Lecture Notes in Computer Science 6849

Commenced Publication in 1973
Founding and Former Series Editors:
Gerhard Goos, Juris Hartmanis, and Jan van Lee

Editorial Board

Dieter Pfoser   Yufei Tao
Kyriakos Mouratidis   Mario A. Nascimento
Mohamed Mokbel   Shashi Shekhar
Yan Huang (Eds.)

Advances in Spatial and Temporal Databases

12th International Symposium, SSTD 2011
Minneapolis, MN, USA, August 24-26, 2011
Proceedings

 Springer

Volume Editors

Dieter Pfoser
Research Center "ATHENA", Athens, Greece
E-mail: pfoser@imis.athena-innovation.gr

Yufei Tao
Chinese University of Hong Kong, China
E-mail: taoyf@cse.cuhk.edu.hk

Kyriakos Mouratidis
Singapore Management University, Singapore
E-mail: kyriakos@smu.edu.sg

Mario A. Nascimento
ATH University of Alberta, Edmonton, AB, Canada
E-mail: mario.nascimento@ualberta.ca

Mohamed Mokbel
Shashi Shekhar
University of Minnesota, Minneapolis, MN, USA
E-mail: {mokbel, shekhar}@cs.umn.edu

Yan Huang
University of North Texas, Denton, TX, USA
E-mail: huangyan@unt.edu

ISSN 0302-9743   e-ISSN 1611-3349
ISBN 978-3-642-22921-3   e-ISBN 978-3-642-22922-0
DOI 10.1007/978-3-642-22922-0
Springer Heidelberg Dordrecht London New York

Library of Congress Control Number: 2011933231

CR Subject Classification (1998): H.2.0, H.2.8, H.2-4, I.2.4

LNCS Sublibrary: SL 3 – Information Systems and Application, incl. Internet/Web
and HCI

Typesetting: Camera-ready by author, data conversion by Scientific Publishing Services, Chennai, India

Printed on acid-free paper

Springer is part of Springer Science+Business Media (www.springer.com)

Preface

SSTD 2011 was the 12th in a series of events that discuss new and exciting research in spatio-temporal data management and related technologies. Previous symposia were successfully held in Santa Barbara (1989), Zurich (1991), Singapore (1993), Portland (1995), Berlin (1997), Hong Kong (1999), Los Angeles (2001), Santorini, Greece (2003), Angra dos Reis, Brazil (2005), Boston (2007), and Aalborg, Denmark (2009). Before 2001, the series was devoted solely to spatial database management, and called SSD. From 2001, the scope was extended in order to also accommodate temporal database management, in part due to the increasing importance of research that considers spatial and temporal aspects jointly.

SSTD 2011 introduced several innovative aspects compared to previous events. In addition to the research paper track, the conference hosted a demonstrations track, and as a novelty, a vision and challenges track focusing on ideas that are likely to guide research in the near future and to challenge prevalent assumptions.

SSTD 2011 received 63 research submissions from 22 countries A thorough review process led to the acceptance of 24 high-quality papers, geographically distributed as follows: USA 9, Germany 3, Greece 2, Canada 2, Switzerland 1, Norway 1, Republic of Korea 1, Japan 1, Italy 1, Hong Kong 1, Denmark 1, China 1. The papers are classified in the following categories, each corresponding to a conference session: (1) Moving Objects and Sensor Networks, (2) Temporal and Streaming Data, (3) Knowledge Discovery, (4) Spatial Networks, (5) Multidimensional Query Processing, (6) Access Methods.

This year's best paper award went to "FAST: A Generic Framework for Flash-Aware Spatial Trees." The paper presents a general technique for converting a traditional disk-oriented structure to an access method that works well on flash-memory devices. Applicable to several well-known structures (including the B- and R-trees), the technique aims at achieving two purposes simultaneously: (a) minimizing the update and query overhead, and (b) preventing the loss of data even in a system crash, thus ensuring data durability. The paper contains several novel ideas, which are of independent interests since they may also be useful in designing other flash-aware algorithms. In addition, the paper features a real system that implements the proposed technique and is demonstrated to have excellent performance in practice through extensive experiments. Besides the best paper, a few other high-quality research papers were selected and the authors were invited to submit extended versions of their work to a special issue of the *Geoinformatica* journal (Springer).

Although the previous symposium in the SSTD series (2009) also included a demonstrations track, submissions were evaluated alongside regular research papers by a single Program Committee. SSTD 2011, for the first time, appointed dedicated Co-chairs to organize an autonomous demonstrations track, who in turn recruited a separate Program Committee comprising 9 members. The purpose of this track was to illustrate engaging systems that showcase underlying solid research and its applicability. The track received 16 submissions from a total of 51 authors or co-authors coming from Germany (21), USA (21), Canada (5), Italy (2), Switzerland (1), and France (1). The selection criteria for the demonstration proposals included novelty, technical advances, and overall practical attractiveness of the demonstrated system. Out of the 16 submissions, 8 were accepted and presented in a special session of the symposium. The best demonstration paper was recognized with "SSTD 2011's Best Demo Award."

Another novelty in SSTD 2011 was the vision and challenges track. The aim of this track was to describe revolutionary ideas that are likely to guide research in the near future, challenge prevalent assumptions in the research community, and identify novel applications and technology trends that create new research directions in the area of spatial and spatiotemporal databases. A separate 12-member Program Committee was formed for this track (coordinated by the same Co-chairs of the demonstrations track). Twenty-one submissions were received from a total of 58 authors and co-authors from USA (19), Germany (7), Italy (2), Greece (2), Brazil (1), U.K. (1), Switzerland (1). Eight of the submissions were accepted and were presented in the symposium in two dedicated sessions. The top three contributions, chosen based on their technical merit as well as their presentation in the symposium, received the Headwaters Awards. The awards were valued at $1,000, $750, and $500 for the three selected contributions (in the form of travel reimbursements), and were kindly sponsored by the Computing Community Consortium (CCC) of the Computing Research Association (CRA).

The keynote address titled "Underexplored Research Topics from the Commercial World" was delivered by Erik Hoel (ESRI). Two panels were held. Panel A titled "Envisioning 2020 Spatial Research Challenges and Opportunities" was chaired by Erwin Gianchandani (CCC) and Panel B titled "Sustainable Energy: Spatial Challenge" was chaired by Ghaleb Abdulla (USDOE LLNL).

To be able to create such a highly attractive SSTD 2011 symposium program, we owe our gratitude to a range of people. We would like to thank the authors, irrespectively of whether their submissions were accepted or not, for their support of the symposium series and for sustaining the high quality of the submissions. We are grateful to the members of the Program Committees (and the external reviewers) for their thorough and timely reviews. In addition, we are grateful to Nikos Mamoulis for his advice and support. We hope the technical program put together for this edition of the SSTD symposium series leads to interesting and fruitful discussions during and after the symposium.

June 2011

Dieter Pfoser
Yufei Tao
Kyriakos Mouratidis
Mario Nascimento
Mohamed Mokbel
Shashi Shekhar
Yan Huang

Organization

Steering Committee

The SSTD Endowment

General Co-chairs

Shashi Shekhar   University of Minnesota, USA
Mohamed Mokbel   University of Minnesota, USA

Program Co-chairs

Dieter Pfoser   IMIS/RC ATHENA, Greece
Yufei Tao   Chinese University of Hong Kong, SAR China

Demo Co-chairs

Mario Nascimento   University of Alberta, Canada
Kyriakos Mouratidis   Singapore Management University, Singapore

Publicity Chair

Jin Soung Yoo   IPFW, USA

Treasurer

Jing (David) Dai   IBM, USA

Sponsorship Chair

Latifur Khan   University of Texas at Dallas, USA

Proceedings Chair

Yan Huang   University of North Texas, USA

Registration Chair

Wei-Shinn Ku   Auburn University, USA

Local Arrangements

Francis Harvey   University of Minnesota, USA

Webmaster

Michael R. Evans   University of Minnesota, USA

Program Committee

Divyakant Agrawal	Feifei Li	Cyrus Shahabi
Walid Aref	Jianzhong Li	Rick Snodgrass
Lars Arge	Xuemin Lin	Yannis Theodoridis
Claudia Bauzer Medeiros	Nikos Mamoulis	Goce Trajcevski
Michela Bertolotto	Test Member	Anthony Tung
Thomas Brinkhoff	Beng Chin Ooi	Agnes Voisard
Bin Cui	Dimitris Papadias	Carola Wenk
Maria Luisa Damiani	Stavros Papadopoulos	Ouri Wolfson
Ralf Hartmut Güting	Chiara Renso	Xiaokui Xiao
Marios Hadjieleftheriou	Dimitris Sacharidis	Xing Xie
Erik Hoel	Simonas Saltenis	Ke Yi
Christian S. Jensen	Markus Schneider	Man Lung Yiu
George Kollios	Bernhard Seeger	Yu Zheng
Bart Kuijpers	Thomas Seidl	Shuigeng Zhou

Vision/Challenge Program Committee

Spiros Bakiras	Erik Hoel	Simonas Saltenis
Thomas Brinkhoff	Vagelis Hristidis	Vassilis Tsotras
Ralf Hartmut Güting	Panos Kalnis	Ouri Wolfson
Marios Hadjieleftheriou	Matthias Renz	Baihua Zheng

Demo Program Committee

Feifei Li	Stavros Papadopoulos	Xiaokui Xiao
Hua Lu	Dimitris Sacharidis	Man Lung Yiu
Apostolos Papadopoulos	Marcos Salles	Karine Zeitouni

External Reviewers

Achakeyev, Daniar
Armenantzoglou, Nikos
Booth, Joel
Bouros, Panagiotis
Buchin, Maike
Cheema, Aamir
Chen, Zaiben
Cheng, Shiwen
Demiryurek, Ugur
Dittrich, Jens
Dziengel, Norman
Fishbain, Barak
Fries, Sergej
Färber, Ines
Giatrakos, Nikos
Green Larsen, Kasper
Hassani, Marwan
Huang, Zengfeng
Hung, Chih-Chieh

Jestes, Jeffrey
Jeung, Hoyoung
Kashyap, Abhijith
Kazemi, Leyla
Kellaris, George
Khodaei, Ali
Kremer, Hardy
Le, Wangchao
Li, Xiaohui
Marketos, Gerasimos
Moelans, Bart
Moruz, Gabriel
Omran, Masoud
Patt-Shamir, Boaz
Pavlou, Kyriacos
Pelekis, Nikos
Ruiz, Eduardo
Sakr, Mahmoud
Shen, Zhitao

Shirani-Mehr, Houtan
Stenneth, Leon
Sun, Guangzhong
Tang, Mingwang
Thomsen, Christian
Trimponias, George
Wang, Lixing
Wang, Lu
Wijsen, Jef
Xu, Bo
Xu, Linhao
Yang, Bin
Yao, Bin
Zhang, Wenjie
Zhang, Ying
Zhang, Zhenjie
Zheng, Wenchen
Zhu, Qijun
Zhu, Yin

Table of Contents

Session 3: Access Methods

Session 4: Moving Objects and Sensor Networks

Session 5: Multidimentional Query Processing

Session 6: Temporal and Streaming Data

Vision and Challenge Papers

Demonstratioions

Keynote Speech: Underexplored Research Topics from the Commercial World

Erik Hoel

Redlands, CA, USA
ehoel@esri.com

Abstract. Active research in the spatio-temporal database domain is approaching fifty years of age, beginning with the early research by Waldo Tobler at Michigan, Roger Tomlinson on CGIS, Edgar Horwood at Washington, Howard Fisher at Harvard, and Donald Cooke at Census. Very significant progress has been made during this period, with spatial data now becoming ubiquitous with the current generation of web applications, imbedded mapping, smartphones, and location-based services. However, many of the most challenging problems being faced by industry in the spatio-temporal domain remain relatively unaddressed by the research community. Many of these problems are related to the devel opment of technology and applications primarily intended for the defense and intelligence worlds. This domain generally involves:

- Prohibitively large quantities of data,
- Real time data fusion,
- Remote sensing (including video and multispectral),
- Large-scale automated and semi-automated feature extraction, and
- Geostreaming, including real-time/continuous analysis and geoprocessing.

In addition, in order to achieve scalability and elasticity, non-traditional architectures and data storage technologies (e.g., NoSQL and multi-tenant) are frequently employed. The guiding mantras of "simple scales, complex fails", as well as "precompute as if your life depends upon it" are key to success in these domains.

Finally, another area that deserves the enhanced attention of the research community involves complex interactions between vast collections of objects in time and space (e.g., migrations, flocking behavior, or communications), with the goal being to infer something about the processes going on. The spatial interaction domain is of particular significance to both social networking as well as intelligence. This talk will provide researchers with a discussion of these topics, presenting additional background and context to these difficult real-world problems along with an overview of what is currently considered the state of the art in framework architectures and production systems.

D. Pfoser et al. (Eds.): SSTD 2011, LNCS 6849, p. 1, 2011.
© Springer-Verlag Berlin Heidelberg 2011

SSCP: Mining Statistically Significant Co-location Patterns

Sajib Barua and Jörg Sander

Dept. of Computing Science, University of Alberta, Edmonton, Canada
{sajib,jsander}@ualberta.ca

Abstract. Co-location pattern discovery searches for subsets of spatial features whose instances are often located at close spatial proximity. Current algorithms using user specified thresholds for prevalence measures may report co-locations even if the features are randomly distributed. In our model, we look for subsets of spatial features which are co-located due to some form of spatial dependency but not by chance. We first introduce a new definition of co-location patterns based on a statistical test. Then we propose an algorithm for finding such co-location patterns where we adopt two strategies to reduce computational cost compared to a naïve approach based on simulations of the data distribution. We propose a pruning strategy for computing the prevalence measures. We also show that instead of generating all instances of an auto-correlated feature during a simulation, we could generate a reduced number of instances for the prevalence measure computation. We evaluate our algorithm empirically using synthetic and real data and compare our findings with the results found in a state-of-the-art co-location mining algorithm.

1 Introduction

Co-location patterns are subsets of Boolean spatial features whose instances are often seen to be located at close spatial proximity [6]. Co-location mining gives important domain related insights for areas such as ecology, biology, epidemiology, earth science, transportation etc. Co-location mining finds frequently co-located patterns among spatial features. For instance, the Nile crocodile and the Egyptian plover (a bird that has a symbiotic relationship with the Nile crocodile) are often seen together giving rise to a co-location pattern {Nile crocodile, Egyptian plover}. In urban areas, we may see co-location patterns such as {shopping mall, parking}, {shopping mall, resturant}.

Existing co-location mining algorithms are motivated by the concept of association rule mining (ARM) [1]. However, in a spatial domain there is no natural notion of a transaction as in market basket data [6]. Yet, most of the co-location mining algorithms [6,16,2] adopt an approach similar to the Apriori algorithm proposed for ARM in [1], introducing some notion of transaction over the space. One such notion is the *feature centric model* [12] where each instance of a spatial feature generates a transaction. Such a transaction includes other feature instances (relevant to the reference feature) appearing in the neighborhood of

D. Pfoser et al. (Eds.): SSTD 2011, LNCS 6849, pp. 2–20, 2011.
© Springer-Verlag Berlin Heidelberg 2011

the instance that defines the transaction. Similar to the *support* measure of the ARM algorithm, a prevalence measure called *Participation Index (PI)* is proposed that is anti-monotonic and thus helps to prune the search space when searching all prevalent co-location patterns. In existing co-location mining algorithms [12,6,14,15], a co-location type C is declared as a prevalent co-location pattern and finally reported, if its PI-value is greater than a user specified threshold. With a small threshold value, meaningless co-location patterns could be reported and with a large threshold value, meaningful co-location patterns could be missed.

To overcome the limitations of the existing approaches when using global prevalence thresholds, we define the notion of a co-location based on some form of dependency among involved features. We also introduce a computationally efficient method for mining such co-locations of different sizes. We argue that the PI threshold should not be global but should be decided based on the distribution and the number of instances of each individual feature involved in a co-location. Instead of a threshold based approach, we use a statistical test in order to decide whether an observed co-location is significant or is likely to have occurred by chance. First, we use a prevalence measure to capture the spatial dependency among features in a co-location. Then we test the null hypothesis H_0 of no spatial dependency against an alternative hypothesis H_1 of spatial dependency among the spatial features in a co-location. We compute the prevalence measure of a co-location in the observed data. Using randomization test, we compute the probability (p) of seeing the same value of prevalence measure or greater under a null model. In the null model, each individual feature maintains a similar spatial distribution as in the observed data but without any interdependency among features. For a given level of significance (α), the prevalence measure value computed from the observed data will be significant if $p \leq \alpha$.

A randomization test poses some computational challenges. The main computation here is the prevalence measure computation for each possible co-location. We propose a pruning technique which is able to detect unnecessary candidate co-locations ahead, prunes them and thus reduces the amount of prevalence measure computations. During the randomization test, we generate instances of each feature based on the null model. This data generation step however takes time. For auto-correlated features, we can avoid generating all instances in many cases and thus save time in the data generation step. Since we use a clustering model for simulating an auto-correlated feature during the randomization test, we can identify those clusters that can not be involved in any co-location without generating the instances of such clusters. For the prevalence measure computation, we also implement a grid based spatial index to find the neighbors of a feature instance. With increasing number of feature instances, our approach shows an increasing speedup compared to a naïve approach.

The rest of the paper is organized as follows. Section 2 describes related work. In Sect. 3, we define the notion of a statistically significant co-location pattern and propose our method SSCP for mining such patterns. In Sect. 4, we show

how to improve the runtime of the baseline algorithm. We conduct experiments to validate our proposed model in Sect. 5. Section 6 concludes the paper.

2   Related Work

In spatial statistics, the co-location mining problem is seen a bit different than in the data mining community. There co-location pattern mining is similar to the problem of finding associations in multi-type spatial point processes. Association or interaction in a spatial point process is known as the second order effect. The second order effect is a result of the spatial dependence and represents the tendency of neighboring values to follow each other in terms of their deviation from their mean. There are several measures used to compute spatial interaction such as Ripley's K-function [11], distance based measures (e.g., F function, G function) [7], and co-variogram function [3]. These measures can summarize a point pattern and are able to detect clustering tendency at different scales. With a large collection of Boolean spatial features, computation of the above measures becomes expensive as the number of candidate subsets increases exponentially in the number of different features. Mane et al. in [9] combine a spatial statistical measure with a data mining tool to find the clusters of female chimpanzees' locations and investigate the dynamics of spatial interaction of a female chimpanzee with other male chimpanzees in the community. There, each female chimpanzee represents a unique mark. Two clustering methods (SPACE-1 and SPACE-2) are proposed which use Ripley's K-function to find clusters among different marked point processes.

In the data mining community, co-location pattern mining approaches are mainly based on spatial relationship such as "close to" proposed by Han and Koperski in [8], which presents a method to mine spatial association rules indicating a strong spatial relationship among a set of spatial and some non-spatial predicates. Morimoto in [10] proposes a method to find groups of different service types originated from nearby locations and report a group if its occurrence frequency is above a given threshold. Such groups can give important insight to improve location based services. Shekhar et al. [12] discuss three models (reference feature centric model, window centric model, and event centric model) that can be used to materialize "transactions" in a continuous spatial domain so that a frequent itemset mining approach can be used. A co-location mining algorithm is developed which utilizes the anti-monotonic property of a proposed participation index (PI) to find all possible co-location patterns.

Many of the follow-up work on the co-location mining approach in [12] have focused on improving the runtime. As an extension of the work in [12], [6] proposes a multi-resolution pruning technique and also compares the PI with the cross K-function, showing that the PI is an upper bound of the cross K-function. To improve the runtime of the algorithm in [6], Yo et al. in [14,15] propose two instance look-up schemes where a transaction is materialized in two different ways. In [15], a transaction is materialized from a star neighborhood and in [14], a transaction is materialized from a clique neighborhood. Xiao et al. in [13] improve the runtime of frequent itemset based methods [10,6,14] by starting from the most dense region of objects and then proceeding to less dense regions. From a dense region, the method counts the number of instances of a feature participating in a candidate co-location. Assuming that all the remaining instances are in co-locations, the method then estimates an upper bound of the prevalence measure and if it is below the threshold the candidate co-location is pruned.

All the above mentioned co-location pattern discovery methods use a predefined threshold to report a candidate co-location as prevalent. Therefore, if thresholds are not selected properly, meaningless co-location patterns could be reported in the presence of spatial auto-correlation and feature abundance, or meaningful co-location patterns could be missed when the threshold is too high.

3   Problem Definition

3.1   Motivating Examples

Consider a scenario (Fig. 1(a)) with two different animal species A and B. Assume there is a true spatial dependency so that As are likely to be seen close to Bs. Assume furthermore that in the given study area, there are only few instances of A but B is abundant. Hence many of the B's instances will not have As in their neighborhood. On the other hand, all As will be close to some Bs and form co-locations. Since many of the Bs are without As, B's participation ratio will be small which results the participation index of $\{A, B\}$ to be rather low (likely lower than a given threshold to avoid reporting meaningless co-locations), and we might not report the pattern.

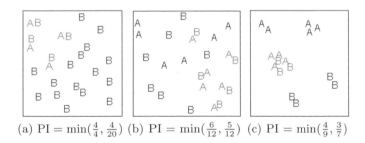

(a) PI $= \min(\frac{4}{4}, \frac{4}{20})$ (b) PI $= \min(\frac{6}{12}, \frac{5}{12})$ (c) PI $= \min(\frac{4}{9}, \frac{3}{7})$

Fig. 1. Three scenarios with features A and B. Red instances are in co-location.

Consider another scenario (Fig. 1(b)) where two spatial features A and B are abundant in a study area and randomly distributed. Even though there is no true spatial dependency between A and B, we might see enough instances of $\{A, B\}$ so that the PI-value of $\{A, B\}$ is above the given threshold. Hence $\{A, B\}$ would be reported as a prevalent co-location pattern.

The presence of spatial auto-correlation of features, i.e., the tendency of clustering among instances of the same type (which is not uncommon in spatial domains) can also lead to reporting meaningless co-location pattern as prevalent if a threshold for the participation index is not set properly. Consider a scenario (Fig. 1(c)) where both feature A and feature B are spatially auto-correlated.

If a cluster of A and a cluster of B happen to overlap by chance, a good number of instances of $\{A, B\}$ will be generated which could result in a high PI-value. Even though no real spatial association or interaction between A and B exists, $\{A, B\}$ could be reported as a prevalent co-location pattern.

3.2   Basic Idea for Finding Significant Co-location Patterns

We suggest that, instead of using a global threshold, we could estimate, for the given number of As and Bs, how rare the observed PI-value is compared to a PI-value when A and B have no spatial relationship. If the observed PI-value is significantly higher than a PI-value under no spatial relationship, we conclude that A and B are spatially related, and $\{A, B\}$ should be reported as a prevalent co-location pattern. Hence, the decision of co-location pattern detection does not depend on a user-defined prevalence measure threshold. In addition, note that such an approach works with any type of prevalence measures to capture spatial dependency among features and is not dependent only on the PI measure.

A measure for spatial dependency among features tries to capture the strength of a co-location; the PI is one such measure that we will adopt in our method. The idea is to estimate the probability of the observed PI-value (computed from the given data set) under some null hypothesis. If features were spatially independent of each other, what is the chance of observing a PI-value equal or higher than the PI-value observed in the given data set? The answer gives us a p-value. If the p-value is low, the observed PI-value is a rare phenomenon under our null assumption, thus indicating a spatial dependency among the features. The observed PI-value is said to be statistically significant at level α, if $p \leq \alpha$.

The current literature does not also consider the spatial auto-correlation property in co-location pattern detection approaches. We study how the existing participation index measure behaves in the presence of spatial auto-correlation.

3.3   Null Model Design

The null hypothesis for our approach is that different features are distributed in the space independently of each other. However, an individual feature may be auto-correlated, and in the given data set we determine how auto-correlated it is by modeling it as in [7] as a cluster process. For instance, in a Poisson cluster process, we start with a Poisson process of parent points. Each parent point then gives rise to a finite set of offspring points according to some distribution and these offsprings are distributed in a predefined neighborhood of their parent point. The offspring sets finally give the final cluster process. The *intensity* of a parent process can be homogenous or non-homogenous. In such a model, auto-correlation can be measured in terms of intensity and type of distribution of a parent process and offspring process around each parent. For instance, in a Thomas process [7], cluster centers are generated from a Poisson process with intensity κ. The spatial distribution of the offsprings of each cluster follows an isotropic Gaussian $N(0, \sigma^2 I)$ displacement from the cluster center. The number of offsprings in each cluster is drawn from a Poisson distribution with mean μ.

If an auto-correlated feature is present, we first estimate parameters using a model fitting technique. The method of Minimum Contrast [4] fits a point process model to a given point data set. This technique first computes a summary statistics from the point data. A theoretically expected value of the model to fit is either derived or estimated from simulation. Then the model is fitted to the given data set by finding optimal parameter values of the model to give the closest match between the theoretical curve and the empirical curve. For the Thomas process cluster model, κ, σ, and μ are the summary statistics that are estimated. Other alternative models such as the Matérn Cluster process [7] and the log Gaussian Cox process [7] can be used for parameter estimation of auto-correlated data.

We will simulate artificial data sets under our null hypothesis. The data sets maintain the following properties of the observed data: (1) same number of instances for each feature, and (2) similar spatial distribution for each individual feature by maintaining the same values of the summary statistics estimated from the given data set. For example, if a feature is spatially auto-correlated in the original data set, in our artificial data sets, conforming to the null hypothesis, the feature will also be clustered in the same degree and the clusters will be randomly distributed over the study area. The distribution of a spatial feature is described in terms of summary statistics, i.e. a set of parameters. For an auto-correlated feature A, the summary statistics are κ, σ, and μ values. If A is randomly distributed, we need to know its Poisson intensity which could be either homogenous (a constant) or non-homogenous (a function of x and y).

3.4   Definition of Co-location

First we state two definitions from the literature [12,6] since we use them as a spatial dependency measure:

Definition 1. Let C be a co-location of k different features $f_1, f_2, \ldots, f_k$. In an instance of C, one instance from each of k features will be present and all these feature instances are neighbors (based on a neighborhood relationship R_d) of each other. The ***Participation Ratio*** of feature f_i in co-location C, $pr(C, f_i)$, is the fraction of instances of f_i participating in any instance of C. Formally, $pr(C, f_i) = \frac{|\text{distinct}(\pi_{f_i}(\text{all instances of } C))|}{|\text{instances of } f_i|}$. Here π is the relational projection.

For instance, let $C = \{P, Q, R\}$ and P, Q, and R have n_P, n_Q, and n_R instances respectively. If n_P^C, n_Q^C, and n_R^C distinct instances of P, Q, and R participate in co-location C, the participation ratio of P, Q, R are $\frac{n_P^C}{n_P}$, $\frac{n_Q^C}{n_Q}$, $\frac{n_R^C}{n_R}$ respectively.

Definition 2. The ***Participation Index*** (PI) of a co-location $C = \{f_1, f_2, \ldots, f_k\}$ is defined as $PI(C) = \min_k\{pr(C, f_k)\}$

For example, let $C = \{P, Q, R\}$ where the participation ratios of P, Q, and R are $\frac{2}{4}$, $\frac{2}{7}$, and $\frac{1}{8}$ respectively. The minimum participation ratio $\frac{1}{8}$ is selected as the participation index of co-location C.

Lemma 1. *The participation ratio and the participation index are monotonically non-increasing with the increase of co-location size that is if $C' \subset C$ then $pr(C', f) \geq pr(C, f)$ where $f \in C'$ and $PI(C') \geq PI(C)$ [12].*

Using PI as a measure of spatial dependency, we can define a statistically significant co-location pattern so that a statistically significant co-location pattern C is the effect of a true spatial dependency among features participating in C.

Definition 3. *We say a co-location pattern $C = \{f_1, f_2, \ldots, f_k\}$ is statistically significant at level α, if the probability (p-value) of seeing the observed PI-value or larger in a data set conforming to a null hypothesis is not greater than α.*

Statistical Significance Test: The main idea of the statistical significance test is to estimate the probability p of seeing the observed PI-value or greater for a co-location $C = \{f_1, f_2, \ldots, f_k\}$ under the null hypothesis. The null hypothesis is an independence assumption H_0 that there is no spatial dependency among the k features of C. Let $PI_{\mathrm{obs}}(C)$ denote the participation index of C in the observed data, and let $PI_0(C)$ denote the participation index of C in a data set generated under our null hypothesis. Then we have to estimate the probability $p = P(PI_0(C) \geq PI_{\mathrm{obs}}(C))$ that $PI_0(C)$ is at least $PI_{\mathrm{obs}}(C)$ if the k features are independent of each other. This p-value is compared to a predefined threshold α. If $p \leq \alpha$, the null hypothesis is rejected and the $PI_{\mathrm{obs}}(C)$-value is significant at level α. Hence the co-location of $f_1, \ldots, f_k$ in the observed data is significant at level α. α is the probability of committing type I error that is rejecting a null hypothesis when the null hypothesis is true, i.e. the probability of accepting a spurious co-location pattern. If a typical value of $\alpha = 0.05$ is used, there is 5% chance that a spurious co-location is reported as a prevalent co-location pattern.

To compute the desired probability p, we can do randomization tests. We generate a large number of simulated data sets that conform to the null hypothesis [see Sect. 3.3]. Then we compute the PI-value of a co-location C, $PI_0^{R_i}(C)$, in each simulation run (or data set) R_i and compute the p value as:

$$p = \frac{R^{\geq PI_{\mathrm{obs}}} + 1}{R + 1} \tag{1}$$

Here the numerator is the number of simulations ($R^{\geq PI_{\mathrm{obs}}}$) plus the observed data set where $PI_0^{R_i}(C)$ is equal or greater than $PI_{\mathrm{obs}}(C)$. The denominator is the total number of simulations (R) plus the observed data.

4   Algorithm

Given a set of spatial features, our objective is to mine all statistically significant co-location patterns of different sizes. For each pattern we have to compute the probability (p-value) of the observed PI-value under the null hypothesis. Computing a p-value using a randomization test is computationally expensive, since we have to compute the PI-value of each co-location in each simulation run. The PI-value computation of a co-location C requires checking the neighborhood of each spatial feature participating in C. For a more accurate estimation of the p-value, we generate at least 999 simulations, which results in a significant amount of computation for a decision on whether a co-location is statistically significant or not. The total computational cost also increases with the number of distinct spatial features. Hence one important challenge is to reduce the total amount of computation of a randomization test.

In each simulation run, we generate instances of each feature. For each auto-correlated feature, we only generate feature instances of those clusters which are close enough to other different features (auto-correlated or not auto-correlated) to be potentially involved in co-locations. In Fig. 2, we show two auto-correlated features ∘ and △. Instances of feature ∘ appear in six clusters and instances of feature △ also appear in six clusters. Each cluster is represented by a cluster center. We sort the cluster centers according to their x-coordinate values. Let $R1$ and $R2$ be the cluster radius of feature ∘ and feature △ respectively and let R_d be the radius of a co-location neighborhood. Two clusters, each from different features, are defined as close enough if the distance between their centers is not more than $R1 + R2 + R_d$. Hence we can avoid generating instances of a cluster whose center is far away ($> R1 + R2 + R_d$) from centers of cluster of different features. In Fig. 3, we show that only a partial amount of instances of each auto-correlated feature has to be generated to compute the PI-value which is same as the PI-value even if it were computed from all the instances as in Fig. 2.

In each simulation run, we need the PI-value of each possible co-location C to compare with its $PI_{\mathrm{obs}}(C)$-value. If the PI-value of C is greater than its PI-value in the observed data, $R^{\geq PI_{\mathrm{obs}}}$ of C is increased by one. The total PI-value computation in a simulation run can be reduced by identifying those candidate co-locations for which the PI-value in the current simulation run is less than the PI-value computed from the observed data; and hence can not increase the $R^{\geq PI_{\mathrm{obs}}}$-value of (1). We apply a pruning technique to identify those candidate co-locations and prune them so that we do not need to compute their PI-values.

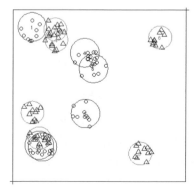

Fig. 2. Instances of all clusters

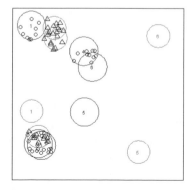

Fig. 3. Generated instances

First, we compute the PI-value of each possible co-location in the observed data. Let C be a co-location and let its PI-value in the observed data be $PI_{\text{obs}}(C)$. In a simulation run R_i, $PI_0^{R_i}(C)$ can not be equal to or greater than $PI_{\text{obs}}(C)$ ($PI_0^{R_i}(C) \not\geq PI_{\text{obs}}(C)$) if the PI-value of any subset C' of C is less than $PI_{\text{obs}}(C)$ that is $PI_0^{R_i}(C') < PI_{\text{obs}}(C)$. If that happens, we can tell C can not contribute to the $R^{\geq PI_{\text{obs}}}$-value of (1). Hence the PI-value computation for C is unnecessary for R_i and C can be pruned. Only if the PI-values of all subsets of C in R_i are greater than or equal to $PI_{\text{obs}}(C)$, there is still a chance that $PI_0^{R_i}(C)$ can be greater than or equal to $PI_{\text{obs}}(C)$. In that case, we compute the PI-value of C in R_i and compare it with $PI_{\text{obs}}(C)$.

Lemma 2. *For a co-location C, its PI-value in simulation R_i, $PI_0^{R_i}(C)$, can not be greater than or equal to its PI-value in the observed data, $PI_{obs}(C)$, if the PI-value of any subset C' of C in R_i, $PI_0^{R_i}(C')$, is less than $PI_{obs}(C)$.*

Proof. If C' is a subset of C for which $PI_0^{R_i}(C') < PI_{\text{obs}}(C)$, then we have to show that $PI_0^{R_i}(C) < PI_{\text{obs}}(C)$. Assume $PI_0^{R_i}(C') < PI_{\text{obs}}(C)$. According to lemma 1, $PI_0^{R_i}(C') \geq PI_0^{R_i}(C)$, then $PI_0^{R_i}(C) < PI_{\text{obs}}(C)$.   □

Using the lemma 2, whether $PI_0^{R_i}(C)$ is computed or not can be decided from the PI-values of C's subsets. For a C of size k ($k > 2$), there are k many subsets of $(k-1)$-size, $\binom{k}{k-2}$ subsets of $(k-2)$-size, $\binom{k}{k-3}$ subsets of $(k-3)$-size, and so on. The decision on whether $PI_0^{R_i}(C)$ will be computed or not can be made by checking C's subsets of any size. Whatever subset size is chosen, the PI-value of all possible subset of that size is required to be pre-computed and available. Using lemma 2, the PI-values of some subsets (of size 3 or more) are not required to be computed. Hence, the total computational cost can not be reduced if we need to compute the PI-values even for such subsets. However computation of PI-values for all subsets of C makes sense if their $PI_0^{R_i}$-value is not less than PI_{obs}-value. This is, however, not very likely. On the other hand, to mine statistically significant 2-size co-location patterns, we compute the PI-value of all possible ($\binom{n}{2}$) 2-size subsets anyway. We just need to store these PI-values so that they can be used for the pruning decision of C of k-size ($k > 2$). Although storing $\binom{n}{2}$ PI-values costs some space, we can reuse it for each simulation run during the randomization test. Thus we save additional PI-value computations for pruned subsets and even save space for storing those. While checking the 2-size subsets of C, if one is found for which the PI-value is less than $PI_{\text{obs}}(C)$, we will stop checking the remaining 2-size subsets (see algorithm 2). Finally after checking 2-size subsets of C, if C is not pruned, we compute the PI-value of C ($PI_0^{R_i}(C)$) for a simulation run R_i. If $PI_0^{R_i}(C) \geq PI_{\text{obs}}(C)$ in R_i, then R_i contributes to the p-value of (1). We maintain a counter $R^{\geq PI_{\text{obs}}}$ which is incremented by one in a simulation run if $PI_0^{R_i}(C) \geq PI_{\text{obs}}(C)$. Finally, C will be reported as statistically significant at level α, if the p-value i.e. $\frac{R^{\geq PI_{\text{obs}}}+1}{R+1} \leq \alpha$.

Here we illustrate the pruning strategy using four features A,B,C, and D. First we compute PI_{obs} for each possible co-location pattern. In each simulation run R_i, we start with computing the $PI_0^{R_i}$-value of each possible 2-size pattern and increment $R^{\geq PI_{\mathrm{obs}}}$ of a pattern by 1 if $PI_0^{R_i} \geq PI_{\mathrm{obs}}$. Now lets consider a 3-size pattern $\{A, B, C\}$. Assume $PI_0^{R_i}\{A, B\} < PI_{\mathrm{obs}}\{A, B, C\}$, then $PI_0^{R_i}\{A, B, C\} < PI_{\mathrm{obs}}\{A, B, C\}$. Hence we do not compute $PI_0^{R_i}\{A, B, C\}$ and $\{A, B, C\}$ can be pruned. The decision for a 4-size pattern $\{A, B, C, D\}$ is done by checking its 2-size subsets in a similar manner. Note that we can not prune it based on the fact that we could prune $\{A, B, C\}$ as $PI_{\mathrm{obs}}\{A, B, C, D\}$ will, in general, be different from $PI_{\mathrm{obs}}\{A, B, C\}$ and it could still be possible that $PI_0^{R_i}\{A, B, C, D\} \geq PI_{\mathrm{obs}}\{A, B, C, D\}$. The PI-value decreases with the increase of the size of a co-location. Hence, if the number of features increases, we will see more pruning effect in co-locations of smaller size than in co-locations of larger size. Algorithm 1 shows the pseudo-code of our approach.

Complexity: In the worst case, there is no pruning in each simulation R_i and we compute the $PI_0^{R_i}$-value of each possible co-location C. Before computing the $PI_0^{R_i}$-value of C, we lookup the stored $PI_0^{R_i}$-values of its 2-size subsets. Hence the cost for C is the sum of the lookup cost and the cost for computing its $PI_0^{R_i}$-value. Assume that a lookup costs β unit of computation. For a co-location C of k-size, the lookup cost for its $\binom{k}{2}$ pairs is $P_1^k = \binom{k}{2}\beta$. For computing $PI_0^{R_i}(C)$, we lookup the neighborhoods of all instances of each feature in C and determine if at least one instance of each feature in C is present in a neighborhood. Hence the cost of PI-value computation for C of k-size is $P_2^k = k \times \max_k\{\#$ of instances of feature $f_k\} \times \beta \approx k\delta\beta$ [assume $\delta = \max_{i=1}^n\{\#$ of instances of feature $f_i\}$]. With n total features, there are $\binom{n}{k}$ different k-size co-locations. Hence the total cost for all different k-size co-locations is $\binom{n}{2}P_2^2 + \sum_{k=3}^n \binom{n}{k}\left(P_1^k + P_2^k\right)$. Using the equalities of $\sum_{k=q}^n \binom{n}{k}\binom{k}{q} = 2^{n-q}\binom{n}{q}$ and $\sum_{k=2}^n k\binom{n}{k} = n(2^{n-1} - 1)$, the above cost is equal to $\binom{n}{2}(2^{n-2} - 1)\beta + n(2^{n-1} - 1)\delta\beta$ which is of $O(2^n)$ when $n > 4$. The worst case is expensive with large n. In many real applications (e.g. ecology) the largest co-location size that typically exists in the data is much smaller than n since a finite co-location neighborhood can typically not accommodate instances of n different features when n is large. While checking neighborhoods of feature instances, we can tell the size of the largest co-location. Note that, we are not looking for a statistically significant co-location pattern with weak spatial dependency. Hence, we do not compute $PI_0^{R_i}$-value of C in a simulation R_i, if its PI_{obs}-value equals 0. All these keep the actual cost in practice less than the cost in the worst case.

5   Experimental Evaluation

5.1   Synthetic Data

Negative Association: Here we show that existing support threshold based co-location mining algorithms can report a set of negatively associated features as a prevalent co-location pattern. We generate a data set of two features $\circ$ and $\triangle$, each of 40 instances and these two features inhibit each other. Such a pairwise inhibition type can be modeled by a Strauss process [7], which has

---

Algorithm 1. SSCP: Mining Statistically Significant Co-location Patterns

---

Input: A Spatial data set $\mathcal{SD}$ with N spatial features $\mathcal{S} = \{f_1, \ldots, f_N\}$. Each feature
  f_i has I_i number of instances. Level of significance α, and total simulation runs R.
Output: Set of statistically significant co-location patterns $\mathcal{C}$.
Variables:
  k: co-location size
  C_O^k: Set of all k-size candidate co-locations. Each candidate co-location is stored
  along with its $PI_{\mathbf{obs}}$-value and $R^{\geq PI_{\mathrm{obs}}}$-value.
  C_0^2: Set of all 2-size co-locations of null model. Each co-location is stored along
  with its $PI_0^{R_j}$-value from a simulation run R_j.
Method:
1: $\mathcal{C} \leftarrow \{\}$
   // Compute $PI_{\mathbf{obs}}$-value of all candidate co-locations from $\mathcal{SD}$
2: **for** $k = 2$ **to** N **do**
3:   **for** $i = 1$ **to** $\binom{N}{k}$ **do**
4:     **Generate** k-size i-th **candidate co-location and store it in**
       $C_O^k[i].$**pattern**
5:     **Compute its** $PI_{\mathbf{obs}}$**-value**
6:     $C_O^k[i].PI_{\mathbf{obs}} \leftarrow PI_{\mathbf{obs}}$; $C_O^k[i].R^{\geq PI_{\mathbf{obs}}} \leftarrow 0$
7: **for** $i = 1$ **to** $\binom{N}{2}$ **do** $C_0^2[i].$**pattern** $\leftarrow C_O^2[i].$**pattern**
   // **Computing** p**-value for all candidate co-locations**
8: **for** $j = 1$ **to** R **do**
9:   **Generate a simulated data set** R_j **under the null model**
10:    **for** $i = 1$ **to** $\binom{N}{2}$ **do**
11:      **Compute its** $PI_0^{R_j}$**-value and** $C_0^2[i].PI \leftarrow PI_0^{R_j}$
12:      **if** $C_0^2[i].PI \geq C_O^2[i].PI_{\mathbf{obs}}$ **then**
13:        $C_0^2[i].R^{\geq PI_{\mathbf{obs}}} \leftarrow C_0^2[i].R^{\geq PI_{\mathbf{obs}}} + 1$
14:    **for** $k = 3$ **to** N **do**
15:      **for** $i = 1$ **to** $\binom{N}{k}$ **do**
16:        **if** (**isPrunedCand**($C_O^k[i].$**pattern**,$C_O^k[i].PI_{\mathbf{obs}}$,$C_0^2, k$)) **then**
17:          **continue** // $PI_0^{R_j}$**-value of** $C_O^k[i].$**pattern is not computed**
18:        **Compute the** $PI_0^{R_j}$**-value of candidate co-location** $C_O^k[i].$**pattern**
19:        **if** $PI_0^{R_j} \geq C_O^k[i].PI_{\mathbf{obs}}$ **then**
20:          $C_O^k[i].R^{\geq PI_{\mathbf{obs}}} \leftarrow C_O^k[i].R^{\geq PI_{\mathbf{obs}}} + 1$
21: **for** $k = 2$ **to** N **do**
22:    **for** $i = 1$ **to** $\binom{N}{k}$ **do**
23:      **Compute** p**-value of** $C_O^k[i].$**pattern as** $\frac{C_O^k[i].R^{\geq PI_{\mathbf{obs}}}+1}{R+1}$
24:      **if** $p \leq \alpha$ **then**
25:        $\mathcal{C} \leftarrow \mathcal{C} \bigcup C_O^k[i].$**pattern**
26: **return** $\mathcal{C}$

---

three parameters $(\beta, \gamma,$ and $r)$. The probability density of the Strauss process
X is $\alpha\beta^{n(X)}\gamma^{s(X)}$, where α is a normalizing constant, $n(X)$ is the total number
of points, and $s(X)$ is the number of pairs in X which lie closer than r units
apart. β is the contributing factor of each point to the density, γ controls the
strength of the interaction between points, and r is the interaction distance.

Algorithm 2 isPrunedCand(CandPattern, PI_{obs}, C_0^2, k)

1: **for** each 2-size subset l of CandPattern **do**
2:      Find index x of l in C_0^2.pattern.
3:      **if** $C_0^2[x].PI < PI_{obs}$ **then**
4:          **return** TRUE
5: **return** FALSE

When $\gamma = 1$, the overall density becomes the density of a Poisson process. With $\gamma > 1$, the point process exhibits clustering, with $\gamma = 0$, points exhibit no interaction within distance r, and with $0 < \gamma < 1$, the points exhibit a negative association (inhibition). Our data is generated from a multi-type Strauss process where the parameter of interaction among similar type of feature instances is 0.43, the parameter of interaction among different types of feature instances is 0.4, and the interaction radius (r) is set to 0.1. β is 210. The study area is a unit square and co-location radius (R_d) is 0.1. Even when imposing a negative association between $\circ$ and $\triangle$, we might still see instances of co-location of $\{\circ, \triangle\}$. The data set is shown in Fig. 4(a). The computed $PI_{obs}(\{\circ, \triangle\})$-value is 0.55 and the existing mining algorithm will report co-location $\{\circ, \triangle\}$ as prevalent if a value less than 0.55 is set as PI threshold. Through randomization test, we find that the probability of seeing $PI_0(\{\circ, \triangle\})$ being at least 0.55 under the null model is 0.931 ($\frac{930+1}{999+1}$) which means observing $PI_{obs}\{\circ, \triangle\}$-value is quite likely under a null model. Hence our method will not report $\{\circ, \triangle\}$ as a significant co-location pattern. This can be validated from the estimation of Ripley's K-function. In Fig. 4(b), we see that estimation of $K_{\circ,\triangle}(r)$ using Ripley's isotropic edge correction (solid line) is always below the theoretical curve (dashed line), which means that the average number of $\triangle$ found in a neighborhood of radius r of a $\circ$ is always less than the expected value (πr^2).

Auto-correlation: In this experiment, we show the effect of spatial auto-correlation in mining a true co-location pattern. We generated a synthetic data

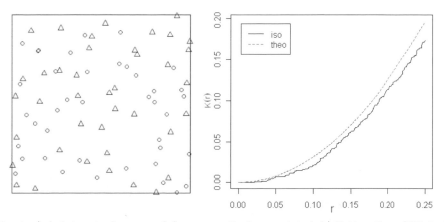

Fig. 4. a) A data set where $\circ$ and $\triangle$ are negatively associated. b) Estimation of Ripley's K- function $K_{\circ,\triangle}(r)$ from the data set.

set with 2 different features ○, and △. Feature ○ has 100 instances which are spatially auto-correlated, and feature △ has 120 instances which are independently and uniformly distributed. The study area is a unit square. As a good number of △ instances are uniformly spread over the space, △s will be found in most of the clusters of ○. In our generated data (Fig. 5), it happens that each cluster of ○s contains some △s which makes a ○ being co-located with at least one △ in a given neighborhood (a circle with a radius of 0.1 unit). Hence the participation ratio of ○ is 1, whereas the participation ratio of △ is 0.49 ($\frac{59}{120}$). Finally the PI-value of {○, △} is 0.49. The spatial distribution of ○ follows the model of Matérn's cluster process [7]. The summary statistics of the Matérn's cluster process has three parameters. Here cluster centers are generated from a Poisson process with intensity κ. Then each cluster center is replaced by a random number of offspring points with a Poisson (μ) which are uniformly and independently distributed inside a disc of radius r centered at the cluster center. The summary statistics of feature ○ are $\kappa = 40$, $\mu = 5$, and $r = 0.05$. During the randomization tests, in each simulation feature ○ is generated using the same summary statistics value for κ, μ, and r. Feature △ is uniformly distributed. The number of instances for each feature are also kept the same as in the observed data. We run 999 simulations leading to a p-value of 0.383 ($\frac{382+1}{999+1}$) for {○, △} which is greater than $\alpha = 0.05$. Hence co-location {○, △} is not statistically significant. Due to the presence of spatial auto-correlation and abundance of feature instances, the PI-value of the co-location {○, △} is 0.49 and high, but not unexpected, which can mislead the existing algorithms when using thresholds less than 0.49. Note that a value of 0.49 would be relatively high and not an unusual threshold.

Multiple Features: Here, we generated a synthetic data set of 5 different feature types ○, △, +, ×, and ◊ (Fig. 6). Features ○, △ and × have 40 instances each. Feature + has 118 instances, and feature ◊ has 30 instances. Our study area is a unit square and the co-location neighborhood radius (R_d) is 0.1. In the data, we impose some pairwise interactions among features. Features ○ and △ (with

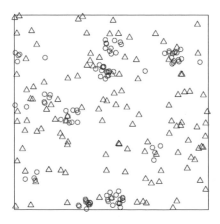

Fig. 5. A data set with two features

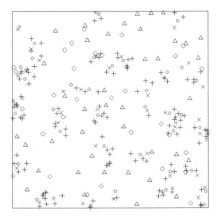

Fig. 6. A data set with five features

individual β value as 300) have a negative association relationship and instances of these two types are generated from an inhibition process (a Multi-Strauss hardcore process [7]), where no two different types of features are seen within a predefined distance threshold (called hardcore distance h, here $h = 0.05$); but an inhibition (negative association) is present at a distance $0.05 < r < 0.1$ where the inhibition parameter γ is 0.3 between ○ and △, and 0.43 among same feature instances. Feature + is spatially auto-correlated ($\kappa = 40$, $\mu = 3$, and $r = 0.05$) and positively associated with feature ○ and feature ×. Hence, we observe a group of + around instances of ○ and ×. Feature ◊ is randomly distributed over the study area. In our null model, both ○ and △ are generated from a Strauss-hardcore process (with the same parameter values as in the synthetic data set) but with no inhibition between them. Feature +, feature ×, and feature ◊ have a similar type of distribution as in the synthetic data set.

From Table 1 we see that $\{○, △\}$ is not reported ($p > 0.05$) as significant due to their inhibitive interaction. Feature ○, feature +, and feature × are strongly associated in the synthetic data and that is captured in the result as we see $\{○, +\}$, $\{○, ×\}$, and $\{+, ×\}$ having a p-value of 0 are reported. $\{○, ◊\}$ is not reported since both features are independent of each other. The same reasoning applies to $\{△, ◊\}$, $\{+, ◊\}$, and $\{×, ◊\}$. Since an inhibition relationship exists between feature ○ and feature △ and a positive association exists among three features ○, +, and ×; feature △ show an inhibition relationship with feature + and feature ×. We find that in our result that $\{△, +\}$, and $\{△, ×\}$ are not reported as significant. It is clear that the existing algorithms using a global threshold for reporting a co-location as prevalent would either miss a true co-location (presence of a relationship) or even report a meaningless co-location (no relationship or inhibition among features), if a proper threshold is not set.

In Table 2, we find that $\{○, △, +\}$, $\{○, △, ×\}$, and $\{○, △, ◊\}$ are not reported. In these 3-size subsets, the presence of ○ and △ prohibits a positive association among 3 features. On the other hand, the positive association among features ○, +, and × is captured in our result. Table 2 reports $\{○, +, ×\}$ as significant. Feature △ shows an inhibition relationship with feature + and feature ×. Hence $\{△, +, ×\}$ can not be a true co-location which is also not reported as significant. Feature ◊ is randomly distributed and independent of other features. $\{△, +, ◊\}$ and $\{△, ×, ◊\}$ are not reported. They can not be true co-locations as each of those 3-size subsets includes two features (such as feature △, and feature +) which are inhibitive to each other. Feature ◊ with a strongly associated pair (such as feature ○, and feature +) could result in a positive association. This is also revealed and our result finds $\{○, +, ◊\}$, $\{○, ×, ◊\}$, and $\{+, ×, ◊\}$ as significant.

Subset $\{○, +, ×, ◊\}$ ($PI_{\text{obs}} = 0.55$ and $p = 0$) is reported as significant. A positive association among ○, +, and × gives rise to this co-location. $\{○, △, +, ×\}$ ($PI_{\text{obs}} = 0.4$, $p = 0.299$), $\{○, △, +, ◊\}$ ($PI_{\text{obs}} = 0.275$, $p = 0.661$), and $\{○, △, ×, ◊\}$ ($PI_{\text{obs}} = 0.25$, $p = 0.665$) are not reported. This is due to the fact that two negatively associated features ○ and △ are present in those 4-size subsets. $\{△, +, ×, ◊\}$ ($PI_{\text{obs}} = 0.25$, $p = 0.668$) is also not reported. The inhibition of △ with + and × and the independence of ◊ with other features prohibits

Table 1. 2-size co-locations

Co-location	PI_{obs}	p-value	Significant
$\{\circ, \triangle\}$	0.525	1	No
$\{\circ, +\}$	1	0	Yes
$\{\circ, \times\}$	1	0	Yes
$\{\circ, \Diamond\}$	0.55	0.697	No
$\{\triangle, +\}$	0.593	0.988	No
$\{\triangle, \times\}$	0.475	0.997	No
$\{\triangle, \Diamond\}$	0.575	0.539	No
$\{+, \times\}$	1	0	Yes
$\{+, \Diamond\}$	0.559	0.63	No
$\{\times, \Diamond\}$	0.6	0.451	No

Table 2. 3-size co-locations

Co-location	PI_{obs}	p-value	Significant
$\{\circ, \triangle, +\}$	0.525	0.669	No
$\{\circ, \triangle, \times\}$	0.4	0.916	No
$\{\circ, \triangle, \Diamond\}$	0.275	0.987	No
$\{\circ, +, \times\}$	1	0	Yes
$\{\circ, +, \Diamond\}$	0.55	0.039	Yes
$\{\circ, \times, \Diamond\}$	0.55	0.013	Yes
$\{\triangle, +, \times\}$	0.425	0.835	No
$\{\triangle, +, \Diamond\}$	0.314	0.927	No
$\{\triangle, \times, \Diamond\}$	0.275	0.944	No
$\{+, \times, \Diamond\}$	0.559	0.011	Yes

a positive association among these 4 features. Finally, the only 5-size pattern $\{\circ, \triangle, +, \times, \Diamond\}$ ($PI_{obs} = 0.25$, $p = 0.185$) is also not reported.

Runtime Comparison: For an auto-correlated feature, we do not need to generate all of its instances and we can also prune candidate co-locations which can not contribute to the p-value computation under certain circumstances (see Sect. 4). In a naïve approach, we do not apply any of these techniques.

We generate a data set with 4 different features $\circ$, $\triangle$, $+$, and $\times$. Features $\circ$, $\triangle$, and $+$ are auto-correlated features; whereas feature $\times$ is randomly distributed. The study area is a square with an area of 100 sq. units and the co-location neighborhood radius R_d is set to 0.1. We impose a strong spatial relationship among $\circ$, $\triangle$, and $\times$. Feature $\circ$, feature $\triangle$, and feature $+$ have 400 instances each and feature $\times$ has 20 instances. In total there are 1220 instances. The significant patterns found are $\{\circ, \triangle\}$ ($PI_{obs} = 0.9125$, $p = 0$), $\{\circ, \times\}$ ($PI_{obs} = 0.17$, $p = 0$), $\{\triangle, \times\}$ ($PI_{obs} = 0.16$, $p = 0$), and ($\{\circ, \triangle, \times\}$ ($PI_{obs} = 0.145$, $p = 0$). We conduct four more, similar experiments, and in each experiment we keep the total cluster number of each auto-correlated feature the same but the total number of instances per cluster is increased by a factor k for all clusters. All these experiments also report the mentioned co-location patterns as significant. Figure 7 shows the runtime. Figure 8 shows that with the increase of total instances, we obtain an increasing speedup growing from 2.5 to 4.5. The experiment is conducted on a 16 Quad-Core AMD Opteron processor machine with a cpu speed of 2.4 Ghz. The main memory size is 62 GB and the OS is Linux. For an auto-correlated feature, if the number of clusters increases, the chance of a cluster being close to other features will be higher. Hence the data generation step might have to generate more instances of each auto-correlated feature in such cases. In another 5 experiments, we increase the number of instances of feature $\times$ and the number of clusters of each auto-correlated feature by the same factor but keep the number of instances per cluster same. Figure 9 shows the runtime and Fig. 10 shows the speedup of the five experiments. We see that with the increase of the number of clusters, after increasing first, the speedup eventually goes down. This happens when more and more instances actually have to be generated, eventually leaving only the speedup due to candidate pruning.

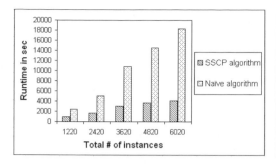

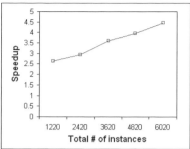

Fig. 7. Runtime comparison   **Fig. 8.** Speedup

5.2 Real Data Set

Ants Data: The nesting behavior of two species of ants (*Cataglyphis bicolor* and *Messor wasman*) is investigated to check if they have any dependency on biological grounds. The Messor ants lives on seeds while the *Cataglyphis* ants collect dead insects for foods which are for the most part dead *Messor* ants. *Messor* ants are killed by *Zodarium frenatum*, a hunting spider. The question is if there is any possible connection we can determine between these two ant species that could be derived from their nest locations. The full data set gives the spatial locations of nests of the two species recorded by Professor R.D. Harkness at a site in northern Greece [5]. It comprises 97 nests (68 *Messor* and 29 *Cataglyphis*) inside an irregular convex polygon polygonal boundary (Fig. 11).

We run our algorithm on the ants data and computed the PI-value. Each of the 24 *Cataglyphis* ant nests is close to at least one *Messor* ant's nest, not more than 50 unit away. Hence the participation ratio of *Cataglyphis* ant is $\frac{24}{29} = 0.8275862$. For *Messor* ants, the participation ratio is $\frac{30}{68} = 0.4411765$. Finally the PI_{obs}-value of co-location $\{Cataglyphis, Messor\}$ is 0.4411765. In the randomization test, we generate 999 simulation runs and find that in 141 simulation runs, the $PI_0^{R_i}$-value is greater than or equal to the PI_{obs}-value. The p-value $= \frac{141+1}{999+1} = 0.142$ is greater than 0.05. Hence we can conclude that the co-location $\{Cataglyphis, Messor\}$ is not statistically significant at level 0.05 and

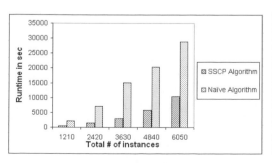

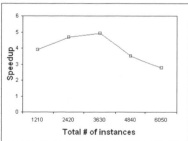

Fig. 9. Runtime comparison   **Fig. 10.** Speedup

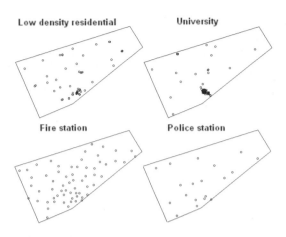

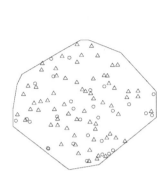

Fig. 11. Ants data - $\circ$ = Cataglyphis ants and $\triangle$ = Messor ants

Fig. 12. Toronto address data (4 features)

there is no spatial dependency between these two types of ants. In fact, clear evidence of a spatial association between these two species is also not found [5].

Toronto Address Repository Data: The Toronto Open Data provides a data set with over $500,000$ addresses within the City of Toronto enclosed in a polygonal area. Each address point has a series of attributes including a feature class with 65 features and coordinates. After removing missing data and removing features with very high frequency (e.g. high density residential), we select 10 features for our experiment: low density residential (66 instances), nursing home (31 instances), public primary school (510 instances), separate primary school (166 instances), college (32 instances), university (91 instances), fire station (63 instances), police station (19 instances), other emergency service (21 instances), and fire/ambulance station (16 instances). Due to space limitations, only some of the feature distributions are shown in Fig. 12. To determine if a feature shows clustering (spatial auto-correlation), regularity (inhibition), randomness (Poisson), we compute the *pair co-relation function* $g(r) = \frac{K'(r)}{2\pi r}$ where K' is the derivative of the K-function [7]. Values of $g(r) > 1$, at smaller distances r suggest clustering; $g(r) = 1$ indicates randomness, and $g(r) < 1$ suggests inhibition. Police stations, fire stations, fire/ambulance stations, separate primary schools show regular distributions, since $g(r) < 1$ at smaller r values. The remaining features are auto-correlated since their $g(r) > 1$ for smaller values of r. The co-location neighborhood radius (R_d) is set to 500.

In Table 3, we show 2-size, 3-size, and 4-size co-locations. Note that the PI_{obs}-values are so low that existing algorithms would return almost every feature combination as a co-location if their global threshold would be set so that the above statistically significant co-locations can be returned.

Table 3. Found co-locations. A feature present in a co-location is shown by $\checkmark$.

Low density resid.	Univ.	Fire Station	Police station	College	Other emerg. service	Nursing home	Public primary school	Separate primary school	PI_{obs}
$\checkmark$	$\checkmark$								0.363
$\checkmark$		$\checkmark$							0.079
$\checkmark$			$\checkmark$						0.157
	$\checkmark$			$\checkmark$					0.263
	$\checkmark$	$\checkmark$							0.095
	$\checkmark$		$\checkmark$						0.120
	$\checkmark$					$\checkmark$			0.022
		$\checkmark$	$\checkmark$						0.095
$\checkmark$			$\checkmark$			$\checkmark$			0.030
$\checkmark$	$\checkmark$	$\checkmark$							0.047
$\checkmark$	$\checkmark$		$\checkmark$						0.087
$\checkmark$	$\checkmark$					$\checkmark$			0.022
	$\checkmark$	$\checkmark$		$\checkmark$					0.0158
	$\checkmark$	$\checkmark$	$\checkmark$						0.0317
$\checkmark$	$\checkmark$	$\checkmark$	$\checkmark$						0.0158
	$\checkmark$	$\checkmark$					$\checkmark$	$\checkmark$	0.012

6   Conclusions

In this paper, we propose a new definition of co-location patterns and a method to detect them. Existing approaches in the literature find co-locations using a predefined threshold value which can lead to not reporting meaningful co-location patterns or reporting meaningless co-location patterns. Our method is based on a statistical test. Such a statistical test is computationally expensive; we improve the runtime by generating a reduced number of instances for an auto-correlated feature in a simulated data generation step and by pruning unnecessary candidate co-location patterns in the PI-value computation of simulation runs. We evaluate our algorithm using synthetic and real data sets. Future work includes studying and comparing alternative prevalence measures in our framework which could also allow additional pruning techniques.

References

1. Agrawal, R., Srikant, R.: Fast Algorithms for Mining Association Rules in Large Databases. In: Proc. VLDB, pp. 487–499 (1994)
2. Celik, M., Shekhar, S., Rogers, J.P., Shine, J.A.: Mixed-Drove Spatiotemporal Co-occurence Pattern Mining. IEEE TKDE 20(10), 1322–1335 (2008)
3. Cressie, N.A.C.: Statistics for Spatial Data. Wiley, Chichester (1993)
4. Diggle, P.J., Gratton, R.J.: Monte Carlo Methods of Inference for Implicit Statistical Models. J. of the Royal Statist. Society, Series B 46(2), 193–227 (1984)
5. Harkness, R.D., Isham, V.: A Bivariate Spatial Point Pattern of Ants' Nests. J. of the Royal Statist. Society, Series C (Appl. Statist.) 32(3), 293–303 (1983)

6. Huang, Y., Shekhar, S., Xiong, H.: Discovering Colocation Patterns from Spatial Data Sets: A General Approach. IEEE TKDE 16(12), 1472–1485 (2004)
7. Illian, J., Penttinen, A., Stoyan, H., Stoyan, D.: Statistical Analysis and Modelling of Spatial Point Patterns. Wiley, Chichester (2008)
8. Koperski, K., Han, J.: Discovery of Spatial Association Rules in Geographic Information Databases. In: Egenhofer, M.J., Herring, J.R. (eds.) SSD 1995. LNCS, vol. 951, pp. 47–66. Springer, Heidelberg (1995)
9. Mane, S., Murray, C., Shekhar, S., Srivastava, J., Pusey, A.: Spatial Clustering of Chimpanzee Locations for Neighborhood Identification. In: Proc. ICDM, pp. 737–740 (2005)
10. Morimoto, Y.: Mining Frequent Neighboring Class Sets in Spatial Databases. In: Proc. SIGKDD, pp. 353–358 (2001)
11. Ripley, B.: The Second-Order Analysis of Stationary Point Processes. J. of Appl. Probability 13(2), 255–266 (1976)
12. Shekhar, S., Huang, Y.: Discovering Spatial Co-location Patterns: A Summary of Results. In: Jensen, C.S., Schneider, M., Seeger, B., Tsotras, V.J. (eds.) SSTD 2001. LNCS, vol. 2121, pp. 236–256. Springer, Heidelberg (2001)
13. Xiao, X., Xie, X., Luo, Q., Ma, W.Y.: Density Based Co-location Pattern Discovery. In: Proc. GIS, pp. 250–259 (2008)
14. Yoo, J.S., Shekhar, S.: A Partial Join Approach for Mining Co-location Patterns. In: Proc. GIS, pp. 241–249 (2004)
15. Yoo, J.S., Shekhar, S.: A Joinless Approach for Mining Spatial Colocation Patterns. IEEE TKDE 18(10), 1323–1337 (2006)
16. Yoo, J.S., Shekhar, S., Kim, S., Celik, M.: Discovery of Co-evolving Spatial Event Sets. In: Proc. SDM, pp. 306–315 (2006)

An Ontology-Based Traffic Accident Risk Mapping Framework

Jing Wang and Xin Wang

Department of Geomatics Engineering,
University of Calgary, Calgary, AB, Canada
{wangjing,xcwang}@ucalgary.ca

Abstract. Road traffic accidents are a social and public challenge. Various spatial concentration detection methods have been proposed to discover the concentration patterns of traffic accidents. However, current methods treat each traffic accident location as a point without consideration of the severity level, and the final traffic accident risk map for the whole study area ignores the users' requirements. In this paper, we propose an ontology-based traffic accident risk mapping framework. In the framework, the ontology represents the domain knowledge related to the traffic accidents and supports the data retrieval based on users' requirements. A new spatial clustering method that takes into account the numbers and severity levels of accidents is proposed for risk mapping. To demonstrate the framework, a system prototype has been implemented. A case study in the city of Calgary is also discussed.

Keywords: Spatial clustering, GIS, Ontology, Traffic accident, Road safety.

1 Introduction

Road traffic accidents that cause injuries and fatalities are a social and public health challenge [1]. The World Health Organization estimates over 1 million people are killed each year in road collisions, which is equal to 2.1% of the annual global mortality, resulting in an estimated social cost of $518 billion [20]. In Canada, about 3,000 people are killed every year [21]. To significantly reduce traffic fatalities and serious injuries on public roads, identification of the hidden patterns behind the accidents' records is critical [1].

In most cases, the occurrences of traffic accidents are seldom random in space and time, but form clusters that indicate accident concentration areas in geographic space [1]. A concentration area is defined as an area or location where there is a higher likelihood for an accident to occur based on historical data and spatial dependency. Over the years, various spatial concentration detection methods have been proposed and applied to discover traffic accidents' concentration patterns, with the final result of a single map [1, 2, 8, 18, 19, 20, 27, 30, 34].

Traffic analysts and the general public, however, are actually interested in accident concentration areas in terms of specific conditions, such as different time intervals, weather conditions, and road surface conditions. For example, a traffic analyst may be

D. Pfoser et al. (Eds.): SSTD 2011, LNCS 6849, pp. 21–38, 2011.

interested in an accident concentration area for the downtown area during workday rush hours, so that he can locate the most vulnerable locations to accidents and analyze the reasons behind these accidents to improve road safety. A new driver may be interested in a map of the northwest part of city during winter weekends, which can help him avoid dangerous areas when practicing driving in the northwest part of the city in winter time. These different conditions reflect different requirements from users. Therefore, risk maps that meet users' manifold requirements are necessary.

Nevertheless, integration of users' requirements into generating different concentration maps is not an easy task. The first subtask is the selection of the proper datasets based on different users' requirements. One naive option is the translation of users' requirements into traditional database queries. For example, in the former example, they need to define "downtown area" or "rush hours", so users can handle the traffic accident data at the data level, which can sometimes be quite challenging for users who have no or only limited knowledge of the study area and dataset. The second option is the handling of users' requirements at the knowledge level. Knowledge of the study area and datasets are well defined and represented. Users do not need to know the details of the area, and the proper datasets can be retrieved based on their requirements.

After selecting the proper datasets, the second task for generating a concentration map is the application of the proper traffic accident concentration detection methods. Existing traffic accident concentration detection methods treat each accident as a point and then apply traditional point pattern analysis methods on the extracted points. When defining the concentration area, they consider only the number of accidents, ignoring the severity levels associated with the accidents. However, accidents have different severity levels, including fatality, injury, and property damage only (PDO) [6]; and, each level should be treated differently. For example, accidents with fatalities and injuries put more strain on the network than PDO accidents. An intersection with frequent fatal accidents may be more dangerous than an intersection with PDO accidents, in cases where both intersections have the same number of accidents.

Ontology is the explicit specification of a conceptualization [11]. It provides domain knowledge relevant to the conceptualization and axioms for reasoning with it. For accident domain ontology, it has a conceptual and taxonomical representation of accident data, providing domain knowledge, including non-spatial and spatial concepts and definitions relate to the traffic accidents. This enables the user to pose semantic queries with a semantic representation of traffic accident concepts. Therefore, it can provide a knowledge source that supplements domain experts and integrates users' goals into the selection procedure.

Spatial clustering is a data mining method that separates objects into groups (called clusters) based on spatial and non-spatial attributes [13]. For density-based clustering methods, clusters are regarded as regions in the data space where the objects are dense. These regions may have an arbitrary shape, and the points inside a region may be arbitrarily distributed. Therefore, it can be used to find accident concentrations in geographical space.

In this paper, we propose a novel traffic accident risk mapping framework. This framework is based on traffic accident domain ontology that is built at a high generic level. This framework includes the interactive input module, an accident domain ontology, an ontology reasoner, datasets, a clustering engine with a risk model, and a map publishing module. With this framework, spatial concepts are well defined; and, each accident record is described by several characteristics, such as crash time, location and environmental factors.

The users' requirements are translated into a set of subtasks by performing reasoning on the ontology. Appropriate datasets can be chosen by a selection procedure guided by the ontology with respect to the user's goals. After identifying the proper datasets, the proposed density-based spatial clustering method with a user selected traffic accident risk model is applied to the dataset to identify the traffic accident risk area. Finally, the different traffic accident risk maps that meet users' requirements can be generated and published. The following paragraphs summarize the contributions of the paper.

First, we propose an ontology-based traffic accident risk-mapping (ONTO_TARM) framework. In ONTO_TARM, the ontology represents the domain knowledge, including the non-spatial and spatial concepts and definitions related to the traffic accidents, and helps to retrieve information based on users' goals. The framework performs reasoning based on the user's input, returns the most suitable dataset from the raw historical dataset to generate the user's own risk map.

Second, a novel spatial clustering method, density-based clustering for traffic accident risk (DBCTAR), is proposed. This new clustering method has been extended from DBSCAN (Density-Based Spatial Clustering of Applications with Noise) [8] by considering both total accident numbers and the severity levels of the accidents. In a simplified version, the value of equivalent property damage only is calculated for each cluster and used as a risk index. The proposed method is also used for the road network environment. The clustering result shows the boundary of each cluster subject to the boundary of the network.

Third, a prototype of the proposed framework has been implemented. Real traffic accident data have been populated into the prototype. With the web-based publishing module, users can view maps through their web browser.

The paper is organized as follows: Section 2 provides a literature review on the methods of identifying accident concentrations and ontology in traffic accidents; Section 3 describes the proposed ONTO_TARM framework and the DBCTAR spatial clustering method; and, Section 4 presents the implementation of the risk mapping prototype system with a case study in the Calgary area. Section 5 concludes the paper and discusses future research directions.

2 Related Works

2.1 Accident Concentration Detection

Identification of an accident concentration area in a road network is usually simplified into a task that detects concentrations of point events in a network. Various methods have been proposed and applied, including spatial autocorrelation methods and kernel density methods.

The autocorrelation methods detect whether a given point distribution differs from a random distribution throughout the study area [4], such as Ripley's K-function, Getis's G-statistic and Moran's I. These methods can be classified as global methods [34] and local methods [8][2], based on whether the methods apply the spatial autocorrelation significance test globally or locally within the study area. For the identification of accident concentrations, positive spatial autocorrelation indicates that accident distribution is clustered, which means the concentration may happen in the study area. A global method cannot reveal the location of clusters. Local methods need to aggregate point-based accidents into basic spatial units (BSUs). There is no unique solution for the division. A different division may lead to different results at different scales.

A kernel density method aims at calculating and producing a density surface from point features. Here, the density is the total number of accidents per unit area. Usually, this method divides the whole area into grid cells. A hump or kernel with a mathematical equation, called a kernel function, is applied to each accident point. A kernel function is a weighting function that is used to estimate variables' density ranging from 1 to 0 with a given radius, depending on its distance to the accident point. All the values from different points at a given cell are then totaled as the density estimation value.

The traditional kernel method is a two-dimensional planar method, which generates a continuous raster surface with equal-sized cells covering the whole area in which the network located. The raster cells with high values indicate the accident concentration areas. The planar method has inherent limitations: First, all of the accidents are only located on streets. The cells that are located outside of the road have risk values that do not match the reality. Second, the density of the road network is ignored [28]. Even if some grid cells have the same density values, they may include different lengths of road sections. The real density values of road network are, therefore, biased. Third, the choice of bandwidth affects the outcome surface.

To overcome these limitations, many studies have attempted to extend the conventional planar method to network spaces. Flahaut et al. [9] developed a kernel density estimation method based on a simple network. Borruso [2] considered the kernel as a density function based on network distances. Xie and Yan [33] pointed out that point events in the network are better measured with density values per linear unit, but they did not consider the bias of their estimator explicitly. Okabe et al. [19] discussed three types of network kernel density estimation. The equal split kernel function and the equal split continuous kernel function have improved the kernel estimation methods. However, no kernel function exists that satisfies a combination of precisely estimating the density of events on a network without bias [28].

Almost all the methods have their own weaknesses in addition to the limitations illustrated. First, all of the above methods handle accident analysis at the data level. They fail to take into consideration users' requirements. As discussed previously, accident distributions are totally different due to many factors, such as time, weather or road surface state. For example, Fig. 1 provides a histogram showing the accident statistics on the 16th Avenue N, Calgary, Alberta, with the same time interval of the day for 4 years (1999-2002). From this chart, it can be seen that the accident numbers vary at different time intervals, meaning that in a specific time range of the day, the concentration of accidents should be different. Thus, it is not difficult to conclude that, given certain factors (such as the weather condition), the risks of the road network should be different. However, current methods do not consider different factors. Although datasets can be generated from database query, the processing remains at the data level, not at the knowledge level. Therefore, current methods cannot satisfy users' needs.

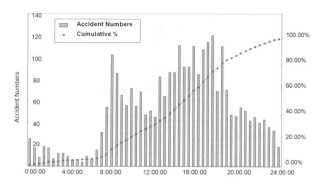

Fig. 1. Accident statistics in the same time intervals on 16th Ave N in Calgary

Second, all of the above methods ignore the severity level of accidents. When users consider the accident risk of the road network, the assumption is that, if an area on the map is marked as high risk, that area should be more vulnerable to accidents. However, the nature of the accidents may be different from one another. One of the obvious distinctions is the severity level. For example, a rear-end accident should not be considered the same as an accident with a fatality. Thus, the risk not only depends on the number of accidents, but also on the severity level of accidents. Unfortunately, most of the previous studies take the accident records as a point without considering the severity levels, and most of the statistical analyses are only based on the number of the accidents.

2.2 Spatial Clustering

Spatial clustering is the process of grouping a set of objects into groups (called clusters) based on their geographical locations and other attributes. Spatial clustering algorithms exploit spatial relationships among the data objects in determining inherent groupings of the input data. Spatial clustering methods can be classified into partitioning methods, hierarchical methods, density-based methods, grid-based methods and model-based methods [13][14]. Among these different types of methods, a density-based clustering method is the most suitable for traffic accident risk analysis, because it can discover arbitrarily shaped clusters based on density. The following paragraphs discuss the classic density-based clustering method on which our research is based.

DBSCAN [8] was the first and is the most classic density-based spatial clustering method. The key concept is the definition of a new cluster or extension of an existing cluster based on a neighborhood. The neighborhood around a point of a given radius (*Eps*) must contain at least a minimum number of points (*MinPts*). Given a dataset (*D*), a distance function (*dist*) and parameters *Eps* and *MinPts*, the following definitions are used to define DBSCAN.

For an arbitrary point p, $p \in D$, the neighborhood of p is defined as $N_{Eps}(p) = \{q \in D \mid dist(p,q) \leq Eps \}$. If $\mid N_{Eps}(p) \mid \geq MinPts$, then p is a core point of a cluster. If p is a core point and q is p's neighbor, q belongs to this cluster, and each of q's neighbors is examined to see if it can be added to the cluster. Otherwise, point q is labeled as noise. The expansion process is repeated for every point in the neighbor-hood. If a cluster cannot be expanded further, DBSCAN chooses another arbitrary unlabelled point and repeats the process. This procedure is iterated until all points in the dataset have been placed in clusters or labeled as noise.

Compared with traditional concentration detection methods, a density-based spatial clustering method can inherently discover the concentrations; dataset segmentation is not needed at the beginning of the process. It can also find arbitrary concentration shapes from the dataset. However, most density-based clustering methods do not consider non-spatial attributes of the data point and cannot be directly applied to data-sets on a spatial network.

2.3 Ontology in Traffic Accidents

Ontology is an explicit representation of knowledge. It is a formal, explicit specifica-tion of shared conceptualizations, representing the concepts and their relations that are relevant for a given domain of discourse [11]. Generally, ontology contains basic modeling primitives such as classes or concepts, relations, functions, axioms and instances [11][12].

The last few years have seen a growing interest [21] in approaches that have do-main ontology add a conceptual level over the data, which is used as a middle layer between the user and the dataset, especially with spatial data. Several ontological approaches are proposed for road accidents. Hwang [14][16] built a high-level con-ceptual framework that includes traffic accident domain ontology. However, this research focused on the task ontology and did not consider the disparity of accidents. Yue et al. [35] presented an ontology-based prototype framework for traffic accident management from a hierarchical structured point of view. However, their ontology was designed only for the traffic management system. It rarely considered the spatial knowledge associated with the accidents.

3   Methodology

3.1   An Ontology-Based Traffic Accident Risk Mapping Framework

As discussed in Section 2, users may demand different road traffic risk maps under various users' requirements and conditions. Fig. 2 shows the proposed ONTO_TARM framework.

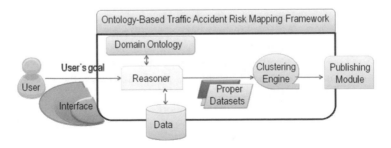

Fig. 2. Ontology-based traffic accident risk mapping (ONTO_TARM) framework

The ONTO_TARM framework consists of 6 components: an interface, an accident domain ontology, an ontology reasoner, traffic accident datasets, a clustering engine, and a publishing module. The users' goals provide the input, which can be represented as natural language, with some predefined words, or it may use a more general format. For example, users can input the time range or select predefined time sections, such as morning or rush hours. The traffic accident domain ontology represents the knowledge of accidents. The ontology reasoner is used to reason the knowledge represented in the ontology. It contains the classification and decomposition rules. Traffic accident datasets contain all the accident records. In this research, we assume the datasets include all the traffic accident data that different users' requirements. The clustering engine uses the proposed spatial clustering method to find the traffic concentration areas based on user' goals. The publishing module is the output part of this framework and is used for the final map generation and publication.

The whole framework works as follows: The interface handles users' goals as inputs, sending them to the reasoner. The reasoner parses the users' goals into tasks based on the domain ontology, conducts queries for each task and returns with the proper dataset. Finally, the proper dataset is sent to the clustering engine, and a risk map is generated and published. In the following subsections, each component is introduced and discussed.

Traffic Accident Domain Ontology

Traffic accident domain ontology (TADO) provides formal descriptions of the classes of concepts and the relationships among those concepts that describe road traffic accidents. The structure of TADO is based on Wang et al. [31][32].

Definition 1 (Domain ontology structure). An ontology structure of a domain is a 7-tuple $O := \{D, C, R, A, H^C, prop, att\}$, where D is the domain context identifier, C is a set called concept, R is the relation identifiers (C and R are disjoint and provide necessary conditions for membership), and A is a set of attributes to describe C and R. H^C, which is a concept hierarchy classification, is a set of hierarchical trees that define the concept taxonomy in the domain. The *prop* function relates concepts non-taxonomically: $R \rightarrow C \times C$. Each attribute in A can be treated as a specific kind of relation, where the function *att* relates literal values to concepts: $A \rightarrow C$. Elements C and R can be regarded as the high-level encapsulation of the analysis and design model for the ontology.

Definition 2 (Classification). H^C is a set directed, transitive relations: $H^C = \{ h^C \subseteq C \times C \}$, where $h^C(C_1, C_2)$ means that C_1 is a sub-concept of C_2 in the relation h^C. Usually, H^C includes a set of classification instances. Depending on the application, the classification constraints may be different. Even the same concept can be categorized into several categories.

Each component of the top-level ontology is discussed in detail.

Domain Context Identifier

In TADO, the domain context identifier D is *TrafficAccidentRecordDomain*.

Concepts

Accident records include spatial and non-spatial information. For example, each accident record has the attributes of location and accident time. Therefore, the concept set *C* of TADO includes three main classes: *GeospatialThing*, *AccidentRecord* and *AccidentCondition* to represent this information.

For the spatial information, we extend the ontology conceptual tree from the Cyc knowledge base [5] by altering the *GeographicalThing* to *GeospatialThing* with customized spatial classes. The Cyc knowledge base was selected because it is the most commonly used ontology and it contains a great quantity of common sense knowledge encoded in formal logic. *GeospatialThing* is defined as an abstract class to provide the basic classes of geospatially related concepts or entities that can be used to describe the locations of accidents. It includes subclasses *GeometricThing*, *FixedStructure* and *GeographicalRegion*. *FixedStructure* presents the facilities related to the accidents, such as the road. *GeographicalRegion* describes the geographical area with a specific boundary. Any geographical region used in TADO is an instance of *GeographicalRegion*. Various geographical regions, such as *Province*, *City*, *County*, *Community* and *CitySection*, are defined. *Province*, *City* and *County* are defined as regions with political boundaries. *Community* is derived from the census subdivisions and can be classified into city sections. *CitySection* is the region in the city that has formed over a long historical period. For example, Calgary is an instance of the class *City*. Within the boundaries of Calgary, there are around 100 communities, with each community belonging to at least one of the five Calgary *CitySections*, which are the NW, SW, SE, NE and downtown areas.

The non-spatial information describes the non-spatial properties of the accidents. It includes two main subclasses, *AccidentRecord* and *AccidentCondition*. *AccidentRecord* represents the class of the available accident record data. Any record used in TADO is an instance of *AccidentRecord*. Non-spatial properties of this class are defined in *AccidentCondition*, such as *TemporalConditions* and *EnvironmentalConditions*. The *TemporalConditions* class includes different abstract classes based on different time scales, from hourly to yearly. The temporal concepts, such as rush hours and slow hours, are also defined. The *EnvironmentalConditions* define various accident related environmental factors. Examples of these classes are *WeatherConditions* and *RoadConditions*.

Relations

Relations consist of the relationships among *GeospatialThing*, *AccidentRecord* and *AccidentCondition*. *Geospatial Relation* and *AccidentCondition Relation* are the two major types of relations. *Geospatial Relation* includes the spatial relationships among *GeospatialThing*. There are three kinds of geospatial relations: direction, distance and topological relations. A direction relation describes the orientation in space of some objects, such as *north*, *south*, *up*, *down*, *behind* and *front*. A distance relation specifies the distance from an object to a reference object. Some examples of distance relations are *far* and *close-to* (near). A topological relation describes the location of an object relative to a reference object [7]. Topological relations include *disjoint*, *contains/insideof, overlap, cover/covered* and *meet. AccidentCondition Relation* defines relations between *AccidentRecord* and *AccidentCondition*. The relationships also include temporal and non-temporal relationships. Examples of temporal relationships are *at time point of, around time, in the range of, early than, later than*. An example of a non-temporal relationship is *with the condition of*.

Attributes

Attributes define the attributes and properties of the above classes and their subclasses. One example of spatial attributes is *location*. Some examples of non-spatial attribute include *hasName, hasValue, hasTime, hasDate*.

Classification

Classification includes the hierarchical classification used for TADO. Fig. 3 shows the top-level ontology defined in TADO and the hierarchical classification of *GeospatialThing* and *AccidentCondition*. As shown in Fig. 4, *GeospatialThing* is the top class of all spatial things in TADO. In the three subclasses, the *GeometricThing* class includes abstract geometric shapes. *FixedStructure* presents the facilities related to the accidents and includes classes such as *Building, Station, Roadway*. The *Roadway* class includes subclasses *Expressway, Highways, Majorroad*, and *Localroad*. Under the class *GeographicalRegion*, we have *EcologicalRegion, GeoculturealRegion, GeopoliticalRegion*. Subclasses *Country, Province, City, County* and *Community* belong to the *GeopoliticalRegion*.

AccidentCondition can be classified into *TemporalConditions* and *Environmental-Conditions*. The *TemporalConditions* include *Instant, Interval* and *DateTimeDescription* class. The *EnvironmentalConditions* include the *WeatherCondition*, the road *RoadSurfaceCondition*, the *RoadCondition, LightCondition*. Each has detailed subclasses. For example, the *WeatherCondition* includes *SevereWeather* and *FairWeather*. The *high_wind, fog_smog_smoke_dust, hail_sleet, raining* and *snow* are all in the severe weather condition class.

Function prop

The function *prop* relates concepts non-taxonomically among the concepts. It can be an instance of geospatial relation or non-geospatial relation, such as *underconditionof()* and *insideof()*. Here, we use *insideof* as an example. In Fig. 4, *City* and *Community* are two classes (concepts) in the ontology, and class *City* is not a super-class of the class *Community*; therefore, *insideof(City, Community)* represents whether a community is inside a city. Thus, *insideof* defines one type of relationship between instances of the two classes.

Function att

The function *att* is used to describe the properties or attributes of a class. For example, *hasName* is used to define the names of instances of each class.

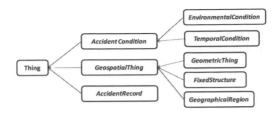

Fig. 3. Top-level conceptual three in TADO

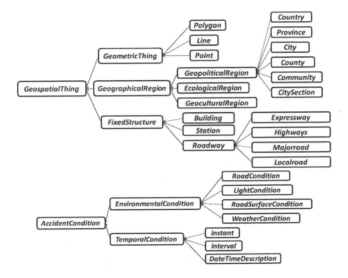

Fig. 4. Classification of TADO

Reasoner

In the framework, the ontology reasoner is used to reason the knowledge represented in the ontology. The input of the reasoner is the user's goals, and the output is a set of proper accident records selected from the raw dataset. After generating a general task from a user's goals, the spatial task identifies the target geographical area. The non-spatial task identifies the proper temporal and environmental factors. For example, if a user's goal is the generation of a risk map for accidents that happened in rush hours on workdays in downtown Calgary, the reasoner first finds "downtown Calgary". A spatial query task is generated as shown in Fig. 5.

The non-spatial task is materialized by the task "accidents that happened in rush hours on workdays with severe weather". This task includes two main components: a temporal condition task and an environmental condition task, as shown in Fig. 6. The non-spatial attributes are both complex tasks that need further decomposition. The temporal condition task is composed of two subtasks: finding "rush hours" and finding "working days". The weather condition task is finding "severe weather" and will return conditions including high_wind, fog_smog_smoke_dust, hail_sleet, raining, and snow.

```
sub-task: findDowntownAreaTask
defgoal find Calgary Downtown Area
Input:
(object (is-a City) (object?ci) (hasName "Calgary"))
(object (is-a CityfSection) (object?cs)
  (hasName "Downtown Area") (insideOf?ci))
(object (is-a community) (object?co) (insideOf?ci)
  (belong-section?cs))
Output:
(object (is-a $?community) (object? co))
```

Fig. 5. Pseudocode of spatial query task `findDowntownAreaTask`

```
sub-task: findAccidentConditionTask
defgoal find Accident Conditions
Input:
(object EnvironmentalCondition?ec
      (RoadSurface-condition "dry"),
      (RoadCondition "straight" || "curve"),
      (WeatherCondition findSevereWeatherTask())
      (LightCondition "artificial"||"nature"))
(object TemporalCondition?tc
      (DateTimeDescription? findWorkingdaysTask())
      (Interval? findRushHoursTask()))
(object (is-a AccidentCondition) (object?ac)
(include?ec & tc))
Output:
(object (is-a $?AccidentCondition) (object?ac))

sub-task: findSevereWeatherTask
defgoal find severe weather conditions
Input:
(object (is-a WeatherCondition) (object?we) include?
"high_wind"||"fog_smog_smoke_dust"||
"hail_sleet"||"raining"||"snow")
Output: (object (is-a WeatherCondition) (object?we))

sub-task: findWorkingdaysTask
defgoal find working days
Input: (object (is-a calendarDay) (object?cd)
is-a?weekday is-not-a?holiday )
Output: (object (is-a calendarDay) (object?cd))

sub-task: findRushHoursTask
defgoal find rush hours
Input: (object (is-a timerange) (object?tr)
equal?TimeofRushHour )
Output: (object (is-a timerange) (object?tr))
```

Fig. 6. Pseudocode of non-spatial query task findAccidentConditionTask

3.2   Density-Based Clustering for Traffic Accident Risk (DBCTAR)

The traffic accident risk in this paper is derived from the accident concentration area. The accumulated accident number is the most common way to reflect the risk level. However, as fatality and injury accidents put more strain on the road network and increase the economic burden on society, these accidents need to be considered differently from PDO accidents, in order to account for their larger effects [25]. Therefore, the risk area should be defined by both the frequency and degree of the severity.

Since the risk areas are arbitrary shapes on the road network, the proposed clustering method is a density-based clustering method for traffic accident risk (DBCTAR). This clustering method is extended from DBSCAN, which is described in Section 2.2.

To consider the severity level of each accident, we propose to assign different weights to accidents with different severity levels. Within a given accident dataset D, we define a variable *RiskIndex* as follows:

$$RiskIndex = \sum_{i=1}^{n} W_i * Count(S_i) \qquad (1)$$

where S_i is the ith severity level, *Count*() is a function to get the total number of accidents at that level, and W_i is the weight assigned to the ith severity level. The *Riskindex* not only considers the number of accidents, but also takes into account the severity level. A new parameter *MinRisk*, which is the threshold of *RiskIndex*, is also defined.

Ideally, no two accidents have the same severity level. However, for the practical cases, assigning unique weights to each accident is not feasible. In road safety research, accident records are usually classified into 3 classes: fatality, injury and PDO. Accident with fatalities and/or injuries can be converted into equivalent property damage only (EPDO) accidents [25].

$$EPDO = W_1 * Count(Fatal) + W_2 * Count(Injury) + W_3 * Count(PDO) \qquad (2)$$

EPDO is calculated by assigning different weighting schemes, as shown in Fig. 7. One of the most commonly used conversion weight settings is recommended by PIARC (Permanent International Association of Road Congresses) with W_1=9.5; W_2=3.5; and W_3=1 [22].

Model	Ratio	Source
1	1:1:1	Simple Total Crash Count
2	9.5:3.5:1	PIARC
3	76.8:8.4:1	North Carolina DOT
4	136.13:4.94:1	Ohio DOT
5	779.9:13.88:1	Transport Canada
6	1300:90:1	Federal Highway Administration

Fig. 7. Different weight models for accident severity level (DOT: Department of Transportation) (taken from [25])

EPDO can be considered as a simplified format of *Riskindex*. To determine the parameter *MinRisk*, we use a method similar to the *k-dist* function [8]. We first build the most significant *k-dist* graph to identify the most suitable *k* value, then calculate the *k*-nearest neighbor's *Riskindex* and sort these values by distance. The threshold *MinRisk* point is located near the first "valley" of the sorted *k*-nearest risk index graph.

With DBCTAR, when the cluster extends an existing cluster from a neighborhood, the neighborhood around a point of a given radius (*Eps*) must contain at least a minimum number of points (*MinPts*) and has a *RiskIndex* larger than *MinRisk*. This algorithm is used in a network environment; therefore, we use road network distance rather than the Euclidian distance. The core point is an accident that has at least *MinPts* accidents within the search distance *Eps*; and, the *RiskIndex* of the accidents within the search distance is larger than *MinRisk*. This core point criteria can be stated as follows: For $p \in D$, the neighborhood of p is defined as $N_{Eps}(p) = \{q \in D \mid network\_dist(p,q) \le Eps \}$. If $\mid N_{Eps}(p) \mid \ge MinPts$ AND $RiskIndex(p) > MinRisk$, then p is a core point of a cluster.

If p is a core point and q is p's neighbor, then q belongs to this cluster; and, each of q's neighbors is examined. Otherwise, if p is not a core point, point q is labeled as noise. The algorithm ends when every point is classified as in a cluster or labeled as noise.

When we implemented the DBCTAR algorithm in the prototype, we simplified the *Riskindex* as follows:

$$RiskIndex = W_1 * Count(\text{Fatal}) + W_2 * Count(\text{Injury}) + W_3 * Count(\text{PDO}) \qquad (3)$$

where W_1, W_2 and W_3 depend on the risk index model that the user selects from Fig. 7. These models use different weighting schemes that reflect different perspectives of the significance of each kind of accident. For example, Transport Canada uses the weight of 13.88 for accidents with injury, which suggests the injury accident is more important than the weight of 3.5 recommended by the PIARC.

4 Implementation and Case Study

To demonstrate the ONTO_TARM framework, a prototype has been developed for the traffic risk map generation and web publication. A graphical user interface has been implemented for setting users' goals. The accident domain ontology is represented with Protégé-OWL 4.0 software [23, 27]. The DBCTAR algorithm and a map generator is implemented using C# with ArcObject 9.3. DBCTAR identifies accident clusters with risk index values. The map generator transforms the clusters into traffic accident concentration areas with different colors for better visualization. The traffic concentration map can be exported as a KML file. An online platform based on Google Map with 3D viewer is also implemented with Apache 4 to publish the KML file.

The testing database included all the reported collisions on the roads in Alberta from 1999 to 2005. The total number of records was more than 770,000. Each record had more than 60 columns of properties, such as date, time, on street, at intersection with street, severity levels and weather conditions. The data from 1999 to 2004 was used to generate traffic accident risk maps and the data from 2005 was used for validation. Locations of the accidents were geocoded with the geocoding service of Google Maps API v3.

Suppose the user's goal is to find the risk map in the downtown area of Calgary at rush hours in the morning. This task refers to the downtown area of Calgary. Without geographical knowledge of Calgary or a definition of rush hours, the traditional method cannot proceed, due to the lack of domain ontology. However, ONTO_TARM first generates the task based on the user's goal and performs the spatial reasoning. The spatial query task is shown in Fig. 5. In the ontology, Calgary is an instance of the *City* class; and, all census units – communities in Calgary – are represented as instances of the *Community* classes. The downtown area is an instance of *CitySection*. This task finds the communities inside of Calgary that belong to the downtown area, returning with five communities – Eau Claire, Chinatown, downtown west end, downtown east village and downtown commercial core.

As the second step, non-spatial reasoning is generated to filter the dataset. The non-spatial task is similar to the task shown in Fig. 6. The final dataset based on the ontology-based query includes 1,032 records. DBCTAR then identifies clusters of this dataset. Maps derived from the clustering results are generated by the map generator.

Fig. 8. Result from the map generator – risk map of morning rush hour (7:30-9:00AM) of Calgary downtown area

Fig. 8 shows one of the traffic accident risk maps. This map has been generated with the risk model parameters recommended by PIARC. The parameter *MinRisk* was set to 8, *Eps* was set to 45 meters, and *MinPts* was set to 3. The validation with the 2005 dataset shows that 66.5% of accidents in 2005 were located in the risk area.

We also compared the risk map generated by our method with the traditional kernel density method. Fig. 9(a) shows the density estimation result when the radius was set to 40 meters and cell size was 10×10 meters. The kernel function was set as a Gaussian function.

(a)   (b)

Fig. 9. Comparison with kernel density method

Figs. 9(a) and 9(b) are two zoomed maps of the same area. Indicated by the arrows, one risk area (left arrow) is around 5 Ave and 5 St SW; and, a second one (right arrow) is around 5 Ave and Macleod Ave SE. According to the 1999-2004 dataset, there were 29 accidents (25 accidents were PDO, 4 were injury) that happened around the first intersection; and, there were 30 accidents (27 accidents were PDO, 3 were injury) happened around the second intersection.

Table 1. Comparison at the intersections

	Intersection 1	Intersection 2
Location	Around 5 Ave & 5 St SW	Around 5 Ave & Macleod Ave SE
Total number of accidents in 1999-2004	29	30
Number of PDO accidents	25	27
Number of accidents with injury	4	3
Risk index with DBCTAR (Transport Canada model)	80.25	68.64
Average density estimation (per 100m^2)	1.4	1.5
Accidents located in the risk area in 2005	5	2

Because the kernel density method only considers the total number of accidents, Intersection 1 has an average density potential of 1.4/100m^2, which is less than Intersection 2, with an average density value of 1.5/100m^2. However, according to our DBCTAR method with the Transport Canada model, Intersection 1 has a higher index (80.25) than Intersection 2 (68.64), as this method also takes into account the severity level of the accidents.

During 2005, there were 5 accidents near Intersection 1, and only 2 accidents were located in Intersection 2. It shows that DBCTAR method is more suitable for determining the accident risk.

The clustering result in the format of a shapefile can be transferred into a KML file, which can be overlaid on Google Map with an online publishing platform (Fig. 10a). To achieve a better visualization result, 3D viewer was also added. Fig. 10b shows another risk map in 3D view.

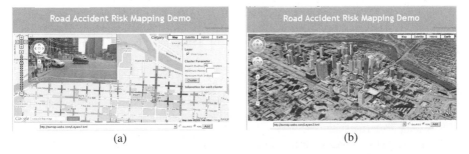

(a)   (b)

Fig. 10. Road accident risk mapping web publishing platform

5 Conclusions and Future Work

This research proposes an ontology-based framework to generate road accident risk maps. In the framework, the ontology represents spatial and non-spatial knowledge about traffic accidents and returns the most suitable dataset from the raw historical datasets based on users' goals. To determine the traffic risk area, a density-based spatial clustering method (DBCTAR) with the ability to handle the accident severity level is also proposed. To demonstrate the system, a prototype of the system has been implemented; and, the preliminary results from the case studies are promising.

This research can be extended in the future in the following ways:

1. A system prototype would benefit from an improved ontology reasoner and a more powerful map generator. The current prototype is still a proof of concept type, using the Jena reasoner actually only works on the level of RDFS(Resource Description Framework Schemas) and needs to be upgraded to handle more complex ontology queries. The map generator needs to be extended with regard to its efficiency. In our experiments, the execution time for more than 10,000 selected records with a complicated road network could take longer than 20 minutes.
2. The current prototype cannot provide recommendations for the weight model selection and clustering parameter settings. We will investigate different risk patterns and collaborate with the experts from civil engineering to find out recommendation values.
3. The risk index can be better defined. For example, the affect of traffic volume should be considered. In addition to the severity level, other properties of the accidents can be adopted in the risk index model.

References

1. Anderson, T.K.: Kernel density estimation and K-means clustering to profile road accident hotspots. Accident Analysis & Prevention 41(3), 359–364 (2009)
2. Black, W.R., Thomas, I.: Accidents on Belgiums motorways: a network autocorrelation analysis. Journal of Transport Geography 6 (March 23-31, 1998)
3. Borruso, G.: Network density estimation: analysis of point patterns over a network. In: Osvaldo, G., Marina, L.G., Vipin, K., Antonio, L., Heow, P.L., Youngsong, M., David, T., Chih Jeng, K.T. (eds.) ICCSA 2005. LNCS, vol. 3482, pp. 126–132. Springer, Heidelberg (2005)
4. Boots, B.N., Getis, A.: Point Pattern Analysis, Sage Newbury Park, CA (1988)
5. CYC, http://www.cyc.com/2003/04/01/cyc
6. Doherty, S.T., Andrey, J.C., MacGregor, C.: The situational risks of young drivers: The influence of passengers, time of day and day of week on accident rates. Accident Analysis & Prevention 30(1), 45–52 (1998)
7. Egenhofer, M.J., Franzosa, R.D.: Point-set topological spatial relations. International Journal of Geographical Information Systems 5(2), 161–174 (1991)
8. Ester, M., Kriegel, H., Sander, J., Xu, X.: A Density-Based Algorithm for Discovering Clusters in Large Spatial Databases with Noise. In: Proc. Second International Conference on Knowledge Discovery and Data Mining, pp. 226–231. AAAI Press, Portland (1996)

9. Flahaut, B., Mouchart, M., Martin, E.S., Thomas, I.: The local spatial autocorrelation and the kernel method for identifying black zones a comparative approach. Accident Analysis & Prevention 35, 991–1004 (2003)
10. Getis, A.: A history of the concept of spatial autocorrelation: a geographer's perspective. Geographical Analysis 40, 297–309 (2008)
11. Gruber, T.R.: A translation approach to portable ontologies. Knowledge Acquisition 5(2), 199–220 (1993)
12. Gómez-Pérez, A., Fernández-López, M., Corcho, O.: Ontological engineering: with examples from the areas of knowledge management, e-commerce and the Semantic Web. Springer, Heidelberg (2004)
13. Han, J., Kamber, M., Tung, A.K.H.: Spatial Clustering Methods in Data Mining: A Survey. In: Miller, H.J., Han, J. (eds.) Geographic Data Mining and Knowledge Discovery. Taylor and Francis, London (2001)
14. Han, J., Kamber, M.: Data Mining: Concepts and Techniques, 2nd edn. Morgan Kaufmann, San Francisco (2006)
15. Hwang, J.: Ontology-based spatial clustering method: case study of traffic accidents. Student Paper Sessions, UCGIS Summer Assembly (2003)
16. Hwang, S.: Using Formal Ontology for Integrated Spatial Data Mining. In: Computational Science and Its Applications – ICCSA 2004, pp. 1026–1035 (2004)
17. Maedche, A., Zacharias, V.: Clustering ontology-based metadata in the semantic web. In: Elomaa, T., Mannila, H., Toivonen, H. (eds.) PKDD 2002. LNCS (LNAI), vol. 2431, pp. 348–360. Springer, Heidelberg (2002)
18. Okabe, A., Yamada, I.: The K-function method on a network and its computational implementation. Geographical Analysis 33(3), 271–290 (2001)
19. Okabe, A., Satoh, T., Sugihara, K.: A kernel density estimation method for networks, its computational method and a GIS-based tool. International Journal of Geographical Information Science 23(1), 7–32 (2009)
20. Peden, M., Scurfield, R., Sleet, D., Mohan, D., Hyder, A.A., Jarawan, E., Mathers, C. (eds.): World Report on Road Traffic Injury Prevention. World Health Organization, Geneva (2004)
21. Peuquet, D.J.: Representations of Space and Time. Guilford, New York (2002)
22. PIARC, Road Safety Manual, World Roads Association Cedex (2003)
23. Protégé, http://protege.stanford.edu/index.html
24. RememberRoadCrashVictims.ca (2009), http://www.RememberRoadCrashVictims.ca
25. Rifaat, S.M., Tay, R.: Effect of Street Pattern on Road Safety: Are Policy Recommendations Sensitive to Different Aggregations of Crashes by Severity? Transportation Research Record: Journal of the Transportation Research Board, 58–65 (2010)
26. Shino, S.: Analysis of a distribution of point events using the network-based quadrat method. Geographical Analysis 40, 380–400 (2008)
27. Smith, M.K., Welty, C., McGuinness, D.L.: OWL Web Ontology Language Guide. W3C (2004), http://www.w3.org/TR/owl-guide/
28. Steenberghen, T., Aerts, K., Thomas, I.: Spatial clustering of events on a network. Journal of Transport Geography 18, 411–418 (2010)
29. Steenberghen, T., Dufays, T., Thomas, I., Flahaut, B.: Intra-urban location and clustering of road accidents using GIS: a Belgian example. International Journal of Geographical Information Science 18(2), 169–181 (2004)
30. Stefanakis, E.: NET-DBSCAN: clustering the nodes of a dynamic linear network. International Journal of Geographical Information Science 21(4), 427–442 (2007)

31. Wang, X., Gu, W., Ziébelin, D., Hamilton, H.: An Ontology-Based Framework for Geo-spatial Clustering. International Journal of Geographical Information Science 24(1), 1601–1630 (2010)
32. Wang, X., Hamilton, H.J.: Towards An Ontology-Based Spatial Clustering Framework. In: Proceedings of the Eighteenth Canadian Artificial Intelligence Conference (AI 2005), Victoria, Canada, pp. 205–216 (2005)
33. Xie, Z., Yan, J.: Kernel density estimation of traffic accidents in a network space. Computers, Environment and Urban Systems 32, 396–406 (2008)
34. Yamada, I., Thill, J.-C.: Comparison of planar and network K-functions in traffic accident analysis. Journal of Transport Geography 12, 149–158 (2004)
35. Yue, D., Wang, S., Zhao, A.: Traffic Accidents Knowledge Management Based on Ontology. In: Proceedings of the 2009 Sixth International Conference on Fuzzy Systems and Knowledge Discovery, vol. 07, pp. 447–449. IEEE Computer Society, Los Alamitos (2009)

Comparing Predictive Power in Climate Data: Clustering Matters

Karsten Steinhaeuser[1,2], Nitesh V. Chawla[1], and Auroop R. Ganguly[2]

[1] Department of Computer Science and Engineering,
Interdisciplinary Center for Network Science and Applications,
University of Notre Dame, Notre Dame IN 46556, USA
{ksteinha,nchawla}@nd.edu
[2] Computational Sciences and Engineering Division,
Oak Ridge National Laboratory, Oak Ridge TN 37831, USA
gangulyar@ornl.gov

Abstract. Various clustering methods have been applied to climate, ecological, and other environmental datasets, for example to define climate zones, automate land-use classification, and similar tasks. Measuring the "goodness" of such clusters is generally application-dependent and highly subjective, often requiring domain expertise and/or validation with field data (which can be costly or even impossible to acquire). Here we focus on one particular task: the extraction of ocean climate indices from observed climatological data. In this case, it is possible to quantify the relative performance of different methods. Specifically, we propose to extract indices with complex networks constructed from climate data, which have been shown to effectively capture the dynamical behavior of the global climate system, and compare their predictive power to candidate indices obtained using other popular clustering methods. Our results demonstrate that network-based clusters are statistically significantly better predictors of land climate than any other clustering method, which could lead to a deeper understanding of climate processes and complement physics-based climate models.

1 Introduction

Cluster analysis is an unsupervised data mining technique that divides data into subsets (called *clusters*) of elements that are – in some way – similar to each other [13]. As such, it is a versatile analysis tool that has been employed in a wide range of application settings including image segmentation [31], text and document analysis [22], and bioinformatics [1]. Clustering has also been applied for mining climate, ecological, and other environmental data. Examples include definition of ecoregions via multivariate clustering [11], automatic classification of land cover from remotely sensed data [16], and the definition of climate zones [5]; for a more complete survey see [10].

In the domain of climate data sciences, clustering is especially useful for discovery or validation of climate indices [23]. A *climate index* summarizes variability at local or regional scales into a single time series and relates these values

D. Pfoser et al. (Eds.): SSTD 2011, LNCS 6849, pp. 39–55, 2011.
© Springer-Verlag Berlin Heidelberg 2011

to other events [33]. Let us consider one particular task: the extraction of ocean climate indices from historical data. For instance, one of the most studied indices is the Southern Oscillation Index (SOI), which is strongly correlated with the El Niño phenomenon and is predictive of climate in many parts of the world [20]; see [35] for other examples. Thus, ocean dynamics are known to have a strong influence over climate processes on land, but the nature of these relationships is not always well understood.

In fact, many climate indices – including the SOI – were discovered through observation, then developed more formally with hypothesis-guided analysis of data. However, given the increasing availability of extensive datasets, climate indices can also be extracted in a data-driven fashion using clustering [23,19]. This approach presents a unique set of challenges including data representation, selection of a clustering method, and evaluation. Because it is such a difficult problem, climate scientists often resort to relatively simple algorithms such as *k*-means [5,16].

For example, as anecdotal evidence of these challenges, Loveland et al. reported that in developing a global land cover dataset, "The number of clusters created for each continent was based on the collective judgment of the project team" [16]. This and similar accounts therefore beg the fundamental question, *What is the best clustering method for climate datasets?* In response we posit that deriving clusters from complex networks, which have been shown to capture the dynamical behavior of the climate system [3,24,26,28,32], may be an effective approach. This raises the issue of evaluation and validity of discovered clusters as climate indices. As typical of any clustering task, evaluation is highly subjective, relying on the judgment of a domain expert as field data for validation can be costly or even impossible to acquire. Instead of evaluating clusters directly, however, one can measure performance in terms of an external criterion, i.e., their predictive power.

Contributions. We combine these two challenges – "choice of clustering" and "evaluation" – in a comprehensive comparative study within the context of climate indices. This paper expands upon the general methodology outlined in our prior work [24,25] but focuses on comparing different clustering algorithms as well as evaluating different regression algorithms for their predictability on climate indices. Specifically, the contributions of the present work can be summarized as follows. We extract ocean climate indices from historical data using traditional clustering methods in addition to network-based clusters (Sections 3 & 4). We then generate predictive models for land climate at representative target regions around the globe using the clusters as predictors (Section 5), using the same process as before [24]. We compare the clustering methods based on their ability to predict climate variability and demonstrate that the network-based indices have significantly more predictive power (Section 6). Finally, we provide domain interpretations of the clustering results for a selected case study to illustrate the potential value of data mining methodologies for climate science (Section 7).

To our knowledge, this is the first study to systematically address the problem of clustering climate data for the purpose of discovering climate indices.

In particular, we cluster a large corpus of ocean climate data using various popular clustering algorithms, in addition to the clusters obtained from complex networks constructed with these data. Each set of clusters then serves as input to a predictive model for land climate of the general form $f : \mathbf{x} \rightarrow y$, where $\mathbf{x}$ represents a set of cluster centroids and y is given by one of two climate variables (temperature and precipitation) at nine target regions around the globe.

Our experimental results demonstrate that the network-based clusters are statistically significantly better predictors (climate indices) than clusters obtained using traditional clustering methods. In comparing different regression algorithms, we also note that more complex methods do not necessarily improve performance for this particular predictive task.

2 Climate Data

In the following, we briefly describe the characteristics of the dataset used in our analysis as well as the pre-processing steps required for the purpose of discovering climate indices [24,25].

2.1 Dataset Description

The climate data stems from the NCEP/NCAR Reanalysis Project [14] (available at [34]). This dataset is constructed by assimilating remote and in-situ sensor measurements from around the world and is widely recognized as one of the best available proxies for global observations (it is obviously impossible to obtain exact data for the entire globe).

Although most climate indices are defined for temperature and/or pressure-related variables, we did not want to constrain ourselves by an *a priori* selection of variables. In fact, one question of interest was whether other variables also have predictive power, and hence we consider a wide range of surface and atmospheric measurements. Specifically, we include the following seven variables (abbreviation, definition in parentheses): *sea surface temperature* (SST, water temperature at the surface), *sea level pressure* (SLP, air pressure at sea level), *geopotential height* (GH, elevation of the 500mbar pressure level above the surface), *precipitable water* (PW, vertically integrated water content over the entire atmospheric column), *relative humidity* (RH, saturation of humidity above the surface), *horizontal wind speed* (HWS, measured in the plane near the surface), and *vertical wind speed* (VWS, measured in the atmospheric column).

This line of research (including [24,25]) is the first to use such a wide range of variables in climate networks studies. The data are available as monthly averages for a period of 60 years (1948-2007), for a total of 720 data points. Measurements are sampled at points (grid cells) on a $5° \times 5°$ latitude-longitude spherical grid. A schematic diagram of the data for a single time step t_i is shown in Fig. 1.

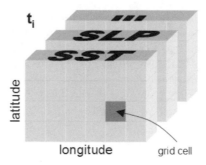

Fig. 1. Schematic depiction of gridded climate data for multiple variables at a single timestep t_i in the rectangular plane

2.2   Seasonality and Autocorrelation

The spatio-temporal nature of climate data poses a number of unique challenges. For instance, the data may be noisy and contain recurrence patterns of varying phase and regularity. Seasonality in particular tends to dominate the climate signal especially in mid-latitude regions, resulting in strong temporal autocorrelation. This can be problematic for prediction, and indeed climate indices are generally defined by the *anomaly series*, that is, departure from the "usual" behavior rather than the actual values. We follow precedent of related work [23,24,25,27] and remove the seasonal component by monthly z-score transformation and de-trending.

At each grid point, we calculate for each month $m = \{1, ..., 12\}$ (i.e., separately for all Januaries, Februaries, etc.) the mean

$$\mu_m = \frac{1}{Y} \sum_{y=1948}^{2007} a_{m,y} \tag{1}$$

and standard deviation

$$\sigma_m = \sqrt{\frac{1}{Y-1} \sum_{y=1948}^{2007} (a_{m,y} - \mu_m)^2} \tag{2}$$

where y is the year, Y the total number of years in the dataset, and $a_{m,y}$ the value of series A at *month = m, year = y*. Each data point is then transformed (a^*) by subtracting the mean and dividing by the standard deviation of the corresponding month,

$$a^*_{m,y} = \frac{a_{m,y} - \mu_m}{\sigma_m} \tag{3}$$

The result of this process is illustrated in Fig. 2(b), which shows that de-seasonalized series has significantly lower autocorrelation than the raw data. In addition, we de-trend the data by fitting a linear regression model and retaining only residuals. For the remainder of this paper, all data used in experiments or discussed hereafter have been de-seasonalized and de-trended as just described.

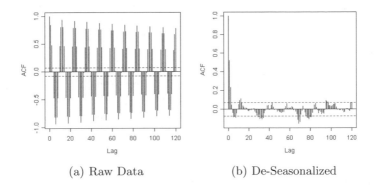

(a) Raw Data   (b) De-Seasonalized

Fig. 2. The de-seasonlized data (b) exhibits significantly lower autocorrelation than the raw data (a)

2.3 Data Representation

In this paper, we employ two distinct representations for the different analysis methods. The first is used strictly to construct climate networks; it considers each grid point as a network vertex and the corresponding data as a time series (Section 3). The second is used for the traditional clustering methods (Sections 4.2-4.5) and consists of a flat-file format, wherein each grid point is considered as an instance (row) and each time step as an attribute (column); the temporal nature of the data as well as certain aspects of the relationships between grid points is lost.

3 Climate Networks

The intuition behind this methodology is that the dynamics in the global climate system can be captured by a complex network [3,24,28]. Vertices represent spatial grid points, and weighted edges are created based on statistical relationships between the corresponding pairs of time series.

3.1 Estimating Link Strength

When dealing with anomaly series we need not consider the mean behavior, only deviations from it. Therefore, Pearson correlation (r) is a logical choice as measure of the edge strength [23,24,25,27], computed for two series A and B of length t as

$$r(A, B) = \frac{\sum_{i=1}^{t}(a_i - \bar{a})(b_i - \bar{b})}{\sqrt{\sum_{i=1}^{t}(a_i - \bar{a})^2 \sum_{i=1}^{t}(b_i - \bar{b})^2}} \tag{4}$$

where a_i is the ith value in A and $\bar{a}$ is the mean of all values in the series. Note that r has a range of $(-1, 1)$, where 1 denotes perfect agreement and - 1 perfect disagreement, with values near 0 indicating no correlation. Since an inverse relationship is equally relevant in the present application we set the edge weight to $|r|$, the absolute value of the correlation.

We should point out here that nonlinear relationships are known to exist within climate, which might suggest the use of a nonlinear correlation measure. However, Donges et al. [3] examined precisely this question and concluded that, "the observed similarity of Pearson correlation and mutual information networks can be considered statistically significant." Thus it is sensible to use the simplest possible measure, namely (linear) Pearson correlation.

3.2   Threshold Selection and Pruning

Computing the correlation for all possible pairs of vertices results in a fully connected network but many (in fact most) edges have a very low weight, so that pruning is desirable. Since there is no universally optimal threshold [21], we must rely on some other criterion. For example, Tsonis and Roebber [27] opt for a threshold of $r \geq 0.5$ while Donges et al. [3] use a fixed edge density ρ to compare networks, noting that "the problem of selecting the exactly right threshold is not as severe as might be thought."

We believe that a significance-based approach is more principled and thus appropriate here. Specifically, we use the *p-value* of the correlation to determine statistical significance. Two vertices are considered connected only if the *p-value* of the corresponding correlation r is less than 1×10^{-10}, imposing a very high level of confidence in that relationship. This may seem like a stringent requirement but quite a large number of edges satisfy this criterion and are retained in the final network.

In [24], we examined the topological and geographic properties of these networks in some detail. Suffice it to say here that for all variables the networks have a high average clustering coefficient and a relatively short characteristic path length, suggesting that there is indeed some community structure; more on this in the following section.

4   Clustering Methods

In this section we provide succinct descriptions of the clustering methods used in this comparative study; for algorithms we defer the reader to the original works, as cited. Sections 4.1 & 4.2 were developed in [24] but are included for completeness. Note that climate networks employ a network-based data representation whereas traditional clustering methods use a flat-file representation of time series at each grid cell, as described in Section 2.3.

4.1   Network Communities

This method is based on the climate networks described in Section 3. There exists a rich body of literature on the theory and applications of clustering in networks, also called *community detection* due to its origins in social network analysis [29]; other examples include discovery of functional modules in protein-protein interactions [1], characterization of transportation networks [8], and many more. However, to our knowledge we are the first to apply community detection in climate networks [24].

In choosing an appropriate algorithm for this study, three constraints guided our selection: *(i)* the ability to utilize edge weights, *(ii)* suitability for relatively dense networks, and *(iii)* overall computational efficiency. The first requirement in particular eliminates a large number of algorithms from consideration as they only work with unweighted networks. Thus, all results presented here were obtained with the algorithm described in [18] using the default parameter settings, which meets all the above criteria (tests with other algorithms produced comparable results). A fringe benefit of this algorithm is an option to determine the number of clusters from the data.

4.2   K-Means Clustering

The k-means algorithm is one of the oldest and well-known methods for cluster analysis, and several refinements have been proposed over the years. We use the implementation described in [12]. Its fundamental aim is to partition the data into k distinct clusters such that each observation belongs to the cluster with the "closest" mean, where closeness is measured by the Euclidean distance function. Due to its simplicity the algorithm is popular and enjoys widespread use, but it also suffers from drawbacks including the need to specify the number of clusters k *a priori* as well as sensitivity to noise, outliers, and initial conditions (mitigated by running the algorithm multiple times).

4.3   K-Medoids Clustering

This algorithm is a variation on k-means clustering in that it also seeks to partition the data into k clusters by minimizing the distance to the cluster centers, except that the data points themselves are chosen as centers (called *medoids*). We use an implementation known as Partitioning Around Medoids (PAM) [15]. It is subject to some of the same problems as k-means but is more robust to outliers and noise in the data.

4.4   Spectral Clustering

This term refers to a class of clustering techniques that utilize the eigenvalues of a similarity matrix constructed from the data (called the *spectrum*, hence the name) for dimensionality reduction and then find clusters in the lower-dimensional space. The method used to compute the similarity matrix is also referred to as kernel function. Data can be partitioned either into two parts – recursively if necessary – or directly into k subsets. We use the algorithm described in [17], which utilizes multi-way partitioning and was shown to yield good results on a wide variety of challenging clustering problems.

4.5   Expectation Maximization

An expectation-maximization (EM) algorithm is a general technique for find-ing maximum likelihood parameter estimates in statistical models, and cluster analysis is one of its most common applications. In general, EM methods are computationally expensive but work well in a variety of application settings. We use the algorithm described in [6], which implements EM for a parameterized mixture of k Gaussians and is reasonably efficient.

5   Experimental Setup

This section explains how the various algorithms are used to obtain potential climate indices and how we compare them in a predictive setting.

5.1   Extracting Candidate Indices

Recall the definition of a climate index, that is, a summary of climate variability over one or more ocean regions, which is related to climate on land. Our first task is to extract potential indices from historical data using clustering. We run each algorithm described in Sec. 4 on all data corresponding to ocean grid points. With the exception of the network-based approach, the number of clusters k must be specified *a priori*. Therefore, we perform a comprehensive set of experiments by running each algorithm for $k = 5$, $k = 10$, and k equal to the number of clusters k_n obtained using community detection in networks to assure the fairest possible comparison (78 clusters total where k differs between variables).

5.2   Evaluating Predictive Power

The upcoming report from the Intergovernmental Panel on Climate Change (expected 2013) calls for attention to regional assessments of climate, so we focus on prediction at regional scale. We use nine target regions covering every continent, illustrated in Figure 3 (consistent with [24] for comparison). Some, like Peru and the Sahel, have known relationships with major climate indices; others were included to provide a representative set of regions around the world.

Moreover, we consider two climate variables in each region, temperature and precipitation, for a total of 18 response variables (9 regions × 2 variables). We chose these variables primarily for their relevance to human interests: they di-rectly influence our health and well-being as well as our environment, infrastruc-tures, and other man-made systems. Precipitation obtained from reanalysis has potential issues but is used here to develop initial insights and to compare with temperature, which is considered more reliable. In the following, we outline the step-by-step procedure used for the predictive modeling [24]:

1. For each of the algorithm-parameter combinations described above, create a corresponding set of predictors ($\mathbf{x}$) consisting of the cluster centroids by averaging the time series for all grid points assigned to each cluster.

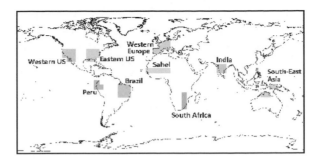

Fig. 3. Target regions for climate indices

2. Similarly, for each target region, create two response variables (y) by computing average temperature / precipitation over all grid points in the region.
3. Divide the data into a 50-year training set (1948-1997) and a 10-year test set (1998-2007).
4. For each of the 18 response variables, build a regression model $f : \mathbf{x} \rightarrow y$ on the training data and generate predictions for the unseen test data using each set of predictors from Step 1 in turn.

While it is conceivable to use any number of machine learning algorithms in Step 4, we start with linear regression in Section 6.1 as it gives us a performance baseline while maintaining interpretability of the model, which is important to domain scientists. In Section 6.3 we then go on to explore alternate prediction algorithms in this context.

To quantify performance we calculate root mean square error (RMSE) between the predictions and the actual (observed) data. Unlike simple correlation, which measures only covariance between two series, RMSE incorporates notions of both variance and estimator bias in a single metric.

6 Experimental Results

In this section, we present our empirical comparison of clustering algorithms and evaluate the predictive power of the derived climate indices.

6.1 Comparing Clustering Algorithms

First we seek to answer the question, *Which clustering method produces the best candidate indices?* The RMSE scores for all prediction tasks are summarized in Table 1; the lowest (best) and highest (worst) score in each row is shown in **bold** and *italic*, respectively.

Examining the results in some detail, we note that network-based clusters achieve the best score on 6 of 18 prediction tasks, more than any other algorithm-parameter combination. This is a first indication that climate networks may be well-suited for the discovery of climate indices. Network clusters also have the lowest mean RMSE across both temperature and precipitation, affirming that they are effective for diverse predictive tasks; even in cases where networks are not the outright best option they seem to offer competitive performance. To support this notion, we evaluate network-based clusters relative to the other methods using the Hochberg procedure of the Friedman test [2] at 95% confidence intervals – a non-parametric way to determine statistical significance of performance rankings across multiple experiments.

The outcomes are included at the bottom of Table 1; a checkmark ($\checkmark$) denotes that the network-based clusters are significantly better than the clusters in that column. Indeed we find this to be unequivocally the case, suggesting that networks capture the complex relationships in climate quite well. It is worth noting that clusters obtained with k-medoids and spectral clustering achieve scores comparable to, or even better than, the much more complex expectation-maximization algorithm. Regardless, we are led to conclude that community detection in climate networks yields the best candidate indices.

6.2 Validating Predictive Skill

Now that we established network-based clusters as having the highest predictability we must address the question, *Do these indices offer any true predictive power?* The answer was provided in [24]. To ascertain that the network clusters indeed contain useful information, we showed that they provide "lift" over random predictions as well as a simple univariate predictor. Our experimental results demonstrated that the network clusters do in fact have some predictive power, improving on the baseline by as much as 35%. Moreover, we can further enhance performance through feature selection [24].

6.3 Prediction Algorithms

Knowing that there is predictive information in the ocean clusters begs yet another question, namely, *What type of model can best harness this predictive power?* As alluded to in Section 5, linear regression merely provided a baseline comparison; it is entirely possible that other machine learning algorithms are better suited for modeling the processes connecting ocean and land climatology, for example, more sophisticated regressors that are able to capture nonlinear relationships.

Therefore, we also compare several fundamentally different prediction algorithms. In particular, we include neural networks (NN), regression trees (RTree) and support vector regression (SVR). The RMSE scores for the corresponding prediction tasks are summarized in Table 2; the lowest (best) score in each row is shown in **bold**.

Most obviously, we find that support vector regression achieves the lowest RMSE in 13 of 15 cases (including one tie), while linear regression comes in close second; the actual scores of these two methods are generally quite close. In contrast, neural networks and regression trees perform notably worse. These observations are confirmed by the statistical significance test (Hochberg procedure

Table 1. Comparison of clustering methods: RMSE scores for predictions of temperature and precipitation using candidate indices obtained via community detection in networks as well as k-means, k-medoids, spectral, and expectation-maximization clustering for $k = 5$, $k = 10$ and $k = k_n$, the number of network clusters for each variable. The best (**bold**) and worst (*italic*) scores in each row are indicated. A checkmark ($\checkmark$) at the bottom of a column denotes that the network-k-based clusters are significantly better according to the Friedman test of ranks at 95% confidence.

	Networks	K-Means			K-Medoids			Spectral			Expectation-Max.		
Region	k_n	$k=5$	$k=10$	$k=k_n$	$k=5$	$k=10$	$k=k_n$	$k=5$	$k=10$	$k=k_n$	$k=5$	$k=10$	$k=k_n$
Air Temperature													
SE Asia	**0.541**	0.629	0.694	0.886	0.826	0.973	*1.009*	0.751	0.731	0.827	0.760	0.653	0.703
Brazil	0.534	0.536	0.532	0.528	**0.509**	0.512	0.539	0.579	0.519	*0.582*	0.577	0.553	0.562
India	0.649	0.784	*1.052*	0.791	0.685	**0.587**	0.627	0.597	0.698	0.618	0.612	0.977	0.796
Peru	**0.468**	0.564	0.623	0.615	0.524	0.578	0.510	0.716	0.585	0.676	0.685	*0.837*	0.793
Sahel	0.685	0.752	0.750	0.793	**0.672**	0.733	0.752	0.866	0.801	0.697	0.820	*0.969*	0.854
S Africa	0.726	0.711	*0.968*	0.734	**0.674**	0.813	0.789	0.692	0.857	0.690	0.761	0.782	0.892
East US	0.815	0.824	0.844	0.811	*0.908*	**0.742**	0.758	0.848	0.839	0.799	0.768	0.753	0.846
West US	0.767	0.805	0.782	0.926	0.784	*1.021*	**0.744**	0.777	0.810	0.755	0.780	0.811	0.766
W Europe	0.936	1.033	0.891	0.915	0.950	1.071	*1.116*	**0.868**	0.898	0.962	0.947	0.975	0.986
Mean	**0.680**	0.737	0.793	0.778	0.726	0.781	0.765	0.744	0.749	0.734	0.746	*0.812*	0.800
±StdDev	0.150	0.152	0.165	0.135	0.155	*0.204*	0.200	**0.109**	0.128	0.117	0.111	0.148	0.120
Precipitation													
SE Asia	**0.665**	0.691	0.700	0.684	0.694	0.695	0.689	0.727	0.673	0.706	*0.739*	0.736	0.719
Brazil	**0.509**	0.778	0.842	0.522	0.817	0.986	1.110	1.272	*1.549*	1.172	1.353	1.351	1.284
India	0.672	0.813	0.823	0.998	0.798	1.072	*1.145*	**0.654**	1.018	0.820	0.714	0.820	0.720
Peru	0.864	1.199	1.095	1.130	1.064	0.934	*1.227*	0.859	0.994	**0.836**	0.872	0.845	0.837
Sahel	**0.533**	0.869	0.856	0.593	1.043	0.847	0.648	0.838	*1.115*	0.846	0.804	1.047	0.963
S Africa	0.697	0.706	0.705	0.703	0.702	0.706	*0.742*	0.704	**0.655**	0.683	0.753	0.729	0.736
East US	0.686	0.750	0.808	0.685	0.814	0.679	*0.851*	0.686	0.789	0.711	**0.645**	0.685	0.745
West US	0.605	0.611	0.648	0.632	0.617	0.610	0.589	**0.587**	0.635	0.645	0.599	0.646	*0.656*
W Europe	**0.450**	0.584	0.549	0.542	*0.720*	0.581	0.665	0.545	0.632	0.493	0.532	0.651	0.673
Mean	**0.631**	0.778	0.781	0.721	0.808	0.790	0.851	0.764	*0.896*	0.768	0.779	0.835	0.815
±StdDev	**0.124**	0.182	0.156	0.207	0.154	0.175	0.247	0.217	*0.307*	0.188	0.239	0.230	0.199
Friedman Test ($\alpha = 0.05$)	$\checkmark$	$\checkmark$	$\checkmark$	$\checkmark$	$\checkmark$	$\checkmark$	$\checkmark$	$\checkmark$	$\checkmark$	$\checkmark$	$\checkmark$	$\checkmark$	$\checkmark$

Table 2. RMSE scores for predictions with network clusters using linear regression (LR), neural networks (NN), regression trees (RTree) and support vector regression (SVR). The best score in each row is indicated in **bold**.

	Region	LR	NN	RTree	SVR
Air Temperature	SE Asia	**0.541**	0.629	0.743	**0.541**
	Brazil	**0.534**	0.568	0.686	0.570
	India	0.649	0.646	0.704	**0.595**
	Peru	0.468	**0.459**	0.616	0.589
	Sahel	0.685	0.866	0.983	**0.662**
	S Africa	0.726	0.838	0.849	**0.714**
	East US	0.815	0.895	1.060	**0.773**
	West US	0.767	0.835	0.860	**0.755**
	W Europe	0.936	1.018	0.014	**0.890**
	Mean	0.680	0.750	0.835	**0.677**
	±StdDev	0.150	0.182	0.159	**0.116**
Precipitation	SE Asia	0.665	0.703	0.791	0.653
	Brazil	**0.509**	0.547	0.771	0.597
	India	0.672	0.809	1.045	**0.646**
	Peru	0.864	1.006	0.960	**0.842**
	Sahel	**0.533**	0.785	0.663	0.542
	S Africa	0.697	0.787	0.767	**0.684**
	East US	0.686	0.684	0.771	**0.649**
	West US	0.605	0.647	0.696	**0.603**
	W Europe	0.450	0.522	0.569	**0.448**
	Mean	0.631	0.721	0.782	**0.629**
	±StdDev	0.124	0.148	0.145	**0.107**
Friedman ($\alpha = 0.05$)			✓	✓	

of the Friedman test [2] at 95% confidence) included at the bottom of Table 2; a checkmark (✓) denotes that linear regression performs significantly better than the algorithm in that column. Note: repeating the significance test relative to the SVR scores does not change the results, i.e., they are significantly better than NN and RTree, but *not* LR.

It is prudent not to draw any general conclusions from these results, but empirical evidence suggests that the more complex regression models *do not* necessarily improve performance. We conjecture that the reason for this is a combination of high-frequency noise and a relatively small number of training samples in the data, which collectively can lead to overfitting with more complex modes. Thus, for the sake of computational efficiency as well as interpretability of results, it is advisable to use a linear regression model.

However, given larger datasets or slightly different prediction tasks, it is possible that the gains from alternate prediction algorithms – including but not limited to those compared in this paper – would indeed be more substantial.

7   Domain Interpretation

Due to space constraints we cannot examine every set of clusters with respect to its climatological interpretation, but we present one case study using Peru as illustrative example (focused only on prediction, a descriptive analysis is provided in our prior work [24]).

Air Temperature in Peru. We chose this region because it is related to the El Niño phenomenon and hence domain knowledge in this area is plentiful. The predictions for temperature using all and "selected" [24] network clusters are shown in Figure 4, along with the actual (observed) data. It is apparent that the predictive model works quite well here, capturing all major variations. In fact, the RMSE score of 0.468 is among the lowest of any prediction task (Table 1).

Examining the nine "selected" clusters in more detail, we find that this particular index is composed of the following variables: 2 SST, 1 GH, 1 PW, 1 HWS and 4 VWS. For reference, the clusters of SST and VWS are depicted in Figure 5. It comes as no surprise that the selected clusters of sea surface temperature are numbers 5 (containing the areas that define several prominent El Niño indices) and 6 (the equatorial Pacific stretching into South-East Asia). However, VWS clusters 1, 11, 12 and 14 are also included. This is curious as vertical wind speed – convective activity over the oceans – is not thought to have any predictive power in climate, yet our findings seem to suggest otherwise.

We contemplate this possibility with a thought experiment: *How do clusters obtained with our data mining approach compare to those supported by domain knowledge?*

To answer this question, we asked a domain expert to narrow down the clusters to only those intuitively expected to be of relevance. SST-5 and SST-6 were chosen based on known relationships, as well as PW-7 due to spatial proximity. We repeat the regression with using only these three clusters and obtain an RMSE of 0.552. This score is lower than most of the methods included in our comparison (Table 1), meaning that traditional clustering methods cannot match current domain knowledge. But *network-based clusters* significantly improve the

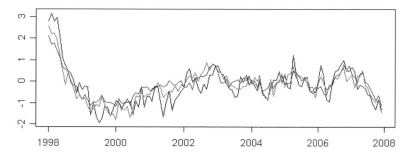

Fig. 4. Prediction of air temperature in Peru with all (red) and "selected" (blue) network clusters compared to observations (black). Best viewed in color

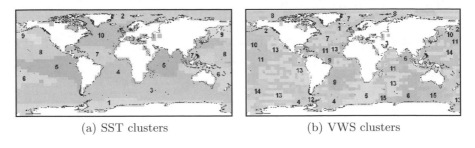

(a) SST clusters   (b) VWS clusters

Fig. 5. Depictions of sample clusters (reproduced from [24]). Best viewed in color.

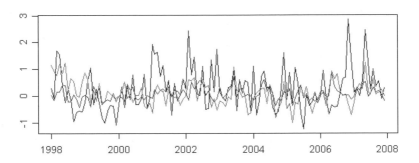

Fig. 6. Prediction of precipitation in Peru with all (red) and "selected" (blue) network clusters compared to observations (black). Best viewed in color.

score, suggesting that climate networks glean additional predictive power from the data. Whether and to what extent this holds in other situations remains an open question for climate scientists.

Precipitation in Peru. The predictions for precipitation using all and "selected" [24] network clusters are shown in Figure 6, along with the actual (observed) data. In contrast to temperature, the RMSE score of 0.864 for this region is on the low end across all prediction tasks (Table 1). While the very low-frequency signal is predicted to some degree, the observed data has much more variability not captured by the model. This is generally true for precipitation, namely, that the mean behavior is represented reasonably well while the model fails to predict the more sporadic short-duration, large-magnitude events. Accordingly, it is relatively more difficult to improve upon baseline methods (Table 2). Nonetheless, in some cases we observed considerable gains, prompting a more thorough investigation of the circumstances under which predictions of precipitation are improved by climate indices.

8   Discussion and Future Work

In this paper, we presented an empirical comparison of clustering methods for climate data on the basis of their ability to extract climate indices. Our experimental results demonstrate that ultimately the choice of algorithm is quite important: clustering matters! More specifically, community detection in climate networks stands out among competing methods as the superior approach across a diverse range of test cases, thereby reinforcing the notion that networks are able to effectively capture the complex relationships within the global climate system. In contrast, the prediction algorithm itself had a relatively smaller impact on quality of the predictions.

Consequently, the application of network-theoretical concepts could have far-reaching implications for climate science, e.g., studying properties of the climate system, detecting changes over time, and complementing the predictive skill of physics-based models with data-guided insights. In addition, complex networks may also prove useful in other applications involving ecological, environmental and/or social data, helping us understand the behavior of and interactions between these systems.

Acknowledgments. The authors would like to thank Prof. Vipin Kumar for the helpful discussions. Prof. Nitesh Chawla and Karsten Steinhaeuser are affiliated with the Interdisciplinary Center for Network Science & Applications (iCeNSA) at the University of Notre Dame. This work was supported in part by the NSF Grant 0826958. The research was performed as part of a project titled "Uncertainty Assessment and Reduction for Climate Extremes and Climate Change Impacts", funded in FY2010 by the "Understanding Climate Change Impacts: Energy, Carbon, and Water Initiative", within the LDRD Program of the Oak Ridge National Laboratory, managed by UT-Battelle, LLC for the U.S. Department of Energy under Contract DEAC05-00OR22725. The United States Government retains a non-exclusive, paid-up, irrevocable, world-wide license to publish or reproduce the published form of this manuscript, or allow others to do so, for Government purposes.

References

1. Asur, S., Ucar, D., Parthasarathy, S.: An ensemble framework for clustering protein-protein interaction graphs. Bioinformatics 23(13), 29–40 (2007)
2. Demšar, J.: Statistical Comparisons of Classifiers over Multiple Data Sets. Mach. Learn. Res. 7, 1–30 (2006)
3. Donges, J.F., Zou, Y., Marwan, N., Kurths, J.: Complex networks in climate dynamics. Eur. Phs. J. Special Topics 174, 157–179 (2009)
4. Floyd, R.W.: Algorithm 97: Shortest Path. Comm. ACM 5(6), 345 (1962)
5. Fovell, R.G., Fovell, M.-Y.C.: Climate Zones of the Conterminous United States Defined Using Cluster Analysis. J. Climate 6(11), 2103–2135 (1993)
6. Fraley, C., Raftery, A.E.: Model-based clustering, discriminant analysis, and density estimation. J. Am. Stat. Assoc. 92, 611–631 (2002)
7. Glantz, M.H., Katz, R.W., Nicholls, N.: Teleconnections linking worldwide climate anomalies: scientific basis and societal impact. Cambridge University Press, Cambridge (1991)

8. Guimerá, R., Mossa, S., Turtschi, A., Amaral, L.A.N.: The worldwide air transportation network: Anomalous centrality, community structure, and cities' global roles. Proc. Nat. Acad. Sci. USA 102(22), 7794–7799 (2005)
9. Hall, M.A., Smith, L.A.: Feature Selection for Machine Learning: Comparing a Correlation-based Filter Approach to the Wrapper. In: Int'l Florida AI Research Society Conf., pp. 235–239 (1999)
10. Han, J., Kamber, M., Tung, A.K.H.: Spatial Clustering in Data Mining: A Survey, pp. 1–29. Taylor and Francis, Abington (2001)
11. Hargrove, W.W., Hoffman, F.M.: Using Multivariate Clustering to Characterize Ecoregion Borders. Comput. Sci. Eng. 1(4), 18–25 (1999)
12. Hartigan, J.A., Wong, M.A.: Algorithm AS 136: A K-means clustering algorithm. Applied Statistics (28), 100–108 (1979)
13. Jain, A.K., Murty, N.N., Flynn, P.J.: Data clustering: A Review. ACM Computing Surveys 31(3), 264–323 (1999)
14. Kalnay, E., et al.: The NCEP/NCAR 40-Year Reanalysis Project. BAMS 77(3), 437–470 (1996)
15. Kaufman, L., Rousseeuw, P.J.: Finding Groups in Data: An Introduction to Clustering Analysis. Wiley, Chichester (1990)
16. Loveland, T.R., et al.: Development of a global land cover characteristics database and IGBP DISCover from 1 km AVHRR data. Int. J. Remote Sensing 21(6-7), 1303–1330 (2000)
17. Ng, A.Y., Jordan, M.I., Weiss, Y.: On Spectral Clustering: Analysis and an algorithm. In: Advances in Neural Information Processing Systems, pp. 849–856 (2001)
18. Pons, P., Latapy, M.: Computing communities in large networks using random walks. J. Graph Alg. App. 10(2), 191–218 (2006)
19. Race, C., Steinbach, M., Ganguly, A.R., Semazzi, F., Kumar, V.: A Knowledge Discovery Strategy for Relating Sea Surface Temperatures to Frequencies of Tropical Storms and Generating Predictions of Hurricanes Under 21st-century Global Warming Scenarios. In: NASA Conf. on Intelligent Data Understanding, Mountain View, CA (2010)
20. Ropelewski, C.F., Jones, P.D.: An Extension of the Tahiti-Darwin Southern Oscillation Index. Mon. Weather Rev. 115, 2161–2165 (1987)
21. Serrano, A., Boguna, M., Vespignani, A.: Extracting the multiscale backbone of complex weighted networks. PNAS 106(16), 8847–8852 (2009)
22. Steinbach, M., Karypis, G., Kumar, V.: A Comparison of Document Clustering Techniques. In: ACM SIGKDD Workshop on Text Mining (2000)
23. Steinbach, M., Tan, P.-N., Kumar, V., Klooster, S., Potter, C.: Discovery of Climate Indices using Clustering. In: ACM SIGKDD Conf. on Knowledge Discovery and Data Mining, pp. 446–455 (2003)
24. Steinhaeuser, K., Chawla, N.V., Ganguly, A.R.: Complex Networks as a Unified Framework for Descriptive Analysis and Predictive Modeling in Climate. Technical Report TR-2010-07. University of Notre Dame (2010)
25. Steinhaeuser, K., Chawla, N.V., Ganguly, A.R.: Complex Networks in Climate Science: Progress, Opportunities and Challenges. In: NASA Conf. on Intelligent Data Understanding, Mountain View, CA (2010)
26. Steinhaeuser, K., Chawla, N.V., Ganguly, A.R.: An Exploration of Climate Data Using Complex Networks. ACM SIGKDD Explorations 12(1), 25–32 (2010)
27. Tsonis, A.A., Roebber, P.J.: The architecture of the climate network. Physica A 333, 497–504 (2004)
28. Tsonis, A.A., Swanson, K.L., Roebber, P.J.: What Do Networks Have to Do with Climate? BAMS 87(5), 585–595 (2006)

29. Wasserman, S., Faust, K.: Social Network Analysis: Methods and Applications. Cambridge University Press, Cambridge (1994)
30. Watts, D.J., Strogatz, S.H.: Collective dynamics of 'small-world' networks. Nature 393, 440–442 (1998)
31. Wu, Z., Leahy, R.: An optimal graph theoretic approach to data clustering: Theory and its application to image segmentation. IEEE T. Pattern Anal. 15(11), 1101–1113 (1993)
32. Yamasaki, K., Gozolchiani, A., Havlin, S.: Climate Networks around the Globe are Significantly Affected by El Niño. Phys. Rev. Lett. 100(22), 157–179 (2008)
33. http://cdiac.ornl.gov/climate/indices/indices_table.html
34. http://www.cdc.noaa.gov/data/gridded/data.ncep.reanalysis.html
35. http://www.cgd.ucar.edu/cas/catalog/climind/

Region of Interest Queries in CT Scans

Alexander Cavallaro[2], Franz Graf[1], Hans-Peter Kriegel[1],
Matthias Schubert[1,*], and Marisa Thoma[1,*]

[1] Institute for Informatics, Ludwig-Maximilians-Universität München,
Oettingenstr. 67, D-80538 München, Germany
{graf,kriegel,schubert,thoma}@dbs.ifi.lmu.de
[2] Imaging Science Institute, University Hospital Erlangen, Maximiliansplatz 1,
D-91054 Erlangen, Germany
Alexander.Cavallaro@uk-erlangen.de

Abstract. Medical image repositories contain very large amounts of
computer tomography (CT) scans. When querying a particular CT scan,
the user is often not interested in the complete scan but in a certain
region of interest (ROI). Unfortunately, specifying the ROI in terms of
scan coordinates is usually not an option because an ROI is usually
specified w.r.t. the scan content, e.g. an example region in another scan.
Thus, the system usually retrieves the complete scan and the user has
to navigate to the ROI manually. In addition to the time to navigate,
there is a large overhead for loading and transferring the irrelevant parts
of the scan.

In this paper, we propose a method for answering ROI queries which
are specified by an example ROI in another scan. An important feature of
our new approach is that it is not necessary to annotate the query or the
result scan before query processing. Since our method is based on image
similarity, it is very flexible w.r.t. the size and the position of the scanned
region. To answer ROI queries, our new method employs instance-based
regression in combination with interpolation techniques for mapping the
slices of a scan to a height model of the human body. Furthermore, we
propose an efficient search algorithm on the result scan for retrieving the
ROI with high accuracy. In the experimental evaluation, we examine the
prediction accuracy and the saved I/O costs of our new method on a
repository of 2 526 CT scans.

1   Introduction

Radiology centers all over the world currently collect large amounts of 3D body
images being generated by various scanners like PET-CT, MRT, x-ray or sono-
graphy. Each of these methods generates a three dimensional image of the human
body by transforming the echo of a different type of signal allowing a radiologist
to examine the inner parts of a human body. In the following, we will particularly
focus on CT body scans. However, the methods proposed in this paper are
generally applicable to other types of scans as well.

---

* Corresponding authors.

D. Pfoser et al. (Eds.): SSTD 2011, LNCS 6849, pp. 56–73, 2011.
© Springer-Verlag Berlin Heidelberg 2011

Technically, the result of a CT scan is stored as a stack of 2D images representing 3D slices of the human body, i.e. each slice is considered to have a certain thickness. The scans in a radiology center are stored in a centralized picture archiving and communication system (PACS) and they are transferred via LAN to the workstation of a physician. In commercial PACS, querying CT scans is currently restricted to retrieving complete scans being annotated with certain meta information like patient name, date and type of the examination. Therefore, each time a CT scan is queried, the complete scan, potentially comprising several thousand high-resolution images, has to be loaded from the image repository. For example, the data volume of a thorax scan being generated by a modern scanner comprises around 1 GB of data. Considering that several physicians will simultaneously query a PACS, the loading time of a single CT scan is up to several minutes depending on network and server traffic.

However, in many cases it is not necessary to display the complete scan. For example, if a physician wants to see whether a certain liver lesion has improved between two scans, the user primarily requires the portion of both scans containing the liver. Therefore, the physician loses up to several minutes by loading unnecessary information and searching for the liver within both scans. Thus, a system retrieving the parts of both scans containing the liver, would save valuable time and network bandwidth.

Parts of a CT scan can be efficiently loaded by raster databases [2] as long as the coordinates of the ROI are specified. However, in the given context, the ROI is rather defined by the image content. In other words, the coordinates of organs and other anatomical regions may strongly vary because of differences in the patients' heights or in the scanned body region. Thus, raster coordinates cannot be used to align to CT scans w.r.t. the image content.

In this paper, we focus on a query-by-example setting. Therefore, the query is posed by selecting a certain body region in a scan. The result contains the part of the scan showing the corresponding body region in one or multiple result scans. For example, a radiologist could select a certain area in the scan being currently under examination. He or she might want to see the corresponding regions in scans of patients having the same disease or earlier examinations of the same patient.

The most established approach to answer this type of queries is based on landmark detection. [14] A landmark is an anatomically unique location in the human body which is well-detectable by pattern recognition methods. To use landmarks for query processing, it is first of all necessary to detect as many landmarks as possible in the example scan and all result scans. Let us note that landmark detection employs pattern recognition methods and thus, there is a classification error, i.e. some of the predicted landmark positions are error prone. Furthermore, it can happen that some of the landmarks are not detectable due to disturbances while recording the scan. However, having detected a sufficiently large number of landmarks, it is possible to align both scans and afterwards select the area from the target scan corresponding to the query.

An important aspect of this approach is that landmark detection should be done as a preprocessing step. Thus, the example scan and the target scans need to be annotated with the landmark position to allow efficient query processing. However, this causes a problem when allowing example scans not being stored in the same PACS. In this case, the query might not have any landmarks or it is not labelled with the same set of landmarks. If the example scan and the result scan are taken by CT scanners from different companies, the positioning systems might not be compatible. Another drawback of the landmark approach is the size of the scan. CT scans are often recorded for only a small part of the body. Thus, it cannot be guaranteed that the scanned body region contains a sufficiently large set of alignable landmarks. To conclude, a fixed and comparably small set of landmarks is often not flexible enough to align arbitrary scans.

In this paper, we propose a more flexible approach being based on similarity search on the particular slices of a CT scan. Our new method does not rely on any time-consuming preproccesing step, but it can be directly applied on any query and result scan. Whereas landmark-based approaches can only align scans with respect to a limited amount of fixed points to be matched, our new approach can generate the positions in the scan to be matched on the fly. Thus, we can even align scans being labelled with different types of landmarks or scans not having any detectable landmarks at all.

The key idea behind our method is to map single slices of a CT scan to a generalized height model describing the relative distances between concepts w.r.t. the height axis of the human body. The height model is independent of the individual size and proportions of a particular patient. Let us note that it is possible to use width and depth axes as well. However, the height axis is the predominantly used navigation axis for CT scans.

By mapping single slices to the model, we can better adjust to limited information about the scan and we are independent from the distribution of predefined landmark positions. Our prediction algorithm employs instance-based regression based on Relevant Component Analysis [1] and the X-Tree [3] for efficiently answering kNN queries.

ROI queries are answered as follows: In the first step, the user selects a certain region of interest in the example scan. Afterwards, we employ instance-based regression to determine the query position in the generalized height model. In the next step, we need to determine the part in each target scan corresponding to the query interval in the height model. Let us note that this second step is more complicated, since we cannot directly determine the slice belonging to a particular height value. One solution to this problem would be to label all available slices with the height value in the model. However, labelling all DICOM images in an average PACS would cause an enormous overhead in preprocessing. Since the majority of images will never be involved in answering an ROI query, we follow a different strategy. Instead of preprocessing each image in the PACS, determining height values for a given slice is done on the fly. To make this type of processing efficient, we propose a query algorithm that alternates regression and interpolation steps until the queried ROI is found in the result scan.

Let us note that although the solutions proposed in this paper are very problem-oriented, the solution principle can be extended to other data as well. For example, a similar processing scheme can be applied to video streams (e.g. procedure timing in surveillance videos) or text mining (e.g. news tickers, twitter streams, age classification in Internet forums).

The rest of the paper is organized as follows. Sect. 2 surveys methods that are related to our approach or parts of it. In Sect. 3, we formalize ROI queries and give an overview of our system. Afterwards, Sect. 4 introduces our method for predicting height values for particular CT slices. Sect. 5 first describes interpolation methods for aligning CT scans to a generalized height model and then presents our new query algorithm. The results of our experimental evaluation are shown in Sect. 6. The paper concludes with a brief summary and ideas for future work in Sect. 7.

2   Related Work

In medical imaging, there are various localization or registration approaches. Most of them are very domain specific, like the Talairach space brain atlas [15] or the MNI space [5]. Nevertheless, as these atlases are very specific to their domain, they were not designed to cover the entire body and they can thus hardly be used for general ROI queries.

Position mapping via landmark-detector-based approaches like the Theseus Medico system presented in [14] are more appropriate for our purpose. This prototype provides an image parsing system which automatically detects 22 anatomically relevant landmarks, i.e. invariant points, and 9 organs. [13] It is thus possible to query the database directly for ROIs which are equivalent to these automatically-annotated image regions. However, general queries for arbitrarily defined ROIs are not yet supported.

A more general, landmark-based interpolation approach for mapping a volume into a standardized height space has been proposed by [7]. However, it is very patient-specific and dependent on the used landmarks. Another approach that uses partial volumes as query is described in [6]. It localizes query volumes with sizes ranging from 4 cm to more than 20 cm by comparing the partial volume with an implicit height atlas based on Haar-like features. In [4], we presented an alternative method such that only a single query slice is needed in order to achieve comparable results.

In Sect. 5.2, we introduce an iterative interpolation and regression approach. In contrast to established regression methods, [9,11] we enhance our model with newly generated information after each iteration in order to refine the final model until convergence is reached.

We experimented with several regression methods from the Weka machine learning package [8]. However, simple approaches like linear regression did not yield a sufficient prediction accuracy and more complicated approaches like support vector regression using non-linear kernel functions could not cope with the enormous amount of training data. Therefore, we decided to employ instance-based regression which is robust and sufficiently fast when employing techniques

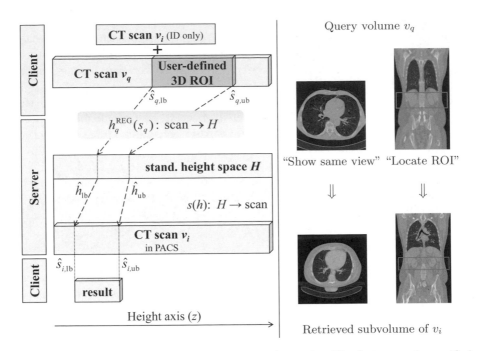

Fig. 1. Workflow of ROI retrieval and two example queries. The first query is specified by a ROI of only one slice, the second is given by a 3D ROI.

of efficiently computing the k-nearest neighbors (k-NN). In particular, we employ k-NN queries being based on the X-Tree [3]. Let us note that there are multiple other index structures [12] for speeding up the same type of query. We decided to employ the X-Tree because it represents an extension of the standard R*-Tree [10] which is better suited for higher dimensionalities.

Current database systems like RasDaMan [2] already support conventional region of interest queries in raster data like CT scans. Nevertheless, the system needs to know the coordinate system in which the query is applied in order to navigate to the requested region. As we do not know the complete coordinate systems of the patients' CT scans in advance and since patients differ in height and body proportions, and thus, locations along the z-axis are not standardized, a globally fixed coordinate system will not be available in our setting. Therefore, our new approach represents a way to bridge the gap between the coordinates in the example scan and the coordinate system of the result scan.

3   Example-Based ROI Queries

In this section, we specify the proposed query process and give an overview of the proposed system. Formally, a dataset consists of n volumes $v_i \in \mathbb{N}^{x(i) \times y(i) \times z(i)}$ with $i \in \{1, \ldots, n\}$ and varying voxel dimensions x, y, z. The height model H is an interval $[h_{\min}, h_{\max}] \in \mathbb{R}_o^+$ representing the extension of the human body in the z-axis. A mapping function $h_i : \mathbb{N} \rightarrow H$ maps slices of volume v_i to a height value $h \in H$. Correspondingly, the reverse mapping function $s_i : H \rightarrow \mathbb{N}$ maps a position h in the height model to a slice number s in v_i. A matching point $p = (\mathbf{s}_p, \mathbf{h}_p, \mathbf{w}_p) \in \mathbb{N} \times H \times \mathbb{R}$ is a triple of a slice number, its corresponding height value in H and a reliability weight w. We use $p_{i,j}$ for naming the jth matching point in scan v_i.

In our system, a region of interest (ROI) query is specified by a set of consecutive slices $(\hat{\mathbf{s}}_{e,\text{lb}}, \dots, \hat{\mathbf{s}}_{e,\text{ub}}) \subseteq \{0, \dots, z(e) - 1\}$ from an example scan v_e and it retrieves a consecutive sequence of CT slices $(\hat{\mathbf{s}}_{i,\text{lb}}, \dots, \hat{\mathbf{s}}_{i,\text{ub}}) \subseteq \{0, \dots, z(i) - 1\}$ from the result scan v_i.

Fig. 1 illustrates the complete workflow of query processing for example-based ROI queries. A user specifies the ROI query on the client computer by marking a region in an example scan v_e. Additionally, the queried scan v_i has to be identified for the server. Let us note that it is not necessary to transfer the complete marked subset of the example scan. Instead it is sufficient to transfer a scale-reduced version of the first and the last slice of the subset. After receiving the slices, the server performs a feature extraction step generating image descriptors for both slices. As an alternative, the client computer might directly compute the required image descriptors and only transfer the descriptors.

In the next step, the server employs a mapping function to predict height values $\hat{h}_{\text{lb}}$ and $\hat{h}_{\text{ub}}$ to describe the borders of the query interval in the height model H. In our system, $h_e^{\text{REG}}(s)$ is implemented by instance-based regression (cf. Sect. 4). Afterwards, our algorithm starts with aligning the result scan v_i to the height model H by employing the algorithm described in Sect. 5.2. In particular, the algorithm employs $h_i^{\text{REG}}(s)$ for generating matching points P_i which are required for the reverse mapping function $s_i(h)_{P_i}$, an interpolation function described in Sect. 5.1. Once the quality of $s_i(h)_{P_i}$ is satisfying, the server selects the sequence of slices $(\hat{\mathbf{s}}_{i,\text{lb}}, \dots, \hat{\mathbf{s}}_{i,\text{ub}}) \subseteq \{0, \dots, z(v_i) - 1\}$ from v_i corresponding to the height interval $[\hat{h}_{\text{lb}}, \hat{h}_{\text{ub}}]$ and returns them to the client as the query result. Let us note that $(\hat{\mathbf{s}}_{i,\text{lb}}, \dots, \hat{\mathbf{s}}_{i,\text{ub}})$ is extended by the amount of slices corresponding to 90 % of the expected prediction error in order to compensate for the inaccuracy of $h_i^{\text{REG}}(s)$.

Table 1 displays an overview of the defined parameters including some additional annotations that will be introduced in the next sections.

4 Efficient Instance-Based Regression

In this section, we introduce our method for mapping a single slice into the standardized height scale H. We already mentioned that there exist methods for landmark and organ detection which mark slices in the scan with the detected landmarks or organs. [13] Using multiple landmarks detected at slices $\mathbf{s}_{i,j}$, which can be mapped to anatomical concepts with known standardized height positions $\mathbf{h}_j$ and reliabilities $\mathbf{w}_j$, we can infer the standardized height of our queried slice. The landmark detector of [13] being used in our experiments was not able to detect landmarks for all available scans, though. The reason why the

Table 1. Notation of frequently used parameters

$v_i \in \mathbb{N}^{x(i) \times y(i) \times z(i)}$	one volume ($i \in \{1, \ldots, n\}$)
$H \in \mathbb{R}_o^+$	standardized height space / height model
$\mathbf{h}_j$	one height value in H
$\mathbf{s}_{i,j}$	one slice number of v_i in $\{0, \ldots, z(i) - 1\}$
$p = (\mathbf{s}_{i,p}, \mathbf{h}_p, \mathbf{w}_p)$	matching point with reliability weight $\mathbf{w}_j$
P_i	set of matching points of v_i
$h_i^{\mathrm{REG}}(s)$	regression function of $\mathbb{N} \to H$
$s_i(h)_{P_i}$	interpolation function of $H \to \mathbb{N}$ using a set P_i
$(\hat{\mathbf{s}}_{i,\mathrm{lb}}, \ldots, \hat{\mathbf{s}}_{i,\mathrm{ub}})$	slice range in v_i
$[h_{\mathrm{lb}}, h_{\mathrm{ub}}]$	interval in H
$F(s_{i,j}) : \mathbb{N} \to \mathcal{F} = \mathbb{R}^d$	image feature transformation of slice j of v_i with $d \in \mathbb{N}$
T_R	training set for regression

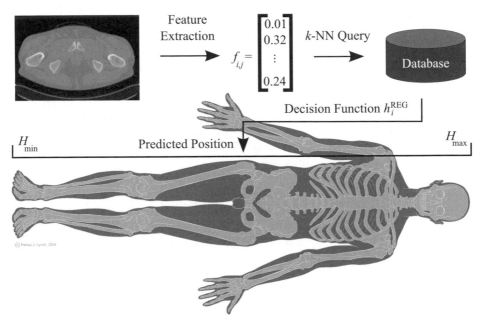

Fig. 2. Overview of content-based matching point generation using instance-based regression on HoGs. (Human model visualization taken from Patrick J. Lynch, medical illustrator and C. Carl Jaffe, MD, cardiologist at http://commons.wikimedia.org/wiki/File:Skeleton_whole_body_ant_lat_views.svg)

detector failed to find landmarks were the following: The image quality is too fuzzy for the detector, the body region covered by the scan is not big enough or only a single slice is available. Further drawbacks of employing landmark detection for generating matching points are the complexity and availability of reliable detectors and that their runtimes are not suitable for interactive query processing. In order to allow instant query processing on arbitrary scans, a faster and more flexible method should be employed that can efficiently generate a matching point for any given slice in the queried scan.

The idea behind our method is to represent the slice of interest $\mathbf{s}_{i,j}$ of the scan v_i by an image descriptor and to employ regression techniques to predict its height value in H. To train the regression function, we employ a training set of height annotated CT scans where each slice is labelled by a height value $h \in H$. An overview of this approach can be see in Fig. 2.

We use a 7 bins histogram of oriented gradients (HoG) which is applied on the cells of a 5x5 grid of the image's sub-regions. In contrast to the image descriptor introduced in [4], we omit the additional global extraction cell. Thus, the concatenated HoGs form a descriptor of size $d = (5 \cdot 5) \cdot 7 = 175$. Since our query algorithm requires multiple online feature extraction steps per query, we down-scale the images to a 100x100 pixels resolution before feature extraction for speeding up feature generation. We denote by $F(\mathbf{s}_{i,j}) : \mathbb{N} \to \mathbb{R}^d$ the feature transformation of the $\mathbf{s}_{i,j}^{\text{th}}$ slice of volume v_i to the final, d-dimensional feature space.

A d-dimensional feature vector $f_{i,j}$ corresponding to the jth slice of the scan v_i can be mapped to the height model H with any given regression function. However, in our experiments the majority of standard regression methods either required extensive training times on our large training datasets of up to 900 000 training examples or they did not yield the required prediction quality. Therefore, we employ an instance-based approach to regression determining the k-nearest neighbors ($k-$NN) of the given feature vector $f_{i,j}$ in the training set T_R, consisting of image features r with existing labels $h(r) \in H$, w.r.t. Euclidean distance. Afterwards, the height of slice $\mathbf{s}_{i,j}$ in scan v_i is predicted using the following decision function:

$$h_i^{\text{REG}}(\mathbf{s}_{i,j}) = \text{median}_{\{r \in T_R \mid r \in k\text{-NN of } F(\mathbf{s}_{i,j})\}} \{h(r)\} \ . \tag{1}$$

Although instance-based regression does not suffer from extensive training times, the cost for large example datasets has to be spent at prediction time. However, the prediction rule does only require to process a single kNN query and thus, allows us to use optimization methods for this well-examined problem.

In order to allow efficient query processing, we transform the high-dimensional feature space of the proposed image features into a lower-dimensional space which can be indexed by suitable spatial index structures. For this paper, we use an X-Tree [3], which is well-suited for data of medium dimension.

We reduce dimensionality d in a supervised way employing Relevant Component Analysis (RCA) [1] with the goal of maintaining the principal information of the original feature vectors $r \in \mathbb{R}^d$. RCA transforms the data into a space minimizing the co-variances within subsets of the data, which are supposed to be similar, the so-called chunklets. Chunklets can be defined by matching a set of class labels or by using clusters. In our setting, we sort the data points used for training the feature transformation according to their height labels and retrieve a pre-defined number (150 chunklets performed well) of equally-sized data subsets. For our datasets, using a 10-dimensional feature representation turned out to be a viable trade-off between prediction time and accuracy. On the average, a query took 22 ms while yielding an average prediction error of only 1.88 cm.

When using positions $\hat{h}_{i,j}^{\text{REG}}$ computed with $h_i^{\text{REG}}(\mathbf{s}_{i,j})$ as matching points for answering ROI queries, we are also interested in how reliable they are. One way to determine a position's reliability is to use the variance of the k-nearest neighbors, with a low variance indicating a reliable prediction. [4] However, in our setting, the best predictions could be observerd with $k = 1$ or $k = 2$. Since building a deviation on 1 or 2 samples does not make any sense, we had to develop an alternative approach for approximating the prediction quality.

We thus perform an additional pre-processing step assigning weights to all instances r in the training database T_R. The weight $w(r)$ of instance r is determined in a leave-one-out run of $h_i^{\text{REG}}(r)$ on T_R. The predicted height value $\hat{h}_r^{\text{REG}}$ is compared to the true position $h(r)$, resulting in the weight $w(r)$:

$$w(r) = 0.1 / \left(0.1 + \left| \hat{h}_r^{\text{REG}} - h(r) \right| \right) . \tag{2}$$

The reliability of a predicted value $\hat{h}_{i,j}^{\text{REG}}$ is now approximated by the average weight $w(x)$ over all k-nearest neighbors x of the queried instance r.

5   Answering ROI Queries

In the following, we define a method for retrieving an ROI in a volume v_i for which no matching points are yet available. As mentioned before, the first step of an ROI query is to determine the query interval $[h_{\text{lb}}, h_{\text{ub}}]$ in the standardized space H corresponding to the ROI in the example scan v_e. We employ instance-based regression as proposed in the previous section for predicting the height of the lower and upper bound of the marked ROI of the example scan.

Once such a query interval is defined, we need to collect a set of matching points $p \in P_i$ for being able to interpolate from the standardized height space H to the volume space of a slice $\mathbf{s}_{i,j} \in \mathbb{N}$ of volume v_i. We will now introduce the interpolation approach used for this purpose.

5.1   Interpolation Using Matching Points

For mapping model positions $h_j \in H$ to slices $\mathbf{s}_{i,j} \in \mathbb{N}$, we use an interpolation approach based on a linear function. However, due to varying body proportions, the patient's position on the scanner table and the imperfect reliability of the used matching points P_i, a strictly linear model is not sufficient. Therefore, we additionally consider a non-linear component in our function which adds an instance-based off-set, comparable to an approach introduced in [7].

The mapping function $s_i(\mathbf{h}_q)_{P_i}$ for mapping the height value $\mathbf{h}_q$ to a slice number $\mathbf{s}_{i,q} \in \mathbb{N}$ is dependent on the scan v_i and the set of matching points P_i. We approximate the slice spacing δ_i describing the thickness of a slice in the target space H as the median slice spacing over all pairs of matching points in P_i as $\hat{\delta}_i = \text{median}_{p,p' \in P_i, \mathbf{s}_{i,p} \neq \mathbf{s}_{i,p'}} |\mathbf{h}_p - \mathbf{h}_{p'}| / |\mathbf{s}_{i,p} - \mathbf{s}_{i,p'}|$.

Let us note that the median is used to achieve a higher stability against outliers caused by unreliable matching points. We define $s_i : H \rightarrow \mathbb{N}$ as:

$$s_i(\mathbf{h}_q)_{P_i} = \frac{\mathbf{h}_q}{\delta_i} - \frac{\sum\limits_{p \in P_i} \mathbf{w}_p \cdot \min\left(1, |\mathbf{h}_q - \mathbf{h}_p|^{-1}\right) \cdot \left(\frac{\mathbf{h}_p}{\delta_i} - \mathbf{s}_{i,p}\right)}{\sum\limits_{p \in P_i} \mathbf{w}_p \cdot \min\left(1, |\mathbf{h}_q - \mathbf{h}_p|^{-1}\right)} . \tag{3}$$

In order to avoid the case of $\mathbf{h}_p = \mathbf{h}_q$, we limit the maximal contribution of a matching point p with the minimum terms. Other, more complex interpolation models usually performed less stable and are thus omitted from this paper.

5.2   Retrieval Algorithm

The quality of the mapping $s_i(\mathbf{h})_{P_i}$ directly depends on the quality of the matching points $p \in P_i$. Having a large set of matching points increases the mapping quality because it increases the likelihood that reliable matching points being close to $[h_{\mathrm{lb}}, h_{\mathrm{ub}}]$ are available. Furthermore, having more matching points decreases the impact of low-quality matching points. However, increasing the amount of matching points is connected with generating costs for feature transformation, dimension reduction and regression.

Thus, we want to employ a minimal number of matching points while achieving high interpolation quality. The core idea of our method is to start with a minimal set of matching points and to measure the quality of the induced mapping function. As long as this quality is significantly increasing, we select slices in the queried scan and induce additional matching points using the regression method proposed in Sect. 4. This process is illustrated in Fig. 3.

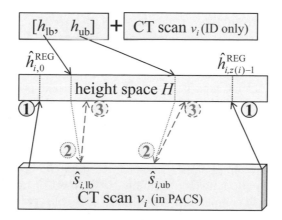

Fig. 3. First steps of Algorithm 1: a query range $[h_{\mathrm{lb}}, h_{\mathrm{ub}}]$ is to be found in a scan v_i. In the initial step (1), the seed slices forming the matching points P_i are selected and mapped to H with h_i^{REG}. In step (2), P_i is used for interpolating a result range $(\hat{s}_{\mathrm{lb}}, \ldots, \hat{s}_{\mathrm{ub}})$ in v_i. Step (3) validates this result range using h_i^{REG} and decides whether a new range should be tested.

Algorithm 1. ROI Query

Input: v_i: Query volume, $[h_{\mathrm{lb}}, h_{\mathrm{ub}}]$: query interval in H, h_i^{REG}: height regression function $\mathbb{N} \to H$, s_i: interpolation function $H \to \mathbb{N}$, ϵ: tolerated result range deviation

1: **function** ROI_QUERY$(v_i, [h_{\mathrm{lb}}, h_{\mathrm{ub}}], h_i^{\mathrm{REG}}, s_i, \epsilon)$
2:     $P_i = (\mathbf{s}_i, \mathbf{h}, \mathbf{w}) \leftarrow \mathrm{INIT}(v_i, h_i^{\mathrm{REG}})$ ▷ Initialize P_i
3:     $\{\mathrm{err}_{\mathrm{lb}}, \mathrm{err}_{\mathrm{ub}}\} \leftarrow \{\infty, \infty\}$ ▷ Errors for lb and ub
4:     $\{\hat{s}_{\mathrm{lb}}, \hat{s}_{\mathrm{ub}}\} \leftarrow \mathrm{NULL}$ ▷ Resulting slice numbers
5:     **while** $\mathrm{err}_{\mathrm{lb}} > \epsilon$ **or** $\mathrm{err}_{\mathrm{ub}} > \epsilon$ **do**
6:        $\{\hat{s}_{\mathrm{lb}}^*, \hat{s}_{\mathrm{ub}}^*\} \leftarrow \{s_i(h_{\mathrm{lb}}), s_i(h_{\mathrm{ub}})\}$ ▷ Interpolation
7:        $\{\hat{h}_{\mathrm{lb}}^{\mathrm{REG}}, \hat{h}_{\mathrm{ub}}^{\mathrm{REG}}\} \leftarrow \{h_i^{\mathrm{REG}}(\hat{s}_{\mathrm{lb}}^*), h_i^{\mathrm{REG}}(\hat{s}_{\mathrm{ub}}^*)\}$ ▷ Regression
8:        $\{\mathrm{err}_{\mathrm{lb}}^*, \mathrm{err}_{\mathrm{ub}}^*\} \leftarrow \left\{\left|\hat{h}_{\mathrm{lb}}^{\mathrm{REG}} - h_{\mathrm{lb}}\right|, \left|\hat{h}_{\mathrm{ub}}^{\mathrm{REG}} - h_{\mathrm{ub}}\right|\right\}$
9:        **if** $\mathrm{err}_{\mathrm{lb}} > \mathrm{err}_{\mathrm{lb}}^*$ **then**   ▷ New lower bound
10:         $\hat{s}_{\mathrm{lb}} \leftarrow \hat{s}_{\mathrm{lb}}^*$; $\mathrm{err}_{\mathrm{lb}} \leftarrow \mathrm{err}_{\mathrm{lb}}^*$
11:        **if** $\mathrm{err}_{\mathrm{ub}} > \mathrm{err}_{\mathrm{ub}}^*$ **then**   ▷ New upper bound
12:         $\hat{s}_{\mathrm{ub}} \leftarrow \hat{s}_{\mathrm{ub}}^*$; $\mathrm{err}_{\mathrm{ub}} \leftarrow \mathrm{err}_{\mathrm{ub}}^*$
13:        Get weights $\hat{w}_{\mathrm{lb}}^{\mathrm{REG}}, \hat{w}_{\mathrm{ub}}^{\mathrm{REG}}$ of new matching points
14:        $P_i.\mathrm{APPEND}\left(\left(\hat{s}_{\mathrm{lb}}^*, \hat{h}_{\mathrm{lb}}^{\mathrm{REG}}, \hat{w}_{\mathrm{lb}}^{\mathrm{REG}}\right), \left(\hat{s}_{\mathrm{ub}}^*, \hat{h}_{\mathrm{ub}}^{\mathrm{REG}}, \hat{w}_{\mathrm{ub}}^{\mathrm{REG}}\right)\right)$ ▷ Extend P_i
15:     **return** $(\hat{s}_{\mathrm{lb}}, \ldots, \hat{s}_{\mathrm{ub}})$

Output: Result range $(\hat{s}_{\mathrm{lb}}, \ldots, \hat{s}_{\mathrm{ub}})$

We use the mechanism of manually generating matching points via regression $h_i^{\mathrm{REG}}(\mathbf{s})$ for measuring the quality of a predicted result range $(\hat{s}_{\mathrm{lb}}, \ldots, \hat{s}_{\mathrm{ub}})$. The error of a prediction $\hat{s}_c$ for a value $\mathbf{h}_c$ is thus defined as: $|h_i^{\mathrm{REG}}(\hat{s}_c) - \mathbf{h}_c|$.

Since $h_i^{\mathrm{REG}}(\hat{s}_c)$ is fixed during query processing, the only possible way to reduce the error is to improve the quality of the matching points. This can happen by either updating their weights $\mathbf{w}_j$ or by adding further matching points. Even though it is sensible to update weights in special cases, the core component of our algorithm involves the second improvement variant.

For a given query interval $[\mathbf{h}_{\mathrm{lb}}, \mathbf{h}_{\mathrm{ub}}]$ our method proceeds as follows (see also Algorithm 1): We select g equally-spaced seed slices $\mathbf{s}_i \subset \{0, \ldots, z(i)-1\}$ to generate an initial set of matching points by predicting their positions as $\hat{\mathbf{h}} \in H^g$ using instance-based regression $h_i^{\mathrm{REG}}(\mathbf{s}_i)$. Using the weights obtained in the regression procedure we can induce an initial set of matching points $P_i = (\mathbf{s}_i, \hat{\mathbf{h}}, \mathbf{w})$. We are now free to make our first prediction of the result range.

We interpolate $\hat{s}_{\mathrm{lb}}^* = s_i(h_{\mathrm{lb}})_{P_i}$ and $\hat{s}_{\mathrm{ub}}^* = s_i(h_{\mathrm{ub}})_{P_i}$ in the queried scan using the current set of matching points P_i. Next, we employ $h_i^{\mathrm{REG}}(\mathbf{s})$ on $(\hat{s}_{\mathrm{lb}}^*, \ldots, \hat{s}_{\mathrm{ub}}^*)$ and determine the prediction error estimate. If the lower or upper bound ($\hat{s}_{\mathrm{lb}}^*$ or $\hat{s}_{\mathrm{ub}}^*$) has been improved compared to the minimal error observed so far, we update the corresponding lower or upper bound ($\hat{s}_{\mathrm{lb}}$ or $\hat{s}_{\mathrm{ub}}$). Finally, we augment the set of matching points P_i by the regression prediction $h_i^{\mathrm{REG}}(\mathbf{s})$ for the boundaries of the target range $\hat{s}_{\mathrm{lb}}^*$ and $\hat{s}_{\mathrm{ub}}^*$. The algorithm terminates if the improvement on both sides of the target range is less than ϵ.

For simplicity reasons, this algorithm omits a number of special cases. Since the derivation of matching points via regression is expensive due to the overhead of feature generation, the algorithm has to ensure that no slice number of v_i is tested multiple times. The search procedure should stop, once there is no more change to the set of matching points P_i because this usually means that the volume is not well enough resolved for perfectly matching the target range. It is also beneficial to test for both bounds whether a new matching point generated for the opposite bound is better suited. Additionally, if only one bound has been established in an acceptable quality, but it remains stable over a couple of iterations, one should refrain from trying to further improve this bound by costly regression calls and only update the opposite bound.

Furthermore, a number of exceptions should be handled: both s_i and h_i^{REG} can be mapped outside of their allowed ranges. In the case of s_i, this may be an indication that the query range is not contained in the volume. Repeated range violations should thus terminate the algorithm with an indication of a mismatch or a partial match. If $h_i^{\text{REG}}(\mathbf{s})$ goes astray, this can either be noise in the regression function or it can be a reason for down-weighting the current set P_i and for seeking further matching points.

6   Experimental Validation

In the following, we present the results of our experimental evaluation by measuring the quality of the retrieval system and by demonstrating the improved query time of our complete system. All of our experiments were performed on subsets of a repository of 4 479 CT scans provided by the Imaging Science Institute of the University Hospital Erlangen for the Theseus Medico project. The scans display various subregions of the human body, starting with the coccyx and ending with the top of the head.

For generating a ground truth of height labels, we used the landmark detector of [13] annotating each scan with up to 22 landmarks. This restricted the dataset to 2 526 scans where landmarks could be detected. The complete repository contains more than a million single CT slices comprising a data volume of 520 GB.

We implemented our prototype in JAVA 1.6 and stored the scans and their annotations in a MySQL database. To simulate the distributed environment of a radiology center, we employed the LAN and the workstations in our lab consisting of common workstations of varying type and configuration being connected by a 100 Mb Ethernet.

6.1   Prediction via Regression

In the following section, we first examine the used image features on their suitability for k-NN regression. Afterwards, we describe the beneficial effects of reducing the original image feature space using RCA.

Regression Quality. For these experiments, we have to provide height labels $h(r)$ as ground truth for all entries of the required regression database, i.e. for each scan in the training dataset T_R. Basically, there are two methods for generating these labels. The first is to manually mark the highest and the lowest point in all scans of a database and to linearly interpolate the height values. [4] We refer to this method as manual labelling.

Since instance-based regression profits from a larger database, we also use an automatic labelling method. It assigns height labels to the slices of a volume with the inverse interpolation approach introduced in Sect. 5.1, using standardized landmark positions as matching points. In our experiments, we use the 22 landmarks of [13], marking meaningful anatomical points, which could be detected in 2 526 of our CT scans. These landmarks are time-expensive to compute and their computation fails in the remaining 1 953 scans of our dataset. We will refer to the height labels generated with this interpolation procedure as automatic labelling.

In our first test, we measure the regression performance of the original image descriptors, which have not yet been transformed by RCA. We first examine a manually annotated dataset of 33 CT scans with a total of 18 654 slices. The average leave-one-out prediction errors are displayed in the first row of Table 2. Let us note that leave-one-out in these experiments means that only slices from other scans than the current query scan are accepted as k-nearest neighbors in order to exclude distorting effects of within-scan similarities. When testing k parameters between 1 and 5, we found $k = 1$ to be the best setting for all experiments using the original slice descriptors.

Table 2. Leave-one-out validation (LOO) errors [in cm] of k-NN slice mapping $h_i^{\text{REG}}(\mathbf{s}_{i,j})$ for two database sizes of n CT scans with m slices

Ground Truth	n	m	Error [cm]	Time / Query [ms]
manual	33	18 654	4.285	18
automatic	33	18 654	3.584	18
automatic	376	172 318	1.465	200

The next row of Table 2 displays the error of the same dataset, which has been labelled automatically. Our experiments show that the average registration error of 4.3 cm of the manual labelling is even lowered to 3.6 cm when using the automatic labelling. Thus, we can safely test our regression method on larger datasets, which have been automatically annotated. This allows to fully exploit the strength of the proposed instance-based regression approach. For smaller databases, alternative regression approaches should be considered, however, with the wealth of information available, our lazy learner is very hard to beat.

We observe a steady improvement of the empirical errors for increasing database sizes, however, this comes at the price of longer runtimes. For a dataset of 376 volumes consisting of 172 318 slices, a single query performed as sequential scan in main memory requires 200 ms. The additional cost of keeping the complete training database in main memory poses an additional drawback. The following section evaluates our method of runtime optimization, by using an efficient indexing scheme.

Speed-up via RCA and Indexing. In order to speed up regression, we index the training data in an X-Tree [3] after reducing the dimensionality via RCA. We tested the target dimensions 5, 10, 25 and 50. Using an index, we could now employ the complete dataset of 2 526 scans. We used a subset of 697 scans (= 163 525 slices) as training set for the RCA and tested the performance on the remaining 2 104 scans (901 326 instances). Table 3 shows the average leave-one-out (LOO) errors and query runtimes (excluding the time for feature generation) for the indexes generated from the test set.

Table 3. LOO regression errors [in cm] for RCA-transformed data with query times [in ms] in an X-Tree representing 2104 scans

Dimension	Error [cm]	Time / Query [ms]
5	2.764	4
10	1.881	22
25	1.343	440
50	1.209	2966

As can be been seen in Table 3, the curse of dimensionality causes the X-Tree to lose much of its effectiveness for increasing dimensions. Additionally, the error does only moderately increase for smaller dimensions. Based on these observations, we consider the 10 dimensional data set as the best trade-off, having a prediction error of 1.88 cm and a query time of 22 ms. We use this dataset for all following experiments. The total runtime required for feature generation is combined from the actual feature generation for a down-scaled version of the query slice (20 ms) and the time required for RCA transformation (0.1 ms). Thus, our selected query configuration results in a total prediction time of 42 ms.

Next, in order to validate the performance of the proposed ROI query workflow we will first analyze the accuracy of the retrieved ROIs and then proceed with an examination of retrieval times.

6.2   Precision of ROI Queries

We could again use automatically detected landmarks for defining a ground truth of lower and upper bounds, however, we cannot guarantee for the correctness of these matching points.

Therefore, we generated a new set of annotation points with five new landmark types: "lower plate of the twelfth thoracic vertebra", "lower bound of coccyx", "sacral promontory", "cranial sternum" and "lower xiphoid process". These landmarks were hand-annotated by a medical expert for providing a set of markers which have been verified visually.

In Table 4, we show the results of predicting all visible intervals with ROI queries formed by pairs of these landmarks in the dataset of 33 manually annotated volumes. As not all landmarks were visible in all volumes, only 158 intervals could be tested. Since the annotation error – the deviation of these markers from their expected positions – is at 2.579 cm, we cannot expect the queries to produce more reliable predictions.

Using Algorithm 1 with varying grid sizes g for the initial matching points P_i provides good predictions. We observe, however, that using a larger number of seed points only mildly improves the accuracy of the predictions, but it greatly increases the number of matching points being generated by regression (q). We conclude that two seed points are sufficient for our simple optimization scheme. Any more sophisticated optimization procedures should rather involve an intelligent screening of the proposed result range $(\hat{s}_{lb}, \ldots, \hat{s}_{ub})$ than use more seed points.

Table 4. Average deviation [in cm] of the result ROI of Algorithm 1 from the manually marked ROIs with the number of regression queries q and the runtime per query

Error Measure [cm]: $\text{err}(\hat{s}_{lb}) + \text{err}(\hat{s}_{ub})$	ROI prediction with Algorithm 1			
	g	Error [cm]	q	Time / Query [ms]
	2	2.655	6.8	1 273
Annotation Error:	5	2.549	9.2	1 951
2.579 cm	10	2.430	15.2	3 032
	25	2.573	30.0	5 946
	50	2.385	55.5	10 081

In Fig. 4 we see the cumulative distribution function $F(\text{error} \leq x\,\text{cm})$ for the analyzed query intervals. The 'Annotation' bars show the performance of the annotated ground truth landmarks, and the 'Algorithm 1' bars represent our ROI query algorithm using two seed points. There is almost no difference between the quality of the ground truth and our algorithm. The probability that the total prediction error $(\text{err}(\hat{s}_{lb}) + \text{err}(\hat{s}_{ub}))$ is at most 2 cm lies at 50 %. Again, with a height spacing of 5 mm, this means that in half of the cases, the retrieved range deviates by only two slices for each the lower and upper bound. When

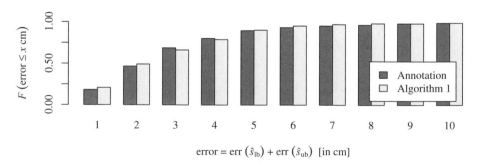

Fig. 4. Cumulative distribution function: $F(\text{error} \leq x\,\text{cm})$ (the steeper the better). It compares the error of Algorithm 1 with the quality of the used annotation points.

thus extending the returned query range by our pre-defined safety range, most returned subvolumes will completely contain the requested ROI.

We thus conclude that ROI queries can be efficiently answered by using Algorithm 1 with two initial matching points. The query time for grid size 2 is 1.5 seconds. Thus, our final experiments will show that the benefit of reducing volume queries to a region of interest strongly outweighs this cost.

6.3   Runtime of ROI Queries

For our last experiment, we chose a random set of 20 volumes from the database and tested them against four ROI queries defined in an example scan. Two queries are aimed at organs ("Left kidney" and "Urinary bladder"), one query ranges from the top of the hip bone to the bottom of Vertebra L5 and the final query is only interested in the view of the arch of aorta. The four hereby defined query ranges have heights of 16.8, 9.6, 4.7 and 0.9 centimeters In Fig. 5, we display the retrieval times of the resulting ROIs and their fraction of the complete dataset of 12 240 slices. Loading the complete 20 volumes from the server takes 1 400 seconds, whereas transferring only the ROIs induced by the given concepts takes 60 to 400 seconds, including the computation overhead for finding the ROI.

To conclude, employing our system for answering ROI queries saved between 77 − 99 % of the loading time compared to the retrieval of the complete scan. Thus, in a clinical routine our system is capable to save valuable time as well as hardware resources.

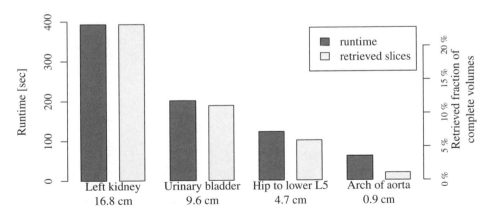

Fig. 5. Average runtimes (ten repetitions) and volume size reduction using ROI queries with Algorithm 1. Each experiment tests 20 volumes with a total of 12 240 slices. Loading the *complete* dataset takes 1 400 seconds.

7 Conclusion

In this paper, we proposed a method for processing region of interest (ROI) queries on a repository of CT scans. An ROI query is specified by giving an example ROI in another CT scan. Since CT scans are usually stored as stacks of 2D images representing a slice in the scan, the answer of an ROI query on a CT scan is a subset of the slices in the target scan representing an ROI which is equivalent to the query ROI.

After query specification, the system maps the ROI to a general height model of the human body. Then, the query region in the height model is mapped to a subregion of the queried scan containing the ROI. Technically, our system is based on an interpolation function using so-called matching points linking a CT scan to the height model. To guarantee the availability of matching points even for unannotated CT scans, we propose a method using content-based image descriptors and regression that can generate matching points for arbitrary slices in a scan. Finally, we propose a query algorithm for finding a stable mapping while deriving a minimal amount of matching points.

In our experimental evaluation, we validated the accuracy of our approach on a large database of 2 526 CT scans and displayed experiments for the reduced transfer volume of ROI queries being processed by our system.

For future work, we plan to extend our system to restrict ROIs in all three dimensions. We also aim to apply our retrieval solution to other types of 3D objects being stored in raster databases and to further, more general regression or interpolation problems.

Acknowledgements. The authors would like to thank Sascha Seifert et.al. for providing the executables of [13] and Robert Forbrig for generating the anatomical annotations used in Sect. 6.2. This research has been supported in part by the Theseus program in the Medico and CTC projects. They are funded by the German Federal Ministry of Economics and Technology under the grant number 01MQ07020. The responsibility for this publication lies with the authors.

References

1. Bar-Hillel, A., Hertz, T., Shental, N., Weinshall, D.: Learning distance functions using equivalence relations. In: Proceedings of the 20th International Conference on Machine Learning (ICML), Washington, DC, pp. 11–18 (2003)
2. Baumann, P., Furtado, P., Ritsch, R., Widmann, N.: The RasDaMan approach to multidimensional database management. In: Proceedings of the 12th ACM Symposium on Applied Computing (ACM SAC), San Jose, CA, pp. 166–173. ACM, New York (1997)
3. Berchtold, S., Keim, D.A., Kriegel, H.P.: The X-Tree: An index structure for high-dimensional data. In: Proceedings of the 22nd International Conference on Very Large Data Bases (VLDB), Bombay, India (1996)
4. Emrich, T., Graf, F., Kriegel, H.P., Schubert, M., Thoma, M., Cavallaro, A.: CT slice localization via instance-based regression. In: Proceedings of the SPIE Medical Imaging 2010: Image Processing (SPIE), San Diego, CA, USA. p. 762320 (2010)

5. Evans, A.C., Collins, D.L., Mills, S.R., Brownand, E.D., Kelly, R.L., Peters, T.M.: 3D statistical neuroanatomical models from 305 MRI volumes. In: IEEE Nuclear Science Symposium and Medical Imaging Conference (1993)
6. Feulner, J., Zhou, S.K., Seifert, S., Cavallaro, A., Hornegger, J., Comaniciu, D.: Estimating the body portion of CT volumes by matching histograms of visual words. In: Proceedings of the SPIE Medical Imaging 2009 Conference (SPIE), Lake Buena Vista, FL, USA, vol. 7259, p. 72591V (2009)
7. Haas, B., Coradi, T., Scholz, M., Kunz, P., Huber, M., Oppitz, U., André, L., Lengkeek, V., Huyskens, D., van Esch, A., Reddick, R.: Automatic segmentation of thoracic and pelvic CT images for radiotherapy planning using implicit anatomic knowledge and organ-specific segmentation strategies. Physics in Medicine and Biology 53(6), 1751–1771 (2008)
8. Hall, M., Frank, E., Holmes, G., Pfahringer, B., Reutemann, P., Witten, I.H.: The WEKA data mining software: an update. ACM SIGKDD Explorations 11(1), 10–18 (2009)
9. Holland, P., Welsch, R.: Robust regression using iteratively reweighted least-squares. Communications in Statistics-Theory and Methods 6(9), 813–827 (1977)
10. Kriegel, H.P., Seeger, B., Schneider, R., Beckmann, N.: The R*-tree: An efficient access method for geographic information systems. In: Proceedings of the International Conference on Geographic Information Systems, Ottawa, Canada (1990)
11. Nguyen, T.: Robust estimation, regression and ranking with applications in portfolio optimization. Ph.D. thesis, Massachusetts Institute of Technology (2009)
12. Samet, H.: Foundations of Multidimensional and Metric Data Structures. Morgan Kaufmann, San Francisco (2006)
13. Seifert, S., Barbu, A., Zhou, S.K., Liu, D., Feulner, J., Huber, M., Suehling, M., Cavallaro, A., Comaniciu, D.: Hierarchical parsing and semantic navigation of full body CT data. In: Proceedings of the SPIE Medical Imaging 2009 Conference (SPIE), Lake Buena Vista, FL, USA. vol. 7259, p. 725902 (2009)
14. Seifert, S., Kelm, M., Möller, M., Mukherjee, S., Cavallaro, A., Huber, M., Comaniciu, D.: Semantic annotation of medical images. In: Proceedings of the SPIE Medical Imaging 2010: Image Processing (SPIE), San Diego, CA, USA. vol. 7628, p. 762808 (2010)
15. Talairach, J., Tournoux, P.: Co-Planar Stereotaxic Atlas of the Human Brain 3-Dimensional Proportional System: An Approach to Cerebral Imaging. Thieme Medical Publishers (1988)

A Critical-Time-Point Approach to All-Start-Time Lagrangian Shortest Paths: A Summary of Results

Venkata M.V. Gunturi, Ernesto Nunes,
KwangSoo Yang, and Shashi Shekhar

Computer Science and Engineering, University of Minnesota, USA
{gunturi,enunes,ksyang,shekhar}@cs.umn.edu

Abstract. Given a spatio-temporal network, a source, a destination, and a start-time interval, the All-start-time Lagrangian Shortest Paths (ALSP) problem determines a path set which includes the shortest path for every start time in the given interval. ALSP is important for critical societal applications related to air travel, road travel, and other spatio-temporal networks. However, ALSP is computationally challenging due to the non-stationary ranking of the candidate paths, meaning that a candidate path which is optimal for one start time may not be optimal for others. Determining a shortest path for each start-time leads to redundant computations across consecutive start times sharing a common solution. The proposed approach reduces this redundancy by determining the critical time points at which an optimal path may change. Theoretical analysis and experimental results show that this approach performs better than naive approaches particularly when there are few critical time points.

1 Introduction

Given a spatio-temporal (ST) network, a source, a destination, and a start-time interval, the All-start-time Lagrangian shortest paths problem (ALSP) determines a path set which includes the shortest path for every start time in the interval. The ALSP determines both the shortest paths and the corresponding set of time instants when the paths are optimal. For example, consider the problem of determining the shortest path between the University of Minnesota and the MSP international airport over an interval of 7:00AM through 12:00 noon. Figure 1(a) shows two different routes between the University and the Airport. The 35W route is preferred outside rush-hours, whereas the route via Hiawatha Avenue is preferred during rush-hours (i.e., 7:00AM - 9:00AM) (see Figure 1(b)). Thus, the output of the ALSP problem may be a set of two routes (one over 35W and one over Hiawatha Avenue) and their corresponding time intervals.

Application Domain: Determining shortest paths is a key task in many societal applications related to air travel, road travel, and other spatio-temporal

D. Pfoser et al. (Eds.): SSTD 2011, LNCS 6849, pp. 74–91, 2011.

Time	Preferred Routes
7:30am	Via Hiwatha
8:30am	Via Hiwatha
9:30am	via 35W
10:30am	via 35W

(a) Different routes between University and Airport [2]

(b) Optimal times

Fig. 1. Problem illustration

networks [9]. Ideally, path finding approaches need to account for the time dependent nature of the spatial networks. Recent studies [6,21] show that indeed, such approaches yield shortest path results that save up to 30% in travel time compared with approaches that assume static network. These savings can potentially play a crucial role in reducing delivery/commute time, fuel consumption, and greenhouse emissions.

Another application of spatio-temporal networks is in the air travel. Maintaining the shortest paths across destinations is important for providing good service to passengers. The airlines typically maintain route characteristics such as average delay, flight time etc, for each route. This information creates a spatio-temporal network, allowing for queries like shortest paths for all start times etc. Figure 2(a) shows the Delta Airlines flight schedule between Minneapolis and Austin (Texas) for different start times [1]. It shows that total flight time varies with the start time of day.

Challenges: ALSP is a challenging problem for two reasons. First, the ranking of alternate paths between any particular source and destination pair in the network is not stationary. In other words, the optimal path between a source and destination for one start time may not be optimal for other start times. Second, many links in the network may violate the property of FIFO behavior. For example, in Figure 2(a), the travel time at 8:30AM is 6 hrs 31mins whereas, waiting 30mins would give a quicker route with 9:10AM flight. This violation of first-in-first-out (FIFO) is called non-FIFO behavior. Surface transportation networks such as road network also exhibit such behavior. For example, UPS [19,7] minimizes the number of left turns in their delivery routes during heavy traffic conditions. This leads to faster delivery and fuel saving.

Related work and their limitations: The related work can be divided on the basis of FIFO vs non-FIFO networks as shown in see Figure 2(b). In a FIFO network, the flow arrives at the destination in the same order as it starts at

Time	Route	Flight Time
8:30am	via Detroit	6 hrs 31 mins
9:10am	direct flight	2 hrs 51 mins
11:00am	via Memphis	4 hrs 38mins
11:30am	via Atlanta	6 hrs 28 mins
2:30pm	direct flight	2 hrs 51 mins

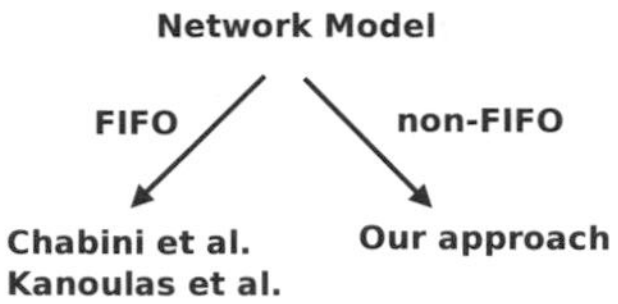

Network Model

FIFO / \ non-FIFO

Chabini et al.   Our approach
Kanoulas et al.

(a) Minneapolis - Austin (TX)   (b) Related work
Flight schedule

Fig. 2. Application Domain and related work

the source. A* based approaches, generalized for non-stationary FIFO behavior, were proposed in [5,15] for solving the shortest path problem in FIFO networks. However, transportation networks are known to exhibit non-FIFO behavior (see Figure 2(a)). Examples of Non-FIFO networks include multi-lane roads, car-pool lanes, airline networks etc. This paper proposes a method which is suitable for both FIFO and non-FIFO networks.

Proposed approach: A naive approach for solving the non-FIFO ALSP problem would involve determining the shortest path for each start time in the interval. This leads to redundant re computation of the shortest path across consecutive start times sharing a common solution. Some efficiencies can be gained using a time series generalization of a label-correcting algorithm [18]. This approach was previously used to find best start time [13], and is generalized for ALSP in Section 5 under the name modified-BEST. However, this approach still entails large number of redundant computations. In this paper, we propose the use of critical time points to reduce this redundancy. For instance, consider again the problem of determining the shortest path between University of Minnesota and MSP international airport over a time interval of 7:30am through 11:00am. Figure 1(b) shows the preferred paths at some time instants during this time interval, and Figure 3 shows the travel-times for all the candidate paths during this interval.

As can be seen, the Hiawatha route is faster for times in the interval [7:30am 9:30am)[1], whereas 35W route is faster for times in the interval [9:30am 11:00am]. This shows that the shortest path changed at 9:30am. We define this time instant as *critical time point*. Critical time points can be determined by computing the earliest intersection points between the functions representing the total travel time of paths. For example, the earliest intersection point of Hiawatha route was at 9:30am (with 35W route function). Therefore, it would be redundant to recompute shortest paths for all times in interval (7:30am 9:30am) and (9:30am 11:00am] since the optimal path for times within each interval did not change. This approach is particularly useful in case when there are a fewer number of critical time points.

---

[1] Note: an interval (a,b) does not include the end points a and b, [a,b] includes the endpoints a and b, and [a,b) includes a but not b.

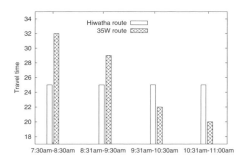

Fig. 3. Total travel time of candidate paths

Contributions: This paper proposes the concept of critical time points, which are the time points at which the shortest path between a source-destination pair changes. Using this idea, we also propose an algorithm, Critical-Time-point-based-ALSP-Solver (CTAS), for solving the ALSP problem. The correctness and completeness of the CTAS is presented. The CTAS algorithm is experimentally evaluated using real datasets. Experiment results show that the critical time point based based approach is particularly useful in cases when the number of critical times is small.

Scope and Outline: This paper models the travel time on any particular edge as a discrete time series. Turn restrictions are not considered in the ST network representation. The paper uses a Dijkstra-like framework. A*-like framework are beyond the scope of the paper. Moreover, we focus on computing the critical time points on the fly rather than precomputing them.

The rest of the paper is organized as follows. A brief description of the basic concepts and a formal problem definition is presented in Section 2. A description of the computational structure of the ALSP problem is presented in Section 3. In Section 4, we propose the CTAS algorithm. Experimental analysis of the proposed methods is presented in Section 5. Finally, we conclude in Section 6 with a brief description of future work.

2   Basic Concepts and Problem Definition

Spatio-temporal Networks: Spatio-temporal networks may be represented in a number of ways. For example, snapshot model, as depicted in Figure 4, views the spatio-temporal network as snapshots of the underlying spatio network at different times. Here, each snapshot shows the edge properties at a particular time instant. For example, travel times in all the edges $((A, B),(A, C),(B, D),(B, C))$ at time $t = 0$ is represented in the top left snapshot in Figure 4. For the sake of simplicity, this example assumes that (B, C) and (C, B) have the same travel time. The same network can also be represented as a time-aggregated graph (TAG) [11] as shown in Figure 5(a). Here, each edge is associated with a time series which represents the total cost of the edge. For example, edge (A, B) is

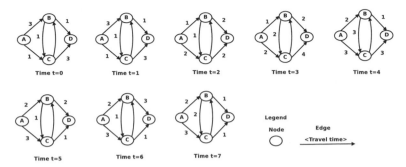

Fig. 4. Snapshot model of spatio-temporal network

associated with the time series [3 3 1 1 2 2 2 2]. This means that the travel time of edge (A, B) at times $t = 0$, $t = 1$, and $t = 2$ is 3, 3 and 1 respectively.

Both TAG and snapshot representations of our sample transportation network use a *Eulerian frame of reference* for describing moving objects. In the Eulerian frame of reference, traffic is observed as it travels past specific locations in the space over a period of time [4]. It is similar to sitting on the side of a highway and watching traffic pass a fixed location.

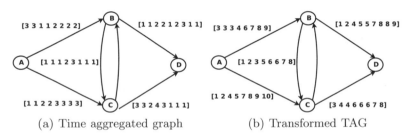

(a) Time aggregated graph   (b) Transformed TAG

Fig. 5. Time aggregated graph example

Traffic movement can also described using a *Lagrangian frame of reference*, in which the observer follows an individual moving object as it moves through space and time [4]. This can be visualized as sitting in a car and driving down a highway. The time-expanded graph (TEG) described later corresponds to this view.

Both Eulerian and Lagrangian-based views represent the same information and one view can be transformed into another view. The choice between the two depends on the problem being studied. In our problem context, it is more logical to use a Lagrangian frame of reference for finding the shortest paths because the cost of the candidate paths should be computed from the perspective of the traveler. TAG can also be used to view the network in Lagrangian frame of reference, however we use TEG for ease of communication. We now define the concept of a Lagrangian path.

Lagrangian Path: *A Lagrangian path P_i is a spatio-temporal path experienced by the traveler between any two nodes in a network.* A Lagrangian path may be viewed as a pair of traditional path in a graph and a schedule specifying arrival time at each node. During traversal of a Lagrangian path, the weight of any particular edge $e = (x, y)$ (where $e \in P_i$) is considered at the time of arrival at node x. For instance, consider the path <A,C,D> for start time $t = 0$. Here, we would start at node A at time $t = 0$. Therefore, the cost of edge (A, C) would be considered at $t = 0$, which is 1 (see Figure 5(a)). Following this edge, we would reach node C at time $t = 1$. Now, edge (C, D) would be considered at time $t = 1$ (because it takes 1 time unit to travel on edge (A, C)).

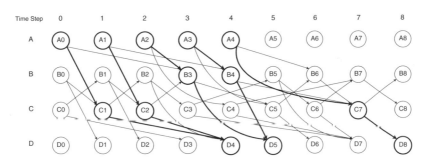

Fig. 6. Time expanded graph example

Conceptually, a Lagrangian path can be visualized as an explicit path in a time-expanded graph. Figure 6 depicts the ST network (shown in Figure 4) represented as a time-expanded graph [17]. Here, each node is replicated across the time instants and edges connect the nodes in a Lagrangian sense. This means that if the travel time on an edge (A, C) is 1 time unit (at time $t = 0$), then there is an edge between node A (at time $t = 0$) and node C (at time $t = 1$). Consider the previous example of Lagrangian path <A,C,D>. This path can be represented as a simple path among nodes A, (at time $t = 0$, which is $A0$ in Figure 6), C (at time $t = 1$, which is $C1$ in Figure 6), and D (at time $t = 4$, which is $D4$ in Figure 6).

As discussed previously, in the case of non-FIFO networks, we may arrive at a destination earlier by waiting at intermediate nodes. For example, in Figure 6 if we start at node A at time $t = 1$ (node $A1$), we would reach node B at time $t = 4$ (node $B4$). However, if we wait at node A for one time unit, then we can reach node B at time $t = 3$ (node $B3$). For simplicity, Figure 6 does not show the wait edges across temporal copies of a physical node e.g. $A0$-$A1$, $A1$-$A2$ etc.

Problem Definition: We define the ALSP problem by specifying input, output, objective function and constraints. Inputs include:
(a) Spatio-temporal network $G = (V, E)$, where V is the set of vertices, and E is the set of edges;
(b) A source s and a destination d pair where $\{s, d\} \in V$;

(c) A discrete time interval λ over which the shortest path between s and d is to be determined;

(d) Each edge $e \in E$ has an associated cost function, denoted as δ. The cost of an edge represents the time required to travel on that edge. The cost function of an edge is represented as a time series.

Output: The output is a set of routes, P_{sd}, from s to d where each route $P_i \in P_{sd}$ is associated with a set of start time instants ω_i, where ω_i is a subset of λ.

Objective function: Each path in P_{sd} is a shortest commuter-experienced travel time path between s and d during its respective time instants.

We assume the following constraints: The length of the time horizon over which the ST network is considered is finite. The weight function δ is a discrete time series.

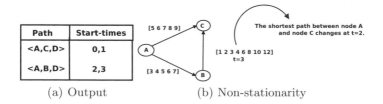

(a) Output   (b) Non-stationarity

Fig. 7. Output of ALSP problem

Example: Consider the sample ST network shown in Figure 5(a). An instance of the ALSP problem on this network with source A destination D and $\lambda = [0, 1, 2, 3]$, is shown in Figure 7(a). Here, path A-C-D is optimal for start times $t = 0$ and $t = 1$, and path A-B-D is optimal for start times $t = 2$ and $t = 3$.

3   Computational Structure of the ALSP Problem

The first challenge in solving the ALSP problem involves capturing the inherent non-stationarity present in the ST network. Due to this non-stationarity, traditional algorithms developed for static graphs cannot be used to solve the ALSP problem because the optimal sub-structure property no longer holds in case of ST networks [13]. On the other hand, algorithms developed for computing shortest paths for a single start time [13,16,17,8,20] are not practical because they would require redundant re-computation of shortest paths across the start times sharing a common solution.

This paper proposes a divide and conquer based approach to handle network non-stationarity. In this approach we divide the time interval over which the network exhibits non-stationarity into smaller intervals which are guaranteed to show stationary behavior[2]. These intervals are determined by computing the critical time points, that is, the time points at which the cost functions representing the total cost of the path as a function of time intersect. Now, within

---

[2] By stationarity, we mean that ranking of the alternate paths between a particular source-destination pair does not change within the interval.

these intervals, the shortest path can be easily computed using a single run of a dynamic programming (DP) based approach [13]. Recall our previous example of determining the shortest paths between the university and the airport over an interval of [7:30am 11:00am]. Here, 9:30am was the critical time point. This created two discrete sub-intervals [7:30 9:30) and [9:30 11:00]. Now, we can compute the ALSP using two runs of a DP based algorithm [13] (one on [7:30 9:30) and another on [9:30 11:00]).

Our second challenge for solving ALSP is capturing the non-FIFO behavior of the network. We do this by converting the travel information associated with an edge into earliest arrival time (at destination) information [12,20]. This is a two step process. First, the travel time information is converted into arrival time information. Second, the arrival time information is converted into earliest arrival time information. The second step captures the possibility of arriving at an end node earlier by waiting (non-FIFO behavior). For example, in the ST network shown in Figure 5(a), the travel time series of edge (A, B) is [3 3 1 1 2 2 2 2]. First, we convert it into arrival time series. This is done by adding the time instant index to the cost. For example, if we leave node A at times $t = 0, 1, 2, 3 \ldots$, we would arrive at node B at times $t = 3+0, 3+1, 1+2, 1+3, 2+4 \ldots$. Therefore, the arrival time series of (A, B) is [3 4 3 4 6 7 8 9]. The second step involves comparing each value of the arrival time series to the value to its right in the arrival time series. A lower value to its right means we can arrive at the end node earlier by waiting. Consider the arrival time series of $(A, B) = $ [3 4 3 4 6 7 8 9]. Here, the arrival time for $t = 1$ is 4 (which is less than the arrival time for $t = 2$). Therefore, the earliest arrival time on edge (A, B) for time $t = 1$ is 3 (by waiting for 1 time unit at node A). This process is repeated for each value in the time series. The earliest arrival time series of edge (A, B) is [3 3 3 4 6 7 8 9]. The earliest arrival time series of the edges are precomputed. Figure 5(b) shows the ST network from Figure 5(a) after the earliest arrival time series transformation.

There are two ways to determine the stationary intervals, either by precomputing all the critical time points, or by determining the critical time points at run time. The first approach was explored in [14] for a different problem. Precomputing the critical time points involves computing intersection points between cost functions of all the candidate paths. Now, in a real transportation network there can be exponential number candidate paths. Therefore, this paper follows the second approach of determining the critical time points at run-time. In this approach only a small fraction of candidate paths and their cost functions are actually considered while computation. Now, we define and describe our method of determining the critical time points.

Critical time point: *A start time instant when the shortest path between a source and destination may change.*

Consider an instance of the ALSP problem on the ST network shown in Figure 7(b), where the source is node A, the destination is C, and $\lambda = $ [0 1 2 3 4]. Here, start time $t = 2$ is a critical time point because the shortest path between node A and C changes for start times greater that $t = 2$. Similarly, $t = 2$ is also a critical time point for the network shown in Figure 5(a), where the source node is A and the destination node is D (see Figure 7(a)). Now, in order to determine these start time instants, we need to model the total cost of the path. This paper proposes using a weight function to capture the total cost of a path. This approach, which associates a weight function to a path, yields a *path-function* which represents the earliest arrival time at the end node of the path. A formal definition of the path-function is given below.

Path Function: *A path function represents the earliest arrival time at the end node of a path as a function of time. This is represented as a time series. A path function is determined by computing the earliest arrival times on its component edges in a Lagrangian fashion.*

For example, consider the path <A,B,C> in Figure 7(b). This path contains two edges, (A, B) and (B, C). The earliest arrival time (EAT) series of edge (A, B) is [3 4 5 6 7], while the EAT of edge (B, C) is [1 2 3 4 6 8 10 12]. Now, the path function of <A,B,C> for start times $[0, 4]$ is determined as follows. If we start at node A at $t = 0$, the arrival time at node B is 3. Now, arrival time at node C through edge (B, C) is considered for time $t = 3$ (Lagrangian fashion), which is 4. Thus, the value of the path function of <A,B,C> for start time $t = 0$ is 4. The value of the path-function for all the start times is computed in similar fashion. This would give the path function of <A,B,C> as [4 6 8 10 12]. This means that if we start at node A at times $t = 0, 1, 2, 3, 4$ then we will arrive at end node C at times $t = 4, 6, 8, 10, 12$. Similarly, the path function of path <A,C> is [5 6 7 8 9] (since it contains only one edge). By comparing the two path-functions we can see that path <A,B,C> has an earlier arrival time (at destination) for start times $t = 0$ and $t = 1$ (ties are broken arbitrarily). However, path <A,C> is shorter for start time $t \geq 2$. Thus, start time $t = 2$ becomes a critical time point. In general, the critical time points are determined by computing the intersection point (with respect to time coordinate) between path functions. In this case, the intersection point between path functions <A,C> and <A,B,C> is at time $t = 2$. Computing this point of intersection is the basis of the CTAS algorithm.

4   Critical Time-Point Based ALSP Solver (CTAS)

This section describes the critical time point based approach for the ALSP problem. This approach reduces the need to re-compute the shortest paths for each start time by determining the critical time points (start times) when the shortest path may change. Although Lagrangian paths are best represented by a time-expanded graph (TEG), these graphs are inefficient in terms of space requirements [13]. Therefore, this paper uses TEG model only for visualizing the ST network. For defining and implementing our algorithm we use TAG [13]. Recall that since the optimal substructure property is not guaranteed in a ST network, the given time interval is partitioned into a set of disjoint sub-intervals, where the shortest path does not change. The *Sub-interval Optimal Lagrangian Path* denotes this shortest path.

Sub-interval optimal Lagrangian path: *is a Lagrangian path, P_i, and its corresponding set of time instants ω_i, where $\omega_i \in \lambda$. P_i is the shortest path between a source and a destination for all the start time instants $t \in \omega_i$.*

The Lagrangian path <A,C,D> shown in Figure 7(a) is an example of sub-interval Optimal Lagrangian Path and its corresponding set $\omega = 0, 1$.

4.1   CTAS Algorithm

The algorithm starts by computing the shortest path for the first start time instant in the set λ. Since the optimal substructure property is not guaranteed in a ST network, the choice of the path to expand at each step made while computing the shortest path for a particular start time may not be valid for later time instants. Therefore, the algorithm stores all the critical time points observed while computing the shortest path for one start time in a data structure called *path-intersection table*. As discussed previously, the critical time point is determined by computing the time instants where the path functions of the candidate paths intersect. The earliest of these critical time points represents the first time instant when the current path no longer remains optimal. The algorithm re-computes the shortest paths starting from this time instant. Since the path functions represent the earliest arrival time (at end node of path) for a given start time, the intersection points represent the start times (at the source) when there is a change in relative ordering of the candidate paths.

The pseudocode for the CTAS algorithm is shown in Algorithm 1. The outer loop of ALSP ensures that a shortest path is computed for each start time instant in the set λ. The inner loop computes a single sub-interval optimal Lagrangian path. There may be several sub-interval optimal Lagrangian paths for the set λ. First, a priority queue is initialized with the source node. The inner loop determines the shortest path for start times greater than the earliest critical time points stored in the path intersection table ($t_{min}s$). For the first iteration, this would just be the first start time instant of our set λ. In each iteration, the algorithm expands along the path which has the minimum weight for time $t = t_{min}$. After a path is expanded the last node of the path is closed for start time $t = t_{min}$. For example, if a path $s-x-y$ is expanded for start time $t = t_{min}$, then node y is closed for start time $t = t_{min}$. This means there cannot be any other shorter path from node s to node y for the start time $t = t_{min}$.

Before expanding a path, the algorithm computes the intersecting points among the path functions available in the priority queue. The earliest intersection point (considering intersection points in the increasing order of their time coordinate) involving the path with minimum weight is the time instant when the stationary ordering among the candidate paths change. After that the path is expanded and path functions to all its neighbors are computed and added to the priority queue. The inner loop terminates when it tries to expand a source-destination path. In the next iteration of the outer loop, the shortest path computation starts from the earliest of the critical time points in path intersection table. Now, the path determined in the previous iteration is closed for all the start times earlier than earliest of the critical time point (and greater

---

Algorithm 1. CTAS Algorithm

1: Determine the earliest arrival time series of each edge
2: Initialize the path intersection table with the first start time in λ
3: **while** a shortest path for each time $t \in \lambda$ waits to be determined **do**
4:    Select the minimum time instants among all time instants at which a shortest path might change and clear path intersection table
5:    Initialize a priority queue with the path functions corresponding to the source node and its neighbors
6:    **while** destination node is not expanded **do**
7:        Choose the path with the minimum weight to expand
8:        Determine the intersection points among the path functions in the priority queue
9:        Delete all the paths ending on that node from the priority queue
10:       Save the time coordinate of the intersection point in the path intersection table
11:       Determine the path functions resulting from expansion of the chosen path
12:       Push the newly determined path functions into the priority queue
13:    **end while**
14: **end while**

---

than t_{min} of previous iteration). In a worst case scenario, the value of the earliest critical time point could be the next time start time instant in the set λ (one more than previous t_{min}). In such a case, the inner loop would have determined the shortest path only for a single start time instant. The algorithm terminates when the earliest of the critical time points is greater than the latest start time instant in the set λ.

Execution Trace: Figure 8 gives an execution trace of the CTAS algorithm on the ST network shown in Figure 5(b). The first iteration of the outer loop starts with initializing the priority queue with the source node. The inner loop builds a shortest path starting from the earliest critical time point. For the first iteration, this would be time $t = 0$ (the first start time instant in λ). First, the source node is expanded for time $t = 0$. Path functions for its immediate neighbors (path <A,B> and <A,C>) are computed and added them to the priority queue. In this case, the path functions just happen to be their edge weight functions.

In the first iteration of the inner loop, the algorithm chooses the path whose path function has minimum weight at the start time instant chosen in step 4 of the algorithm. This would be $t = 0$ for the first case. At this point the algorithm has two choices, path function <A,C> and <A,B>(see Figure 8). The algorithm chooses <A,C> because it has lowest cost for $t = 0$. This path is expanded and path functions for its neighbors, <A,C,D> and <A,C,B>, are computed and added to the priority queue. Again the choice of path <A,C> may not be valid in later times, therefore, the algorithm closes the node C only for the start time $t = 0$. The algorithm stores the intersection point between the path <A,B> and <A,C> (which is at $t = 2$) in the path intersection table.

The next iteration expands path <A,C,B>. There is no intersection between the path functions of <A,C,D> and <A,C,B> (in our interval λ). Again, node

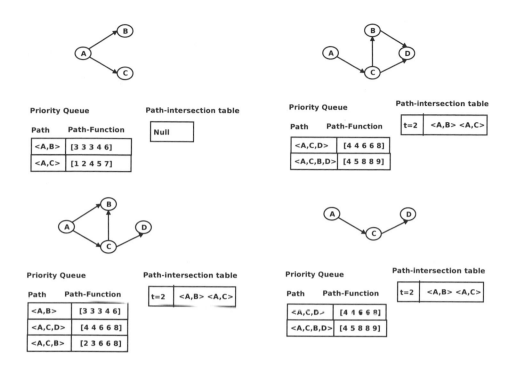

Fig. 8. Execution trace of CTAS algorithm

B is closed only for $t = 0$. At this point, all the paths which end in node B are deleted from the priority queue. Now, the priority queue contains two paths, <A,C,D> and <A,C,B,D>. This time the algorithm tries to expand along path <A,C,D> because it has least cost. Again the intersection point between the path functions is computed. Here, paths <A,C,D> and <A,C,B,D> do not intersect[3]. This time a complete source to destination path was expanded. This is the terminating condition of the inner loop. This completes one iteration of the outer loop. Next iteration of the outer loop builds the shortest path starting at the earliest of the critical time points stored in the previous iteration. This happens to be $t = 2$ for our example. The next iteration of the inner loop starts for start time $t = 2$. At this point the path intersection table is cleared. We see that the algorithm did not compute the shortest path for start time $t = 1$. Figure 6 also showed that the shortest path did not change start time $t = 1$ (shortest paths between node A and node D for start-times $t = 0, 1, 2, 3$ are shown in bold). The fact that the next iteration of the CTAS algorithm starts with $t = 2$ shows that it saves computation. The algorithm terminates when shortest paths of all time instants in the set λ have been determined. Figure 7(a) shows the final result of the algorithm.

---

[3] Note that here path <A,C,B,D> may also be expanded.

4.2   Analysis of the CTAS Algorithm

The CTAS algorithm divides a given interval (over which shortest paths have to be determined) into a set of disjoint sub-intervals. Within these intervals, the shortest path does not change. The correctness of the CTAS algorithm requires the path returned for a particular sub-interval to be optimal and the completeness of the algorithm guarantees that none of the critical time points are missed. The correctness of the CTAS can be easily argued on the basis greedy nature of the algorithm. Lemma 1 shows that the CTAS algorithm does not miss any critical time point.

Lemma 1. *The CTAS algorithm recomputes the shortest path for all those start times $t_i \in \lambda$, where the shortest path for previous start time t_{i-1} can be different from the shortest path for start time t_i.*

Proof. Consider the sample network shown in Figure 9, where a shortest path has to be determined between node s and d for discrete time interval $\lambda = [1, 2 \ldots T]$. First, the source node is expanded for the start time instant $t = 1$. As a result, path functions for all the neighbors <s,2>, <s,1>, <s,x_i> are added to the priority queue. With loss of generality assume that path <s,1> is chosen in the next iteration of the inner loop. Also assume that the earliest intersection point between path functions <s,x_i > and <s,1> is at $t = \alpha$ between <s,1> and <s,2>. Now, the queue contain paths <s,1,d>, <s,2>, <s,x_i>. Here we have two cases. First, path <s,1,d> has lower cost for start time $t = 1$. Second, path <s,2> has lower cost for start time $t = 1$.

Considering the first case, without loss of generality assume that the earliest intersection point between the path functions <s,1,d> and <s,2> is at $t = \beta$. Note that both $t = \alpha$ and $t = \beta$ denote the start times at the source node. Consider the case when $\beta \leq \alpha$ (Note that β cannot be greater than α as all the edges have positive weights). In such a case the shortest path is recomputed for starting time $t = \beta$ and the path <s,1,d> is closed for all start times $1 \leq t < \beta$. Assume for the sake of contradiction, that there is a shortest path P_x from source to destination that is different from path <s,1,d> which is optimal for start time $t_x \in [1, \beta)$. Assume that $P_x = < s, x_1, x_2, x_3, \ldots, d >$. This means that path $< s, x_1 >$ had least for time $t = t_x$. However, by the nature of the algorithm, this path would have been expanded instead of path $< s, 1 >$ (a contradiction). Moreover, as all the travel times positive, if sub path <s,x_1 > was not shorter than <s,1,d> for start times earlier than $t = \beta$. Any positive weight addition to the path function (through other edges) cannot make P_x shorter than <s,1,d>. Consider the second case when <s,2> had lower cost for start time $t = 1$. Now, path <s,2> would be expanded and path function <s,2,d> would be added to priority queue. Again, assume that path functions <s,1,d> and <s,2,d> intersect at time $t = \beta$. A similar argument can be given for this case as well.

Theorem 1. *CTAS algorithm is complete.*

Proof. There may be several sub-interval optimal Lagrangian paths P_i over set λ. Each P_i is associated with a set of time instants ω_i, where $\bigcup_{\forall i \in |P_{sd}|} \omega_i = \lambda$.

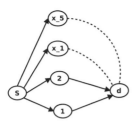

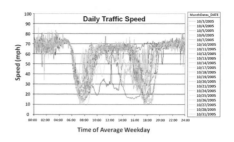

Fig. 9. Network for Lemma 1

Fig. 10. Speed profiles for one highway segment

The completeness proof of the CTAS algorithm is presented in two parts. First, using Lemma 1 we can conclude that the CTAS algorithm does not miss any start time instant when the shortest path changes. Secondly, the outer loop iterates until the algorithm determines a shortest path for all the time instants in set λ. This happens when the earliest of the path-order change times t_{min} falls outside λ. This proves the completeness of the algorithm.

Discussion: The CTAS algorithm shows better performance than a naive approach which, determines the shortest paths for each start time in the user specified time interval. However, this happens only when the ratio of the number of critical time points to that of start time instants is low. This ratio can be denoted as the change probability shown by Equation 1.

$$change\_probability = \frac{\#critical\ time\ points}{\#start\_time\ instants} \tag{1}$$

When the change probability is nearly 1, there would be a different shortest path for each start time in an interval. In this worst case scenario, the CTAS approach would also have to recompute the shortest path for each start time. A theoretical bound of $n^{\Theta(\log n)}$ (where n is the number nodes in the graph) on the number of critical time points was given in [10] for the case of piecewise linear cost functions. In our case, the number of critical time points is bounded by length of start time interval ($|\lambda|$) due to discrete nature of the input.

5   Experimental Evaluation

Experiments were conducted to evaluate the performance of CTAS algorithm as different parameters were varied. The experiments were carried on a real dataset containing the highway road network of Hennepin county, Minnesota, provided by NAVTEQ [3]. The dataset contained 1417 nodes and 3754 edges. The data set also contained travel times for each edge at time quanta of 15mins. Figure 10 shows the speed profiles for a particular highway segment in the dataset over a period of 30days. As can be seen, the speed varies with the time of day. For experimental purposes, the travel times were converted into time quanta of 1mins.

This was done by replicating the data inside time interval. The experiments were conducted on an Intel Quad core Linux workstation with 2.83Ghz CPU and 4GB RAM. The performance of CTAS algorithm was compared against a modified version of the existing BEST start time algorithm proposed in [13].

Modified-BEST (MBEST) algorithm: The MBEST algorithm consists of two main parts. First, the shortest path between source and destination is determined for all the desired start time instants. Second, the computed shortest paths are post-processed and a set of distinct paths is returned. The MBEST algorithm uses a label correcting approach similar to that of BEST algorithm, proposed in [13] to compute the shortest paths between source and destination. The algorithm associates two lists viz, the arrival time list and ancestor list, with each node. The arrival time array, $C_v[t]$, represents the earliest arrival time at node v for the start time t at the source. The ancestor array, $An_v[t]$, represents the previous node in the path from source for time t. These lists are updated using Equation 2, where γ_{uv} represents the earliest arrival time series of the edge (u, v). The algorithm terminates when there are no more changes in the arrival time list of any node.

$$C_v[t] = \min\{C_v[t], \gamma_{uv}[C_u[t]]\}, uv \in E \qquad (2)$$

Experimental setup: The experimental setup is shown in Figure 11. The first step of the experimental evaluation involved combining the travel time information along with the spatial road network to represent the ST network as a Time-aggregated graph. A set of different queries (each with different parameters) was run on the CTAS and the MBEST algorithms. The following parameters were varied in the experiments: (a) length of the time interval over which shortest paths were desired ($|\lambda|$), (b) total travel time of the route, (c) time of day (rush hour vs non-rush hours). A speedup ratio, given by Equation 3, was computed for each run. The total number of re-computations avoided by CTAS was also recorded for each run. In worst case, the shortest paths may have to be re-computed for each time instant in the interval λ. The total number of re-computations saved by CTAS is the difference between $|\lambda|$ and re-computations performed.

$$speed - up \; ratio = \frac{MBEST \; runtime}{CTAS \; runtime} \qquad (3)$$

Number of re-computations saved in CTAS: Figure 12 shows the number of re-computations saved by the CTAS algorithm. The experiments showed that more saving was gained where paths were shorter. Similarly, fewer number of re-computations were performed in case of Non-rush hours. This is because there were fewer intersections among the path-functions.

Effect of length of start time interval ($|\lambda|$): This experiment was performed to evaluate the effect of length of start time interval (λ) over which the shortest path was desired. Figure 13(a) shows the speed-up ratio for Non-rush hours and Figure 13(b) shows the speed-up ratio for Rush hours. The speed ratio was calculated for paths with travel time 30 and travel time 40. These travel times

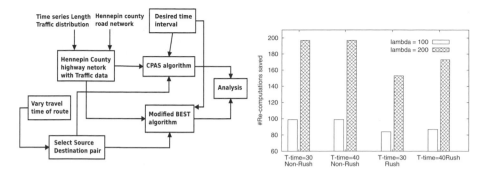

Fig. 11. Experimental setup

Fig. 12. Re-computations saved

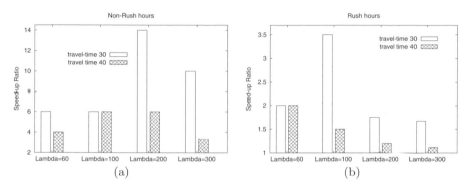

Fig. 13. Effect of Lambda on Speed-up

indicate the time required to travel on these paths during non-rush hours when there is no traffic. The experiments showed that run-time of both CTAS and MBEST increased with increase in the lambda. Runtime of MBEST increases steadily whereas CTAS increases very slowly with *lambda* for a travel time of 30. However, the run-time of CTAS increases rapidly for travel time of 40.

Effect of total travel time of a path: This experiment was performed to evaluate the effect of total travel time of a path on the candidate algorithms. Figure 14 shows the speed-up ratio as the total travel time of the path was varied. The experiments showed that the runtime of CTAS algorithm increased with a corresponding increase in the total travel time of the path, whereas the runtime of the MBEST algorithm remained the same. This is because, CTAS algorithm follows a Dijkstra's like approach and expands the paths, but MBEST is follows a label correcting approach and terminates when there no more changes in the arrival time array of any node.

Effect of different start times: Experiments showed that better speed-up was obtained for Non-rush hours than the rush hours (see Figure 14 and Figure 13). This is because there are fewer number of intersection points in case of non-Rush hours.

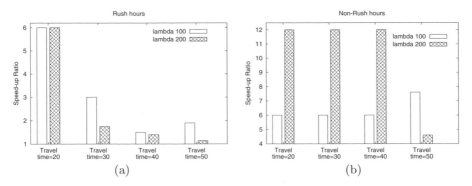

Fig. 14. Effect of travel time on Speed-up

6 Conclusions and Future Work

The All-start-time shortest Lagrangian shortest path problem (ALSP) is a key component of applications in transportation networks. ALSP is a challenging problem due to the non-stationarity of the spatio-temporal network. Traditional A* and Dijkstra's based approaches incur redundant computation across the time instants sharing a common solution. The proposed Critical Time-point based ALSP Solver (CTAS), reduces this redundant re computation by determining the time points when the ranking among the alternate paths between the source and destination change. Theoretical and experimental analysis show that this approach is more efficient than naive particularly in case of few critical time points.

In future the we plan to extend the CTAS algorithm by developing an A* based approach. We plan to reduce the overhead of computing the earliest arrival time series for all the time instants by doing the computation at runtime. We also plan to explore alternative approaches similar to [15] for solving non-FIFO ALSP. Performance of CTAS would also be evaluated on larger datasets.

Acknowledgment. We are particularly grateful to the members of the Spatial Database Research Group at the University of Minnesota for their helpful comments and valuable suggestions. We would also like to extend our thanks to Kim Koffolt for improving the readability of this paper. This work was supported by NSF grant (grant number NSF III-CXT IIS-0713214) and USDOD grant (grant number HM1582-08-1-0017 and HM1582-07-1-2035). The content does not necessarily reflect the position or the policy of the government and no official endorsement should be inferred.

References

1. Delta airlines, http://www.delta.com/
2. Google maps, http://www.maps.google.com/
3. Navteq, http://www.navteq.com/

4. Batchelor, G.K.: An introduction to fluid dynamics. Cambridge Univ. Press, Cambridge (1973)
5. Chabini, I., Lan, S.: Adaptations of the a* algorithm for the computation of fastest paths in deterministic discrete-time dynamic networks. IEEE Transactions on Intelligent Transportation Systems 3(1), 60–74 (2002)
6. Demiryurek, U., Banaei-Kashani, F., Shahabi, C.: A case for time-dependent shortest path computation in spatial networks. In: Proc. of the SIGSPATIAL Intl. Conf. on Advances in GIS, GIS 2010, pp. 474–477 (2010)
7. Deutsch, C.: Ups embraces high-tech delivery methods. NY Times (July 12, 2007), http://www.nytimes.com/2007/07/12/business/12ups.html
8. Ding, B., Yu, J.X., Qin, L.: Finding time-dependent shortest paths over large graphs. In: Proc. of the Intl. Conf. on Extending Database Technology (EDBT), pp. 205–216 (2008)
9. Evans, M.R., Yang, K., Kang, J.M., Shekhar, S.: A lagrangian approach for storage of spatio-temporal network datasets: a summary of results. In: Proc. of the SIGSPATIAL Intl. Conf. on Advances in GIS, GIS 2010, pp. 212–221 (2010)
10. Foschini, L., Hershberger, J., Suri, S.: On the complexity of time-dependent shortest paths. In: SODA. pp. 327–341 (2011)
11. George, B., Shekhar, S.: Time-aggregated graphs for modelling spatio-temporal networks. J. Semantics of Data XI, 191 (2007)
12. George, B., Shekhar, S., Kim, S.: Spatio-temporal network databases and routing algorithms. Tech. Rep. 08-039, Univ. of Minnesota - Comp. Sci. and Engg. (2008)
13. George, B., Kim, S., Shekhar, S.: Spatio-temporal network databases and routing algorithms: A summary of results. In: Papadias, D., Zhang, D., Kollios, G. (eds.) SSTD 2007. LNCS, vol. 4605, pp. 460–477. Springer, Heidelberg (2007)
14. Gunturi, V., Shekhar, S., Bhattacharya, A.: Minimum spanning tree on spatio-temporal networks. In: Bringas, P.G., Hameurlain, A., Quirchmayr, G. (eds.) DEXA 2010. LNCS, vol. 6262, pp. 149–158. Springer, Heidelberg (2010)
15. Kanoulas, E., Du, Y., Xia, T., Zhang, D.: Finding fastest paths on a road network with speed patterns. In: Proceedings of the 22nd International Conference on Data Engineering (ICDE), p. 10 (2006)
16. Kaufman, D.E., Smith, R.L.: Fastest paths in time-dependent networks for intelligent vehicle-highway systems application. I V H S Journal 1(1), 1–11 (1993)
17. Köhler, E., Langkau, K., Skutella, M.: Time-expanded graphs for flow-dependent transit times. In: Möhring, R.H., Raman, R. (eds.) ESA 2002. LNCS, vol. 2461, pp. 599–611. Springer, Heidelberg (2002)
18. Kleinberg, J., Tardos, E.: Algorithm Design. Pearson Education, London (2009)
19. Lovell, J.: Left-hand-turn elimination. NY Times (December 9, 2007), http://www.nytimes.com/2007/12/09/magazine/09left-handturn.html
20. Orda, A., Rom, R.: Shortest-path and minimum-delay algorithms in networks with time-dependent edge-length. J. ACM 37(3), 607–625 (1990)
21. Yuan, J., Zheng, Y., Zhang, C., Xie, W., Xie, X., Sun, G., Huang, Y.: T-drive: driving directions based on taxi trajectories. In: Proc. of the SIGSPATIAL Intl. Conf. on Advances in GIS, GIS 2010, pp. 99–108 (2010)

Online Computation of Fastest Path in Time-Dependent Spatial Networks*

Ugur Demiryurek[1], Farnoush Banaei-Kashani[1], Cyrus Shahabi[1],
and Anand Ranganathan[2]

[1] University of Southern California- Department of Computer Science
Los Angeles, CA USA
{demiryur,banaeika,shahabi}@usc.edu
[2] IBM T.J. Watson Research Center
Hawthorne, NY USA
aranganaus@ibm.com

Abstract. The problem of point-to-point fastest path computation in static spatial networks is extensively studied with many precomputation techniques proposed to speed-up the computation. Most of the existing approaches make the simplifying assumption that travel-times of the network edges are constant. However, with real-world spatial networks the edge travel-times are time-dependent, where the arrival-time to an edge determines the actual travel-time on the edge. In this paper, we study the online computation of fastest path in time-dependent spatial networks and present a technique which speeds-up the path computation. We show that our fastest path computation based on a bidirectional time-dependent A* search significantly improves the computation time and storage complexity. With extensive experiments using real data-sets (including a variety of large spatial networks with real traffic data) we demonstrate the efficacy of our proposed techniques for online fastest path computation.

1   Introduction

With the ever-growing popularity of online map applications and their wide deployment in mobile devices and car-navigation systems, an increasing number of users search for point-to-point fastest paths and the corresponding travel-times. On static road networks where edge costs are constant, this problem has been extensively studied and many efficient speed-up techniques have been developed to compute the fastest path in a matter of milliseconds (e.g., [27,31,28,29]). The static fastest path approaches make the simplifying assumption that the travel-time for each edge of the road network is constant (e.g., proportional to the length of the edge). However, in real-world the actual travel-time on a road segment heavily depends on the traffic congestion and, therefore, is a function of time i.e., *time-dependent*. For example, Figure 1 shows the variation

---

* This research has been funded in part by NSF grants IIS-0238560 (PECASE), IIS-0534761,IIS-0742811 and CNS-0831505 (CyberTrust), and in part from CENS and METRANS Transportation Center, under grants from USDOT and Caltrans.Any opinions, findings, and conclusions or recommendations expressed in this material are those of the author(s) and do not necessarily reflect the views of the National Science Foundation.

D. Pfoser et al. (Eds.): SSTD 2011, LNCS 6849, pp. 92–111, 2011.

of travel-time (computed by averaging two-years of historical traffic sensor data) for a particular road segment of I-10 freeway in Los Angeles as a function of arrival-time to the segment. As shown, the travel-time changes with time (i.e, the time that one arrives at the segment entry determines the travel-time), and the change in travel-time is significant. For instance, from 8AM to 9AM the travel-time of the segment changes from 32 minutes to 18 minutes (a 45% decrease). By induction, one can observe that the time-dependent edge travel-times yield a considerable change in the actual fastest path between any pair of nodes throughout the day. Specifically, the fastest between a source and a destination node varies depending on the departure-time from the source. Unfortunately, all those techniques that assume constant edge weights fail to address the fastest path computation in real-world time-dependent spatial networks.

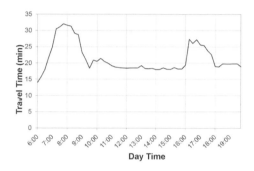

Fig. 1. Real-world travel-time on a segment of I-10 in LA

The time-dependent fastest path problem was first shown by Dreyfus [10] to be polynomially solvable in FIFO networks by a trivial modification to Dijkstra algorithm where, analogous to shortest path distances, the arrival-time to the nodes is used as the labels that form the basis of the greedy algorithm. The FIFO property, which typically holds for many networks including road networks, suggests that moving objects exit from an edge in the same order they entered the edge[1]. However, the modified Dijkstra algorithm [10] is far too slow for online map applications which are usually deployed on very large networks and require almost instant response times. On the other hand, there are many efficient precomputation approaches that answer fastest path queries in near real-time (e.g., [27]) in static road networks. However, it is infeasible to extend these approaches to time-dependent networks. This is because the input size (i.e., the number of fastest paths) increases drastically in time-dependent networks. Specifically, since the length of a s-d path changes depending on the departure-time from s, the fastest path is not unique for any pair of nodes in time-dependent networks. It has been conjectured in [3] and settled in [11] that the number of fastest paths between any pair of nodes in time-dependent road networks can be super-polynomial. Hence, an algorithm which considers the every possible path (corresponding to every possible departure-time from

---

[1] The fastest path computation is shown to be NP-hard in non-FIFO networks where waiting at nodes is not allowed [23]. Violation of the FIFO property rarely happens in real-world and hence is not the focus of this study.

the source) for any pair of nodes in large time-dependent networks would suffer from exponential time and prohibitively large storage requirements. For example, the time-dependent extension of Contraction Hierarchies (CH) [1] and SHARC [5] speed-up techniques (which are proved to be very efficient for static networks) suffer from the impractical precomputation times and intolerable storage complexity (see Section 3).

In this study, we propose a bidirectional time-dependent fastest path algorithm (B-TDFP) based on A* search [17]. There are two main challenges to employ bidirectional A* search in time-dependent networks. First, finding an admissible heuristic function (i.e., lower-bound distance) between an intermediate v_i node and the destination d is challenging as the distance between v_i and d changes based on the departure-time from v_i. Second, it is not possible to implement a backward search without knowing the arrival-time at the destination. We address the former challenge by partitioning the road network to non-overlapping partitions (an off-line operation) and precompute the intra (node-to-border) and inter (border-to-border) partition distance labels with respect to *Lower-bound Graph $\underline{G}$* which is generated by substituting the edge travel-times in G with minimum possible travel-times. We use the combination of intra and inter distance labels as a heuristic function in the online computation. To address the latter challenge, we run the backward search on the lower-bound graph ($\underline{G}$) which enables us to filter-in the set of the nodes that needs to be explored by the forward search.

The remainder of this paper is organized as follows. In Section 2, we explain the importance of time-dependency for accurate and useful path planning. In Section 3, we review the related work on time-dependent fastest path algorithms. In Section 4, we formally define the time-dependent fastest path problem in spatial networks. In Section 5, we establish the theoretical foundation of our proposed bidirectional algorithm and explain our approach. In Section 6, we present the results of our experiments for both approaches with a variety of spatial networks with real-world time-dependent edge weights. Finally, in Section 7, we conclude and discuss our future work.

2   Towards Time-Dependent Path Planning

In this section, we explain the difference between fastest computation in time-dependent and static spatial networks. We also discuss the importance and the feasibility of time-dependent route planning.

To illustrate why classic fastest path computations in static road networks may return non-optimal results, we show a simple example in Figure 2 where a spatial network is modeled as a time-dependent graph and edge travel-times are function of time. Consider the snapshot of the network (i.e., a static network) with edge weights corresponding to travel-time values at $t=0$. With classic fastest path computation approaches that disregard time-dependent edge travel-times, the fastest path from s to d goes through v_1, v_2, v_4 with a cost of 13 time units. However, by the time when v_2 is reached (i.e., at $t=5$), the cost of edge $e(v_2, v_4)$ changes from 8 to 12 time units, and hence reaching d through v_2 takes 17 time units instead of 13 as it was anticipated at $t=0$. In contrast, if the time-dependency of edge travel-times are considered and hence the path going through v_3 was taken, the total travel-cost would have been 15 units which is the actual optimal fastest path. We call this shortcoming of the classic fastest path computation techniques as *no-lookahead* problem. Unfortunately, most of the existing state of the art path planning applications (e.g., Google Maps, Bing Maps) suffer from the no-lookahead shortcoming and, hence, their fastest path recommendation remains the same throughout the day regardless of the departure-time from the source (i.e., query time). Although some of these applications provide alternative paths under traffic conditions (which may seem similar to time-dependent planning at first), we observe that the recommended alternative paths and their corresponding travel-times still remain unique during the day, and hence no time-dependent planning. To the best of our knowledge, these applications compute $top\text{-}k$ fastest paths (i.e., k alternative paths) and their corresponding travel-times with and without taking into account the traffic conditions. The travel-times which take into account the traffic conditions are simply computed by considering increased edge weights (that corresponds to traffic congestion) for each path. However, our time-dependent path planning results in different optimum paths for different departure-times from the source. For example, consider Figure 3(a) where Google Maps offer two alternative paths (and their travel-times under no-traffic and traffic conditions) for an origin and destination pair in Los Angeles road network. Note that the path recommendation and the travel-times remain the same regardless of when the user submits the query. On the other hand, Figure 3(b) depicts the time-dependent path recommendations (in different colors for different departure times) for the same origin and destination pair where we computed the time-dependent fastest paths for 38 consecutive departure-times between 8AM and 5:30PM, spaced 15 minutes apart[2]. As shown, the optimal paths change frequently during the course of the day.

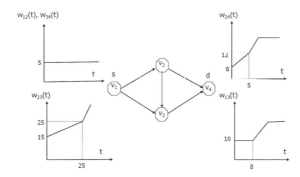

Fig. 2. Time-dependent graph

One may argue against the feasibility of time-dependent path planning algorithms due to a) unavailability of the time-dependent edge travel-times, or b) negligible gain of time-dependent path planning (i.e., how much time-dependent planning can improve the travel-time) over static path planning. To address the first argument, note that recent advances in sensor networks enabled instrumentation of road networks in major cities for collecting real-time traffic data, and hence it is now feasible to accurately model

---

[2] The paths are computed using the algorithm presented in Section 5 where time-dependent edge travel-times are generated based on the two-years of historical traffic sensor data collected from Los Angeles road network.

the time-dependent travel-times based on the vast amounts of historical data. For instance, at our research center, we maintain a very large traffic sensor dataset of Los Angeles County that we have been collecting and archiving the data for past two years (see Section 6.1 for the details of this dataset). As another example, PeMS [24] project developed by UC Berkeley generates time-varying edge travel-times using historical traffic sensor data throughout California. Meanwhile, we also witness that the leading navigation service providers (such as Navteq [22] and TeleAtlas [30]) started releasing their time-dependent travel-time data for road networks at high temporal resolution. With regards to the second argument, several recent studies showed the importance of time-dependent path planning in road networks where real-world traffic datasets have been used for the assessment. For example, in [7] we report that the fastest path computation that considers time-dependent edge travel-times in Los Angeles road network decreases the travel-time by as much as 68% over the fastest path computation that assumes constant edge travel-times. We made the similar observation in another study [15] under IBM's Smart Traffic Project where the time-dependent fastest path computation in Stockholm road network can improve the travel-time accuracy up to 62%. Considering the availability of high-resolution time-dependent travel-time data for road networks, and the importance of time-dependency for accurate and useful path planning, the need for efficient algorithms to enable next-generation time-dependent path planning applications becomes apparent and immediate.

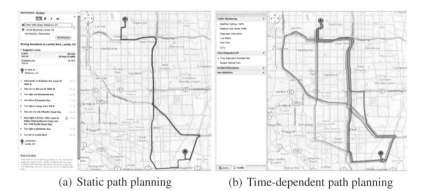

(a) Static path planning   (b) Time-dependent path planning

Fig. 3. Static vs Time-dependent path planning

3   Related Work

In the last decade, numerous efficient fastest path algorithms with precomputation methods have been proposed (see [29,27] for an overview). However, there are limited number of studies that focus on efficient computation of time-dependent fastest path (TDFP) problem.

Cooke and Halsey [2] first studied TDFP computation where they solved the problem using Dynamic Programming in discrete time. Another discrete-time solution to TDFP problem is to use time-expanded networks [19]. In general, time-expanded network (TEN) and discrete-time approaches assume that the edge weight functions are defined over a finite discrete window of time $t \in t_0, t_1, .., t_n$, where t_n is determined by the total duration of time interval under consideration. Therefore, the problem is reduced to the problem of computing minimum-weight paths over a static static network per time window. Hence, one can apply any static fastest path algorithms to compute TDFP. Although these algorithms are easy to design and implement, they have numerous shortcomings. First, TEN models create a separate instance of network for each time instance hence yielding a substantial amount of storage overhead. Second, such approaches can only provide approximate results because the model misses the state of the network between any two discrete-time instants. Moreover, the difference between the shortest path obtained using TEN approach and the optimal shortest path is *unbounded*. This is because the query time can be always between any two of the intervals which are not captured by the model, and hence the error is is accumulated on each edge along the path. In [12], George and Shekhar proposed a time-aggregated graph approach where they aggregate the travel-times of each edge over the time instants into a time series. Their model requires less space than that of the TEN and the results are still approximate with no bounds.

In [10], Dreyfus showed that TDFP problem can be solved by a generalization of Dijkstra's method as efficiently as for static fastest path problems. However, Halpern [16] proved that the generalization of Dijkstra's algorithm is only true for FIFO networks. If the FIFO property does not hold in a time-dependent network, then the problem is NP-Hard. In [23], Orda and Rom introduced Bellman-Ford based algorithm where they determine the path toward destination by refining the arrival-time functions on each node in the whole time interval T. In [18], Kanoulas et al. proposed Time-Interval All Fastest Path (allFP) approach in which they maintain a priority queue of all paths to be expanded instead of sorting the priority queue by scalar values. Therefore, they enumerate all the paths from the source to a destination node which incurs exponential running time in the worst case. In [9], Ding et al. used a variation of Dijkstra's algorithm to solve the TDFP problem. With their TDFP algorithm, using Dijkstra like expansion, they decouple the path-selection and time-refinement (computing earliest arrival-time functions for nodes) for a given starting time interval T. Their algorithm is also shown to run in exponential time for special cases (see [4]). The focus of both [18] and [9] is to find the fastest path in time-dependent road networks for a given start time-interval (e.g., between 7:30AM and 8:30AM).

The ALT algorithm [13] was originally proposed to accelerate fastest path computation in static road networks. With ALT, a set of nodes called landmarks are chosen and then the shortest distances between all the nodes in the network and all the landmarks are computed and stored. ALT employs triangle inequality based on distances to the landmarks to obtain a heuristic function to be used in A* search. The time-dependent variant of this technique is studied in [6] (unidirectional) and [21] (bidirectional A* search) where heuristic function is computed w.r.t lower-bound graph. However, the landmark selection is very difficult (relies on heuristics) and the size of the search space is severely affected by the choice of landmarks. So far no optimal strategy with respect to landmark selection and random queries has been found. Specifically, landmark selection is NP-hard [26] and ALT does not guarantee to yield the smallest search spaces with respect to fastest path computations where source and destination nodes are chosen at random. Our experiments with real-world time-dependent travel-times show that our approach consumes much less storage as compared to ALT based approaches and yields faster response times (see Section 6). In two different studies, The Contraction Hierarchies (CH) and SHARC methods (also developed for static networks) were augmented to time-dependent road networks in [1] and [5], respectively. The main idea of these techniques is to remove unimportant nodes from the graph without changing the fastest path distances between the remaining (more important) nodes. However, unlike the static networks, the importance of a node can change throughout the time under consideration in time-dependent networks, hence the importance of the nodes are time varying. Considering the super-polynomial input size (as discussed in Section 1), and hence the super-polynomial number of important nodes with time-dependent networks, the main shortcomings of these approaches are impractical preprocessing times and extensive space consumption. For example, the precomputation time for SHARC in time-dependent road networks takes more than 11 hours for relatively small road networks (e.g. LA with 304,162 nodes) [5]. Moreover, due to the significant use of arc flags [5], SHARC does not work in a dynamic scenario: whenever an edge cost function changes, arc flags should be recomputed, even though the graph partition need not be updated. While CH also suffers from slow preprocessing times, the space consumption for CH is at least 1000 bytes per node for less varied edge-weights where the storage cost increases with real-world time-dependent edge weights. Therefore, it may not be feasible to apply SHARC and CH to continental size road networks which can consist of more than 45 million road segments (e.g., North America road network) with possibly large varied edge-weights.

4   Problem Definition

There are various criteria to define the cost of a path in road networks. In our study we define the cost of a path as its travel-time. We model the road network as a *time-dependent weighted graph* as shown in Figure 2 where time-dependent travel-times are provided as a function of time which captures the typical congestion pattern for each segment of the road network. We use piecewise linear functions to represent the time-dependent travel-times in the network.

Definition 1. Time-dependent Graph. *A Time-dependent Graph is defined as $G(V, E, T)$ where $V = \{v_i\}$ is a set of nodes and $E \subseteq V \times V$ is a set of edges representing the network segments each connecting two nodes. For every edge $e(v_i, v_j) \in E$, and $v_i \neq v_j$, there is a cost function $c_{v_i,v_j}(t)$, where t is the time variable in time domain T. An edge cost function $c_{v_i,v_j}(t)$ specifies the travel-time from v_i to v_j starting at time t.*

Definition 2. Time-dependent Travel Cost. *Let $\{s = v_1, v_2, ..., v_k = d\}$ denotes a path which contains a sequence of nodes where $e(v_i, v_{i+1}) \in E$ and $i = 1, ..., k - 1$. Given a $G(V, E, T)$, a path $(s \leadsto d)$ from source s to destination d, and a departure-time at the source t_s, the time-dependent travel cost $TT(s \leadsto d, t_s)$ is the time it takes to travel the path. Since the travel-time of an edge varies depending on the arrival-time to that edge, the travel-time of a path is computed as follows:*

$$TT(s \rightsquigarrow d, t_s) = \sum_{i=1}^{k-1} c_{v_i,v_{i+1}}(t_i) \text{ where } t_1 = t_s, t_{i+1} = t_i + c_{(v_i,v_{i+1})}(t_i), i = 1,..,k.$$

Definition 3. Lower-bound Graph. *Given a $G(V, E, T)$, the corresponding Lower-bound Graph $\underline{G}(V, E)$ is a graph with the same topology (i.e, nodes and edges) as graph G, where the weight of each edge c_{v_i,v_j} is fixed (not time-dependent) and is equal to the minimum possible weight c_{v_i,v_j}^{min} where $\forall\ e(v_i, v_j) \in E, t \in T\ c_{v_i,v_j}^{min} \leq c_{v_i,v_j}(t).$*

Definition 4. Lower-bound Travel Cost. *The lower-bound travel-time $LTT(s \rightsquigarrow d)$ of a path is less than the actual travel-time along that path and computed w.r.t $\underline{G}(V, E)$ as*

$$LTT(s \rightsquigarrow d) = \sum_{i=1}^{k-1} c_{v_i,v_{i+1}}^{min}, i = 1,..,k.$$

It is important to note that for each source and destination pair (s, d), $LTT(s \rightsquigarrow d)$ is *time-independent* constant value and hence t is not included in its definition. Given the definitions of TT and LTT, the following property always holds for any path in $G(V, E, T)$: $LTT(s \rightsquigarrow d) \leq TT(s \rightsquigarrow d, t_s)$ where t_s is an arbitrary departure-time from s. We will use this property in subsequent sections to establish some properties of our proposed solution.

Definition 5. Time-dependent Fastest Path (TDFP). *Given a $G(V, E, T)$, s, d, and t_s, the time-dependent fastest path $TDFP(s, d, t_s)$ is a path with the minimum travel-time among all paths from s to d for starting time t_s.*

In the rest of this paper, we assume that $G(V, E, T)$ satisfies the First-In-First-Out (FIFO) property. We also assume that moving objects do not wait at any node. In most real-world applications, waiting at a node is not realistic as it means that the moving object must interrupt its travel by getting out of a road (e.g., exit freeway), and finding a place to park and wait.

5   Time-Dependent Fastest Path Computation

In this section, we explain our bidirectional time-dependent fastest path approach that we generalize bidirectional A* algorithm proposed for static spatial networks [25] to time-dependent road networks. Our proposed solution involves two phases. At the pre-computation phase, we partition the road network into non-overlapping partitions and precompute lower-bound distance labels within and across the partitions with respect to $\underline{G}(V, E)$. Successively, at the online phase, we use the precomputed distance labels as a heuristic function in our bidirectional time-dependent A* search that performs simultaneous searches from source and destination. Below we elaborate on both phases.

5.1   Precomputation Phase

The precomputation phase of our proposed algorithm includes two main steps in which we partition the road network into non-overlapping partitions and precompute lower-bound border-to-border, node-to-border, and border-to-node distance labels.

5.1.1   Road Network Partitioning

Real-world road networks are built on a well-defined hierarchy. For example, in United States, highways connect large regions such as states, interstate roads connect cities within a state, and multi-lane roads connect locations within a city. Almost all of the road network data providers (e.g., Navteq [22]) include road hierarchy information in their datasets. In this paper, we partition the graph to non-overlapping partitions by exploiting the predefined edge class information in road networks. Specifically, we first use higher level roads (e.g., interstate) to divide the road network into large regions. Then, we subdivide each large region using the next level roads and so on. We adopt this technique from [14] and note that our proposed algorithm is independent of the partitioning method, i.e., it yields correct results with all non-overlapping partitioning methods.

With our approach, we assume that the class of each edge $class(e)$ is predefined and we denote the class of a node $class(v)$ by the lowest class number of any incoming or outgoing edge to/from v. For instance, a node at the intersection of two freeway segments and an arterial road (i.e., the entry node to the freeway) is labeled with class of the freeway rather than the class of the arterial road. The input to our hierarchical partitioning method is the road network and the level of partitioning l. For example, if we like to partition a particular road network based on the interstates, freeways, and arterial roads in sequence, we set $l = 2$ where interstate edges represent the class 0. The road network partitions can be conceptually visualized as the areas after removal the nodes with $class(v) \leq l$ from $G(E, V)$.

Definition 6. *Given a graph $G(V, E)$, the partition of $G(V, E)$ is a set of subgraphs $\{S_1, S_2, ..., S_k\}$ where $S_i = (V_i, E_i)$ includes node set V_i where $V_i \cap V_j = \emptyset$ and $\cup_{i=1}^{k} V_i = V$, $i \neq j$.*

Given a $G(E, V)$ and level of partitioning l, we first assign to each node an empty set of partitions. Then, we choose a node v_i that is connected to edges other than the ones used for partitioning (i.e., a node with $class(v_i) > l$) and add partition number (e.g., S_1) to v_i's partition set. For instance, continuing with our example above, a node v_i with $class(v_i) > 2$ represent a particular node that belongs a less important road segment than an arterial road. Subsequently, we expand a shortest path tree from v_i to all it's neighbor nodes reachable through the edges of the classes greater than l, and add S_1 to their partition sets. Intuitively, we expand from v_i until we reach the roads that are used for partitioning. At this point we determine all the nodes that belong to S_1. Then, we select another node v_j with an empty partition set by adding the next partition number (e.g., S_2) to v_j's partition set and repeat the process. We terminate the process when all nodes are assigned to at least one partition. With this method we can easily find the border nodes for each partition, i.e., those nodes which include multiple partitions in their partition sets. Specifically, a node v, with $class(v) \leq l$ belongs to all partitions such that there is an edge e (with $class(e) > l$) connecting v to v' where $v' \in S_i$ and $i = 1, ..., k$, is the border node of the partitions that it connects to. Note that l is a tuning parameter in our partitioning method. Hence, one can arrange the size of the partitions by increasing or decreasing l.

Figure 4 shows the partitioning of San Joaquin (California) network based on the road classes. As shown, higher level edges are depicted with different (thicker) colors.

Fig. 4. Road network partitioning

Each partition is numbered starting from the north-west corner of the road network. The border nodes between partitions S_1 and S_4 are shown in the circled area. We remark that the number of border nodes (which can be potentially large depending on the density of the network) in the actual partitions have a negligible influence on the storage complexity. We explain the effect of the border nodes on the storage cost in the next section.

5.1.2 Distance Label Computation

In this step, for each pair of partitions (S_i, S_j) we compute the lower-bound fastest path cost w.r.t $\underline{G}$ between each border in S_i to each border node in S_j. However, we only store the minimum of all border-to-border fastest path distances. As an example, consider Figure 5 where the lower-bound fastest path cost between b_1 and b_3 (shown with straight line) is the minimum among all border-to-border distances (i.e., b_1-b_4, b_2-b_4, b_2-b_3) between S_1 and S_2. In addition, for each node v_i in a partition S_i, we compute the lower-bound fastest path cost from v_i to all border nodes in S_i w.r.t. $\underline{G}$ and store the minimum among them. We repeat the same process from border nodes in S_i to v_i. For example, border nodes b_1 and b_4 in Figure 5 are the nearest border nodes to s and d, respectively. We will use the precomputed node-to-border, border-to-border, and border-to-node lower-bound travel-times (referred to as distance labels) to construct our heuristic function for online time-dependent A* search. We used a similar distance label precomputation technique to expedite shortest path computation between network Voronoi polygons in static road networks [20].

We maintain the distance labels by attaching three attributes to each node representing a) the partition S_i that contains the node, b) minimum of the lower-bound distances from the node to border nodes, and c) minimum of the lower-bound distances from border nodes to the node (this is necessary for directed graphs). We keep border-to-border distance information in a hash table. Since we only store one distance value for each partition pair, the storage cost of the border-to-border distance labels is negligible.

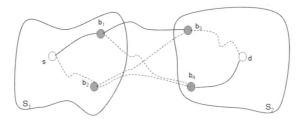

Fig. 5. Lower-bound distance computation

Another benefit of our proposed lower-bound computation is that the lower-bounds need to be updated when it is necessary. Specifically, we update the intra and inter distance labels only when the minimum travel-time of an edge changes, otherwise, the travel-time updates are discarded. Note that intra distance label computation is local, i.e., we only update the intra distance labels for the partitions in which the minimum travel-time of an edge changes.

5.2   Online B-TDFP Computation

As showed in [10], the time-dependent fastest path problem (in FIFO networks) can be solved by modifying Dijkstra algorithm. We refer to modified Dijkstra algorithm as time-dependent Dijkstra (TD-Dijkstra). TD-Dijkstra visits all network nodes reachable from s in every direction until destination node d is reached. On the other hand, a time-dependent A* algorithm can significantly reduce the number of nodes that have to be traversed in TD-Dijkstra algorithm by employing a heuristic function $h(v)$ that directs the search towards destination. To guarantee optimal results, $h(v)$ must be admissible and consistent (a.k.a, monotonic). The admissibility implies that $h(v)$ must be less than or equal to the actual distance between v and d. With static road networks where the length of an edge is constant, Euclidian distance between v and d is used as $h(v)$. However, this simple heuristic function cannot be directly applied to time-dependent road networks, because, the optimal travel-time between v and d changes based on the departure-time t_v from v. Therefore, in time-dependent road networks, we need to use an estimator that never overestimates the travel-time between v and d for any possible t_v. One simple lower-bound estimator is $d_{euc}(v, d)/max(speed)$, i.e., the Euclidean distance between v and d divided by the maximum speed among the edges in the entire network. Although this estimator is guaranteed to be a lower-bound, it is a very loose bound, and hence yields insignificant pruning.

With our approach, we obtain a much tighter bound by utilizing the precomputed distance labels. Assuming that an on-line time-dependent fastest path query requests a path from source s in partition S_i to destination d in partition S_j, the fastest path must pass through from one border node b_i in S_i and another border node b_j in S_j. We know that the time-dependent fastest path distance passing from b_i and b_j is greater than or equal to the precomputed lower-bound border-to-border (e.g., $LTT(b_l, b_t)$) distance for S_i and S_j pair. We also know that a time-dependent fastest path distance from s to b_i

is always greater than or equal to the precomputed lower-bound fastest path distance of s to its nearest border node b_s. Analogously, same is true from the border node b_d (i.e., nearest border node) to d in S_j. Thus, we can compute a lower-bound estimator of s by $h(s) = LTT(s, b_s) + LTT(b_l, b_t) + LTT(b_d, d)$.

Lemma 1. *Given an intermediate node v_i in S_i and destination node d in S_j, the estimator $h(v_i)$ is admissible, i.e., a lower-bound of time-dependent fastest path distance from v_i to d passing from border nodes b_i and b_j in S_i and S_j, respectively.*

Proof. Assume $LTT(b_l, b_t)$ is the minimum border-to-border distance between S_i and S_j, and b'_i, b'_j are the nearest border nodes to v_i and d in $\underline{G}$, respectively. By definition of $\underline{G}(V, E)$, $LTT(v_i, b'_i) \leq TDFP(v_i, b_i, t_{v_i})$, $LTT(b_l, b_t) \leq TDFP(b_i, b_j, t_{b_i})$, and $LTT(b'_j, d) \leq TDFP(b_j, d, t_{b_j})$ Then, we have $h(v_i) = LTT(v_i, b'_i) + LTT(b_l, b_t) + LTT(b'_j, d) \leq TDFP(v_i, b_i, t_{v_i}) + TDFP(b_i, b_j, t_{b_i}) + TDFP(b_j, d, t_{b_j})$

We can use our $h(v)$ heuristic with unidirectional time-dependent A* search in road networks. The time-dependent A* algorithm is a best-first search algorithm which scans nodes based on their time-dependent cost label (maintained in a priority queue) to source similar to [10]. The only difference to [10] is that the label within the priority queue is not determined only by the time-dependent distance to source but also by a lower-bound of the distance to d, i.e., $h(v)$ introduced above.

To further speed-up the computation, we propose a bidirectional search that simultaneously searches forward from the source and backwards from the destination until the search frontiers meet. However, bidirectional search is challenging in time-dependent road networks for two following reasons. First, it is essential to start the backward search from the arrival-time at the destination t_d and exact t_d cannot be evaluated in advance at the query time (recall that arrival-time to destination depends on the departure-time from the source in time-dependent road networks). We address this problem by running a backward A* search that is based on the reverse lower-bound graph $\overleftarrow{G}$ (the lower-bound graph with every edge reversed). The main idea with running backward search in $\overleftarrow{G}$ is to determine the set of nodes that will be explored by the forward A* search. Second, it is not straightforward to satisfy the consistency (the second optimality condition of A* search) of $h(v)$ as the forward and reverse searches use different distance functions. Next, we explain bidirectional time-dependent A* search algorithm (Algorithm 1) and how we satisfy the consistency.

Given $G = (V, E, T)$, s and d, and departure-time t_s from s, let Q_f and Q_b represent the two priority queues that maintain the labels of nodes to be processed with forward and backward A* search, respectively. Let F represent the set of nodes scanned by the forward search and N_f is the corresponding set of labeled vertices (those in its priority queue). We denote the label of a node in N_f by d_{fv}. Analogously, we define B, N_b, and d_{fv} for the backward search. Note that during the bidirectional search F and B are disjoint but N_f and N_b may intersect. We simultaneously run the forward and backward A* searches on $G(V, E, T)$ and $\overleftarrow{G}$, respectively (Line 4 in Algorithm 1). We keep all the nodes visited by backward search in a set H (Line 5). When the search frontiers meet, i.e., as soon as N_f and N_b have a node u in common (Line 6), the cost of the time-dependent fastest path ($TDFP(s, u, t_s)$) from s to u is determined. At this

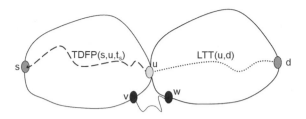

Fig. 6. Bidirectional search

point, we know that $TDFP(u, d, t_u) > LTT(u, d)$ for the path found by the backward search. Hence, the time-dependent cost of the paths (found so far) passing from u is the upper-bound of the time-dependent fastest path from s to d, i.e., $TDFP(s, u, t_s) + TDFP(u, d, t_u) \geq TDFP(s, d, t_s)$.

If we stop the searches as soon as a node u is scanned by both forward and backward searches, we cannot guarantee finding the time-dependent fastest path from u to d within the set of nodes in H. This is due to inconsistent potential function used in bidirectional search that relies on two independent potential functions for two inner A* algorithms. Specifically, let $h_f(v)$ (estimated distance from node v to target) and $h_b(v)$ (estimated distance from node v to source) be the potential functions used in the forward and backward searches, respectively. With the backward search, each original edge $e(i, j)$ considered as $e(j, i)$ in the reverse graph where h_b used as the potential function, and hence the reduced cost[3] of $e(j, i)$ w.r.t. h_b is computed by $c_{h_b}(j, i)=c(i, j)-h_b(j)+h_b(i)$ where $c(i, j)$ is the cost in the original graph. Note that h_f and h_b are consistent if, for all edges (i, j), $c_{h_f}(i, j)$ in the original graph is equal to $c_{h_b}(j, i)$ in the reverse graph. If h_f and h_b are not consistent, there is no guarantee that the shortest path can be found when the search frontiers meet. For instance, consider Figure 6 where the forward and backward searches meet at node u. As shown, if v is scanned before u by the forward search, then $TDFP(s, u, t_s) > TDFP(s, v, t_s)$. Similarly if w is scanned before u by the backward search, the $LTT(u, d) > LTT(w, d)$ and hence $TDFP(u, d, t_u) > TDFP(w, d, t_w)$. Consequently, it is possible that $TDFP(s, u, t_s) + TDFP(u, d, t_u) \geq TDFP(s, v, t_s) + TDFP(w, d, t_w)$. To address this challenge, one needs to find a) a consistent heuristic function and stop the search when the forward and backward searches meet or b) a new termination condition. In this study, we develop a new termination condition (the proof of correctness is given below) in which we continue both searches until the Q_b only contains nodes whose labels exceed $TDFP(s, u, t_s) + TDFP(u, d, t_u)$ by adding all visited nodes to H (Line 9-11). Recall that the label (denoted by d_{bv}) of node v in the backward search priority queue Q_b is computed by the time-dependent distance from the destination to v plus the lower-bound distance from v to s, i.e., $d_{bv} = TDFP(v, d, t_v) + h(v)$. Hence, we stop the search when $d_{bv} > TDFP(s, u, t_s) + TDFP(u, d, t_u)$. As we explained, $TDFP(s, u, t_s) + TDFP(u, d, t_u)$ is the length of the fastest path seen so far (not necessarily the actual fastest path) and is updated during the search when a new

---

[3] A* search is equivalent to Dijkstra's algorithm on a transformed network in which the cost of each edge $c(i, j)$ is equal to $c(i, j)-h(i)+h(j)$.

common node u' found with $TDFP(s, u', t_s) + TDFP(u', d, t_{u'}) < TDFP(s, u, t_s) + TDFP(u, d, t_u)$. Once both searches stop, H will include all the candidate nodes that can possibly be part of the time-dependent fastest path to d. Finally, we continue the forward search considering only the nodes in H until we reach d (Line 12).

Algorithm 1. B-TDFP Algorithm

1: //Input: G_T, $\overleftarrow{G}$, s:source, d:destination,t_s:departure time
2: //Output: a (s, d, t_s) fastest path
3: //$FS()$:forward search, $BS()$:backward search, N_f/N_b: nodes scanned by FS()/BS(), d_{bv}:label of the minimum element in BS queue
4: $FS(G_T)$ and $BS(\overleftarrow{G})$ //start searches simultaneously
5: $N_f \leftarrow FS(G_T)$ and $N_b \leftarrow BS(\overleftarrow{G})$
6: $If \ N_f \cap N_b \neq \emptyset \ then \ u \leftarrow N_f \cap N_b$
7: $M = TDFP(s, u, t_s) + TDFP(u, d, t_u)$
8: $end \ If$
9: $While \ d_{bv} \geq M$
10: $N_b \leftarrow BS(\overleftarrow{G})$
11: $End \ While$
12: $FS(N_b)$
13: $return \ (s, d, l_s)$

Lemma 2. *Algorithm 1 finds the correct time-dependent fastest path from source to destination for a given departure-time l_s.*

Proof. We prove Lemma 2 by contradiction. The forward search in Algorithm 1 is the same as the unidirectional A* algorithm and our heuristic function $h(v)$ is a lower-bound of time-dependent distance from u to v. Therefore, the forward search is correct. Now, let $P(s, (u), d, t_s)$ represent the path from s to d passing from u where forward and backward searches meet and ω denotes the cost of this path. As we showed ω is the upper-bound of actual time-dependent fastest path from s to d. Let ϕ be the smallest label of the backward search in priority queue Q_b when both forward and backward searches stopped. Recall that we stop searches when $\phi > \omega$. Suppose that Algorithm 1 is not correct and yields a suboptimal path, i.e., the fastest path passes from a node outside of the corridor generated by the forward and backward searches. Let $P*$ be the fastest path from s to d for departure-time t_s and cost of this path is α. Let v be the first node on $P*$ which is going to be explored by the forward search and not explored by the backward search and $h_b(v)$ is the heuristic function for the backward search. Hence, we have $\phi \leq h_b(v) + LTT(v, d)$, $\alpha \leq \omega < \phi$ and $h_b(v) + LTT(v, d) \leq LTT(s, v) + LTT(v, d) \leq TDFP(s, v, t_s) + TDFP(v, t, t_v) = \alpha$, which is a contradiction. Hence, the fastest path will be found in the corridor of the nodes labeled by the backward search.

6 Experimental Evaluation

6.1 Experimental Setup

We conducted extensive experiments with different spatial networks to evaluate the performance of our proposed bidirectional time-dependent fastest path (B-TDFP) approach. As of our dataset, we used California (CA), Los Angeles (LA) and San Joaquin County (SJ) road network data (obtained from Navteq [22]) with approximately 1,965,300, 304,162 and 24,123 nodes, respectively. We conducted our experiments on a server with 2.7 GHz Pentium Core Duo processor with 12GB RAM memory.

6.1.1   Time-Dependent Network Modeling

At our research center, we maintain a very large-scale and high resolution (both spatial and temporal) traffic sensor (i.e., loop detector) dataset collected from entire LA County highways and arterial streets. This dataset includes both inventory and real-time data for 6300 traffic sensors covering approximately 3000 miles. The sampling rate of the streaming data is 1 reading/sensor/min. We have been continuously collecting and archiving the traffic sensor data for the past two years. We use this real-world dataset to create time varying edge weights; we spatially and temporally aggregate sensor data by assigning interpolation points (for each 5 minutes) that depict the travel-times on the network segments. Based on our observation, all roads are un-congested between 9PM and 6AM, and hence we assume static edge weights during this interval. In order to create time-dependent edge weights for the local streets in LA, CA and SJ, we developed a traffic modeling approach [8] that synthetically generates the edge travel-time profiles. Our approach uses spatial (e.g., locality, connectivity) and temporal (e.g., rush hour, weekday) characteristics to generate travel-time for network edges that does not have readily available sensor data.

6.2   Results

In this section, we report the experimental results from our fastest path queries in which we determine the s and d nodes uniformly at random. We also pick our departure-time randomly and uniformly distributed in time domain T. The average results are derived from 1000 random s-d queries. We only present the results for LA and CA, the experimental results for both SJ and LA are very similar.

6.2.1   Comparison with ALT

In this set of experiments we compare our algorithm with time-dependent ALT (TD-ALT) approaches [6,21] with respect to storage and response time. We run our proposed algorithm both unidirectionally and bidirectionally (in CA network) and compare with [6] and [21], respectively. As we mentioned, selecting good landmarks that lead to good performance is very difficult and hence several heuristics have been proposed for landmark selection. Among these heuristics, we use the best known technique; maxCover (see [6]) with 64 landmarks. We computed travel-times between each node and the landmarks with respect to G. Under this setting, to store the precomputed distances, TD-ALT attaches to each node an array of 64 elements corresponding to the number of landmarks. Assuming that each array element takes 2 bytes of space, the additional storage requirement of TD-ALT is 63 Megabytes. On the other hand, with our algorithm, we divide CA network to 60 partitions and store the intra and inter distance labels. The total storage requirement of our proposed solution is 8.5 Megabytes where we consume, for each node, an array of 2 elements (corresponding to *from* and *to* distances to the closest border node) plus the border-to-border distance labels. Since the experimental results for both unidirectional and bidirectional searches differ insignificantly and due to space limitations, we only present the results from unidirectional search below. As shown in Figure 7(a) the response time of our unidirectional time-dependent A* search (U-TDFP) is approximately three times better than that of TD-ALT for all times. This is because the search space of TD-ALT is severely affected by the quality of the landmarks which are selected based on a heuristic. Specifically, TD-ALT may yield very loose bounds based on the randomly selected s and d, and hence the large search space. In addition, with each iteration, TD-ALT needs to find the best landmark (among 64 landmarks) which yields largest triangular inequality distance for better pruning; it seems that the overhead of this operation is not negligible. On the other hand, U-TDFP yields a more directional search with the help of intra and inter distance labels with no additional computation.

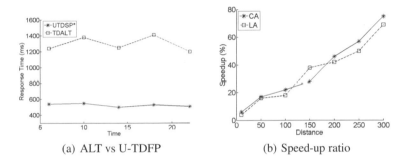

(a) ALT vs U-TDFP   (b) Speed-up ratio

Fig. 7. TD-ALT Comparison and Speed-up Ratio Analysis

6.2.2 Performance of B-TDFP

In this set of experiments, we compare the performance of our proposed approach to other existing TDFP methods w.r.t to a) preprocessing time, b) storage (byte per node), c) the average number of relaxed edges, and d) average query time. Table 1 shows the preprocessing time (Pre Processing), storage (Storage), number of scanned nodes (#Nodes), and response time (Res. Time) of time-dependent Dijkstra (TD-Dijkstra) implemented based on [10], unidirectional (U-TDFP) and bidirectional (B-TDFP) time-dependent A* search implemented using our proposed heuristic function, time-dependent Contraction Hierarchies (TD-CH) [1], and time-dependent SHARC (TD-SHARC) [5]. To implement U-TDFP and B-TDFP, we divide CA and LA network to 60 (which roughly correspond to counties in CA) and 25 partitions, respectively. Comparing TD-Dijkstra with our approach, we observe a very high trade-off between the query results and precomputation in both LA and CA networks. Our proposed B-TDFP performs 23 times better than TD-Dijkstra depending on the network while preprocessing and storage overhead is relatively small. As shown, the preprocessing time and storage complexity is directly proportional to network size.

Comparing the time-dependent variant of SHARC (TD-SHARC) and CH (TD-CH) with our approach, we observe B-TDFP outperforms TD-SHARC and TD-CH in preprocessing and response time. We also observe that as the graph gets bigger or more

Table 1. Experimental Results

	Algorithm	PreProcessing [h:m]	Storage [B/node]	#Nodes	Res. Time [ms]
CA	TD-Dijkstra	0:00	0	1162323	4104.11
	U-TDFP	1:13	6.82	90575	310.17
	B-TDFP	1:13	6.82	67172	182.06
	TD-SHARC	19:41	154.10	75104	227.26
	TD-CH	3:55	1018.33	70011	209.12
LA	TD-Dijkstra	0:00	0	210384	2590.07
	U-TDFP	0:27	3.51	11115	197.23
	B-TDFP	0:27	3.51	6681	101.22
	TD-SHARC	11:12	68.47	9566	168.11
	TD-CH	1:58	740.88	7922	140.25

edges are time-dependent, the preprocessing time of TD-SHARC increases drastically. The preprocessing of TD-SHARC takes very long for both road networks, i.e., up to 20 times more than B-TDFP. The reason for the performance gap is that TD-SHARC's contraction routine cannot bypass the majority of the nodes in time-dependent road networks as in the static road networks. Recall that the importance of a node can change throughout the time under consideration in time-dependent road networks. In addition, TD-SHARC is very sensitive to edge cost function changes, i.e. whenever cost function of an edge changes, the preprocessing phase needs to be repeated to determine the by-pass nodes. While TD-CH tend to have better response times than TD-SHARC, the space consumption of TD-CH is significantly high (approximately 1000 bytes per node in CA network). For this reason, TD-CH is not feasible for very large road networks such as North America and Europe. We note that, to improve the response and preprocessing time, several variations of TD-SHARC and TD-CH algorithms are implemented in the literature. These variations trade-off between the optimality of the solution and the response time. For example, the response time of Heuristic TD-SHARC [5] is shown much better than that of original TD-SHARC algorithm. However, the path found by the Heuristic TD-SHARC is not optimal and the error rate is not bounded. As another example, the performance of TD-SHARC can be improved by combining with another technique called Arc-Flags [5]. Similar performance improvements can be applied to our proposed approach. For instance, we can terminate the search when the search frontiers meet and report the combination of path found by the forward and backward search as the result. However, as mentioned in Section 5.2, we cannot guarantee the optimal solution in this setting. Moreover, based on our initial observation and implementation, we can also integrate our algorithm with Arc-Flags. However, the focus of our study is to develop a technique that yields exact solutions. Hence, for the sake of simplicity and fair comparison, we only compare the original algorithms that yields exact results and do not consider integrating different methods.

6.2.3   Quality of Lower-Bounds

As discussed, the performance of time-dependent A* search depends on the lower-bound distance. In this set of experiments, we analyze the quality of our proposed lower-bound computed based on the Distance Labels explained in Section 5.1.2. We define the lower-bound quality by $lg = \frac{\delta(u,v)}{d(u,v)}$, where $\delta(u,v)$ and $d(u,v)$ represent the estimated and actual travel-times between nodes u and v, respectively. Table 2 reports lg based on three different heuristic function, namely Naive, ALT, and DL (i.e., our heuristic function computed based on Distance Labels). Similar to other experiments, the values in Table 2 are obtained by selecting s, d and t_s uniformly at random between 6AM and 9PM. We compute the naive lower-bound estimator by $\frac{d_{euc}(u,v)}{max(speed)}$, i.e., the Euclidean distance between u and v is divided by the maximum speed among the edges in the entire network. We obtain the ALT lower-bounds based on $\underline{G}$ and the maxCover ([6]) technique with 64 landmarks. As shown, DL provides better heuristic function in both LA and CA. The reason is that the ALT's lg relies on the distribution of the landmarks, and hence depending on the location of s and d it is possible to get very loose bounds. On the other hand, the lower-bounds computed based on Distance Labels are more directional. Specifically, with our approach the s and d nodes must reside in one of the partitions and the (border-to-border) distance between these partitions is always considered for the lower-bound computation.

Table 2. Lower-bound Quality

Network	Naive (%)	ALT (%)	DL (%)
CA	21	42	63
LA	33	46	66

6.2.4   Bidirectional vs. Unidirectional Search

In another set of experiments, we study the impact of path length (i.e., distance from s to d) on the speed-up of bidirectional search. Hence, we measure the performance of B-TDFP and U-TDFP with respect to distance by varying the path distance (1 to 300 miles) between s and d. Figure 7(b) shows the speed-up with respect to distance. We observe that the speed-up is significantly more especially for long distance queries. The reason is that for short distances the computational overhead incurred by B-TDFP is not worthwhile as U-TDFP visits less number of nodes anyway.

7   Conclusion and Future Work

In this paper, we proposed a time-dependent fastest path algorithm based on bidirectional A*. Unlike the most path planning studies, we assume the edge weights of the road network are time varying rather than constant. Therefore, our approach yield a much more realistic scenario, and hence, applicable to the to real-world road networks. We also compared our approaches with those handful of time-dependent fastest path studies. Our experiments with real-world road network and traffic data showed that our proposed approaches outperform the competitors in storage and response time significantly. We intend to pursue this study in two different directions. First, we plan to investigate new data models for effective representation of spatiotemporal road networks. This is critical in supporting development of efficient and accurate time-dependent algorithms, while minimizing the storage and computation costs. Second, to support rapid changes of the traffic patterns (that may happen in case of accidents/events, for example), we intend to study incremental update algorithms for both of our approaches.

References

1. Batz, G.V., Delling, D., Sanders, P., Vetter, C.: Time-dependent contraction hierarchies. In: ALENEX (2009)
2. Cooke, L., Halsey, E.: The shortest route through a network with timedependent internodal transit times. Journal of Mathematical Analysis and Applications (1966)
3. Dean, B.C.: Algorithms for min-cost paths in time-dependent networks with wait policies. Networks (2004)
4. Dehne, F., Omran, M.T., Sack, J.-R.: Shortest paths in time-dependent fifo networks using edge load forecasts. In: IWCTS (2009)
5. Delling, D.: Time-dependent SHARC-routing. In: Halperin, D., Mehlhorn, K. (eds.) Esa 2008. LNCS, vol. 5193, pp. 332–343. Springer, Heidelberg (2008)
6. Delling, D., Wagner, D.: Landmark-based routing in dynamic graphs. In: Demetrescu, C. (ed.) WEA 2007. LNCS, vol. 4525, pp. 52–65. Springer, Heidelberg (2007)
7. Demiryurek, U., Kashani, F.B., Shahabi, C.: A case for time-dependent shortest path computation in spatial networks. In: ACM SIGSPATIAL (2010)
8. Demiryurek, U., Pan, B., Kashani, F.B., Shahabi, C.: Towards modeling the traffic data on road networks. In: SIGSPATIAL-IWCTS (2009)
9. Ding, B., Yu, J.X., Qin, L.: Finding time-dependent shortest paths over large graphs. In: EDBT (2008)
10. Dreyfus, S.E.: An appraisal of some shortest-path algorithms. Operations Research 17(3) (1969)
11. Foschini, L., Hershberger, J., Suri, S.: On the complexity of time-dependent shortest paths. In: SODA (2011)
12. George, B., Kim, S., Shekhar, S.: Spatio-temporal network databases and routing algorithms: A summary of results. In: Papadias, D., Zhang, D., Kollios, G. (eds.) SSTD 2007. LNCS, vol. 4605, pp. 460–477. Springer, Heidelberg (2007)
13. Goldberg, A.V., Harellson, C.: Computing the shortest path: A* search meets graph theory. In: SODA (2005)
14. Gonzalez, H., Han, J., Li, X., Myslinska, M., Sondag, J.P.: Adaptive fastest path computation on a road network: A traffic mining approach. In: VLDB (2007)
15. Guc, B., Ranganathan, A.: Real-time, scalable route planning using stream-processing infrastructure. In: ITS (2010)
16. Halpern, J.: Shortest route with time dependent length of edges and limited delay possibilities in nodes. Mathematical Methods of Operations Research (1969)
17. Hart, P., Nilsson, N., Raphael, B.: A formal basis for the heuristic determination of minimum cost paths. IEEE Transactions on Systems Science and Cybernetics (1968)

18. Kanoulas, E., Du, Y., Xia, T., Zhang, D.: Finding fastest paths on a road network with speed patterns. In: ICDE (2006)
19. Kohler, E., Langkau, K., Skutella, M.: Time-expanded graphs for flow-dependent transit times. In: Proc. 10th Annual European Symposium on Algorithms (2002)
20. Kolahdouzan, M., Shahabi, C.: Voronoi-based k nearest neighbor search for spatial network databases. In: VLDB (2004)
21. Nannicini, G., Delling, D., Liberti, L., Schultes, D.: Bidirectional a* search for time-dependent fast paths. In: McGeoch, C.C. (ed.) WEA 2008. LNCS, vol. 5038, pp. 334–346. Springer, Heidelberg (2008)
22. NAVTEQ, http://www.navteq.com (accessed in May 2010)
23. Orda, A., Rom, R.: Shortest-path and minimum-delay algorithms in networks with time-dependent edge-length. J. ACM (1990)
24. PeMS, https://pems.eecs.berkeley.edu (accessed in May 2010)
25. Pohl, I.: Bi-directional search. In: Machine Intelligence. Edinburgh University Press, Edinburgh (1971)
26. Potamias, M., Bonchi, F., Castillo, C., Gionis, A.: Fast shortest path distance estimation in large networks. In: CIKM (2009)
27. Samet, H., Sankaranarayanan, J., Alborzi, H.: Scalable network distance browsing in spatial databases. In: SIGMOD (2008)
28. Sanders, P., Schultes, D.: Highway hierarchies hasten exact shortest path queries. In: Brodal, G.S., Leonardi, S. (eds.) ESA 2005. LNCS, vol. 3669, pp. 568–579. Springer, Heidelberg (2005)
29. Sanders, P., Schultes, D.: Engineering fast route planning algorithms. In: Demetrescu, C. (ed.) WEA 2007. LNCS, vol. 4525, pp. 23–36. Springer, Heidelberg (2007)
30. TELEATLAS, http://www.teleatlas.com (accessed in May 2010)
31. Wagner, D., Willhalm, T.: Geometric speed-up techniques for finding shortest paths in large sparse graphs. In: Di Battista, G., Zwick, U. (eds.) ESA 2003. LNCS, vol. 2832, pp. 776–787. Springer, Heidelberg (2003)

Dynamic Pickup and Delivery with Transfers

Panagiotis Bouros[1], Dimitris Sacharidis[2,*], Theodore Dalamagas[2],
and Timos Sellis[1,2]

[1] National Technical University of Athens, Greece
pbour@dblab.ece.ntua.gr
[2] Institute for the Management of Information Systems, R.C. "Athena", Greece
{dsachar,dalamag,timos}@imis.athena-innovation.gr

Abstract. In the dynamic Pickup and Delivery Problem with Transfers
(dPDPT), a set of transportation requests that arrive at arbitrary times
must be assigned to a fleet of vehicles. We use two cost metrics that capture both the company's and the customer's viewpoints regarding the
quality of an assignment. In most related problems, the rule of thumb is
to apply a two-phase local search algorithm to heuristically determine a
good requests-to-vehicles assignment. This work proposes a novel solution based on a graph-based formulation of the problem that treats each
request independently. Briefly, in this conceptual graph, the goal is to
find a shortest path from a node representing the pickup location to that
of the delivery location. However, we show that efficient Bellman-Ford or
Dijkstra-like algorithms cannot be applied. Still, our method is able to
find dPDPT solutions significantly faster than a conventional two-phase
local search algorithm, while the quality of the solution is only marginally
lower.

Keywords: Pickup and delivery problem, dynamic shortest path.

1 Introduction

The family of pickup and delivery problems covers a broad range of optimization
problems that appear in various logistics and transportation scenarios. Broadly
speaking, these problems look for an assignment of a set of transportation requests to a fleet of vehicles in a way that satisfies a number of constraints and at
the same time minimizes a specific cost function. In this context, a transportation request is defined as *picking up* an object (e.g., package, person, etc.) from
one location and *delivering* it to another; hence the name.

In its simplest form, the *Pickup and Delivery Problem* (PDP) only imposes
two constraints. The first, termed *precedence*, naturally states that pickup should
occur before delivery for each transportation request. The second, termed *pairing*, states that both the pickup and the delivery of each transportation request

---

* The author is supported by a Marie Curie Fellowship (IOF) within the European
Community FP7.

D. Pfoser et al. (Eds.): SSTD 2011, LNCS 6849, pp. 112–129, 2011.

should be performed by the same vehicle. The *Pickup and Delivery Problem with Transfers* (PDPT) [11,26] is a PDP variant that eliminates the pairing constraint. In PDPT, objects can be transferred between vehicles. *Transfers* can occur in predetermined locations, e.g., depots, or in arbitrary locations as long as the involved vehicles are in close proximity to each other at some time. We refer to the latter case as *transfer with detours*, since the vehicles may have to deviate from their routes.

Almost every pickup and delivery problem comes in two flavors. In *static*, all requests are known in advance and the goal is to come up with the best vehicle routes from scratch. On the other hand, in *dynamic*, a set of vehicle routes, termed the *static plan*, has already been established. Then, additional requests arrive ad hoc, i.e., at arbitrary times, and the plan must be modified to satisfy them. While algorithms for static problems can also solve the dynamic counterpart, they are rarely used as they take a lot of time to execute. Instead, common practice is to apply two-phase local search algorithms. In the first phase, a quick solution is obtained by assigning each standing request to the vehicle that results in the smallest cost increase. In the second phase, the obtained solution is improved by reassigning requests.

This paper proposes an algorithm for the *dynamic Pickup and Delivery Problem with Transfers* (dPDPT). Although works for the dynamic PDP can be extended to consider transfers between vehicles, to the best of our knowledge, this is the first work targeting dPDPT. Our solution processes requests independently, and does not follow the two-phase paradigm. Satisfying a request is treated as a shortest path problem in a conceptual graph. Intuitively, the object must travel from the pickup to the delivery location following the vehicles' routes and schedules.

The primary goal in pickup and delivery problems is to minimize the total operational cost required to satisfy the requests, i.e., the company's expenses. Under our dPDPT formulation, a satisfied request is represented as a path p. We define its *operational cost* O_p as the additional cost (total delay), with respect to the static plan, incurred by the vehicles in order to accommodate the solution p. In addition, we consider the promptness of satisfying the request. We define the *customer cost* C_p of a path p as the delivery time of the object. These costs are in general conflicting, as they represent two distinct views. For example, the path with the earliest delivery time may require significant changes in the schedule of the vehicles and cause large delays on the static plan. In contrast, the path with the smallest operational cost could result in late delivery. Our algorithm can operate under any monotonic combination of the two costs. However, in this work, we consider operational cost as more important; customer cost is used to solve ties.

Finding the shortest path (according to the two costs) in the conceptual graph is not straightforward. The reason is that the weights of the edges depend on both the operational and customer cost of the path that led to this edge. In fact, an important contribution of this paper is that we show, contrary to other time-dependent networks, that the conceptual graph does not exhibit the *principle of optimality*, which is necessary to apply efficient Bellman-Ford or Dijkstra-like algorithms. Hence, one has to enumerate all possible paths. However, despite this fact, extensive experimental results show that our method is significantly faster than a two-phase local search algorithm adapted for dPDPT.

The remainder of this paper is organized as follows. Section 2 reviews related work and Section 3 formally introduces dPDPT. Section 4 presents our graph-based formulation and algorithm. Then, Section 5 presents an extensive experimental evaluation and Section 6 concludes this work.

2   Related Work

This work is related to pickup and delivery, and shortest path problems.

Pickup and Delivery Problems
In the *Pickup and Delivery Problem* (PDP) objects must be transported by a fleet of vehicles from a pickup to a delivery location with the minimum cost, under two constrains: (1) pickup occurs before delivery (*precedence*), and (2) pickup and delivery of an object is performed by the same vehicle (pairing). PDP is NP-hard since it generalizes the well-known Traveling Salesman Problem (TSP). Exact solutions employ column generation [14,34,36], branch-and-cut [10,32] and branch-and-cut-and-price [31] methodologies. On the other hand, the heuristics for the approximation methods take advantage of local search [2,23,26,33].

Other PDP variations introduce additional constraints. For instance, in the *Pickup and Delivery Problem with Time Windows* (PDPTW), pickups and deliveries are accompanied with a time window that mandates when the action can take place. In the *Capacitated Pickup and Delivery Problem* (CPDP), the amount of objects a vehicle is permitted to carry at any time is bounded by a given capacity. In the *Pickup and Delivery Problem with Transfers* (DPDT), studied in this paper, the pairing constraint is lifted. [11] proposes a branch-and-cut strategy for DPDT. [26] introduces the Pickup and Delivery Problem with Time Windows and Transfers and employs a local search optimization approach. In all the above problems, the transportation requests are known in advance, hence they are characterized as *static*. A formal definition of static PDP and its variants can be found in [4,9,28].

Almost all PDP variants also have a *dynamic* counterpart. In this case, a set of vehicle routes, termed the *static plan*, has already been established, and additional requests arrive ad hoc, i.e., at arbitrary times. Thus, the plan must be modified to satisfy them. A survey on dynamic PDP can be found in [3]. Typically, two-phase local search methods are applied for the dynamic problems. The first phase applies an insertion heuristic [30], whereas the second employs tabu search [15,24,25]. To the best of our knowledge our work is the first to address the *dynamic Pickup and Delivery Problem with Transfers* (dPDPT).

Shortest Path Problems
Bellman-Ford and Dijkstra are the most well-known algorithms for finding the *shortest path* between two nodes in a graph. The ALT algorithms [16,17,29]

perform a bidirectional A* search and exploit a lower bound of the distance between two nodes to direct the search. To compute this bound they construct an embedding on the graph. There exist a number of materialization techniques [1,20,21] or encoding/labeling schemes [6,7] that can be used to efficiently compute the shortest path. Both the ALT algorithms and the materialization and encoding methods are mostly suitable for graphs that are not frequently updated, since they require expensive precomputation.

In *multi-criteria shortest path problems* the quality of a path is measured by multiple metrics, and the goal is to find all paths for which no better exists. Algorithms are categorized into three classes. The methods of the first class (e.g., [5]) apply a user preference function to reduce the original multi-criteria problem to a conventional shortest path problem. The second class contains the interactive methods (e.g., [18]) that interact with a decision maker to come up with the answer path. Finally, the third class includes label-setting and label-correcting methods (e.g., [19,22,35]). These methods construct a label for every path followed to reach a graph node. Then, at each iteration, they select the path with the minimum cost, defined as the combination of the given criteria, and expand the search extending this path.

In *time-dependent shortest path problems* the cost of traveling from node n_i to n_j in a graph (e.g., the road network of a city) depends on the departure time t from n_i. [8] is the first attempt to solve this problem using a Bellman-Ford based solution. However, as discussed in [13], Dijkstra can also be applied for this problem, as long as the earliest possible arrival time at a node is considered. In the context of transportation systems, the *FIFO* (a.k.a. *non-overtaking*) property of a road network is considered as a necessity in order to achieve an acceptable level of complexity. According to this property delaying the departure from a graph node n_i to reach n_j cannot result in arriving earlier at n_j. However, even when the FIFO property does not hold it is possible to provide an efficient solution [12,27] by properly adjusting the weights in graph edges [12].

3   Problem Formulation

Section 3.1 provides basic definitions and introduces the dynamic Pickup and Delivery Problem with Transfers. Section 3.2 details the actions allowed for satisfying a request and their costs.

3.1   Definitions

Assume that a company has already scheduled its fleet of vehicles to service a number of requests. We refer to this schedule as the *static plan*, since we assume that it is given as input. The static plan consists of a set of vehicles following some routes; we overload notation r_a to refer to both a vehicle and its route. The *route* of a vehicle r_a is a sequence of distinct spatial locations, where each location n_i is associated with an *arrival time* A_i^a and a *departure time* D_i^a. Note that the requirement for distinct locations within a route is introduced to simplify notation and avoid ambiguity when referring to a particular location. Besides, if a vehicle visits a location multiple times, its route can always be represented as a set of distinct-locations routes. The difference $D_i^a - A_i^a$ is a non-negative number; it may be zero indicating that vehicle r_a just passes by n_i, or a positive number corresponding to some service at n_i, e.g., pickup, delivery, mandatory stop, etc. For two consecutive locations n_i and n_j on r_a, the difference $A_j^a - D_i^a \geq 0$ corresponds to the travel time from n_i to n_j.

An ad-hoc dPDPT *request* is a pair of locations n_s and n_e, signifying that a package must be picked up at n_s and be delivered at n_e. In order to complete a request, it is necessary to perform a series of modifications to the static plan. There are five types of modifications allowed, termed *actions*: pickup, delivery, transport, transfer without detours, and transfer with detours. Each action, described in detail later, results in the package changing location and/or vehicle. A sequence of actions is called a *path*. If the initial and final location of the package in a path are n_s and n_e, respectively, the path is called a *solution* to the request.

There are two costs associated with an action. The *operational cost* measures the time spent by vehicles in order to perform the action, i.e., the delay with respect to the static plan. The *customer cost* represents the time when the action is completed. Furthermore, the operational cost O_p of a path p is defined as the sum of operational costs for each action in the path, and the customer cost C_p is equal to the customer cost of the final action in p. Therefore, for a solution p, O_p signifies the company's cost in accommodating the request, while C_p determines the delivery time of the package according to p.

Any monotonic combination (e.g., weighted sum, product, min, max, average etc.) of the two costs could be a meaningful measure for assessing the quality of a solution. In the remainder of this paper, we assume that the operational cost is more important, and that the customer cost is of secondary importance. Therefore, a path p is preferred over q, if $O_p < O_q$, or $O_p = O_q \wedge C_p < C_q$. Equivalently, we may define the *combined cost* of a path p as:

$$cost(p) = M \cdot O_p + C_p , \qquad (1)$$

where M is a sufficiently large number (greater than the largest possible customer cost divided by the smaller possible operational cost) whose sole purpose is to assign greater importance to the operational cost. Based on this definition, the *optimal solution* is the one that has the lowest combined cost, i.e., the minimum customer cost among those that have the least operational cost. The *dynamic Pickup and Delivery with Transfers* (dPDPT) problem is to find the optimal solution path.

3.2   Actions

It is important to note that, throughout this paper, we follow the convention that an action is completed by vehicle r_a at a location n_i just before r_a departs from n_i. Since r_a can have multiple tasks to perform at n_i according to the static plan, this convention intuitively means that we make no assumptions about the order in which a vehicle performs its tasks. In any case, the action will have concluded by the time r_a is ready to depart from n_i.

Consider a path p with operational and customer costs O_p and C_p, respectively. Further, assume that the last action in p results in the package being onboard vehicle r_a at location n_i. Let p' denote the path resulting upon performing an additional action E on p. In the following, we detail each possible action E, and the costs of the resulting path p', denoted as $O_{p'}$, $C_{p'}$, which may depend on the current path costs O_p, C_p.

Pickup

The pickup action involves a single vehicle, r_a, and appears once as the first action in a solution path. Hence, initially the package is at the pickup location n_s of the request, p is empty, and $O_p = C_p = 0$.

We distinguish two cases for this action. First, assume that n_s is included in the vehicle's route, and let A_s^a, D_s^a denote the arrival and departure times of r_a at n_s according to the static plan. In this case, the pickup action is denoted as E_s^a. No modification in r_a's route is necessary, and thus there is zero additional operational cost for executing E_s^a. The customer cost for the resulting path p' becomes equal to the scheduled (according to the static plan) departure time D_s^a from n_s; without loss of generality, we make the assumption that the request arrives at time 0. Therefore,

$$\left.\begin{array}{l} O_{p'} = 0 \\ C_{p'} = D_s^a \end{array}\right\} \text{ for } p' = E_s^a. \tag{2a}$$

In the second case, the pickup location n_s is not in the r_a route. Let n_i be a location in the r_a route that is *sufficiently close* to n_s; then, r_a must take a detour from n_i to n_s. A location is sufficiently close to n_s if the detour is short, i.e., its duration, denoted as T_{si}^a, is below some threshold (a system parameter). Hence, it is possible that a sufficiently close location does not exists for route r_a; clearly, if no such location exists for any route, then the request is unsatisfiable. When such a n_i exists, the pickup action is denoted as E_{si}^a. Figure 1(a) shows a pickup action with detour. The solid line in the figure denotes the vehicle route r_a and the dashed line denotes the detour performed by r_a from n_i to n_s to pickup the package. The operational cost of a pickup action with detour is equal to the delay T_{si}^a due to the detour. The customer cost of p' is the scheduled departure time from n_i incremented by the delay. Therefore,

$$\left.\begin{array}{l} O_{p'} = T_{si}^a \\ C_{p'} = D_i^a + T_{si}^a \end{array}\right\} \text{ for } p' = E_{si}^a. \tag{2b}$$

Delivery

The delivery action involves a single vehicle, r_a, and appears once as the last action in a solution path. Similar to pickup, two cases exist for this action. In the first case, n_e appears in the route r_a, and delivery is denoted as E_e^a. The costs for path p' are shown in Equation 3a. In the second case, a detour of length T_{ie}^a at location n_i is required, and delivery is denoted as E_{ie}^a. Figure 1(b) presents

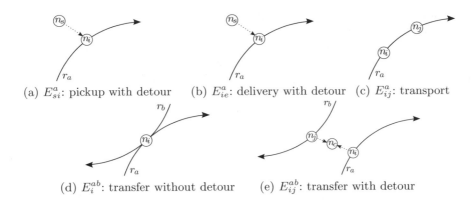

(a) E_{si}^a: pickup with detour   (b) E_{ie}^a: delivery with detour (c) E_{ij}^a: transport

(d) E_i^{ab}: transfer without detour   (e) E_{ij}^{ab}: transfer with detour

Fig. 1. Actions allowed for satisfying a dPDPT request

an E_{ie}^a delivery action with detour. The costs for p' are shown in Equation 3b, where we make the assumption that it takes $T_{ie}^a/2$ time to travel from n_i to n_e.

$$\left.\begin{array}{l} O_{p'}= O_p \\ C_{p'}= C_p \end{array}\right\} \text{ for } p' = pE_e^a. \tag{3a}$$

$$\left.\begin{array}{l} O_{p'}= O_p + T_{ie}^a \\ C_{p'}= C_p + T_{ie}^a/2 \end{array}\right\} \text{ for } p' = pE_{ie}^a. \tag{3b}$$

Transport

The transport action involves a single vehicle, r_a, and corresponds to the carrying of the package by a vehicle between two successive locations on its route. Figure 1(c) illustrates such a transportation action from location n_i to n_j onboard vehicle r_a. As assumed, path p results in the package being onboard r_a at location n_i. The transport action, denoted as E_{ij}^a, has zero operational cost, as the vehicle is scheduled to move from n_i to n_j anyway. The customer cost is incremented by the time required by vehicle r_a to travel from n_i to n_j and finish its tasks at n_j. Therefore,

$$\left.\begin{array}{l} O_{p'}= O_p \\ C_{p'}= C_p + D_j^a - D_i^a \end{array}\right\} \text{ for } p' = pE_{ij}^a. \tag{4}$$

Transfer without detours

The transfer without detours action, denoted as E_i^{ab}, involves two vehicles, r_a and r_b, and corresponds to the transfer of the package from r_a to r_b at a common location n_i, e.g., a depot, drop-off/pickup point, etc. For example, Figure 1(d) shows such a transfer action via the common location n_i. Let A_b^i, D_b^i be the arrival and departure times of vehicle r_b at location n_i. We distinguish three cases.

In the first, the last action in path p concludes after vehicle r_b arrives and before it departs from n_i, i.e., $A_i^b \le C_p \le D_i^b$. Since there is no delay in the schedule of vehicles, the action's operational cost is zero, while the customer cost of the resulting path p' becomes equal to the scheduled departure time of r_b from n_i. Therefore,

$$\left.\begin{array}{l} O_{p'} = O_p \\ C_{p'} = D_i^b \end{array}\right\} \text{ for } p' = pE_i^{ab}, \text{ if } A_i^b \leq C_p \leq D_i^b. \tag{5a}$$

In the second case, the last action in path p concludes before vehicle r_b arrives at n_i, i.e., $C_p < A_i^b$. For the transfer to proceed, vehicle r_a must wait at n_i until r_b arrives. The operational cost is incremented by the delay, which is equal to $A_i^b - C_p$. On the other hand, the customer cost becomes equal to the scheduled departure time of r_b from n_i. Therefore,

$$\left.\begin{array}{l} O_{p'} = O_p + A_i^b - C_p \\ C_{p'} = D_i^b \end{array}\right\} \text{ for } p' = pE_i^{ab}, \text{ if } C_p < A_i^b. \tag{5b}$$

In the third case, the last action in p concludes after vehicle r_b is scheduled to depart from n_i, i.e., $C_p > D_i^b$. This implies that r_b must wait at n_i until the package is ready for transfer. The delay is equal to $C_p - D_i^b$, which contributes to the operational cost. The customer cost becomes equal to the delayed departure of r_b from n_i, which coincides with C_p. Therefore,

$$\left.\begin{array}{l} O_{p'} = O_p + C_p - D_i^b \\ C_{p'} = C_p \end{array}\right\} \text{ for } p' = pE_i^{ab}, \text{ if } C_p > D_i^b. \tag{5c}$$

Transfer with detours

Consider distinct locations n_i on r_a and n_j on r_b. Assume that short detours from n_i and n_j are possible, i.e., the detour durations are below some threshold, and that they have a common rendezvous point. The transfer with detours action, denoted as E_{ij}^{ab}, involves the two vehicles, r_a and r_b, and corresponds to the transportation of the package on vehicle r_a via the n_i detour to the rendezvous location, its transfer to vehicle r_b, which has taken the n_j detour, and finally its transportation to n_j. Figure 1(e) illustrates a transfer action between vehicles r_a and r_b via a detour to their common rendezvous point n_c. Notice the difference with Figure 1(d) where the transfer action occurs without a detour. To keep the notation simple, we make the following assumptions: (1) the n_i detour travel time of r_a is equal to that of the n_j detour of r_b, denoted as T_{ij}^{ab}; and (2) it takes $T_{ij}^{ab}/2$ time for both r_a and r_b to reach the rendezvous location.

Similar to transferring without detours, we distinguish three cases. In the first, the package is available for transfer at the rendezvous location, at time $C_p + T_{ij}^{ab}/2$, after the earliest possible and before the latest possible arrival of r_b, i.e., $A_j^b + T_{ij}^{ab}/2 \leq C_p + T_{ij}^{ab}/2 \leq D_j^b + T_{ij}^{ab}/2$. Both vehicles incur a delay in their schedule by T_{ij}^{ab}. Therefore,

$$\left.\begin{array}{l} O_{p'} = O_p + 2 \cdot T_{ij}^{ab} \\ C_{p'} = D_j^b + T_{ij}^{ab} \end{array}\right\} \text{ for } p' = pE_{ij}^{ab}, \text{ if } A_j^b \leq C_p \leq D_j^b. \tag{6a}$$

In the second case, the package is available for transfer before the earliest possible arrival of r_b at the rendezvous location, i.e., $C_p + T_{ij}^{ab}/2 < A_j^b + T_{ij}^{ab}/2$. Vehicle r_a must wait for $A_j^b - C_p$ time. Therefore,

$$\left.\begin{array}{l} O_{p'} = O_p + A_j^b - C_p + 2 \cdot T_{ij}^{ab} \\ C_{p'} = D_j^b + T_{ij}^{ab} \end{array}\right\} \quad \text{for } p' = pE_{ij}^{ab}, \text{ if } C_p < A_j^b. \tag{6b}$$

Finally, in the third case, the package is available for transfer after the latest possible arrival of r_b at the rendezvous location, i.e., $C_p + T_{ij}^{ab}/2 > D_j^b + T_{ij}^{ab}/2$. Vehicle r_b must wait for $C_p - D_j^b$ time. Therefore,

$$\left.\begin{array}{l} O_{p'} = O_p + C_p - D_j^b + 2 \cdot T_{ij}^{ab} \\ C_{p'} = C_p + T_{ij}^{ab} \end{array}\right\} \quad \text{for } p' = pE_{ij}^{ab}, \text{ if } C_p > D_j^b. \tag{6c}$$

4   Solving Dynamic Pickup and Delivery with Transfers

Section 4.1 models dynamic Pickup and Delivery with Transfers as a dynamic shortest path graph problem. Section 4.2 introduces the SP algorithm that identifies the solution to a dPDPT request.

4.1   The Dynamic Plan Graph

We construct a weighted directed graph, termed *dynamic plan graph*, in a way that a sequence of actions corresponds to a simple path on this graph. A vertex of the graph corresponds to a spatial location. In particular, a vertex V_i^a represents the spatial location n_i of route r_a. Additionally, there exist two special vertices, V_s and V_e, which represent the request's pickup and delivery, respectively, locations. Therefore, five types of edges exist:

(1) A *pickup edge* E_{si}^a connects V_s to V_i^a, and represents a pickup action by vehicle r_a with a detour at n_i. Edge E_{ss}^a from V_s to V_s^a (two distinct vertices that correspond to the same spatial location n_s) represents the case of pickup with no detour.

(2) A *delivery edge* E_{ie}^a connects V_i^a to V_e, and represents a delivery action by vehicle r_a with a detour at n_i. Edge E_{ee}^a from V_e to V_e^a represents the case of pickup with no detour.

(3) A *transport* edge E_{ij}^a connects V_i^a to V_j^a, and represents a transport action by r_a from n_i to its following location n_j on the route.

(4) A *transfer without detours* edge E_i^{ab} connects V_i^a to V_i^b, and represents a transfer from r_a to r_b at common location n_i.

(5) A *transfer with detours* edge E_{ij}^{ab} connects V_i^a to V_j^b, and represents a transfer from r_a to r_b at a rendezvous location via detours at n_i and n_j.

Based on the above definitions, a simple path on the graph is a sequence of distinct vertices that translates into a sequence of actions. Further, a solution for the request is a path that starts from V_s and ends in V_e.

Table 1. Edge weights

Pickup	Delivery	Transport
$w(E_{si}^a) = \langle T_{si}^a, D_i^a + T_{si}^a \rangle$ (7)	$w(E_{ie}^a) = \langle T_{ie}^a, T_{ie}^a/2 \rangle$ (8)	$w(E_{ij}^a) = \langle 0, D_j^a - D_i^a \rangle$ (9)

Transfer		
$w(E_{ij}^{ab}) = \begin{cases} \langle 2 \cdot T_{ij}^{ab}, D_j^b - C_p + T_{ij}^{ab} \rangle, & \text{if } A_j^b \leq C_p \leq D_j^b \\ \langle A_j^b - C_p + 2 \cdot T_{ij}^{ab}, D_j^b - C_p + T_{ij}^{ab} \rangle, & \text{if } C_p < A_j^b \\ \langle C_p - D_j^b + 2 \cdot T_{ij}^{ab}, T_{ij}^{ab} \rangle, & \text{if } C_p > D_j^b. \end{cases}$		(10)

The final issue that remains is to define the weights $\mathcal{W}$ of the edges. We assign edge E a pair of weights $w(E) = \langle w_O(E), w_C(E) \rangle$, so that $w_O(E)$ (resp. $w_C(E)$) corresponds to the operational (resp. customer) cost of performing the action associated with the edge E. Recall from Section 3.2 that the costs of the last action in a sequence of actions depends on the total costs incurred by all previous actions. Consequently, the weights of an edge E from V to V' are *dynamic*, since they depend on the costs of the path p that lead to V. Assuming O_p and C_p are the costs of p, and $O_{p'}$ and $C_{p'}$ those of path $p' = pE$ upon executing E, we have that $w(E) = \langle O_{p'} - O_p, C_{p'} - C_p \rangle$. Table 1 summarizes the formulas for the weights of all edge types; note that the weights for actions with no detours are obtained by setting the corresponding T value to zero. In the formulas, A_j^b, D_i^a, D_j^a and D_j^b have fixed values determined by the static plan. On the other hand, C_p depends on the path p that leads to V_i^a.

Clearly, a path from V_s to V_e that has the lowest combined cost according to Equation 1 is an optimal solution.

Proposition 1. *Let R be a collection of vehicles routes and (n_s, n_e) be a dPDPT request over R. The solution to the request is the shortest path from vertex V_s to V_e on the dynamic plan graph G_R with respect to cost() of Equation 1.*

Example 1. Figure 2(a) pictures a collection of vehicle routes $R = \{r_a(n_1, n_3), r_b(n_2, n_6), r_c(n_4, n_8, n_9)\}$, and the pickup n_s and the delivery location n_e of a dPDPT request. Locations n_1 on route r_a and n_2 on r_b are sufficiently close to location n_s and thus, pickup actions E_{s1}^a and E_{s2}^b are possible. Similar, the E_{9e}^c delivery action is possible at location n_9 on route r_c. Finally, we also identify two transfer actions, E_{34}^{ac} and E_{68}^{bc}, as locations n_3, n_4 and n_6, n_8 have common rendezvous points n_5 and n_7, respectively.

To satisfy the dPDPT request (n_s, n_e) we define the dynamic plan graph G_R in Figure 2(b) containing vertices $V_s, V_1^a, V_2^b, \ldots, V_e$. Notice that the graph does not include any vertices for the rendezvous points n_5 and n_7. Dynamic plan graph G_R contains two paths from V_s to V_e which means that there two different ways to satisfy the dPDPT request: $p_1(V_s, V_1^a, V_3^a, V_4^c, V_8^c, V_9^c, V_e)$ and

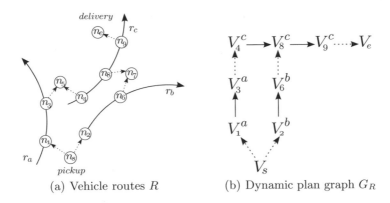

(a) Vehicle routes R (b) Dynamic plan graph G_R

Fig. 2. A collection of vehicles routes R, a dPDPT request (n_s, n_e) over R, and the dynamic plan graph G_R. The solid lines denote the vehicle routes/transport edges while the dashed lines denote the pickup, delivery and transfer with detour actions/edges.

$p_2(V_s, V_2^b, V_6^b, V_8^c, V_9^c, V_e)$. Next, assume, for simplicity, that the detour cost is equal to T for all possible actions. Further, consider paths $p'_1(V_s, V_1^a, V_3^a)$ and $p'_2(V_s, V_2^b, V_6^b)$, i.e., just before the transfer of the package takes place, and assume that $A_4^c < C_{p'_1} < D_4^c$ and $C_{p'_2} > D_8^c$ hold. Note, that the operational cost of the two paths is exactly the same, i.e., $O_{p'_1} = O_{p'_2} = T$, coming from the pickup of the package at n_s. Now, according to Equation 10 and after the transfers E_{34}^{ac} and E_{68}^{bc} take place, we get path $p''_1 = p'_1 E_{68}^{bc}$ and $p''_2 = p'_2 E_{68}^{bc}$ with $O_{p''_1} = 3 \cdot T < O_{p''_2} = 3 \cdot T + C_{p'_2} - D_8^c$, and therefore, $cost(p''_1) < cost(p''_2)$. Finally, since no other transfer incurs in order to delivery the package, this holds also for paths p_1 and p_2, i.e., $cost(p_1) < cost(p_2)$, and thus, the solution to the dPDPT request (n_s, n_e) is path $p_1(V_s, V_1^a, V_3^a, V_4^c, V_8^c, V_9^c, V_e)$ with $O_{p_1} = 4 \cdot T$ and $C_{p_1} = D_9^c + \frac{3 \cdot T}{2}$.

4.2   The SP Algorithm

According to Proposition 1, the next step is to devise an algorithm that computes the two-criterion shortest path w.r.t. $cost()$ on the dynamic plan graph. Unfortunately, the dynamic weights of the edges in the graph violate the *subpath optimality*; that is, the lowest combined cost path from V_s to V_e that passes through some vertex V may not contain the lowest combined cost path from V_s to V. The following theorem supports this claim.

Theorem 1. *The dynamic plan graph does not have subpath optimality for any monotonic combination of the operational and customer costs.*

Proof. Let p, q be two paths from V_s to V_i^a, with costs O_p, C_p and O_q, C_q, respectively, such that $O_p < O_q$ and $C_p < C_q$, which implies that for any monotonic combination of the operational and customer costs, p has lower combined cost than q. Let p' and q' be the paths resulting after traversing a transfer without detours edge E_i^{ab}.

Assume that $C_p < C_q < A_i^b$, so that the second case of Equation 10 applies for the weight $w(E_i^{ab})$ (setting $T_{ij}^{ab} = 0$). Then, $O_{p'} = O_p + A_i^b - C_p$, $O_{q'} = O_q + A_i^b - C_q$, and $C_{p'} = C_{q'} = D_i^b$. Assuming that $O_p - C_p > O_q - C_q$, we obtain that $O_{q'} < O_{p'}$. Therefore, for any monotonic combination that considers both costs, q''s combined cost is lower that that of p''s.   □

As a result, efficient algorithms based on the subpath optimality, e.g., Dijkstra and Bellman-Ford cannot be employed to compute the shortest path on the dynamic plan graph. In contrast, an exhaustive enumeration of all paths from V_s to V_e is necessary, and for this purpose we introduce a label-setting algorithm called SP. Note that, in the sequel, we only discuss the case when actions occur with detours as it is more general.

The SP algorithm has the following key features. First, similar to all algorithms for multi-criteria shortest path, it may visit a vertex V_i^a more than once following multiple paths from the initial vertex V_s. For each of these paths p, the algorithm defines a *label* in the form of $\langle V_i^a, p, O_p, C_p \rangle$, where O_p is the operational and C_p the customer cost of path p as introduced in Section 4.1. Second, at each iteration, SP selects the label $\langle V_i^a, p, O_p, C_p \rangle$ of the most "promising" path p, in other words, the path with the lowest $cost(p)$, and expands the search considering the outgoing edges from V_i^a on the dynamic plan graph G_R. If vertex V_i^a has an outgoing delivery edge E_{ie}^a, SP identifies a path from initial vertex V_s to final V_e called *candidate* solution. The candidate solution is an *upper bound* to the final solution and it is progressively improved until it becomes equal to the final. The role of a candidate solution is twofold; it triggers the termination condition and prunes the search space. Finally, the algorithm terminates the search when $cost(p)$ of the most "promising" path p is equal to or higher than $cost(p_{cand})$ of the current candidate solution p_{cand} which means that neither p or any other path at future iterations can be better than current p_{cand}.

Figure 3 illustrates the pseudocode of the SP algorithm. SP takes as inputs: a dPDPT request (n_s, n_e) and the dynamic plan graph G_R of a collection of vehicle routes R. It returns the shortest path from V_s to V_e on G_R with respect to $cost()$. The algorithm uses the following data structures: (1) a priority queue $\mathcal{Q}$, (2) a path p_{cand}, and (3) a list $\mathcal{T}$. The priority queue $\mathcal{Q}$ is used to perform the search by storing every label $\langle V_i^a, p, O_p, C_p \rangle$ to be checked, sorted by $cost(p)$ in ascending order. The target list $\mathcal{T}$ contains entries of the form $\langle V_i^a, 0, T_{ie}^a \rangle$, where V_i^a is a vertex of the dynamic plan graph involved in a delivery edge E_{ie}^a, List $\mathcal{T}$ is used to construct or improve the candidate solution p_{cand}.

The execution of the SP algorithm involves two phases: the *initialization* and the *core* phase. In the initialization phase (Lines 1–4), SP first creates the pickup E_{si}^a and delivery edges E_{ie}^a on the dynamic plan graph G_R. For this purpose it identifies each location n_i on every vehicle route r_a that is sufficiently close to pickup location n_s (resp. delivery n_e), i.e., the duration T_{si}^a (resp. T_{ie}^a) of the detour from n_s to n_i (resp. n_i to n_e) is below some threshold (a system parameter). Then, the algorithm initializes the priority queue $\mathcal{Q}$ adding every

Algorithm SP
Input: dPDPT request (n_s, n_e), dynamic plan graph G_R
Output: shortest path from V_s to V_e w.r.t. $cost()$
Parameters:

priority queue $\mathcal{Q}$: the search queue sorted by $cost()$ in ascending order
path p_{cand}:   the candidate solution to the dPDPT request
list $\mathcal{T}$:   the target list

Method:

1. **construct** pickup edges E_{si}^a;
2. **construct** delivery edges E_{ie}^a;
3. **for** each pickup edge $E_{si}^a(V_s, V_i^a)$ **do**
        push label $\langle V_i^a, E_{si}^a, T_{si}^a, D_i^a + T_{si}^a \rangle$ in $\mathcal{Q}$;
4. **for** each delivery edge $E_{ie}^a(V_i^a, V_e)$ **do**
        insert $\langle V_i^a, T_{ie}^a, T_{ie}^a/2 \rangle$ in $\mathcal{T}$;
5. **while** $\mathcal{Q}$ is not empty **do**
6.   **pop** label $\langle V_i^a, p, O_p, C_p \rangle$ from $\mathcal{Q}$;
7.   **if** $cost(p) \geq cost(p_{cand})$ **then return** p_{cand};
8.   ImproveCandidateSolution$(p_{cand}, \mathcal{T}, \langle V_i^a, p, O_p, C_p \rangle)$;
9.   **for** each outgoing transport E_{ij}^a **or** transfer edge E_{ij}^{ab} on G_R **do**
10.   **extend** path p and **create** p';
11.   **compute** $O_{p'}$ and $C_{p'}$;
12.   **if** $cost(p') < cost(p_{cand})$ **then**
                ignore path p';
13.   **else**
14.   **push** label $\langle V', p', O_{p'}, C_{p'} \rangle$ in $\mathcal{Q}$ **where** V' is the last vertex in p';
15.   **end if**
16.   **end for**
17. **end while**
18. **return** p_{cand} **if exists, otherwise null**;

Fig. 3. The SP algorithm

vertex V_i^a involved in a pickup edge E_{si}^a on G_R and constructs the target list $\mathcal{T}$. In the core phase (Lines 5–17), the algorithm performs the search. It proceeds iteratively popping, first, the label $\langle V_i^a, p, O_p, C_p \rangle$ from $\mathcal{Q}$ on Line 6. Path p has the lowest $cost(p)$ value compared to all others paths in $\mathcal{Q}$. Next, SP checks the termination condition (Line 7). If the check succeeds, i.e., $cost(p) \geq cost(p_{cand})$, then current candidate p_{cand} is returned as the final solution.

If the termination condition fails, the algorithm first tries to improve candidate solution p_{cand} calling the ImproveCandidateSolution$(p_{cand}, \mathcal{T}, \langle V_i^a, p, O_p, C_p \rangle)$ function on Line 8. The function checks if the target list $\mathcal{T}$ contains an entry $\langle V_i^a, T_{ie}^a, T_{ie}^a/2 \rangle$ for the vertex V_i^a of the current label and constructs the path pE_{ie}^a from V_s to V_e. If $cost(pE_{ie}^a) < cost(p_{cand})$ then a new improved candidate solution is identified and thus, $p_{cand} = pE_{ie}^a$. Finally, SP expands the search considering all outgoing transport and transfer edges from V_i^a on G_R (Lines 9–17). Specifically, the path p of the current label is extended to $p' = pE_{ij}^a$ (transport edge) or to $p' = pE_{ij}^{ab}$ (transfer edge), and the operational $O_{p'}$ and the customer cost $C_{p'}$ of the new path p' are computed according to Equations 9 and 10. Then, on Line 12, the algorithm determines whether p' is a "promising" path and thus, it must be extended at a future iteration, or it must be discarded. The algorithm discards path p' if $cost(p') \geq cost(p_{cand})$ which means that p' cannot produce a better solution than current p_{cand}. Otherwise, p' is a "promising" path, and SP inserts label $\langle V', p', O_{p'}, C_{p'}\rangle$ in $\mathcal{Q}$ where V' is the last vertex in path p'.

Example 2. We illustrate the SP algorithm using Figure 2. To carry out the search we make the following assumptions, similar to Example 1. The detour cost is equal to T for all edges. For the paths $p'_1(V_s, V_1^a, V_3^a)$ and $p'_2(V_s, V_2^b, V_6^b)$, i.e., just before the transfer of the package takes place, $A_4^c < C_{p'_1} < D_4^c$ and $C_{p'_2} > D_8^c$ hold. Finally, we also assume that $D_1^a < D_3^a < D_2^b < D_6^b$.

First, SP initializes the priority queue $\mathcal{Q} = \{\langle V_1^a, (V_s, V_1^a), T, D_1^a + T\rangle, \langle V_2^b, (V_s, V_2^b), T, D_2^b + T\rangle\}$ and constructs the target list $\mathcal{T} = \{\langle V_9^c, T, T/2\rangle\}$. Note that the leftmost label in $\mathcal{Q}$ always contains the path with the lowest $cost()$ value. At the first iteration, the algorithm pops label $\langle V_1^a, (V_s, V_1^a), T, D_1^a + T\rangle$, considers transport edge E_{13}^a, and pushes $\langle V_3^a, p'_1(V_s, V_1^a, V_3^a), T, D_3^a + T\rangle$ to $\mathcal{Q}$. Next, at the second iteration, SP pops label $\langle V_3^a, p'_1(V_s, V_1^a, V_3^a), T, D_3^a + T\rangle$ from $\mathcal{Q}$, considers the transfer edge E_{34}^{ac}, and pushes $\langle V_4^c, p''_1(V_s, V_1^a, V_3^a, V_4^c), 3 \cdot T, D_4^c + T\rangle$ to $\mathcal{Q}$ (remember $A_4^c < C_{p'_1} < D_4^c$). The next two iterations are similar, and thus, after the fourth iteration we have:

$$\mathcal{Q} = \{\langle V_4^c, p''_1(V_s, V_1^a, V_3^a, V_4^c), 3 \cdot T, D_4^c + T\rangle, \qquad \text{and } p_{cand} = \textbf{null}$$
$$\langle V_8^c, p''_2(V_s, V_2^b, V_6^b, V_8^c), 3 \cdot T + C_{p'_2} - D_8^c, C_{p'_2} + T\rangle\}$$

Now, at the next two iterations, SP expands path p''_1 considering transport edges E_{48}^c and E_{89}^c as $O_{p''_1} < O_{p''_2}$. Therefore, at the seventh iteration, the algorithm pops label $\langle V_9^c, (V_s, V_1^a, V_3^a, V_4^c, V_8^c, V_9^c), 3 \cdot T, D_9^c + T\rangle$ from $\mathcal{Q}$. Since the target list $\mathcal{T}$ contains an entry for vertex V_9^c, SP identifies candidate solution $p_{cand} = p_1(V_s, V_1^a, V_3^a, V_4^c, V_8^c, V_9^c, V_e)$ with $O_{p_1} = 4 \cdot T$ and $C_{p_1} = D_9^c + \frac{3 \cdot T}{2}$. Finally, assuming without loss of generality that $D_6^b > D_8^c$ also holds and therefore, $C_{p'_2} - D_8^c = D_6^b + T - D_8^c > T$, at the eighth iteration, the algorithm pops $\langle V_8^c, p''_2(V_s, V_2^b, V_6^b, V_8^c), 3 \cdot T + C_{p'_2} - D_8^c, C_{p'_2} + T\rangle$ and terminates the search because $O_{p''_2} = 3 \cdot T + C_{p'_2} - D_8^c > 4 \cdot T = O_{p_1}$ and thus, $cost(p''_2) > cost(p_1)$. The solution to the dPDPT request (n_s, n_e) is $p_1(V_s, V_1^a, V_3^a, V_4^c, V_8^c, V_9^c, V_e)$.

5   Experimental Evaluation

In this section, we present an experimental study of our methodology for solving dynamic Pickup and Delivery Problem with Transfers (dPDPT). We compare SP against HT, a method inspired by [15,25] that combines an insertion heuristic with tabu search. All methods are written in C++ and compiled with gcc. The evaluation is carried out on a 3Ghz Core 2 Duo CPU with 4GB RAM running Debian Linux.

5.1   The HT Method

Satisfying dPDPT requests with HT involves two phases. In the first phase, for every new dPDPT request, the method employs the cheapest insertion heuristic to include the pickup n_s and the delivery location n_e in a vehicle route. The idea is the following. HT examines every vehicle route r_a and for each pair of consecutive locations n_i and n_{i+1} in r_a (forming an insertion "slot"), it computes the detour cost $DS = dist(n_i, n_s) + dist(n_s, n_{i+1}) - dist(n_i, n_{i+1})$ for inserting pickup n_s (resp. delivery n_e) in between n_i and n_{i+1}. The detour cost DS signifies the extra time vehicle r_a must spend and therefore, it increases the total operational cost. Then, HT selects the best overall insertion, i.e., the route r_a and the "slots", such that the combined detour cost for inserting both pickup n_s and delivery location n_e is minimized.

The second phase of HT takes place periodically after k requests are satisfied with the cheapest insertion. It involves a tabu search procedure that reduces the total operating cost. At each iteration, the tabu search considers every satisfied request and calculates what would be the change (increase or decrease) in the total operational cost removing the request from its current vehicle route r_a and inserting it to another r_b. Then, the tabu search selects the request with the best combination of removal and insertion, and performs these actions. Finally, the selected combination is characterized as tabu and cannot be executed for a number of future iterations.

5.2   Experiments

To conduct our experiments, we consider the road networks of two cities; Oldernburg (OL) with 6,105 spatial locations (Figure 4), and Athens (ATH) with 22,601 locations (Figure 5). First, we generate random pickup and delivery requests at each network and exploit the HT method to construct collections of vehicle routes varying either the number of routes $|R|$, from 100 to 1000, or the number of requests $|Reqs|$ involved, from 200 to 2000. Then, for each of these route collections, we generate 500 random dPDPT requests and employ the SP and the HT method to satisfy them. For HT, we introduce three variations HT1, HT3 and HT5 such that the tabu search is invoked once (after 500 requests are satisfied), three times (after 170) and five times (after 100), respectively. In addition, each time the tabu search is invoked, it performs 10 iterations. For each method, we measure (1) the increase in the total operational cost of the vehicles after all 500 requests are satisfied (sub-figures (a) and (c)) and (2) the total time needed to satisfy the requests. Finally, note that we store both the road network and the vehicle routes on disk and that, we consider a main-memory cache mechanism capable of retaining 10% of the total space occupied on disk.

Examining Figures 4 and 5 we make the following observations. The SP method requires significantly less time to satisfy the 500 ad-hoc dPDPT requests, for all the values of the $|Reqs|$ and $|R|$ parameters, and for both the underlying road networks. In fact, when varying $|R|$, SP is always one order of magnitude faster than all three HT variants. In contrast, SP results in slightly increased total operational cost compared to HT, in most of the cases, and especially for large road networks as ATH. However, this advantage of HT comes with a unavoidable trade-off between the increase of the total operational cost and the time needed to satisfy the ad-hoc dPDPT requests. The more often HT employs the tabu search, the lower the increase of the total operational cost of the vehicles is. But, on the other hand, since each iteration of the tabu search needs

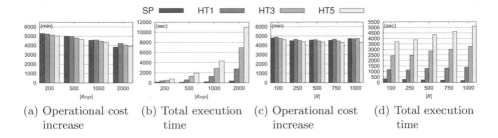

(a) Operational cost increase

(b) Total execution time

(c) Operational cost increase

(d) Total execution time

Fig. 4. City of Oldenburg (OL) Road Network

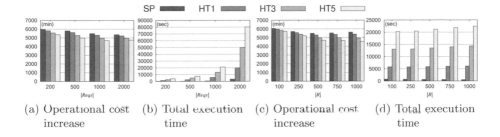

(a) Operational cost increase

(b) Total execution time

(c) Operational cost increase

(d) Total execution time

Fig. 5. City of Athens (ATH) Road Network

to examine every route and identify the best reassignment for all the existing requests, the total time of HT5 is higher than the time of HT3 and HT1.

Finally, we notice that as the number of pickup and delivery requests $|Reqs|$ involved in the initial static plan increases, satisfying the 500 ad-hoc dPDPT requests, either with HT or SP, results in a lower increase of the total operational cost but the total time needed to satisfy these requests increases. Notice that this is true regardless of the size of the underlying road network. As $|Reqs|$ increases and while $|R|$ remains fixed, the vehicle routes contain more spatial locations. This provides more insertion "slots" and enables both HT and SP to include the pickup and the delivery location of a dPDPT request with a lower cost. On the other hand, HT slows down since it has to examine the reassignment of more requests during the tabu search, and SP needs more time because the dynamic plan graph is larger. Similar observations can be made in case of varying the number of routes $|R|$.

6   Conclusions

This work studies the dynamic Pickup and Delivery Problem with Transfers (dPDPT). This is the first work addressing the dynamic flavor of the problem. We propose a methodology that formulates dPDPT as a graph problem and identifies the solution to a request as the shortest path from a node representing the pickup location to that of the delivery location. Our experimental analysis shows that our method is able to find dPDPT solutions significantly faster than a conventional two-phase local search algorithm, while the quality of the solution is only marginally lower.

References

1. Agrawal, R., Jagadish, H.V.: Materialization and incremental update of path information. In: ICDE, pp. 374–383 (1989)
2. Bent, R., Hentenryck, P.V.: A two-stage hybrid algorithm for pickup and delivery vehicle routing problems with time windows. Computers and Operations Research 33(4), 875–893 (2006)
3. Berbeglia, G., Cordeau, J.F., Laporte, G.: Dynamic pickup and delivery problems. European Journal of Operational Research 202(1), 8–15 (2010)
4. Berbeglia, G., Cordeau, J.F., Gribkovskaia, I., Laporte, G.: Static pickup and delivery problems: a classification scheme and survey. TOP 15, 1–31 (2007)
5. Carraway, R.L., Morin, T.L., Moskowitz, H.: Generalized dynamic programming for multicriteria optimization. European Journal of Operational Research 44(1), 95–104 (1990)
6. Cheng, J., Yu, J.X.: On-line exact shortest distance query processing. In: EDBT, pp. 481–492 (2009)
7. Cohen, E., Halperin, E., Kaplan, H., Zwick, U.: Reachability and distance queries via 2-hop labels. In: SODA, pp. 937–946 (2002)
8. Cooke, K.L., Halsey, E.: The shortest route through a network with time-dependent internodal transit times. Journal of Mathematical Analysis and Applications 14(3), 493–498 (1966)
9. Cordeau, J.F., Laporte, G., Ropke, S.: Recent Models and Algorithms for One-to-One Pickup and Delivery Problems. In: Vehicle Routing: Latest Advances and Challenges, pp. 327–357. Kluwer, Dordrecht (2008)
10. Cordeau, J.F.: A branch-and-cut algorithm for the dial-a-ride problem. Operations Research 54(3), 573–586 (2006)
11. Cortés, C.E., Matamala, M., Contardo, C.: The pickup and delivery problem with transfers: Formulation and a branch-and-cut solution method. European Journal of Operational Research 200(3), 711–724 (2010)
12. Ding, B., Yu, J.X., Qin, L.: Finding time-dependent shortest paths over large graphs. In: EDBT, pp. 205–216 (2008)
13. Dreyfus, S.E.: An appraisal of some shortest-path algorithms. Operations Research 17(3) (1969)
14. Dumas, Y., Desrosiers, J., Soumis, F.: The pickup and delivery problem with time windows. European Journal of Operational Research 54(1), 7–22 (1991)
15. Gendreau, M., Guertin, F., Potvin, J.Y., Séguin, R.: Neighbourhood search heuristics for a dynamic vehicle dispatching problem with pick-ups and deliveries. Transportation Research Part C: Emerging Technologies 14(3), 157–174 (2006)
16. Goldberg, A.V., Harrelson, C.: Computing the shortest path: A search meets graph theory. In: SODA 2005: Proceedings of the sixteenth annual ACM-SIAM symposium on Discrete algorithms, pp. 156–165 (2005)
17. Goldberg, A.V., Kaplan, H., Werneck, R.F.: Reach for A*: Efficient point-to-point shortest path algorithms. In: Proc. of the 8th WS on Algorithm Engineering and Experiments (ALENEX), pp. 129–143. SIAM, Philadelphia (2006)

18. Granat, J., Guerriero, F.: The interactive analysis of the multicriteria shortest path problem by the reference point method. European Journal of Operational Research 151(1), 103–118 (2003)
19. Guerriero, F., Musmanno, R.: Label correcting methods to solve multicriteria shortest path problems. Journal of Optimization Theory and Applications 111(3), 589–613 (2001)
20. Jing, N., Huang, Y.W., Rundensteiner, E.A.: Hierarchical encoded path views for path query processing: An optimal model and its performance evaluation. In: TKDE, vol. 10(3), pp. 409–432 (1998)
21. Jung, S., Pramanik, S.: An efficient path computation model for hierarchically structured topographical road maps. In: TKDE, vol. 14(5), pp. 1029–1046 (2002)
22. Kriegel, H.P., Renz, M., Schubert, M.: Route skyline queries: A multi-preference path planning approach. In: ICDE, pp. 261–272 (2010)
23. Li, H., Lim, A.: A metaheuristic for the pickup and delivery problem with time windows. In: ICTAI, pp. 333–340 (2001)
24. Mitrović-Minić, S., Krishnamurti, R., Laporte, G.: Double-horizon based heuristics for the dynamic pickup and delivery problem. Transportation Research Part B: Methodological 38(7), 669–685 (2004)
25. Mitrović-Minić, S., Laporte, G.: Waiting strategies for the dynamic pickup and delivery problem with time windows. Transportation Research Part B: Methodological 38(7), 635–655 (2004)
26. Mitrović-Minić, S., Laporte, G.: The pickup and delivery problem with time windows and transshipment. INFOR 44(3), 217–228 (2006)
27. Orda, A., Rom, R.: Shortest-path and minimum-delay algorithms in networks with time-dependent edge-length. J. ACM 37(3), 607–625 (1990)
28. Parragh, S., Doerner, K., Hartl, R.: A survey on pickup and delivery problems, Part II: Transportation between pickup and delivery locations. Journal für Betriebswirtschaft 58, 81–117 (2008)
29. Pohl, I.: Bi-directional search. Machine Intelligence 6, 127–140 (1971)
30. Popken, D.A.: Controlling order circuity in pickup and delivery problems. Transportation Research Part E: Logistics and Transportation Review 42(5), 431–443 (2006)
31. Ropke, S., Cordeau, J.F.: Branch and cut and price for the pickup and delivery problem with time windows. Transportation Science 43(3), 267–286 (2009)
32. Ropke, S., Cordeau, J.F., Laporte, G.: Models and branch-and-cut algorithms for pickup and delivery problems with time windows. Networks 49(4), 258–272 (2007)
33. Ropke, S., Pisinger, D.: An adaptive large neighborhood search heuristic for the pickup and delivery problem with time windows. Transportation Science 40(4), 455–472 (2006)
34. Sigurd, M., Pisinger, D., Sig, M.: Scheduling transportation of live animals to avoid spread of diseases. INFORMS Transportation Science 38, 197–209 (2004)
35. Tian, Y., Lee, K.C.K., Lee, W.C.: Finding skyline paths in road networks. In: GIS, pp. 444–447 (2009)
36. Xu, H., Chen, Z.L., Rajagopal, S., Arunapuram, S.: Solving a practical pickup and delivery problem. Transportation Science 37(3), 347–364 (2003)

Finding Top-k Shortest Path Distance Changes in an Evolutionary Network

Manish Gupta[1,*], Charu C. Aggarwal[2], and Jiawei Han[1]

[1] University of Illinois at Urbana-Champaign, IL, USA
gupta58@illinois.edu, hanj@cs.uiuc.edu
[2] IBM T.J. Watson Research Center, NY, USA
charu@us.ibm.com

Abstract. Networks can be represented as evolutionary graphs in a variety of spatio-temporal applications. Changes in the nodes and edges over time may also result in corresponding changes in structural garph properties such as shortest path distances. In this paper, we study the problem of detecting the top-k most significant shortest-path distance changes between two snapshots of an evolving graph. While the problem is solvable with two applications of the all-pairs shortest path algorithm, such a solution would be extremely slow and impractical for very large graphs. This is because when a graph may contain millions of nodes, even the storage of distances between all node pairs can become inefficient in practice. Therefore, it is desirable to design algorithms which can directly determine the significant changes in shortest path distances, without materializing the distances in individual snapshots. We present algorithms that are *up to two orders of magnitude* faster than such a solution, while retaining comparable accuracy.

1 Introduction

The problem of network evolution [1,7,12,17] has seen increasing interest in recent years of the dynamic nature of many web-based, social and information networks which continuously change over time. The evolution of such networks may also result in changes in important structural properties such as pairwise shortest-path distances. In this paper, we will study the problem of finding the top k shortest path distance changes in an evolutionary network. This problem may be interesting in the context of a number of practical scenarios:

- Social and information networks are inherently dynamic, and the change in shortest paths between nodes is critical in understanding the changes in connections between different entities. It can also be helpful for tasks such as dynamic link prediction modeling with the use of shortest-path recommendation models or in providing insights about new events in the social graph. For example, the distance between tags in an inter-tag correlation graph may change because of tagging events, or the distance between actors in IMDB[1] may change because of the introduction of new movies. Similarly, change in distances on word co-occurrence graphs for micro-blogs (such as Twitter[2] tweets) can result in better detection and summarization of events.

---

* Work was partially done during employment at IBM T.J. Watson Research Center.
[1] http://www.imdb.com/
[2] http://twitter.com

D. Pfoser et al. (Eds.): SSTD 2011, LNCS 6849, pp. 130–148, 2011.

- Many developing countries have witnessed a rapid expansion in their road networks. An example would be the well-known *Golden Quadrilateral (GQ)*[3] project
in India. The detection of important distance changes may provide further insights
about connectivity implications. For example, given multiple plans to set up a road
network, one can measure the utility of a proposed plan in a particular marketing
scenario.
- The launch of collaborative programs changes the structure of virtual collaboration
networks. The underlying changes provide an understanding of critical connections
between collaborative entities (e.g., authors in *DBLP*) and their evolution.

The detection of interesting hot-spots for which average distance to other parts of the
network has changed suddenly is an interesting problem. This is closely related to the
problem of finding the top-k maximum distance change node pairs. We will present a
number of algorithms for this problem in this paper. We compare our algorithms on a
number of real data sets.

We note that the problem of shortest path change can be solved directly by running
well known all-pair shortest path algorithms and simply comparing the distance changes
between all pairs. However, this is not a practical solution for very large graphs. For
example, for a graph containing 10^8 nodes, the number of possible node pairs would
be 10^{16}. The complexity of the all-pairs shortest path computation increases at least
quadratically [2] with the number of nodes. Furthermore, the storage of such pairwise
paths can be impractical. While this has not been a problem with the small memory-
resident graphs which are frequently used with conventional algorithms, it is much more
challenging for the large-scale networks which arise in social and information network
scenarios. In fact, in our experiments, we found it virtually impossible to use such brute-
force algorithms in any meaningful way. Therefore, the goal of the paper is to enable
practical use of such algorithms in large-scale applications.

The remainder of this paper is organized as follows. In Section 2, we provide an
overview of the related work. We introduce our basic algorithm, the Incidence Algorithm and a randomized algorithm to estimate importance of an edge in a graph in
Section 3. In Section 4, we discuss various algorithms for ranking of nodes which can
potentially be a part of the top-k node pairs. We present our experiments on large graphs
in Section 5 and finally conclude with a summary of the work in Section 6.

2   Related Work

The problem of finding the top-k node pairs with maximum shortest path distance
change can be solved by a straightforward applications of two instances of the all-pairs
shortest path (APSP) problem. Clearly, the running time is sensitive to the method used
for APSP computation. Consider a graph containing n nodes and m edges. One can use
a variety of methods such as Shimbel's algorithm [20], Dijkstra's algorithm [11], Johnson's algorithm [15], or the Floyd and Warshall [14,21] algorithms, all which require at
least $O(n \cdot m)$ time. Such running times are not very practical for very large graphs containing millions of nodes. A randomized algorithm by Cohen [10] allows us to compute
the number of nodes at a distance d from each of the nodes in the graph. While such an
approach can be used to approximately determine a superset of the relevant node pairs,
a part of the method requires $O(mlog(n) + nlog^2(n))$ time. This is quite inefficient.

---

[3] http://en.wikipedia.org/wiki/Golden_Quadrilateral

Our problem is also related to that of finding time-dependent shortest paths [25,26] in a dynamic graph. A number of algorithms in computer networks [22,8] also solve the problem of recomputing shortest path trees when edges are added to or removed from the graph. Some related work [6,3,19,4] for this problem proposes methods for exact computation of dynamic shortest paths in a variety of graph settings and in parallel or distributed scenarios. Our problem is however that of finding the maximum shortest path *change* between pairs of nodes, rather than that of designing incremental algorithms for *maintaining* shortest paths.

The problem of shortest path distance change is also related to that of outlier detection, since unusually large changes in distances can be considered abnormal behavior. Some work has also been done on node outlier detection in *static* weighted networks [5]. Anomalies which are caused by *dynamic behavior* such as label modifications, vertex/edge insertions and vertex/edge deletions have been studied in [13]. In [18], the authors detect anomalies such as missing connected subgraphs, missing random vertices and random topological changes over web crawl snapshots by measuring the amount and the significance of changes in consecutive web graphs.

As we will see later, we need to design methods for measuring the betweenness of edges in order to determine key changes in the network. A number of betweenness measures have been proposed in [23,24], though the methods for computing such measures are too slow to be of practical use in very large scale applications.

3   Shortest Path Evolution: Model and Algorithms

Before discussing the problem further, we will introduce some formal notations and definitions. Consider an undirected connected graph G with snapshots $G_1(V_1, E_1)$ at time t_1 and $G_2(V_2, E_2)$ at time t_2. For the purpose of this paper, we only consider the case where new nodes and edges are added to the graph, and they do not get removed. This is quite often the case in many natural information networks such as *IMDB* movie network or *DBLP co-authorship* network in which objects are constantly added over time. Each edge e can be expressed as a three-tuple (u, v, w) where u and v are the nodes on which the edge is incident and w is the weight of the edge. The edge weights denote the distance between two nodes and can only decrease over time. Let $d_1(u, v)$ and $d_2(u, v)$ denote the shortest path distances between nodes u and v in snapshots G_1 and G_2. Let the distance change between nodes u and v be denoted by $\Delta d(u, v)$. We aim to find these top-k node pairs (u, v) with the largest $\Delta d(u, v)$, so that there exists no pair (u', v') where $u \neq u'$ and/or $v \neq v'$, s.t. $\Delta d(u', v') > \Delta d(u, v)$.[4]

One of the keys to determining the node pairs with the most change is to determine the *critical edges* which lie on the shortest paths between many pairs of nodes. Clearly the addition of such edges can lead to tremendous changes in the shortest path distances. Therefore, we define the concept of *edge importance* as the probability that an edge will belong to some shortest path tree. In this section, we first propose a randomized algorithm for edge importance estimation. Then, we will leverage this notion to propose the

---

[4] Suppose that $\Delta(u, v)$ is large, as a result of addition of a new set S of edges. It may superficially seem that pairs of the form (node near u, node near v) would also have huge distance changes (since they are related in a somewhat similar way to S) and may swamp the top-k. However, when we consider the effect of multiple new edges in a dense real graph, the swamping effect is not quite as explicit. This suggests that subtle structural effects play an important role in defining the solutions of the underlying problem.

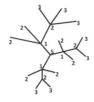

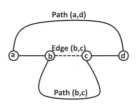

Fig. 1. Distance labeling on SPT_S for a unit weighted undigraph

Fig. 2. Small example

Incidence Algorithm. This is followed by a method to improve its accuracy. Finally, we propose node ranking algorithms to improve the efficiency of the Incidence Algorithm.

3.1 Edge Importance Estimation Algorithm

We first define some concepts which relate to edge-importance.

Definition 1 (Importance number of an edge). *The Importance Number of an edge* $e \in E$ *in a graph* $G(V, E)$ *is the probability that the edge will lie on a randomly chosen shortest path tree in the graph.*

Let us denote the shortest path tree rooted at any node x as the source node by SPT_x. The *Edge importance* $I(e)$ for an edge e can be accurately estimated by using the edge importance measured over a random sample of shortest path trees for the same graph. We will present detailed results in Section 5 which show that a random sampling approach works quite well. For estimating the edge importance numbers, we first choose a set of α nodes randomly into a set S and initialize the importance numbers for all edges to 0. For each of the nodes x in set S, we run the well known Dijkstra algorithm over the graph G by using x as the source node.

Algorithm 1. Edge importance number estimation

1: Input: Graph $G(V, E)$ with n nodes and m edges.
2: Randomly sample α nodes from the graph into set S.
3: Initialize importance number $I(e)$ of every edge $e \in E$ as 0.
4: **for** each node $x \in S$ **do**
5:     Run Dijkstra with x as the source node in $O(m + n)$ time to get a shortest path tree SPT_x
6:     Label the nodes of SPT_x with the distance from the source node x. Figure 1 shows the labeling for a shortest path tree on a unit weighted undirected graph.
7:     For each edge $(i, j) \in E$, identify if (i, j) is an alternative tight edges for SPT_x. Tight edges can be discovered in $O(m)$ time.
8:     Choose β random shortest path trees by perturbing the tree SPT_x by replacing γ number of edges in SPT_x each by one of the alternative tight edges.
9:     **for** every alternative shortest path tree SPT_{xy} chosen in Step 8 **do**
10:         Estimate importance number $I(e, SPT_{xy})$ for every edge e wrt SPT_{xy} as the number of its descendants normalized by n. This takes $O(m)$ time.
11:         Add $I(e, SPT_{xy})$ to $I(e)$.
12:     **end for**
13: **end for**
14: Compute avg. importance numbers for each edge by normalizing $I(e)$ by $\alpha\beta$.
15: **return** Average importance for each edge $e \in E$

Definition 2 (Distance Label). *The distance label $d^x(i)$ for a node i in a shortest path tree SPT_x is defined as the shortest path distance of the node from the source node x. Note that distance labels are valid if $d^x(j) \leq d^x(i) + w(i,j)$ or $d^x(i) \leq d^x(j) + w(i,j)$ where $d^x(i)$ and $d^x(j)$ are the distance labels for nodes i and j in SPT_x.*

Definition 3 (Tight edge). *An edge (i,j) is a tight edge with respect to shortest path tree SPT_x, if $d^x(j) = d^x(i) + w(i,j)$ or $d^x(i) = d^x(j) + w(i,j)$.*

Definition 4 (Alternative tight edge). *An edge (i,j) is an alternative tight edge if it is a tight edge, but it is not a part of the current shortest path tree.*

After constructing the shortest path trees, we perform distance labeling on SPT_x (as shown in Figure 1). Further, we can generate alternative shortest path trees for SPT_x by replacing tree edges with alternative tight edges. Step 7 in Algorithm 1 iterates over the entire edge set of the graph and checks if it is an alternative tight edge. Step 8 generates random shortest path trees by replacing one or more edges in SPT_x by one of the alternative tight edges.

Observation 1. *For every alternative tight edge, there is a corresponding edge in SPT_x, which can be replaced to create a new perturbed shortest path tree.*

We note that the edge to be replaced can be determined by adding the alternative tight edges, and removing one of the tree edges from the cycle thus created.

We generate a random perturbation of original SPT_x by selecting alternative tight edges for some randomly selected edges of SPT_x. This can be achieved easily by considering the subgraph of all tight edges, and picking one which forms a shortest path tree. We consider β such shortest path trees for each SPT_x.

Definition 5 (Descendant of an edge). *A descendant in the context of an edge (i,j) and a shortest path tree SPT_x is defined as any node that lies in the subtree rooted at the end-point of the edge farther from the source node x.*

The concept of edge descendant is useful, because it measures the importance of the edge (i,j) in connecting nodes contained in SPT_x. Specifically, the number of descendants of edge (i,j) provides a clear understanding of its connectivity level from the source node x to other nodes of the tree. Step 10 computes edge importance as the normalized number of descendants. Let $I(e, SPT_{xy})$ denote the importance of an edge for the yth SPT which is obtained by a random perturbation of SPT_x. This is computed as the ratio of number of descendants for edge (i,j) with respect to SPT_{xy} to $|V|$. $0 \leq I(e, SPT_{xy}) \leq 1$. Finally, in Step 14, average importance numbers are computed for every edge by computing average across all the sampled shortest path trees. Thus, the average importance of an edge is estimated as $I(e) = \frac{\sum_x \sum_y I(e, SPT_{xy})}{\alpha\beta}$. The Importance estimation algorithm runs in $O(\alpha\beta m)$ time. We will use these importance numbers later to improve the Incidence Algorithm described in the next subsection and also for node ranking algorithms.

3.2   The Incidence Algorithm

In this subsection, we present the Incidence Algorithm (Algorithm 2).

Definition 6 (Active node). *A node is **active** if new edges or edges with changed weights are incident on it.*

The intuition for the *Incidence Algorithm* is that the maximum distance change node pairs will include at least one node as an active node with high probability.

Observation 2. *Node pairs containing at least one active node can cover most of the top-k node pairs with maximum distance change with high probability for small values of k. This is particularly true for more dense networks.*

Example: Consider a path $a - b - c - d$ as shown in Figure 2. The solid lines show the shortest paths between the nodes in snapshot G_1. (a, b) and (c, d) are the edges in graph G_1 while (b, c) appears as a new edge in graph G_2. Let $Path(a, d)$ and $Path(b, c)$ be the shortest paths between the corresponding nodes in graph G_1.

Let us assume that the new shortest path between nodes b and c as well as the one between nodes a and d passes through the edge (b, c). Then $d_1(b, c)$ should follow the inequality: $d_1(a, d) - w(a, b) - w(c, d) \leq d_1(b, c) \leq d_1(a, d) + w(a, b) + w(c, d)$. Note that $\Delta(a, d) = d_1(a, d) - w(a, b) - w(c, d) - w(b, c)$ while for $\Delta(b, c)$, we have $d_1(a, d) - w(a, b) - w(c, d) - w(b, c) \leq \Delta(b, c) \leq d_1(a, d) + w(a, b) + w(c, d) - w(b, c)$. Thus, we observe that $\Delta(b, c) \geq \Delta(a, d)$. The distance change would be equal only if shortest path between a and d contained the shortest path between nodes b and c in graph G_1. The changed shortest distance paths in graph G_2 would make use of one or more of the new edges. Using the above example, we can show that the distance change between the endpoints of any new edge would always be greater than or equal to distance change between any other node pair. As shown in Section 5 this is generally true in a wide variety of scenarios, though it may not be true across all network structures.

Algorithm 2. Incidence Algorithm

1: Input: Graphs $G_1(V_1, E_1)$ and $G_2(V_2, E_2)$, Active node set V'.
2: HEAP h ← ϕ
3: **for** every node $n \in V'$ **do**
4:     Run Dijkstra Algorithm from n on G_1 and G_2.
5:     **for** every node $v \in V_1 \cap V_2$ **do**
6:         h.insert($(n, v), \Delta(n, v)$) (Regularly clean heap to control size)
7:     **end for**
8: **end for**
9: **return** top-k pairs (u, v) with maximum $\Delta(u, v)$

While the *Incidence Algorithm* provides a decent first-approximation, it is rather naive in its approach. To improve the accuracy, we can consider active node set as the seed set and try to expand from this seed set to include neighbors of nodes in active node set. This expanded seed set is used for determining the source nodes for the different runs of the Dijkstra shortest path algorithm. The goal is to consider only promising neighbors of promising nodes from the active node set for expansion. We will discuss below how this selective expansion is performed.

3.3 Selective Expansion of Active Node Set V'

Without any expansion from the active node set, we just detect the epicenters of shortest path distance changes. However, we may wish to find out node pairs that surround these epicenters too. We use the following method for selective expansion. We first run Dijkstra algorithm from each of the currently active nodes and compute the current top-k shortest path distance change node pairs. Within these top-k, we look at the active nodes and can expand from them. However, we would like to expand one node at a time. Hence, we need to rank the neighbors of the currently active nodes. We would select the neighbor with the highest rank as the source node for the next run of the Dijkstra algorithm. The rank of a neighbor node (say a in Figure 2 which is a neighbor of node b where $a \notin V'$ and $b \in V'$) should depend on the probability that a large number of shortest paths from this node would use the edge (a, b).

Thus, the rank of a neighbor a would be computed as $rank(a) = \frac{I(edge(a,b))}{\sum_{x \in nbr(a)} I(edge(a,x))}$. Then we simply choose the node with the maximum rank and use it as the source node for running Dijkstra algorithm. We update the top-k node pairs using the new node pairs obtained from the latest Dijkstra run. Node a also becomes active. If the top-k node pair list changes, we choose a new node again, else we terminate the selective expansion.

4   Node Ranking for Improved Efficiency

The *Incidence Algorithm* with selective expansion helps us to obtain the top-k node pairs with high accuracy. However, it is still not quite efficient. If the snapshots of the graph are taken after long intervals, new edges would be incident on a large percentage of the nodes in the graph. The efficiency of our *Incidence Algorithm* is dependent upon the size of this set. Hence, when solving the problem over snapshots of a graph taken over longer time intervals, the *Incidence Algorithm* would be computationally expensive. In this section, we discuss strategies to rank the nodes in the active node set so that the order helps to select the top few nodes which can be used to run single-source based shortest path algorithms and capture the maximum distance change pairs. The trade-off is between the number of nodes selected and accuracy. The goal is to rank the nodes, so that by processing them one by one we obtain more accurate results by running shortest path algorithms from a very small number of source nodes. In the following, we discuss some of the ranking strategies that we used in order to achieve this goal.

4.1   Edge Weight Based Ranking (EWBR)

A node has a higher probability of contributing to the top-k node pairs with maximum shortest distance path change if a large number of *new* low-weight edges are incident on this node. The higher the number of new edges incident on the node, the greater the likelihood that the distances of the other nodes to this node have changed. Of course, such an approach is quite simple, and may have its own drawbacks. For example, an edge contributes only a small part to the shortest path length. So, for graphs with longer shortest path distances, a path would typically consist of a large number of edges and hence the greedy approach of just considering the first edge in that path may not be sufficient. This leads to low node ranking accuracy.

4.2   Edge Weight Change Based Ranking (EWCBR)

A node with a large number of edges whose weight has changed by relatively larger amounts is more important. If the edge weight decreases by a large amount, this edge would naturally become a part of more number of shortest paths. We note that the weight of an edge corresponds to the distance along it. Of course, this distance could be defined differently for different applications. For example, for a co-authorship application, the distance could be the inverse of the collaboration frequency. For edges where one of the nodes was not present in the previous snapshot, change in similarity (1/weight) is set to similarity in the new snapshot. This essentially implies that the similarity in the old snapshot is considered as 0.

4.3 Importance Number Based Ranking (INBR)

The previous algorithm does not distinguish between the different kinds of edges. As discussed earlier, the edge importance is estimated as its likelihood of occurring along a shortest path. A node has a higher probability of contributing to the top-k node pairs with maximum shortest distance path change, if a large number of new (or weight-changed) important edges are incident on this node. The importance is measured on the new snapshot. Thus, ranking nodes in this order and considering the top few nodes would ensure that we are capturing the effect of most of the important edges. If importance numbers follow a power law, then contributions to distance changes by the tail of this edge ordering should be minimal. Therefore, the consideration of only a top few should provide us high accuracy.

4.4 Importance Number Change Based Ranking (INCBR)

If an edge has a lower importance number in the old graph snapshot and now its importance number has increased a lot, it implies that the edge is important with respect to our task. Similarly, edges with high edge importance scores in old snapshot and low scores for the new snapshot are also interesting. Active nodes with large number of such new or weight-changed edges become important. Note that the importance numbers in the old snapshots for edges that are completely new (i.e., not just weight changes) is considered to be 0.

4.5 Ranking Using Edge Weight and Importance Numbers (RUEWIN)

To rank a node, we can use both the number of important edges and the weight of the new or weight-changed edges incident on a node. Apart from the absolute values of the two quantities, we can also use the actual change in the quantities for node ranking. RUEWIN uses change in weight multiplied by absolute values of importance numbers in the new snapshot for ranking, while RUEWINC uses change in edge weight multiplied by change in importance numbers for ranking.

4.6 Clustering Based Ranking (CBR)

Intuitively a new inter-cluster edge with low weight would be more important in reducing the shortest path distance between a pair of nodes compared to an intra-cluster edge. This is because nodes within a cluter are already well connected, and nodes across clusters have high likelihood to be used in a shortest path. In other words, an edge which connects two regions of high density is more important than an edge which connects nodes within a high density region.

In this scheme of ranking, we first partition the graph using a minimum cut based algorithm. We use METIS [16] for this purpose. From each of the partitions, we randomly choose one of the nodes (called the representative node of the cluster) on which at least one new edge is incident. If we do not perform partitioning, we can randomly select initial nodes from the entire set of nodes on which new edges are incident. We call the approach with partitioning as CBRP and the one without partitioning as CBR. The Dijkstra algorithm is run with the representative node as the source node. Other nodes on which new edges are incident are assigned to clusters whose representative nodes are the closest to the current node.

Now, we need to estimate the change in distance that occurs between two nodes because of a new edge and use this measure as the importance of the new edge. Inter-cluster distance is computed as the distance between the representative nodes of the two clusters. Distance change is computed as the estimated old distance between the pair of nodes on which the new edge is incident minus the weight of the new edge. The old distance between the two nodes corresponding to an inter-cluster edge is estimated as the sum of the distances of the nodes from the representative nodes of their respective clusters and the inter-cluster distance between the corresponding clusters. We compute the old distance corresponding to an intra-cluster edge in the same way, except that both the nodes belong to the same cluster. Finally, the cluster based score of a node is determined as the sum of the importance of the new edges incident on the node. Note that in this method, the estimation of the old distance for an intra-cluster edge can be very inaccurate. The accuracy of the estimate depends on the size of the cluster. If the cluster has a high radius, the estimate would be bad. However, the intra-cluster edges are not really important for finding the top-k node pairs for which maximum shortest path distance change has occurred, unless they belong to clusters with very large radius. We experimented and noticed that relative estimates of shortest path distances are quite similar to actual distances. Also, we observed that the radius of the clusters are comparatively smaller when we use partitioning based method to choose the initial set of source nodes rather than making a random choice from all over the graph. Lower radius of clusters means better estimates of shortest path distances between nodes in the old graph snapshot, which leads to better accuracy.

The preprocessing steps involved in this form of ranking are: (1) A graph partitioning of the old snapshot using METIS (2) c Dijkstra algorithm runs where c is the number of clusters. Also note that, for an edge that connects an old node to a new node, we cannot use the above logic to estimate the distance saved. Hence, we set their cluster based scores to 0.

5   Experimental Results

The goal of the section is to show the practical usability and efficiency of the method in large scale systems at the expense of a modest loss of accuracy. In order to provide further insights, we also present several intermediate results which show that in many real data sets, a recurring theme is that there are often only few node pairs which share the bulk of the changes in shortest path distances. This suggests that our ranking approach for finding these nodes is likely to be efficient and accurate.

5.1 Datasets

We used the *DBLP co-authorship graphs*[5], *IMDB co-starring graphs*[6] and *Ontario road network graphs*[7]. The nodes for the *DBLP co-authorship graphs* are authors and the edges denote the collaborations. The edge weight is the reciprocal of the co-authorship frequency. We used five pairs of the co-authorship graphs from 1980 to 2000 for testing purposes. The nodes for the *IMDB co-starring graphs* are actors and actresses and edges denote the collaborations. The edge weight is the reciprocal of the co-starring frequency. We consider the co-starring graphs from 1950 to 1952. The nodes for the *Ontario Road Network* are road intersections and edges denote the road segments. The Geography Markup Language (GML) files for this road network for 2005 and 2008 were obtained from the Canada Statistics website[8]. The dataset are defined in terms of latitudes and longitudes, which were converted into edge lengths for the purpose of the algorithm. Edges from 2005, which were missing from the 2008 snapshot, were added to the data set.

We provide details of the characteristics of these graphs in Tables 1, 2 and 3. In the table, the set of edges with change in weight is denoted by E_{1C}(**c=change**). The set of edges which are absent in G_1 and for which one of the nodes was in G_1 is denoted by E_{2ON}(**o=old, n=new**). The set of edges which were absent in G_1 but both the nodes were present in G_1 is denoted by E_{2OO}.

Table 1. *DBLP* Details

Year	Nodes in LCC	Edges	E_{1C}	E_{2ON}	E_{2OO}	$|V'|$	Max freq
1980	4295	8096					45
1981	5288	10217	607	953	305	895	46
1984	10598	21592					52
1985	13025	27260	1272	2271	616	2128	56
1988	22834	50255					61
1989	27813	61704	3282	5016	1921	4963	63
1993	61592	148287					87
1994	74794	183738	10995	15446	6569	14633	111
1999	163203	456954					221
2000	188048	539957	32876	39542	21004	35723	231

Table 2. *IMDB* Details

Year	Nodes in LCC	Edges	E_{1C}	E_{2ON}	E_{2OO}	$|V'|$	Max freq
1950	144991	8.2M					739
1951	149260	8.6M	111097	74757	179698	16364	745
1952	154719	9.1M	118276	146661	207859	17596	745

Table 3. *Ontario RN* Details

Year	Nodes in LCC	Edges	E_{1C}	E_{2ON}	E_{2OO}	$|V'|$	Max freq
2005	348236	455804					250
2008	367628	494067	2280	11870	9041	29539	250

We note that the *DBLP* and *IMDB* graphs are naturally temporal, and therefore a *cumulative graph* can be defined which aggregates all edges upto time t in order to create a graph. The change is then computed between two such cumulative graphs. For example, a *DBLP* graph at year 1980 captures all the co-authorship relationships in

---

[5] http://www.informatik.uni-trier.de/~ley/db/

[6] http://www.imdb.com/

[7] http://geodepot.statcan.gc.ca/

[8] http://tinyurl.com/yfsoouu

DBLP until the year 1980. In the case of the *Ontario road network*, the two snapshots in 2005 and 2008 were used in order to determine the change.

The ground truth (also referred to as the "golden set") in terms of the node-pairs with maximum change in distance value was determined. We used the fast implementation mentioned in [9] to generate this golden set. As mentioned earlier, this is the alternative (but brute-force method) for determine the precise node pairs with the maximum distance changes. It is important to note that *this process required several days on multiple CPUs to compute the exact distance changes using direct applications of shortest path algorithms.* Our computational challenges in determining the ground truth is itself evidence of the difficulty of the problem.

5.2   Evaluation Methodology

We compute the accuracy of the results by computing the top-k pairs with the greatest change in the original data (also known as the ground truth or golden set), and comparing it with the results of our algorithm. We refer to the fraction of such matching node pairs as the top$K$ Accuracy. We will show the tradeoffs between the running time and the top$K$ Accuracy achieved by the method. Since the core of our algorithm is to rank and select (the top-ranking) nodes for use as source nodes, one proxy for the time spent is the number of nodes from which the shortest path algorithm needs to be executed. Hence, for every ranking method, we plot the top$K$ Accuracy against the number of source nodes. The *area under such a curve* is also a proxy for the average lift achieved with the use of a small number of source nodes. Thus, we compare various ranking algorithms with respect to these areas. Before, describing these results in detail, we will provide some intermediate results which provide some interesting insights.

5.3   Typical Distribution of Maximum Distance Change Values

We first show that much of the change in distance values is *large for a small number of node pairs* in the real data sets tested. Figure 3 shows the distribution of the top 1000 distance change values for each of the five *DBLP* snapshots. We plot the distance change values on the Y-axis, whereas the rank of the node pair (with greatest change) is illustrated on the X-axis. Figure 4 shows the same results for the *IMDB* data set for snapshot changes across the years 1950-1951 and 1951-1952. We note that the distribution of the change values is quite skewed. Only the top few distance change values are very large. This suggests that there are only a few node pairs across which most of the changes were concentrated. This was a recurring theme through the real data sets encountered. The rarity of source nodes reflecting high changes makes it possible to discover them with the use of ranking methods. In the next subsection, we will examine this issue from the perspective of specific nodes (as opposed to node pairs).

5.4   Contributions by Individual Nodes

Next, we examine the *involvement of specific nodes* in these pairs of nodes between which most of the distance change is concentrated. Table 4 shows the number of unique nodes in top 1000 maximum distance change node pairs. The results show that a small number of nodes ($\sim$250 on the average, and not increasing significantly with graph size) are a part of the top 1000 maximum distance change pairs. This provides evidence that the ranking approach should approximate the top-k maximum shortest distance

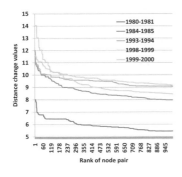
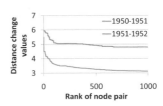

Fig. 3. Variation of shortest path distance changes for *DBLP*

Fig. 4. Variation of shortest path distance changes for *IMDB*

Table 4. Number of nodes in top 1000 maximum distance change node pairs

DBLP Snapshots	#nodes
1980-1981	294
1984-1985	267
1988-1989	112
1993-1994	150
1999-2000	179

IMDB Snapshots	#nodes
1950-1951	343
1951-1952	152

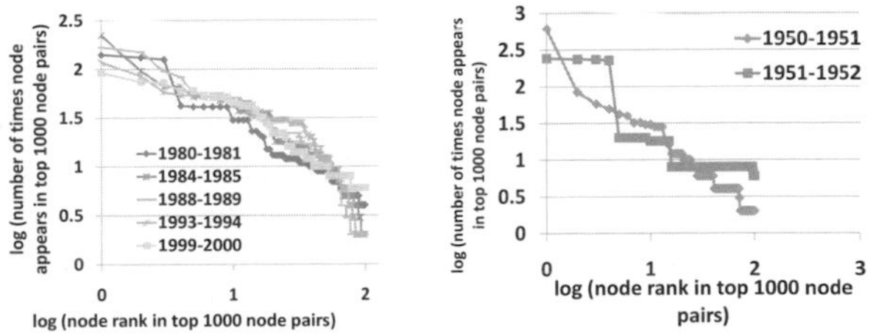

(a) DBLP   (b) IMDB

Fig. 5. A few nodes are responsible for the top few maximum distance change node pairs

change node pairs well, because most of the distances are concentrated in a few nodes. The trends in this distribution are also illustrated graphically in Figures 5(a) and 5(b) respectively.

5.5 Accuracy Using the Incidence Algorithm and Selective Expansion

Table 5 and Table 6 show the accuracy of our Incidence Algorithm for the *DBLP* and *IMDB* graphs respectively. Each row shows the accuracy for a particular data set over different values of k. Note that even without any selective expansion, the algorithm performs quite well and almost always determines the top five node pairs accurately.

Table 7 shows the accuracy of our Incidence Algorithm with selective seed set expansion for the *DBLP* graphs. Note that selective expansion improves accuracy significantly compared to the Incidence Algorithm, and the difference is especially significant for large values of k.

Table 5. Accuracy of Incidence Algorithm (*DBLP*)

Snapshots	K=1	K=5	K=10	K=50	K=100	K=500
1980-1981	1	0.8	0.8	0.8	0.86	0.638
1984-1985	1	1	0.9	0.92	0.8	0.776
1988-1989	1	1	1	0.76	0.84	0.784
1993-1994	1	1	1	1	0.76	0.734
1999-2000	1	1	0.8	0.62	0.69	0.86

Table 6. Accuracy of Incidence Algorithm (*IMDB*)

Snapshots	K=1	K=5	K=10	K=50	K=100	K=500
1950-1951	1	1	1	1	0.93	0.982
1951-1952	1	1	1	0.94	0.8	0.57

Table 7. Accuracy of the Incidence Algorithm with selective expansion (*DBLP*)

Snapshots	K=1	K=5	K=10	K=50	K=100	K=500
1980-1981	1	0.8	0.8	0.8	0.88	0.84
1984-1985	1	1	1	1	1	1
1988-1989	1	1	1	0.76	0.84	0.896
1993-1994	1	1	1	1	0.76	0.734
1999-2000	1	1	1	0.8	0.69	0.86

5.6   Accuracy of Edge Importance Numbers

We note that the edge importance numbers are *estimated* in order to design the change detection algorithm. For the change detection algorithm to work well, the edge importance numbers must also be estimated accurately. Therefore, in this section we will show that this intermediate result on importance numbers is estimated accurately as well. Later, we will show the final quality of the algorithm with the use of these importance numbers. We note that the precise value of the importance numbers can be estimated accurately, provided that we are willing to spend the time required to run the shortest path algorithm from all the nodes. These precise values are compared with the estimated values of the importance numbers. In order to do so, we compute the importance numbers with a varying level of sampling. Specifically, we vary the number of source nodes from 100 to 1000 and also the number of shortest path trees per source node as 100 and 1000. We report comparisons of some of these with the precise values of the importance numbers.

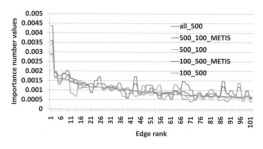

Fig. 6. Convergence of the top 100 edges

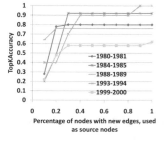

Fig. 7. Performance of CBRP for *k*=50 (*DBLP*). Nodes with new edges ≈ 20% of total nodes

In our algorithm, the importance number was estimated in two ways: (a) The source nodes were sampled randomly from the entire graph (b) The graph was partitioned with the METIS graph partitioning algorithm [16] into a number of different partitions, which was varied from 100 to 500. A node is then randomly selected from each partition as a source node for edge importance number computation.

We note that only the top few important edges are used in the computation process. Therefore, we are mostly interested in the convergence of the importance numbers of the top few edges. Therefore, we will show the results for the top 100 most important edges. The results are illustrated in Figure 6. Here, we plot the edge importance numbers on Y-axis for each of the edges on the X-axis. Note that all of the curves follow the trend which is quite similar to that of the curve corresponding to precise computations (all_500). It is evident from the results that the importance number values converge quite well for the top few edges. This is especially true, when the clustering-based algorithm is used for the sampling process.

5.7   Accuracy Comparison of Ranking Algorithms

We note that a ranking algorithm is effective, if we can run Dijkstra algorithm from a very small number of top ranked nodes and still achieve good accuracy for the maximum shortest path distance change node pairs. We executed each of our ranking algorithms for the different data sets, for values of k ranging from 1 to 500. We varied the number of source nodes used for Dijkstra runs in order to test the sensitivity. An example of such a curve is shown in Figure 7, in which we show the curves for $k=50$ for the *DBLP datasets* using the *CBRP* ranking algorithm. On the X-axis, we use the percentage of nodes from the original graph (from all nodes having at least one new edge incident on them), which are used as a source node for Dijkstra runs. We note that the area under such a curve is good proxy for the effectiveness of our algorithm, because a more effective algorithm would provide a higher accuracy for a small percentage of samples nodes, and provide a considerable amount of "lift". Therefore, we can compute the average area under the curve as a measure of the goodness of the algorithm. For importance number estimation, we used the METIS-based selection of 100 random nodes and 100 random shortest path trees per source node for all the graphs except for the road network. In the case of the *Ontario RN* data set, we used 20 random nodes and just one shortest path tree per source node as the road network is quite sparse.

Figure 8 shows[9] the accuracy comparisons of our different algorithms for the *DBLP data set*. In addition, we added a few baselines. Specifically, the algorithm labeled as *Random* refers to the case were nodes were randomly chosen from the entire graph for running the Dijkstra algorithm. The algorithm labeled as *RandomNWNE* refers to the algorithm in which the nodes were randomly chosen only from the set of nodes on which new edges were incident between two snapshots. We ran each of the algorithms ten times and reported the average values in order to provide statistical stability to the results. From the results, it is evident that the most simplistic algorithm (or *EWBR* algorithm), which considers only the weight of the new edges incident on a node, performs quite poorly. The algorithm which uses importance numbers (or *INBR* algorithm) turns out to be much more accurate. However, the algorithm which considers change in importance numbers (or *INCBR* algorithm) as well is more effective than either of the

---

[9] The acronym for each algorithm in the figures may be found in the section in which these algorithms were introduced.

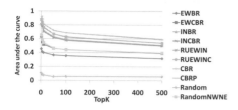
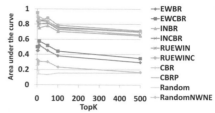

Fig. 8. Accuracy Comparison of Ranking Algorithms:averaged across five pairs of *DBLP* snapshots

Fig. 9. Accuracy Comparison of Ranking Algorithms (*IMDB*): averaged across two pairs of snapshots

previous two algorithms. However, the most superior performance is achieved by the clustering based algorithm (*CBR*). It is interesting to see that the initial selection of the nodes with the use of graph partitioning in *CBRP* provides marginal improvement in accuracy over the variant *CBR* which does not.

Figures 9 and 10 shows the accuracy comparisons of our different algorithms for the *IMDB* and the *Ontario RN* data sets. The results are quite similar to the previous case with some exceptions as noted well.

One difference is that for graphs such as the road network, our clustering based approach does not work well. This is because a number of the edge changes involve new *nodes*. The impact of such nodes is hard to estimate with the use of the clustering algorithm.

5.8   Time Comparison of Ranking Algorithms

In this section, we will show the time-comparison of the different ranking algorithms. We note that each ranking algorithm provides a tradeoff between accuracy and efficiency, since the ranking process provides an *order* to the sampling of these nodes. By picking more nodes along this order, it is possible to improve the effectiveness of the algorithm at the expense of efficiency. In this section, we will explore this tradeoff. Results for the *DBLP* data set (1988-1989) are illustrated in Figure 11. We used $k = 50$ to test the accuracy.

Notice that the individual points on each curve are plotted at intervals of 5% of the nodes with new edges. Finally, all the algorithms reach a maximum accuracy of 0.76 reachable by our Incidence Algorithm on this data set for $k = 50$. The importance number based algorithms (*INBR* and *INCBR*) take some time for the initial estimation of the importance numbers. *INCBR* computes importance numbers for both the old and new snapshots and so its initialization time is approximately twice that of *INBR* which estimates importance numbers for the new snapshot only. The same relationship is true between the *RUEWIN* and *RUEWINC* algorithms as well. The clustering based algorithms (*CBR* and *CBRP*) require some additional time for the initial Dijkstra runs from the representative nodes of the clusters and for finding the nearest cluster for all other nodes with new edges. The *CBRP* algorithm requires some extra time for the initial selection of the nodes with the use of graph partitioning.

We notice that the clustering based ranking technique is the most effective from an overall perspective, and is able to provide almost the same amount of accuracy with the use of just 10% of the time required by the Incidence Algorithm. This is because

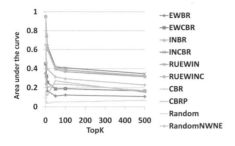

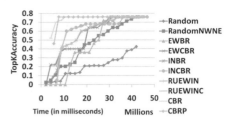

Fig. 10. Accuracy Comparison of Ranking Algorithms (Road network)

Fig. 11. Time vs Accuracy Comparison of Ranking Algorithms (*DBLP*)

the ranking process is superior to the incidence approach from an algorithmic design perspective, especially when most of the changes are concentrated in a few nodes.

5.9 Case Studies

In this section, we provide a number of insights gained from exploration of the maximum distance change pairs. One interesting observation was that *highly ranked nodes often also had a high level of change in the structural centrality*. For the *DBLP* dataset, Table 8 shows the authors that appeared the maximum number of times in the top 1000 node pairs with maximum shortest path distance change. Such authors are the ones who never published with popular authors before, and were therefore at the fringes of the co-authorship graph. However, their (or their neighbor's) subsequent participation in publications with popular authors resulted in increasing their structural centrality. In some cases, these were authors who were working in a single research area who recently started collaborating with authors in other research areas.

For example, let us consider the node (author) *Nuno F. Paulino*. The author published with two other authors in 1993. The corresponding publications were also early publications for these authors. In 1994, Nuno published a paper with five other authors including *Gabor C. Temes* who had been publishing since 1971. As a result of this new paper in 1994, Nuno became more central in the graph, and also appeared in the list of top ranking nodes which were among the sources of greatest change in the graph.

Table 8. Nodes with Greatest Change (*DBLP*)

Graphs	Author
1980-1981	Seppo Sippu
1984-1985	W. Eric L. Grimson
1988-1989	Michael Walker
1993-1994	Nuno F. Paulino
1999-2000	R. Böhm

Table 9. Author-pairs with Greatest Change

Graphs	Author1	Author2
1980-1981	B. Krishnamoorthi	Reginald Meeson
1984-1985	W. Eric L. Grimson	Zachary C. Fluhr
1988-1989	Michael Walker	Cees J. A. Jansen
1993-1994	A. D. Bray	Michael Smyth
1999-2000	Jitka Dupacová	Tiecheng Yan

We also present node pairs with greatest change over different graph snapshots in Table 9. These are typically pairs of authors who belong to two different research areas but recently published a paper together, or published with the same common cluster of nodes. E.g., the nodes *Tiecheng Yan* and *Jitka Dupacová* were typically on the fringes of the co-authorship graph. However, *Tiecheng Yan* published a paper with *Stein W.*

Table 10. Star pairs with the highest change in shortest path distance

Graphs	Actor1	Actor2
1950-1951	Niculescu-Bruna, Ion	Conabie, Gheorghe
1950-1951	Onea, Arion	Conabie, Gheorghe
1950-1951	Niculescu-Bruna, Ion	Scarlatescu, N.
1951-1952	Sanders, Kerry (II)	Hoffman, Dustin
1951-1952	Sanders, Kerry (II)	Gordon, Mark (I)

Table 11. Nodes with greatest change over time period '05-'08

Longitude,Latitude	Location
78.7559W,45.3429N	Clayton Lake, near Dorset
78.7343W,45.3464N	Bear Lake Road, Near Livingstone Lake
78.7024W,45.3112N	Kawagama Lake, near Bear island
78.667W,45.3273N	Near Kimball Lake
78.6956W,45.3216N	Near Kawagama Lake

Table 12. Loc pairs with greatest change ('05-'08)

Location1	Location2
78.704W,45.3321N	78.692W,45.3173N
78.6995W,45.3222N	78.692W,45.3173N
78.6995W,45.3222N	78.692W,45.3173N
78.7033W,45.3347N	78.6956W,45.3216N
78.704W,45.3321N	78.692W,45.3173N

Fig. 12. Representation of locations in Table 11 and 12

Wallace in 1993, and one with Jitka Dupacová in 2000. While the two authors Jitka and Tiecheng were originally located on different corners of the graph, this authorship behavior brought them close to one another.

The results for the *IMDB data set* are illustrated in Table 10. The *IMDB* graph was quite dense with many cliques. This is because many actors may work in the same movie, and this adds a large clique of edges in the graph. For example, *Sanders Kerry (II)* acted as a correspondent for the TV series "Today" in 1952. IMDB lists 938 people as related to this TV series. Furthermore, in 1947, he was involved in the "Meet the Press" show. These factors, resulted in his becoming a node with a large change in centrality. This was also reflected in the fact that it was identified as a top-ranked change node.

The results for *Ontario Road Network* are illustrated in Tables 11 and 12. Locations[10] in Table 11 are marked on the map (Google Maps[11]) in Figure 12 (left). From the results, we can see that the shortest path distance between many locations changed in the area situated in the south west part of Algonquin Provincial Park. We looked back at the road networks and discovered that this happened because of the construction of an extension to the Bear Lake Road during the time period from 2005 to 2008. We mark

---

[10] http://tinyurl.com/mg-mapOne

[11] http://maps.google.com

the corresponding area[12] in Figure 12 (right). Here points A, E correspond to column 3 of Table 12 and points B, C and D correspond to column 2 of Table 12.

From the above case studies, it is evident that our approach can discover interesting node pairs for which shortest path distance change was maximum. Quite often these changes are because of the combinatorial effect of multiple new edges that come in the network. Our algorithms can also discover the most critical edges which result in a significant part of the distance changes. This also helps provide a better understanding of the causality of the changes in the distances.

6   Conclusions and Future Work

In this paper, we presented several fast algorithms to compute top-k node pairs with the greatest evolution in shortest path distances. We experimented using large graphs and showed the efficiency and scalability of our methods. We observed that the change in edge importance number and clustering based methods work quite well for the task. One advantage of the approach is that it was able to create effective results over extremely large data sets in which it was not possible to use traditional methods efficiently. In future work, we will perform investigation of specific structural aspects of graphs which are related to considerable distance change. We will also exploit these algorithms for a variety of event-detection applications in networks.

Acknowledgements.   Research was sponsored in part by the U.S. National Science Foundation under grant IIS-09-05215, and by the Army Research Laboratory under Co-operative Agreement Number W911NF-09-2-0053 (NS-CTA). The views and conclusions contained in this document are those of the authors and should not be interpreted as representing the official policies, either expressed or implied, of the Army Research Laboratory or the U.S. Government. The U.S. Government is authorized to reproduce and distribute reprints for Government purposes notwithstanding any copyright notation here on.

We thank Geography Div, Stats Canada, 2005/ 2008 Road Network File (RNF), 92-500-XWE/XWF. We thank the anonymous reviewers for their insightful comments.

References

1. Aggarwal, C.C. (ed.): Social Network Data Analytics. Springer, Heidelberg (2011)
2. Ahuja, R.K., Magnanti, T.L., Orlin, J.B.: Network Flows: Theory, Algorithms, and Applications. Prentice Hall, Englewood Cliffs (1993)
3. Ahuja, R.K., Orlin, J.B., Pallottino, S., Scutella, M.G.: Dynamic shortest paths minimizing travel times and costs. Networks 41, 205 (2003)
4. Henzinger, M., King, V.: Maintaining minimum spanning forests in dynamic graphs. SIAM Journal on Computing 31(2), 364–374 (2001)
5. Akoglu, L., McGlohon, M., Faloutsos, C.: oddball: Spotting Anomalies in Weighted Graphs. In: Zaki, M.J., Yu, J.X., Ravindran, B., Pudi, V. (eds.) PAKDD 2010. LNCS, vol. 6119, pp. 410–421. Springer, Heidelberg (2010)
6. Chabini, I., Ganugapati, S.: Parallel algorithms for dynamic shortest path problems. International Transactions in Operational Research 9(3), 279–302 (2002)

---

[12] http://tinyurl.com/mg-mapTwo

7. Chakrabarti, D., Kumar, R., Tomkins, A.: Evolutionary clustering. In: KDD, pp. 554–560 (2006)
8. Chakraborty, D., Chakraborty, G., Shiratori, N.: A dynamic multicast routing satisfying multiple qos constraints. Int. J. Network Management 13(5), 321–335 (2003)
9. Cherkassky, B.V., Goldberg, A.V., Silverstein, C.: Buckets, heaps, lists, and monotone priority queues. In: SODA (1997)
10. Cohen, E.: Estimating the size of the transitive closure in linear time. In: FOCS (1994)
11. Dijkstra, E.W.: A note on two problems in connexion with graphs. Numerische Mathematik 1, 269–271 (1959)
12. Doreian, P., Stokman, F.: Evolution of social networks. Gordon and Breach Publishers, New York (1997)
13. Eberle, W., Holder, L.: Discovering structural anomalies in graph-based data. In: ICDMW, pp. 393–398 (2007)
14. Floyd, R.W.: Algorithm 97: Shortest path. Commun. ACM 5(6), 345 (1962)
15. Johnson, D.B.: Efficient algorithms for shortest paths in sparse networks. J. ACM 24(1), 1–13 (1977)
16. Karypis, G., Kumar, V.: A fast and high quality multilevel scheme for partitioning irregular graphs. SIAM J. Sci. Comput. 20(1), 359–392 (1998)
17. Leskovec, J., Kleinberg, J., Faloutsos, C.: Graph evolution: Densification and shrinking diameters. ACM TKDD 1(1), 2 (2007)
18. Papadimitriou, P., Dasdan, A., Garcia-Molina, H.: Web graph similarity for anomaly detection (poster). In: WWW, pp. 1167–1168 (2008)
19. Roditty, L., Zwick, U.: On dynamic shortest paths problems. In: Albers, S., Radzik, T. (eds.) ESA 2004. LNCS, vol. 3221, pp. 580–591. Springer, Heidelberg (2004)
20. Shimbel, A.: Structural parameters of communication networks. Bulletin of Mathematical Biology 15, 501–507 (1953)
21. Warshall, S.: A theorem on boolean matrices. J. ACM 9(1), 11–12 (1962)
22. Zhu, S., Huang, G.M.: A new parallel and distributed shortest path algorithm for hierarchically clustered data networks. IEEE Trans. Parallel Dist. Syst. 9(9), 841–855 (1998)
23. Newman, M.E.J., Girvan, M.: Finding and evaluating community structure in networks. Physical Review E 69 (2004)
24. Brandes, U.: A faster algorithm for betweenness centrality. Journal of Mathematical Sociology 25(2), 163–177 (2001)
25. Ding, B., Yu, J.X., Qin, L.: Finding time-dependent shortest paths over large graphs. In: EDBT, pp. 205–216 (2008)
26. Chon, H.D., Agrawal, D., Abbadi, A.E.: FATES: Finding A Time Dependent Shortest path. Mobile Data Management, 165–180 (2003)

FAST: A Generic Framework for Flash-Aware Spatial Trees

Mohamed Sarwat[1,*], Mohamed F. Mokbel[1,*], Xun Zhou[1], and Suman Nath[2]

[1] Department of Computer Science and Engineering, University of Minnesota
[2] Microsoft Research
{sarwat,mokbel,xun}@cs.umn.edu, sumann@microsoft.com

Abstract. Spatial tree index structures are crucial components in spatial data management systems, designed with the implicit assumption that the underlying external memory storage is the conventional magnetic hard disk drives. This assumption is going to be invalid soon, as flash memory storage is increasingly adopted as the main storage media in mobile devices, digital cameras, embedded sensors, and notebooks. Though it is direct and simple to port existing spatial tree index structures on the flash memory storage, that direct approach does not consider the unique characteristics of flash memory, i.e., slow write operations, and erase-before-update property, which would result in a sub optimal performance. In this paper, we introduce FAST (i.e., Flash-Aware Spatial Trees) as a generic framework for flash-aware spatial tree index structures. FAST distinguishes itself from all previous attempts of flash memory indexing in two aspects: (1) FAST is a generic framework that can be applied to a wide class of data partitioning spatial tree structures including R-tree and its variants, and (2) FAST achieves both *efficiency* and *durability* of read and write flash operations through smart memory flushing and crash recovery techniques. Extensive experimental results, based on an actual implementation of FAST inside the GiST index structure in PostgreSQL, show that FAST achieves better performance than its competitors.

1 Introduction

Data partitioning spatial tree index structures are crucial components in spatial data management systems, as they are mainly used for efficient spatial data retrieval, hence boosting up query performance. The most common examples of such index structures include R-tree [7], with its variants [4,9,22,24]. Data partitioning spatial tree index structures are designed with the implicit assumption that the underlying external memory storage is the conventional magnetic hard disk drives, and thus has to account for the mechanical disk movement and its seek and rotational delay costs. This assumption is going to be invalid soon, as flash memory storage is expected to soon prevail in the storage market replacing the magnetic hard disks for many applications [6,21]. Flash memory storage is increasingly adopted as the main storage media in mobile devices and as a storage alternative in laptops, desktops, and enterprise class servers (e.g.,

---

* The research of these authors is supported in part by the National Science Foundation under Grants IIS-0811998, IIS-0811935, CNS-0708604, IIS-0952977, by a Microsoft Research Gift, and by a seed grant from UMN DTC.

D. Pfoser et al. (Eds.): SSTD 2011, LNCS 6849, pp. 149–167, 2011.

in forms of SSDs) [3,13,15,18,23]. Recently, several data-intensive applications have started using custom flash cards (e.g., ReMix [11]) with large capacity and access to underlying raw flash chips. Such a popularity of flash is mainly due to its superior characteristics that include smaller size, lighter weight, lower power consumption, shock resistance, lower noise, and faster read performance [10,12,14,19].

Flash memory is block-oriented, i.e., pages are clustered into a set of blocks. Thus, it has fundamentally different characteristics, compared to the conventional page-oriented magnetic disks, especially for the write operations. First, write operations in flash are slower than read operations. Second, random writes are substantially slower than sequential writes. In devices that allow direct access to flash chips (e.g., ReMix [11]), a random write operation updates the contents of an already written part of the block, which requires an expensive block erase operation[1], followed by a sequential write operation on the erased block; an operation termed as *erase-before-update* [5,12]. SSDs, which emulate a disk-like interface with a *Flash Translation Layer* (FTL), also need to internally address flash's erase-before-update property with logging and garbage collection, and hence random writes, especially *small* random writes, are significantly slower than sequential writes in almost all SSDs [5].

Though it is direct and simple to port existing tree index structures (e.g., R-tree and B-tree) on FTL-equipped flash devices (e.g., SSDs), that direct approach does not consider the unique characteristics of flash memory and therefore would result in a suboptimal performance due to the random writes encountered by these index structures. To remedy this situation, several approaches have been proposed for flash-aware index structures that either focus on a specific index structure, and make it a flash-aware, e.g., flash-aware B-tree [20,26] and R-tree [25], or design brand new index structures specific to the flash storage [2,16,17].

Unfortunately, previous works on flash-aware search trees suffer from two major limitations. First, these trees are specialized—they are not flexible enough to support new data types or new ways of partitioning and searching data. For example, FlashDB [20], which is designed to be a B-Tree, does not support R-Tree functionalities. RFTL [25] is designed to work with R-tree, and does not support B-tree functionalities. Thus, if a system needs to support many applications with diverse data partitioning and searching requirements, it needs to have multiple tree data structures. The effort required to implement and maintain multiple such data structures is high.

Second, existing flash-aware designs often show trade-offs between efficiency and durability. Many designs sacrifice strict durability guarantee to achieve efficiency [16,17,20,25,26]. They buffer updates in memory and flush them in batches to amortize the cost of random writes. Such buffering poses the risk that in-memory updates may be lost if the system crashes. On the other hand, several designs achieve strict durability by writing (in a sequential log) all updates to flash [2]. However, this increases the cost of search for many log entries that need to be read from flash in order to access each tree node [20]. In summary, no existing flash-aware tree structure achieves both strict durability and efficiency.

---

[1] In a typical flash memory, the cost of *read*, *write*, and *erase* operations are 25, 200, and 1500 μs, respectively [3].

In this paper, we address the above two limitations by introducing FAST; a framework for Flash-Aware Spatial Tree index structures. FAST distinguishes itself from all previous flash-aware approaches in two main aspects: (1) Rather than focusing on a specific index structure or building a new index structure, FAST is a generic framework that can be applied to a wide variety of tree index structures, including B-tree, R-tree along with their variants. (2) FAST achieves both efficiency *and* durability in the same design. For efficiency, FAST buffers all the incoming updates in memory while employing an intelligent *flushing policy* that smartly evicts selected updates from memory to minimize the cost of writing to the flash storage. In the mean time, FAST guarantees durability by sequentially logging each in-memory update and by employing an efficient *crash recovery* technique.

FAST mainly has four modules, *update*, *search*, *flushing*, and *recovery*. The *update* module is responsible on buffering incoming tree updates in an in-memory data structure, while writing small entries sequentially in a designated flash-resident log file. The *search* module retrieves requested data from the flash storage and updates it with recent updates stored in memory, if any. The *flushing* module is triggered once the memory is full and is responsible on evicting flash blocks from memory to the flash storage to give space for incoming updates. Finally, the *recovery* module ensures the durability of in-memory updates in case of a system crash.

FAST is a generic system approach that neither changes the structure of spatial tree indexes it is applied to, nor changes the search, insert, delete, or update algorithms of these indexes. FAST only changes the way these algorithms reads, or updates the tree nodes in order to make the index structure flash-aware. We have implemented FAST within the GiST framework [8] inside PostgrSQL. As GiST is a generalized index structure, FAST can support any spatial tree index structure that GiST is supporting, including but not restricted to R-tree [7], R*-tree [4], SS-tree [24], and SR-tree [9], as well as B-tree and its variants. In summary, the contributions of this paper can be summarized as follows:

- We introduce FAST; a general framework that adapts existing spatial tree index structures to consider and exploit the unique properties of the flash memory storage.
- We show how to achieve efficiency and durability in the same design. For efficiency, we introduce a *flushing policy* that smartly selects parts of the main memory buffer to be flushed into the flash storage in a way that amortizes expensive random write operations. We also introduce a *crash recovery* technique that ensures the durability of update transactions in case of system crash.
- We give experimental evidence for generality, efficiency, and durability of FAST framework when applied to different data partitioning tree index structures.

The rest of the paper is organized as follows: Section 2 gives an overview of FAST along with its data structure. The four modules of FAST, namely, *update*, *search*, *flushing*, and *recovery* are discussed in Sections 3 to 6, respectively. Section 7 gives experimental results. Finally, Section 8 concludes the paper.

2   Fast System Overview

Figure 1 gives an overview of FAST. The original tree is stored on persistent flash memory storage while recent updates are stored in an in-memory buffer. Both parts need to be combined together to get the most recent version of the tree structure. FAST has four main modules, depicted in bold rectangles, namely, *update, search, flushing,* and *crash recovery*. FAST is optimized for both SSDs and raw flash devices. SSDs are the dominant flash device for large database applications. On the other hand, raw flash chips are dominant in embedded systems and custom flash cards (e.g., ReMix [11]), which are getting popular for data-intensive applications.

2.1   FAST Modules

In this section, we explain FAST system architecture, along with its four main modules; (1) Update, (2) Search, (3) Flushing, and (4) Crash recovery. The actions of these four modules are triggered through three main events, namely, *search queries, data updates,* and *system restart*.

Update Module. Similar to some of the previous research for indexing in flash memory, FAST buffers its recent updates in memory, and flushes them later, in bulk, to the persistent flash storage. However, FAST update module distinguishes itself from previous research in two main aspects: (1) FAST does not store the update operations in memory, instead, it stores the *results* of the update operations in memory, and (2) FAST ensures the *durability* of update operations by writing small log entries to the persistent storage. These log entries are written sequentially to the flash storage, i.e., very small overhead. Details of the update module will be discussed in Section 3.

Search Module. The search module in FAST answers point and range queries that can be imposed to the underlying tree structure. The main challenge in the search module is that the actual tree structure is split between the flash storage and the memory. Thus, the main responsibility of the search module

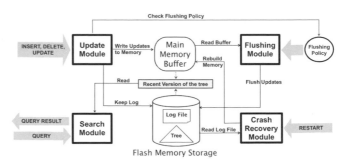

Fig. 1. Tree Modifications Table

is to construct the recent image of the tree by integrating the stored tree in flash with the tree updates in memory that did not make it to the flash storage yet. Details of the search module will be discussed in Section 4.

Flushing Module. As the memory resource is limited, it will be filled up with the recent tree updates. In this case, FAST triggers its flushing module that employs a *flushing* *policy* to smartly select some of the in-memory updates and write them, in bulk, into the flash storage. Previous research in flash indexing flush their in-memory updates or log file entries by writing *all* the memory or log updates once to the flash storage. In contrast, the flushing module in FAST distinguishes itself from previous techniques in two main aspects: (1) FAST uses a *flushing policy* that smartly selects *some* of the updates from memory to be flushed to the flash storage in a way that amortizes the expensive cost of the block erase operation over a large set of random write operations, and (2) FAST logs the flushing process using a single log entry written sequentially on the flash storage. Details of the flushing module will be discussed in Section 5.

Crash Recovery Module. FAST employs a crash recovery module to ensure the durability of update operations. This is a crucial module in FAST, as only because of this module, we are able to have our updates in memory, and not to worry about any data losses. This is in contrast to previous research in flash indexing that may encounter data losses in case of system crash, e.g., [25,26,16,17]. The crash recovery module is mainly responsible on two operations: (1) Once the system restarts after crash, the crash recovery module utilizes the log file entries, written by both the update and flushing modules, to reconstruct the state of the flash storage and in-memory updates just before the crash took place, and (2) maintaining the size of the log file within the allowed limit. As the log space is limited, FAST needs to periodically compact the log entries. Details of this module will be discussed in Section 6.

2.2 FAST Design Goals

FAST avoids the tradeoff of durability and efficiency by using a combination of buffering and logging. Unlike existing efficient-but-not-durable designs [16,17,20,25,26], FAST uses write-ahead-logging and crash recovery to ensure strict system durability. FAST makes tree updates efficient by buffering write operations in main memory and by employing an intelligent flushing policy that optimizes I/O costs for both SSDs and raw flash devices. Unlike existing durable-but-inefficient solutions [2], FAST does not require reading in-flash log entries for each search/update operation, which makes reading FAST trees efficient.

2.3 FAST Data Structure

Other than the underlying index tree structure stored in the flash memory storage, FAST maintains two main data structures, namely, the *Tree Modifications Table*, and *Log File*, described below.

Tree Modifications Table. This is an in-memory hash table (depicted in Figure 2) that keeps track of recent tree updates that did not make it to the flash storage yet. Assuming no hashing collisions, each entry in the hash table represents the modification applied to a unique node identifier, and has the form (*status, list*)

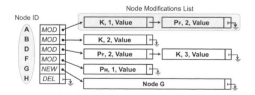

Fig. 2. Tree Modifications Table

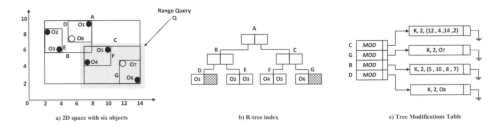

a) 2D space with six objects   b) R-tree index   c) Tree Modifications Table

Fig. 3. An illustrating example for all FAST operations

where *status* is either *NEW*, *DEL*, or *MOD* to indicate if this node is newly created, deleted, or just modified, respectively, while *list* is a pointer to a new node, null, or a list of node modifications based on whether the *status* is *NEW*, *DEL*, or *MOD*, respectively. For *MOD* case, each modification in the list is presented by the triple (*type*, *index*, *value*) where *type* is either K, P_F, or P_M, to indicate if the modified entry is the key, a pointer to a flash node, or a pointer to an in-memory node, respectively, while *index* and *value* determines the index and the new value for the modified node entry, respectively. In Figure 2, there are two modifications in nodes A and D, one modification in nodes B and F, while node G is newly created and node H is deleted.

Log File. This is a set of flash memory blocks, reserved for recovery purposes. A log file includes short logs, written *sequentially*, about insert, delete, update, and flushing operations. Each log entry includes the triple (*operation*, *node_list*, *modification*) where *operation* indicates the type of this log entry as either insert, delete, update, or flush, *node_list* includes the list of affected nodes by this operation in case of a flush operation, or the only affected node, otherwise, *modification* is similar to the triple (*type*, *index*, *value*), used in the *tree modifications table*. All log entries are written *sequentially* to the flash storage, which has a much lower cost than *random* writes that call for the erase operation.

2.4   Running Example

Throughout the rest of this paper, we will use Figure 3 as a running example where six objects O_1 to O_6, depicted by small black circles, are indexed by an R-tree. Then, two objects O_7 and O_8, depicted by small white circles, are to be inserted in the same R-tree. Figure 3a depicts the eight objects in the two-dimensional space domain, while Figure 3b gives the flash-resident R-tree with only the six objects that made it to the flash memory. Finally, Figure 3c gives the in-memory buffer (*tree modifications table*) upon the insertion of O_7 and O_8 in the tree.

3   Tree Updates in FAST

This section discusses the update operations in FAST, which include inserting a new entry and deleting/updating an existing entry. An update operation to any tree in FAST may result in creating new tree nodes as in the case of splitting operations (i.e., when inserting an element in the tree leads to node overflow), deleting existing tree nodes as in the case of merging operations (i.e., when deleting an element from the tree leads to node underflow), or just modifying existing node keys and/or pointers.

Main idea. For any update operation (i.e., insert, delete, update) that needs to be applied to the index tree, FAST does not change the underlying insert, delete, or update algorithm for the tree structure it represents. Instead, FAST runs the underlying update algorithm for the tree it represents, with the only exception of writing any changes caused by the update operation in memory instead of the external storage, to be flushed later to the flash storage, and logging the result of the update operation. A main distinguishing characteristic of FAST is that what is buffered in memory, and also written in the log file, is the *result* of the update operation, not a log of this operation.

Algorithm. Algorithm 1 gives the pseudo code of inserting an object *Obj* in FAST. The algorithms for deleting and updating objects are similar in spirit to the insertion algorithm, and thus are omitted from the paper. The algorithm mainly has two steps: (1) *Executing the insertion in memory* (Line 2 in Algorithm 1). This is basically done by calling the insertion procedure of the underlying tree, e.g., R-tree insertion, with two main differences. First, The insertion operation calls the search operation, discussed later in section 4, to find where we need to insert our data based on the most recent version of the tree, constructed from main memory recent updates and the in-flash tree index structure. Second, the modified or newly created nodes that result back from the insertion operation are *not* written back to the flash storage, instead, they will be returned to the algorithm in a list $\mathcal{L}$. Notice that the insertion procedure may result in creating new nodes if it encounters a split operation. (2) *Buffering and logging the tree updates* (Lines 3 to 22 in Algorithm 1). For each modified node $\mathcal{N}$ in the list $\mathcal{L}$, we check if there is an entry for $\mathcal{N}$ in our in-memory buffer, *tree modifications table*. If this is the case, we first add a corresponding log entry that records the changes that took place in $\mathcal{N}$. Then, we either add the changes in $\mathcal{N}$ to the list of changes in its entry in the *tree modifications table* if this entry status is *MOD*, or update $\mathcal{N}$ entry in the *tree modifications table*, if the entry status is *NEW*. On the other hand, if there is no entry for $\mathcal{N}$ in the *tree modifications table*, we create such entry, add it to the log file, and fill it according to whether $\mathcal{N}$ is a newly created node or a modified one.

Example. In our running example of Figure 3, inserting O_7 results in modifying two nodes, G and C. Node G needs to have an extra key to hold O_7 while node C needs to modify its minimum bounding rectangle that points to G to accommodate its size change. The changes in both nodes are stored in the *tree modifications table* depicted in Figure 3c. The log entries for this operation are depicted in the first two entries of the log file of Figure 4a. Similarly, inserting O_8 results in modifying nodes, D and B.

4   Searching in FAST

Given a query Q, the *search* operation returns those objects indexed by FAST and satisfy Q. The search query Q could be a point query that searches for objects with a specific (point) value, or a range query that searches for objects within a specific range.

Algorithm 1. Insert an Object in the Tree

1: Function INSERT(*Obj*)
/* STEP 1: *Executing the Insertion in Memory only* */
2:   $\mathcal{L}$ ← List of modified nodes from the in-memory execution of inserting *Obj* in the underlying tree
/* STEP 2: *Buffering and Logging the Updates* */
3: **for each** Node $\mathcal{N}$ in $\mathcal{L}$ **do**
4:      *HashEntry* ← $\mathcal{N}$ entry in the *Tree Modifications Table*
5:      **if** *HashEntry* **is not** NULL **then**
6:          Add the triple (*MOD*, $\mathcal{N}$, updates in $\mathcal{N}$) to the log file
7:          **if** the status of *HashEntry* is *MOD* **then**
8:              Add the changes in $\mathcal{N}$ to the list of changes of *HashEntry*
9:          **else**
10:              Apply the changes in $\mathcal{N}$ to the new node of *HashEntry*
11:          **end if**
12:      **else**
13:          *HashEntry* ← Create a new entry for $\mathcal{N}$ in the *Tree Modifications Table*
14:          **if** $\mathcal{N}$ is a newly created node **then**
15:              Add the triple (*NEW*, $\mathcal{N}$, updates in $\mathcal{N}$) to the log file
16:              Set *HashEntry* status to *NEW*, and its pointer to $\mathcal{N}$
17:          **else**
18:              Add the triple (*MOD*, $\mathcal{N}$, updates in $\mathcal{N}$) to the log file
19:              Set *HashEntry* status to *MOD*, and its pointer to the list of changes that took place in $\mathcal{N}$
20:          **end if**
21:      **end if**
22: **end for**

Algorithm 2. Searching for an Object indexed by the Tree

1: Function SEARCH(Query Q, Tree Node R)
/* STEP 1: Constructing the most recent version of R */
2:   $\mathcal{N}$ ← *RetrieveNode(R)*
/* STEP 2: Recursive search calls */
3: **if** $\mathcal{N}$ *is non-leaf node* **then**
4:      Check each entry E in $\mathcal{N}$. If E satisfies the query Q, invoke *Search(Q, E.NodePointer)* for the subtree below E
5: **else**
6:      Check each entry E in $\mathcal{N}$. If E satisfies the search query Q, return the object to which E is pointing
7: **end if**

An important promise of FAST is that it does not change the main search algorithm for any tree it represents. Instead, FAST complements the underlying searching algorithm to consider the latest tree updates stored in memory.

Main idea. As it is the case for any index tree, the *search* algorithm starts by fetching the root node from the secondary storage, unless it is already buffered in memory. Then, based on the entries in the root, we find out which tree pointer to follow to fetch another node from the next level. The algorithm goes on recursively by fetching nodes from the secondary storage and traversing the tree structure till we either find a node that includes the objects we are searching for or conclude that there are no objects that satisfy the search query. The challenging part here is that the retrieved nodes from the flash storage do not include the recent in-memory stored updates. FAST complements this *search* algorithm to apply the recent tree updates to each retrieved node from the flash storage. In particular, for each visited node, FAST constructs the latest version of the node by merging the retrieved version from the flash storage with the recent in-memory updates for that node.

Algorithm. Algorithm 2 gives the pseudo code of the search operation in FAST. The algorithm takes two input parameters, the query Q, which might be a point or range

---

Algorithm 3. Retrieving a tree node

1: Function RETRIEVENODE(Tree Node R)
2:    *FlashNode* ← Retrieve node R from the flash-resident index tree
3:    *HashEntry* ← R's entry in the *Tree Modifications Table*
4:    **if** *HashEntry* is NULL **then**
5:        **return** *FlashNode*
6:    **end if**
7:    **if** the status of *HashEntry* is *MOD* **then**
8:        *FlashNode* ← *FlashNode* ∪ All the updates in *HashEntry* list
9:        **return** *FlashNode*
10:   **end if**
       /* We are trying to retrieve either a new or a deleted node */
11:   **return** the node that *HashEntry* is pointing to

---

query, and a pointer to the root node R of the tree we want to search in. The output of the algorithm is the list of objects that satisfy the input query Q. Starting from the root node and for each visited node R in the tree, the algorithm mainly goes through two main steps: (1) *Constructing the most recent version of R* (Line 2 in Algorithm 2). This is mainly to integrate the latest flash-residant version of R with its in-memory stored updates. Algorithm 3 gives the detailed pseudo code for this step, where initially, we read R from the flash storage. Then, we check if there is an entry for R in the *tree modifications table*. If this is not the case, then we know that the version we have read from the flash storage is up-to-date, and we just return it back as the most recent version. On the other hand, if R has an entry in the *tree modifications table*, we either apply the changes stored in this entry to R in case the entry status is *MOD*, or just return the node that this entry is pointing to instead of R. This return value could be null in case the entry status is *DEL*. (2) *Recursive search calls* (Lines 3 to 7 in Algorithm 2). This step is typical in any tree search algorithm, and it is basically inherited from the underlying tree that FAST is representing. The idea is to check if R is a leaf node or not. If R is a non-leaf node, we will check each entry E in the node. If E satisfies the search query Q, we recursively search in the subtree below E. On the other hand, if R is a leaf node, we will also check each entry E in the node, yet if E satisfies the search query Q, we will return the object to which E is pointing to as an answer to the query.

Example. Given the range query Q in Figure 3a, FAST search algorithm will first fetch the root node A stored in flash memory. As there is no entry for A in the *tree modifications table* (Figure 3c), then the version of A stored in flash memory is the most recent one. Then, node C is the next node to be fetched from flash memory by the searching algorithm. As the *tree modifications table* has an entry for C with status *MOD*, the modifications listed in the *tree modifications table* for C will be applied to the version of C read from the flash storage. Similarly, the search algorithm will construct the leaf nodes F and G Finally, the result of this query is $\{O_4, O_5, O_6, O_7\}$.

5   Memory Flushing in FAST

As memory is a scarce resource, it will eventually be filled up with incoming updates. In that case, FAST triggers its flushing module to free some memory space by evicting a selected part of the memory, termed a *flushing unit*, to the flash storage. Such flushing is done in a way that amortizes the cost of expensive random write operations over a high number of update operations. In this section, we first define the flushing unit. Then, we discuss the flushing policy used in FAST. Finally, we explain the FAST flushing algorithm.

5.1   Flushing Unit

An important design parameter, in FAST, is the size of a *flushing unit*, the granularity of consecutive memory space written in the flash storage during each flush operation. Our goal is to find a suitable *flushing unit* size that minimizes the average cost of flushing an update operation to the flash storage, denoted as C. The value of C depends on two factors: $C_1 = \frac{average\ writing\ cost}{number\ of\ written\ bytes}$; the average cost per bytes written, and $C_2 = \frac{number\ of\ written\ bytes}{number\ of\ updates}$; the number of bytes written per update. This gives $C = C_1 \times C_2$.

Interestingly, the values of C_1 and C_2 show opposite behaviors with the increase of the *flushing unit* size. First consider C_1. On raw flash devices (e.g., ReMix [11]), for a *flushing unit* smaller than a flash block, C_1 decreases with the increase of the flushing unit size (see [19] for more detail experiments). This is intuitive, since with a larger *flushing unit*, the cost of erasing a block is amortized over more bytes in the flushing unit. The same is also true for SSDs since small random writes introduce large garbage collection overheads, while large random writes approach the performance of sequential writes. Previous work has shown that, on several SSDs including the ones from Samsung, MTron, and Transcend, random write latency per byte increases by $\approx$ $32\times$ when the write size is reduced from 16KB to 0.5KB [5]. Even on newer generation SSDs from Intel, we observed an increase of $\approx 4\times$ in a similar experimental setup. This suggests that a flushing unit should *not* be very small, as that would result in a large value of C_1. On the other hand, the value of C_2 increases with increasing the size of the *flushing unit*. Due to non-uniform updates of tree nodes, a large flushing unit is unlikely to have as dense updates as a small flushing unit. Thus, the larger a *flushing unit* is, the less the number of updates per byte is (i.e., the higher the value of C_2 is). Another disadvantage of large *flushing unit* is that it may cause a significant pause to the system. All these suggest that the *flushing unit* should *not* be very large.

Deciding the optimal size of a *flushing unit* requires finding a sweet spot between the competing costs of C_1 and C_2. Our experiments show that for raw flash devices, a *flushing unit* of one flash block minimizes the overall cost. For SSDs, a *flushing unit* of size 16KB is a good choice, as it gives a good balance between the values of C_1 and C_2.

5.2   Flushing Policy

The main idea of FAST *flushing policy* is to minimize the average cost of writing each update to the underlying flash storage. To that end, FAST flushing policy aims to flush the in-memory tree updates that belong to the *flushing unit* that has the highest number of in-memory updates. In that case, the cost of writing the *flushing unit* will be amortized among the highest possible number of updates. Moreover, since the maximum number of updates are being flushed out, this frees up the maximum amount of

---

Algorithm 4. Flushing Tree Updates

1: Function FLUSHTREEUPDATES()
   /* STEP 1: Finding out the list of flushed tree nodes */
2:   *FlushList* ← {ϕ}
3:   *MaxUnit* ← Extract the Maximum from *FlushHeap*
4:   **for each** Node $\mathcal{N}$ in *tree modifications table* **do**
5:      **if** $\mathcal{N} \in MaxUnit$ **then**
6:         $\mathcal{F}$ ← *RetrieveNode*($\mathcal{N}$)
7:         *FlushList* ← *FlushList* $\cup$ $\mathcal{F}$
8:      **end if**
9:   **end for**
   /* STEP 2: Flushing, logging, and cleaning selected nodes */
10:   Flush all tree updates $\in$ *FlushList* to flash memory
11:   Add (*Flush*, All Nodes in *FlushList*) to the log file
12:   **for each** Node $\mathcal{F}$ in *FlushList* **do**
13:      Delete $\mathcal{F}$ from the *Tree Modifications Table*
14:   **end for**

---

memory used by buffered updates. Finally, as done in the update operations, the flushing operation is logged in the log file to ensure the *durability* of system transactions.

Data structure. The flushing policy maintains an in-memory max heap structure, termed *FlushHeap*, of all *flushing units* that have at least one in-memory tree update. The max heap is ordered on the number of in-memory updates for each *flushing unit*, and is updated with each incoming tree update.

5.3   Flushing Algorithm

Algorithm 4 gives the pseudo code for flushing tree updates. The algorithm has two main steps: (1) *Finding out the list of flushed tree nodes* (Lines 2 to 9 in Algorithm 4). This step starts by finding out the victim *flushing unit*, *MaxUnit*, with the highest number of in-memory updates. This is done as an O(1) heap extraction operation. Then, we scan the *tree modifications table* to find all updated tree nodes that belong to *MaxUnit*. For each such node, we construct the most recent version of the node by retrieving the tree node from the flash storage, and updating it with the in-memory updates. This is done by calling the *RetrieveNode*($\mathcal{N}$) function, given in Algorithm 3. The list of these updated nodes constitute the list of to be flushed nodes, *FlushList*. (2) *Flushing, logging, and cleaning selected tree nodes* (Lines 10 to 14 in Algorithm 4). In this step, all nodes in the *FlushList* are written once to the flash storage. As all these nodes reside in one *flushing unit*, this operation would have a minimal cost due to our careful selection of the *flushing unit* size. Then, similar to update operations, we log the flushing operation to ensure *durability*. Finally, all flushed nodes are removed from the *tree modifications table* to free memory space for new updates.

Example. In our running example given in Figure 3, assume that the memory is full, hence FAST triggers its flushing module. Assume also that nodes B, C, and D reside in the same *flushing unit* B_1, while nodes E, F, and G reside in another *flushing unit* B_2. The number of updates in B_1 is three as each of nodes B, C and D has been updated once. On the other hand, the number of updates in B_2 is one because nodes E and F has no updates at all, and node G has only a single update. Hence, *MaxUnit* is set to B_1, and we will invoke *RetrieveNode* algorithm for all nodes belonging to B_1 (i.e., nodes B, C, and D) to get the most recent version of these nodes and flush them to flash memory. Then, the log entry ($Flush$; Nodes B, C, D) is added to the log file (depicted as the last log entry in Figure 4a). Finally, the entries for nodes B, C, and D are removed from the *tree modifications table*.

6 Crash Recovery and Log Compaction in FAST

As discussed before, FAST heavily relies on storing recent updates in memory, to be flushed later to the flash storage. Although such design efficiently amortizes the expensive random write operations over a large number of updates, it poses another challenge where memory contents may be lost in case of system crash. To avoid such loss of data, FAST employs a crash recovery module that ensures the *durability* of in-memory updates even if the system crashed. The crash recovery module in FAST mainly relies on the log file entries, written sequentially upon the update and flush operations.

6.1 Recovery

The recovery module in FAST is triggered when the system restarts from a crash, with the goal of restoring the state of the system just before the crash took place. The state of the system includes the contents of the in-memory data structure, *tree modifications table*, and the flash-resident tree index structure. By doing so, FAST ensures the *durability* of all non-flushed updates that were stored in memory before crash.

Main Idea. The main idea of the *recovery* operation is to scan the log file bottom-up to be aware of the flushed nodes, i.e., nodes that made their way to the flash storage. During this bottom-up scanning, we also find out the set of operations that need to be replayed to restore the *tree modifications table*. Then, the *recovery* module cleans all the flash blocks, and starts to replay the non-flushed operations in the order of their insertion, i.e., top-down. The replay process includes insertion in the *tree modifications table* as well as a new log entry. It is important

Log#	Operation	Node	Modification
1	MOD	C	$K, 2, (12,4,14,2)$
2	MOD	G	$K, 2, O_7$
3	MOD	B	$K, 2, (5,10, 8, 7)$
4	MOD	D	$K, 2, O_8$
5	FLUSH	B, C, D	*

(a) FAST Log File

Log#	Operation	Node	Modification
2	MOD	G	$K, 2, O_7$

(b) FAST Log File after Crash Recovery

Fig. 4. FAST Logging and Recovery

here to reiterate our assumption that there will be no crash during the recovery process, so, it is safe to keep the list of operations to be replayed in memory. If we will consider a system crash during the recovery process, we might just leave the operations to be replayed in the log, and scan the whole log file again in a top-down manner. In this top-down scan, we will only replay the operations for non-flushed nodes, while writing the new log entries into a clean flash block. The result of the crash recovery module is that the state of the memory will be stored as it was before the system crashes, and the log file will be an exact image of the *tree modifications table*.

Algorithm. Algorithm 5 gives the pseudo code for crash recovery in FAST, which has two main steps: (1) *Bottom-Up scan* (Lines 2 to 12 in Algorithm 5). In this step, FAST

Algorithm 5. Crash Recovery

1: Function RECOVERFROMCRASH()
   /* STEP 1: Bottom-Up Cleaning */
2: *FlushedNodes* ← φ
3: **for each** Log Entry $\mathcal{L}$ **in** the log file in a reverse order **do**
4:      **if** the operation of $\mathcal{L}$ is `Flush` **then**
5:          *FlushedNodes* ← *FlushedNodes* ∪ the list of nodes in $\mathcal{L}$
6:      **else**
7:          **if** the node in entry $\mathcal{L}$ ∉ *FlushedNodes* **then**
8:              Push $\mathcal{L}$ into the stack of updates *RedoStack*
9:          **end if**
10:     **end if**
11: **end for**
12: Clean all the log entries by erasing log flash blocks
    /* Phase 2: Top-Down Processing */
13: **while** *RedoStack* **is not** Empty **do**
14:     *Op* ← Pop an update operation from the top of *RedoStack*
15:     Insert the operation *Op* into the *tree modifications table*
16:     Add a log entry for *Op* in the log file
17: **end while**

---

scans the log file bottom-up, i.e., in the reverse order of the insertion of log entries. For each log entry $\mathcal{L}$ in the log file, if the operation of $\mathcal{L}$ is *Flush*, then we know that all the nodes listed in this entry have already made their way to the flash storage. Thus, we keep track of these nodes in a list, termed *FlushedNodes*, so that we avoid redoing any updates over any of these nodes later. On the other side, if the operation of $\mathcal{L}$ is not *Flush*, we check if the node in $\mathcal{L}$ entry is in the list *FlushedNodes*. If this is the case, we just ignore this entry as we know that it has made its way to the flash storage. Otherwise, we push this log entry into a stack of operations, termed *RedoStack*, as it indicates a non-flushed entry at the crash time. At the end of this step, we erase the log flash blocks, and pass the *RedoStack* to the second step. (2) *Top-Down processing* (Lines 13 to 17 in Algorithm 5). This step basically goes through all the entries in the *RedoStack* in a top-down way, i.e., the order of insertion in the log file. As all these operations were not flushed by the crash time, we just add each operation to the *tree modifications table* and add a corresponding log entry. The reason of doing these operations in a top-down way is to ensure that we have the same order of updates, which is essential in case one node has multiple non-flushed updates. At the end of this step, the *tree modifications table* will be exactly the same as it was just before the crash time, while the log file will be exactly an image of the *tree modifications table* stored in the flash storage.

Example. In our running example, the log entries of inserting Objects O_7 and O_8 in Figure 3 are given as the first four log entries in Figure 4a. Then, the last log entry in Figure 4a corresponds to flushing nodes B, C, and D. We assume that the system is crashed just after inserting this flushing operation. Upon restarting the system, the *recovery* module will be invoked. First, the *bottom-up scanning* process will be started with the last entry of the log file, where nodes B, C, and D are added to the list *FlushedNodes*. Then, for the next log entry, i.e., the fourth entry, as the node affected by this entry D is already in the *FlushedNodes* list, we just ignore this entry, since we are sure that it has made its way to disk. Similarly, we ignore the third log entry for node B. For the second log entry, as the affected node G is not in the *FlushedNodes* list, we know that this operation did not make it to the storage yet, and we add it to the

RedoStack to be redone later. The *bottom-up scanning* step is concluded by ignoring the first log entry as its affected node C is already flushed, and by wiping out all log entries. Then, the *top-down processing* step starts with only one entry in the *RedoStack* that corresponds to node G. This entry will be added to the *tree modifications table* and log file. Figure 4b gives the log file after the end of the *recovery* module which also corresponds to the entries of the *tree modifications table* after recovering from failure.

6.2   Log Compaction

As FAST log file is a limited resource, it may eventually become full. In this case, FAST triggers a *log compaction* module that organizes the log file entries for better space utilization. This can be achieved by two space saving techniques: (a) Removing all the log entries of flushed nodes. As these nodes have already made their way to the flash storage, we do not need to keep their log entries anymore, and (b) Packing small log entries in a larger writing unit. Whenever a new log entry is inserted, it mostly has a small size that may occupy a flash page as the smallest writing unit to the flash storage. At the time of compaction, these small entries can be packed together to achieve the maximum possible space utilization.

The main idea and algorithm for the *log compaction* module are almost the same as the ones used for the *recovery* module, with the exception that the entries in the *RedoStack* will not be added to the *tree modifications table*, yet they will just be written back to the log file, in a more compact way. As in the *recovery* module, Figures 4a and 4b give the log file before and after *log compaction*, respectively. The *log compaction* have similar expensive cost as the *recovery* process. Fortunately, with an appropriate size of log file and memory, it will not be common to call the *log compaction* module.

It is unlikely that the *log compaction* module will not really compact the log file much. This may take place only for a very small log size and a very large memory size, as there will be a lot of non-flushed operations in memory with their corresponding log entries. Notice that if the memory size is small, there will be a lot of flushing operations, which means that *log compaction* can always find log entries to be removed. If this unlikely case takes place, we call an *emergency flushing* operation where we force flushing all main memory contents to the flash memory persistent storage, and hence clean all the log file contents leaving space for more log entries to be added.

7   Experimental Evaluation

This section experimentally evaluates the performance of FAST, compared to the state-of-the-art algorithms for one-dimensional and multi-dimensional flash index structures: (1) Lazy Adaptive Tree (LA-tree) [2]: LA-tree is a flash friendly one dimensional index structure that is intended to replace the B-tree. LA-tree stores the updates in cascaded buffers residing on flash memory and, then empties these buffers dynamically based on the operations workload. (2) FD-tree [16,17]: FD-tree is a one-dimensional index structure that allows small random writes to occur only in a small portion of the tree called the head tree which exists at the top level of the tree. When the capacity of the head tree is exceeded, its entries are merged in batches to subsequent tree levels. (3) RFTL [25]: RFTL is a mutli-dimensional tree index structure that adds a buffering layer on top of the flash translation layer (FTL) in order to make R-trees work efficiently on flash devices.

All experiments are based on an actual implementation of FAST, LA-tree, FD-tree, and RFTL inside PostgreSQL [1]. We instantiate B-tree and R-tree instances of FAST, termed FAST-Btree and FAST-Rtree, respectively, by implementing FAST inside the GiST generalized index structure [8], which is already built inside PostgreSQL. In our experiments, we use two synthetic workloads: *(1) Lookup intensive workload (W_L)*: that includes 80% search operations and 20% update operations (i.e., insert, delete, or update). *(2) Update intensive workload, (W_U)*: that includes 20% search operations and 80% update operations.

Unless mentioned otherwise, we set the number of workload operations to 10 million operations, main memory size to 256 KB (i.e., the amount of memory dedicated to main memory buffer used by FAST), tree index size to 512 MB, and log file size to 10 MB, which means that the default log size is $\approx$2% of the index size.

The experiments in this section mainly discuss the effect of varying the memory size, log file size, index size, and number of updates on the performance of FAST-Btree, FAST-Rtree, LA-tree, FD-tree, and RFTL. Also, we study the performance of flushing, log compaction, and recovery operations in FAST. In addition, we compare the implementation cost between FAST and its counterparts. Our performance metrics are mainly the number of flash memory erase operations and the average response time. However, in almost all of our experiments, we got a similar trend for both performance measures. Thus, for brevity, we only show the experiments for the number of flash memory erase operations, which is the most expensive operation in flash storage. Although we compare FAST to its counterparts from a performance point of view, however we believe the main contribution of FAST is not in the performance gain. The generic structure and low implementation cost are the main advantages of FAST over specific flash-aware tree index structures.

All experiments were run on both raw flash memory storage, and solid state drives (SSDs). For raw flash, we used the raw NAND flash emulator described in [2]. The emulator was populated with exhaustive measurements from a custom-designed Mica2 sensor board with a Toshiba1Gb NAND TC58DVG02A1FT00 flash chip. For SSDs, we used a 32GB MSP-SATA7525032 SSD device. All the experiments were run on a machine with Intel Core2 8400 at 3Ghz with 4GB of RAM running Ubuntu Linux 8.04.

7.1   Effect of Memory Size

Figures 5(a) and 5(b) give the effect of varying the memory size from 128 KB to 1024 KB (in a log scale) on the number of erase operations, encountered in FAST-Btree, LA-tree, and FD-tree, for workloads W_L and W_U, respectively. For both workloads and for all memory sizes, FAST-Btree consistently has much lower erase operations than that of the LA-tree. More specifically, Fast-Btree results in having only from half to one third of the erase operations encountered by LA-tree. This is mainly due to the smart choice of *flushing unit* and *flushing policy* used in FAST that amortize the block erase operations over a large number of updates.

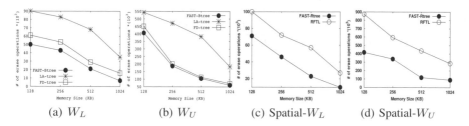

| (a) W_L | (b) W_U | (c) Spatial-W_L | (d) Spatial-W_U |

Fig. 5. Effect of Memory Size

The performance of FAST-Btree is slightly better than that of FD-tree, because FD-tree does not employ a crash recovery technique (i.e., no logging overhead). FAST still performs better than FD-tree due to FAST flushing policy that smartly selects the best block to be flushed to flash memory. Although the performance of FD-tree is close to FAST-Btree, however FAST has the edge of being a generic framework which is applied to many tree index structures and needs less work and overhead (in terms of lines of code) to be incorporated in the database engine.

Figures 5(c) and 5(d) give similar experiments to that of Figures 5(a) and 5(b), with the exception that we run the experiments for two-dimensional search and update operations for both the Fast-Rtree and RFTL. To be able to do so, we have adjusted our workload W_L and W_U to Spatial-W_L and Spatial-W_U, respectively, which have two-dimensional operations instead of the one-dimensional operations used in W_L and W_U. The result of these experiments have the same trend as the ones done for one-dimensional tree structures, where FAST-Rtree has consistently better performance than RFTL in all cases, with around one half to one third of the number of erase operations encountered in RFTL.

7.2 Effect of Log File Size

Figure 6 gives the effect of varying the log file size from 10 MB (i.e., 2% of the index size) to 25 MB (i.e., 5% of the index size) on the number of erase operations, encountered in FAST-Btree, LA-tree, and FD-tree for workload W_L (Figure 6(a)) and FAST-Rtree and RFTL for workload Spatial-W_U (Figure 6(b)). For brevity, we do not show the experiments of FAST-Btree, LA-tree, and FD-tree for workload W_U nor the experiment of FAST-Rtree and RFTL for workload Spatial-W_L. As can be seen from the figures, the performance of both LA-tee, FD-tree, and RFTL is not affected by the change of the log file size. This is mainly because these three approaches rely on buffering incoming updates, and hence does not make use of any log file. It is interesting, however, to see that the number of erase operations in FAST-Btree and FAST-Rtree significantly decreases with the increase of the log file size, given that the memory size is set to its default value of 256 KB in all experiments. The justification for this is that with the increase of the log file size, there will be less need for FAST to do log compaction.

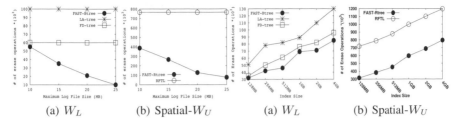

(a) W_L (b) Spatial-W_U (a) W_L (b) Spatial-W_U

Fig. 6. Effect of FAST log file size   **Fig. 7.** Effect of Tree index Size

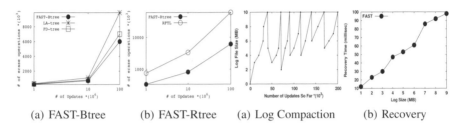

(a) FAST-Btree   (b) FAST-Rtree   (a) Log Compaction   (b) Recovery

Fig. 8. Effect of Number of Updates   **Fig. 9.** Log Compaction and recovery

7.3 Effect of Index Size

Figure 7 gives the effect of varying the index size from 128 MB to 4 GB (in a log scale) on the number of erase operations, encountered in FAST-Btree, LA-tree, and FD-tree for workload W_L (Figure 7(a)) and FAST-Rtree and RFTL for workload Spatial-W_U (Figure 7(b)). Same as in Section 7.2, we omit other workloads for brevity. In all cases, FAST consistently gives much better performance than its counterparts. Both FAST and other index structures have similar trend of a linear increase of the number of erase operations with the increase of the index size. This is mainly because with a larger index, an update operation may end up modifying more nodes in the index hierarchy, or more overlapped nodes in case of multi-dimensional index structures.

7.4 Effect of Number of Updates

Figure 7 gives the effect of varying the number of update operations from one million to 100 millions (in a log scale) on the number of erase operations for both one-dimensional (i.e., FAST-Btree, LA-tree, and FD-tree in Figure 8(a)) and multi-dimensional index structures (i.e., FAST-Rtree and RFTL in Figure 8(b)). As we are only interested in update operations, the workload for the experiments in this section is just a stream of incoming update operations, up to 100 million operations. As can be seen from the figure, FAST scales well with the number of updates and still maintains its superior performance over its counterparts from both one-dimensional (LA-tree) and multi-dimensional index structures (RFTL). FAST performs slightly better than FD-tree; this is because FD-tree (one dimensional index structure) is buffering some of the tree updates in memory and flushes them when needed, but FAST applies a smart flushing policy, which flushes only the block with the highest number of updates.

7.5 Log Compaction

Figure 9(a) gives the behavior and frequency of log compaction operations in FAST when running a sequence of 200 thousands update operations for a log file size of 10 MB. The Y axis in this figure gives the size of the filled part of the log file, started as empty. The size is monotonically increasing with having more update operations till it reaches its maximum limit of 10 MB. Then, the log compaction operation is triggered to compact the log file. As can be seen from the figure, the log compaction operation may compact the log file from 20 to 60% of its capacity, which is very efficient compaction. Another take from this experiment is that we have made only seven log compaction operations for 200 thousands update operations, which means that the log compaction process is not very common, making FAST more efficient even with a large amount of update operations.

7.6 Recovery Performance

Figure 9(b) gives the overhead of the recovery process in FAST, which serves also as the overhead of the log compaction process. The overhead of recovery increases linearly with the size increase of the log file contents at the time of crash. This is intuitive as with more log entries in the log file, it will take more time from the FAST recovery module to scan this log file, and replay some of its operations to recover the lost main memory contents. However, what we really want to emphasize on in this experiment is that the overhead of recovery is only about 100 msec for a log file that includes 9 MB of log entries. This shows that the recovery overhead is a low price to pay to ensure transaction *durability*.

8 Conclusion

This paper presented FAST; a generic framework for flash-aware spatial tree index structures. FAST distinguishes itself from all previous attempts of flash memory indexing in two aspects: (1) FAST is a generic framework that can be applied to a wide class of spatial tree structures, and (2) FAST achieves both *efficiency* and *durability* of read and write flash operations. FAST has four main modules, namely, *update*, *search*, *flushing*, and *recovery*. The *update* module is responsible on buffering incoming tree updates in an in-memory data structure, while writing small entries sequentially in a designated flash-resident log file. The *search* module retrieves requested data from the flash storage and updates it with recent updates stored in memory, if any. The *flushing* module is responsible on evicting flash blocks from memory to the flash storage to give space for incoming updates. Finally, the *recovery* module ensures the durability of in-memory updates in case of a system crash.

References

1. PostgreSQL, http://PostgreSQL, http://www.postgresql.org
2. Agrawal, D., Ganesan, D., Sitaraman, R.K., Diao, Y., Singh, S.: Lazy-adaptive tree: An optimized index structure for flash devices. In: PVLDB (2009)
3. Agrawal, N., Prabhakaran, V., Wobber, T., Davis, J., Manasse, M., Panigrahy, R.: Design Tradeoffs for SSD Performance. In: Usenix Annual Technical Conference, USENIX (2008)
4. Beckmann, N., Kriegel, H.-P., Schneider, R., Seeger, B.: The R*-Tree: An Efficient and Robust Access Method for Points and Rectangles. In: SIGMOD (1990)
5. Bouganim, L., Jónsson, B., Bonnet, P.: uFLIP: Understanding Flash IO Patterns. In: CIDR (2009)
6. Gray, J., Fitzgerald, B.: Flash Disk Opportunity for Server Applications. ACM Queue (2008)
7. Guttman, A.: R-Trees: A Dynamic Index Structure For Spatial Searching. In: SIGMOD (1984)
8. Hellerstein, J.M., Naughton, J.F., Pfeffer, A.: Generalized search trees for database systems. In: VLDB (1995)
9. Katayama, N., Satoh, S.: The sr-tree: An index structure for high-dimensional nearest neighbor queries. In: SIGMOD (1997)
10. Kim, H., Ahn, S.: BPLRU: A Buffer Management Scheme for Improving Random Writes in Flash Storage. In: FAST (2008)
11. Lavenier, D., Xinchun, X., Georges, G.: Seed-based Genomic Sequence Comparison using a FPGA/FLASH Accelerator. In: ICFPT (2006)
12. Lee, S., Moon, B.: Design of Flash-Based DBMS: An In-Page Logging Approach. In: SIGMOD (2007)
13. Lee, S.-W., Moon, B., Park, C., Kim, J.-M., Kim, S.-W.: A case for Flash memory SSD in Enterprise Database Applications. In: SIGMOD (2008)
14. Lee, S.-W., Park, D.-J., sum Chung, T., Lee, D.-H., Park, S., Song, H.-J.: A Log Buffer-Based Flash Translation Layer Using Fully-Associate Sector Translation. In: TECS (2007)
15. Leventhal, A.: Flash Storage Today. ACM Queue (2008)
16. Li, Y., He, B., Luo, Q., Yi, K.: Tree indexing on Flash Disks. In: ICDE (2009)
17. Li, Y., He, B., Yang, R.J., Luo, Q., Yi, K.: Tree indexing on solid state drives. In: PVLDB, vol. 3(1) (2010)
18. Moshayedi, M., Wilkison, P.: Enterprise SSDs. ACM Queue (2008)
19. Nath, S., Gibbons, P.B.: Online Maintenance of Very Large Random Samples on Flash Storage. In: VLDB (2008)
20. Nath, S., Kansal, A.: Flashdb: Dynamic self-tuning database for nand flash. In: IPSN (2007)
21. Reinsel, D., Janukowicz, J.: Datacenter SSDs: Solid Footing for Growth (January 2008), http://www.samsung.com/us/business/semiconductor/news/downloads/210290.pdf
22. Sellis, T.K., Roussopoulos, N., Faloutsos, C.: The R+-Tree: A Dynamic Index for Multi-Dimensional Objects. In: VLDB (1987)
23. Shah, M.A., Harizopoulos, S., Wiener, J.L., Graefe, G.: Fast Scans and Joins using Flash Drives. In: International Workshop of Data Managment on New Hardware, DaMoN (2008)
24. White, D.A., Jain, R.: Similarity indexing with the ss-tree. In: ICDE (1996)
25. Wu, C., Chang, L., Kuo, T.: An Efficient R-tree Implementation over Flash-Memory Storage Systems. In: GIS (2003)
26. Wu, C., Kuo, T., Chang, L.: An Efficient B-tree Layer Implementation for Flash-Memory Storage Systems. In: TECS (2007)

MIDAS: Multi-attribute Indexing for Distributed Architecture Systems

George Tsatsanifos[1], Dimitris Sacharidis[2,*], and Timos Sellis[1,2]

[1] National Technical University of Athens, Greece
gtsat@dblab.ece.ntua.gr
[2] Institute for the Management of Information Systems, R.C. "Athena", Greece
{dsachar,timos}@imis.athena-innovation.gr

Abstract. This work presents a pure multidimensional, indexing infrastructure for large-scale decentralized networks that operate in extremely dynamic environments where peers join, leave and fail arbitrarily. We propose a new peer-to-peer variant implementing a virtual distributed k-d tree, and develop efficient algorithms for multidimensional point and range queries. Scalability is enhanced as each peer has only partial knowledge of the network. The most prominent feature of our method, is that in expectance each peer maintains $O(\log n)$ state and requests are resolved in $O(\log n)$ hops with respect to the overlay size n. In addition, we provide mechanisms for handling peer failures and improving fault tolerance as well as balancing the load of peers. Finally, our work is complemented by an experimental evaluation, where MIDAS is shown to outperform existing methods in spatial as well as in higher dimensional settings.

Keywords: Peer-to-peer systems, kd-trees.

1 Introduction

Peer-to-peer (P2P) systems have emerged as a popular technique for exchanging information among a set of distributed machines. Recently, structured peer-to-peer systems gain momentum as a general means to decentralize various applications, such as lookup services, file systems, content delivery, etc. This work considers the case of a structured distributed storage and index scheme for multidimensional information, termed MIDAS, for **M**ulti-**A**ttribute **I**ndexing for **D**istributed **A**rchitecture **S**ystems. The important feature of MIDAS is that it is capable of efficiently processing the most important types of multi-attribute queries, such as point and range queries, in arbitrary dimensionality.

While a lot of research has been devoted to structured peer-to-peer networks, only a few of them are capable of indexing multidimensional data. We distinguish three categories. The first includes solutions based on a single-dimensional P2P method. The most naive method is to select a single attribute and ignore all others for indexing, which clearly has its disadvantages. A more attractive alternative is to index each dimension separately, e.g., [4], [6]. However, these approaches still have to resort to only one of the

---

* The author is supported by a Marie Curie Fellowship (IOF) within the European Community FP7.

D. Pfoser et al. (Eds.): SSTD 2011, LNCS 6849, pp. 168–185, 2011.

dimensions for processing queries. The most popular approach, [9], [18], [5], within this category is to map the original space into a single dimension using a space filling curve, such as Hilbert or z-curve, and then employ any standard P2P system. These techniques suffer, especially in high dimensionality, as locality cannot be preserved. For instance, a rectangular range in the original space corresponds to multiple non-contiguous ranges in the mapped space.

The second category contains P2P systems that were explicitly designed to store multidimensional information, e.g., [15], [9]. The basic idea in these methods is that each peer is responsible for a rectangular region of the space and it has knowledge of its neighbors in adjacent regions. Being multidimensional in nature, allows them to feature sublinear to the network size cost for most queries. Their main weakness, however, is that they cannot take advantage of a hierarchical indexing structure. As a result, lookups for remote (in the multidimensional space) peers are unavoidably routed through many intermediate node, i.e., jumps cannot be made.

The last category includes methods, e.g., [10], [11], that decentralize a conventional hierarchical multidimensional index, such as the R-tree. The basic idea is that each peer corresponds to a node (internal or leaf) of the index, and establishes link to its parent, children and selected nodes at the same depth of the tree but in different subtrees. Queries are processed similar to the centralized approach, i.e., the index is traversed starting from the root. As a result, these methods inherit nice properties like logarithmic search cost, but face a serious limitation. Peers that correspond to nodes high in the tree can quickly become overloaded as query processing must pass through them. While this was a desirable property in centralized indices in order to minimize the number of I/O operations by maintaining these nodes in main memory, it is a limiting factor in distributed settings leading to bottlenecks. Moreover, this causes an imbalance in fault tolerance: a peer high in the tree that fails requires a significant amount of effort from the system to recover. Last but not least, R-trees are known to suffer in high dimensionality settings, which carries over to their decentralized counterparts. For example, the experiments in [11] showed that for dimensionality close to 20, this method was outperformed by the non-indexed approach of [15].

Motivated by these observations, MIDAS takes a different approach. First, it employs a hierarchical multidimensional index structure, the k-d tree. This has a series of benefits. Being a binary tree, it allows for simple and efficient routing, in a manner reminiscent of Plaxton's algorithm [14] for single dimensional tree-like structures. Unlike other multidimensional index techniques, e.g., [11], peers in MIDAS only correspond to leaf nodes of the k-d tree. This, alleviates bottlenecks and increases scalability as no single peer is burdened with routing multiple requests. Moreover, MIDAS is compatible with conventional techniques for load balancing and replication-based fault tolerance.

In summary, MIDAS is an efficient method for indexing multi-attribute data. We prove that in expectance point queries and range queries are performed in $O(\log n)$ hops; these bounds are smaller than non-indexed multidimensional P2P systems, e.g., $O(d \sqrt[d]{n})$ of [15]. A thorough experimental study on real spatial data as well as on synthetic data of varying dimensionality validates this claim.

The remainder of this paper is organized as follows. Section 2 compares MIDAS to related work. Section 3 describes our index scheme and basic operations including load balancing and fault tolerance mechanisms. Section 4 discusses multidimensional query processing. Section 5 presents an extensive experimental evaluation of all MIDAS' features. Section 6 concludes and summarizes our contributions.

2 Related Work

Structured peer-to-peer networks employ a globally consistent protocol to ensure that any peer can efficiently route a search to the peer that has the desired content, regardless of how rare it is or where it is located. Such a guarantee necessitates a structured overlay pattern. The most prominent class of approaches is *distributed hash tables* (DHTs). A DHT is a decentralized, distributed system that provides a lookup service similar to a hash-table. DHTs employ a consistent hashing variant [12] that is used to assign ownership of a (key, value) pair to a particular peer of an overlay network. Because of their structure, they offer certain guarantees when retrieving a key (e.g., worst-case logarithmic number of hops for lookups, i.e., point queries, with respect to network size). DHTs form a reliable infrastructure for building complex services, such as distributed file systems, content distribution systems, cooperative web caching, multicast, domain name services, etc.

Chord [19] uses a consistent hashing variant to associate unique (single-dimensional) identifiers with resources and peers. A key is assigned to the first peer whose identifier is equal to, or follows the key, in the identifier space. Each peer in Chord has $\log n$ state, i.e., number of neighbors, and resolves lookups in $\log n$ hops, where n is the size of the overlay network, i.e., the number of peers.

Another line of work involves tree-like structures, such as P-Grid [1], Kademlia [13], Tapestry [?] and Pastry [17]. Peer lookup in these systems is based on Plaxton's algorithm [14]. The main idea is to locate the neighbor whose identifier shares the longest common prefix with the requested (single-dimensional) key, and repeat this procedure recursively until the owner of the key is found. Lookups cost $O(\log n)$ hops and each peer has $O(\log n)$ state. MIDAS is similar to these works in that it has a tree-like structure with logarithmic number of neighbors at each peer, but differs in that it is able to perform multidimensional lookups in $O(\log n)$ hops.

We next discuss various structured peer-to-peer systems that natively index multi-attribute keys. In CAN [15], each peer is responsible for its *zone*, which is an axis-parallel orthogonal region of the d-dimensional space. Each peer holds information about a number of adjacent zones in the space, which results in $O(d)$ state. A d-dimensional key lookup is greedily routed towards the peer whose zone contains the key and costs $O(d \sqrt[d]{n})$ hops. Analogous results hold for MURK [9], where the space is a d-dimensional torus. The main concern with these approaches is that their cost (although sublinear to n) is considerate for large networks.

Several approaches, e.g., SCRAP [9], ZNet [18], employ a space filling curve to map the multidimensional space to a single dimension and then use a conventional system to index the resulting space. For instance, [5] uses the z-curve and P-Grid to support multi-attribute range queries. The problem with such methods is that the locality of the original space cannot be preserved well, especially in high dimensionality. As a result a single range query is decomposed to multiple range queries in the mapped space, which increases the processing cost.

MAAN [6] extends Chord to support multidimensional range queries by mapping attribute values to the Chord identifier space via uniform locality preserving hashing. MAAN and Mercury [4] can support multi-attribute range queries through single-attribute query resolution. They do not feature pure multidimensional schemes, as they treat attributes independently. As a result, a range query is forwarded to the first value appearing in the range and then it is spread along neighboring peers exploiting the contiguity of the range. This procedure is very costly particularly in MAAN, which prunes the search space using only one dimension.

The VBI-tree [11] is a distributed framework based on balanced multidimensional tree structured overlays, e.g., R-tree. It provides an abstract tree structure on top of an overlay network that supports any kind of hierarchical tree indexing structures, i.e., when the region managed by a node covers those managed by its children. However, it was shown in [5] that for range queries the VBI-tree suffers in scalability in terms of throughput. Furthermore, it can cause unfairness as peers corresponding to nodes high in the tree are heavily hit.

3 MIDAS Architecture

This section presents the information stored in each peer and details the basic operations in the MIDAS overlay network. In particular, Section 3.1 introduces the distributed index structure, Section 3.2 discusses the information stored within each peer in MIDAS. Section 3.4, 3.3 and 3.5 elaborates on the actions taken when a peer departs, joins, and fails, respectively. Section 3.6 discusses load balancing and fault tolerance.

3.1 Index Structure

The distributed index of MIDAS is an instance of an adaptive k-d tree [3]. Consider a D-dimensional space $I = [\ell_I, h_I]$, defined by a low ℓ_I and a high h_I D-dimensional point. The k-d tree T is a binary tree, in which each node $T[i]$ corresponds to an axis parallel (hyper-) rectangle I_i; the root $T[1]$ corresponds to the entire space, i.e., $I_1 = I$. Each internal node $T[i]$ has always two children, $T[2i]$ and $T[2i+1]$, whose rectangles are derived by splitting I_i at some value s_i along some dimension d_i; the splitting criterion (i.e., the values of s_i and d_i) are discussed in Section 3.3. Note that d_i represents the splitting dimension of node $T[i]$ and not the i-th dimension of the space.

Consider node $T[i]$'s two children, $T[2i]$, $T[2i + 1]$, and their rectangles $I_{2i} = [\ell_{2i}, h_{2i}]$, $I_{2i+1} = [\ell_{2i+1}, h_{2i+1}]$. Assuming that the left child ($T[2i]$) is assigned the lower part of I_i, it holds that (1) $\ell_{2i}[d_j] = \ell_{2i+1}[d_j]$ and $h_{2i}[d_j] = h_{2i+1}[d_j]$ on every dimension $d_j \neq d_i$, and (2) $h_{2i}[d_i] = \ell_{2i+1}[d_i] = s_i$ on dimension d_i. We write $I_{2i} \uplus^{d_i} I_{2i+1}$ to denote that the above properties hold for the two rectangles.

Each node of the k-d tree is associated with a binary identifier corresponding to its path from the root, which is defined recursively. The root has the empty id $\varnothing$; the left (resp. right) child of an internal node has the id of its parent augmented with 0 (resp. 1). Figure 1a depicts a k-d tree of eleven nodes obtained from five splits; next to each node its id is shown. Due to the hierarchical splits, the rectangles of the leaf nodes in a k-d tree constitute a non-overlapping partition of the entire space I. Figure 1b draws the

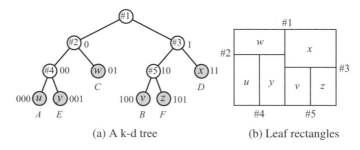

| (a) A k-d tree | (b) Leaf rectangles |

Fig. 1. An example of a two-dimensional k-d tree

rectangles corresponding to the leaves of Figure 1a; the splits are numbered and shown next to the corresponding axis parallel cuts.

A tuple with D attributes is represented as a point in the D-dimensional space I indexed by a k-d tree. A leaf of a k-d tree stores all tuples that fall in its rectangle. The hierarchical structure of the k-d tree allows for efficient methods to process queries, such as range queries, which retrieve all tuples within a range.

3.2   MIDAS Peers

It is important to distinguish the concepts of a physical and a virtual peer. A virtual peer, or simply a *peer*, is the basic entity in MIDAS. On the other hand, a *physical peer* is an actual machine that takes part in the distributed overlay. A physical peer can be responsible for several peers due to node departures or failures (Section 3.4, 3.5), or for load balancing and fault tolerance purposes (Section 3.6).

A peer in MIDAS corresponds to a leaf of the k-d tree, and stores/indexes all tuples that reside in the leaf's rectangle, which is called its *zone*. A peer is denoted with small letters, e.g., u, v, w, etc., whereas a physical peer with capital letters, e.g., A, B, C, etc. For example, in Figure 1a, physical peer C acts as the single peer w corresponding to leaf 01. We emphasize that internal k-d tree nodes, e.g., the non-shaded nodes in Figure 1a, do not correspond to peers and of course not to physical peers. An important property of peers in MIDAS is the following *invariant*.

Lemma 1. *For any point in space I, there exists exactly one peer in MIDAS responsible for it.*

Proof. Each peer corresponds to a k-d tree leaf. The lemma holds because the leaves constitute an non-overlapping partition of the entire space I.   □

A peer u in MIDAS contains only partial information about the k-d tree, which however is sufficient to perform complex query processing discussed in Section 4. In particular, peer u contains the following state. (1) $u.id$ is a bitmap representing the leaf's binary id; $u.id[j]$ is the j-th most significant bit. (2) $u.depth$ is the depth of the leaf in the k-d tree, or equivalently the number of bits in $u.id$. (3) $u.sdim$ is an array of length $u.depth$ so that $u.sdim[j]$ is the splitting dimension of the parent of the j-th node on the path from the root to u. (4) $u.split$ is an array of length $u.depth$ so that $u.split[j]$ is the splitting value of the parent of the j-th node on the path from the root to u. (5) $u.link$ is an array of length $u.depth$ that corresponds to u's routing table, i.e., it contains the peers u has a link to. (6) $u.backlink$ is a list that contains all peers that have u in their $link$ array.

In the following, we explain the contents of $u.link$, which define the routing table of peer u. First, we define an important concept. Consider the prefixes of u's identifier; there are $u.depth$ of them. Each prefix corresponds to a subtree of the k-d tree that contains the leaf u (more accurately the leaf that has id $u.id$) and identifies a node on the path from the root to u. In the example of Figure 1a, $u.id = 000$ has three prefixes: 0, 00 and 000, corresponding to the subtrees rooted at the internal k-d tree nodes with these ids. If we invert the least significant bit of a prefix, we obtain a *maximal sibling subtree*, i.e., a subtree for which there exists no larger subtree that contains it and also not contain the leaf u. Figure 2a shows the maximal sibling subtrees of $u.id = 000$, which are rooted at nodes 1, 01 and 001, as shaded triangles.

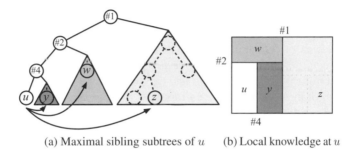

(a) Maximal sibling subtrees of u (b) Local knowledge at u

Fig. 2. Links of peer u

For each maximal sibling subtree, u establishes a link to a peer that resides in it. Note that a subtree may contain multiple leafs and thus multiple peers; MIDAS requires that peer u knows just any one of them. For example, Figure 2a shows the peers in each maximal sibling subtree that u is connected to. Observe that each peer has only partial knowledge about the k-d tree structure. Figure 2b depicts this local knowledge for u, which is only aware about the splits (#1, #2 and #4) along its path to the root. The shaded rectangles corresponds to the subtrees of the same shade in Figure 2a. Peer u knows exactly one other peer within each rectangle. Observe, however, that these rectangles cover the entire space I; this is necessary to ensure that u can locate any other peer, as explained in Section 4.1, and process queries, as discussed in Section 4.

Array $u.link$ defines the routing table. Entry $u.link[j]$ contains the address of a peer that resides in the maximal sibling subtree obtained from the j-length prefix of $u.id$. Continuing the example, u connects to three peers, i.e., $u.link = \{z, w, y\}$. Table 1 depicts the $link$ array for each peer. The notation $u(000)$ indicates that peer u corresponds to k-d tree leaf with id 000. The notation 01: $w(01)$ signifies that peer w with leaf id 01 is located at the subtree rooted at node 01. The first row of Table 1 indicates that u has three links z, w and y in its maximal sibling subtrees rooted at k-d tree nodes with ids 1, 01 and 001, respectively.

Table 1. Routing tables example

Peer	link entries		
$u(000)$	1: $z(101)$	01: $w(01)$	001: $y(001)$
$y(001)$	1: $z(101)$	01: $w(01)$	000: $u(000)$
$w(01)$	1: $v(100)$	00: $u(000)$	
$v(100)$	0: $w(01)$	11: $x(11)$	101: $z(101)$
$z(101)$	0: $y(001)$	11: $x(11)$	100: $v(100)$
$x(11)$	0: $u(000)$	10: $v(100)$	

3.3 Peer Joins

When a new physical peer joins MIDAS, it becomes responsible for a single peer. Initially, the newly arrived physical peer chooses a uniformly random point p in the space I and locates the peer v responsible for it; Section 4.1 details points query processing. There are two scenarios depending on the status of the physical peer responsible for v.

In the *first scenario*, the physical peer responsible for v has no other peers. Then, the k-d tree leaf node with id $v.id$ is *split* and two new leaves are created. The splitting dimension $sdim$ of the node is chosen uniformly at random among all possible dimensions, while the splitting value $split$ is the value of the random point p on the $sdim$ dimension. Peer v now corresponds to the left child. Finally, a new peer w is created for the right child and is assigned to the newly arrived physical peer.

To ensure proper functionality, MIDAS takes the following actions. (1) v sends to w the tuples that fall in w's zone. (2) Peer v: (2a) appends 0 to $v.id$; (2b) increments $v.depth$ by one; (2c) appends w as the last entry in $v.link$; (2d) appends to $v.sdim$ and $v.split$ the new splitting dimension and value. (3) Peer w: (3a) copies v's state; (3b) changes the least significant bit of $w.id$ to 1; (3c) changes the last entry in $w.link$ to v. (4) v keeps one half of its $v.backlink$. (5) w keeps the other half of its $w.backlink$. (4) w notifies its backlinks about its address.

The *second scenario* applies when the physical peer responsible for v has multiple peers, that is v is just one of them. In this case, v simply migrates to the newly arrived physical peer, which has the responsibility to notify the backlinks of v about its address.

We present an example of how the network of Figure 1 was constructed. Assume initially that there is a single physical peer A responsible for peer u, whose zone is the entire space, as shown in Figure 3a. Then, physical peer B joins and causes a split of the k-d tree root along the first dimension (Split #1 in Figure 1). Peer u is now responsible for the leaf with id 0. A new peer v is created with the id 1 and is assigned to the newly arrived physical peer B. Figure 3b depicts the resulting k-d tree; the split node is drawn with a bold line.

Assume next that physical peer C joins and chooses a random point that falls in peer u's zone. Therefore, leaf 0 splits, along the second dimension (Split #2). Peer u becomes responsible for the left child and has the id 00, while a new peer w with id 01 is created and assigned to physical peer C. Figure 3c depicts the resulting k-d tree. Then, physical peer D joins selecting a random point inside v's zone. As a result, leaf

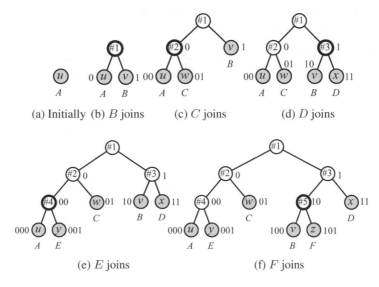

(a) Initially   (b) B joins   (c) C joins   (d) D joins

(e) E joins   (f) F joins

Fig. 3. Network creation

1 splits along $v.sdim[2]$ (Split #3), v is assigned leaf 10, and a new peer x with id 11 is assigned to D; see Figure 3d.

Physical peer E arrives and splits leaf 00 along $u.sdim[3]$ (Split #4). Peer u becomes responsible for the left child and obtains the id 000, while a new peer y with id 001 is created and assigned to E. The resulting k-d tree is shown in Figure 3e. Finally, F joins causing a split of leaf 10 along $v.sdim[3]$ (Split #5). A new peer z is assigned to F with id 101, while peer v gets the id 100. Figure 3f shows the k-d tree after the last join.

The following lemma shows that peer joins in MIDAS are *safe*, that is, Lemma 1 continues to hold.

Lemma 2. *After a physical peer joins, the MIDAS invariant holds.*

Proof. Assume that the MIDAS invariant initially holds. In the first scenario, a physical peer join causes a k-d tree leaf to split. Let u be the peer responsible for the leaf that splits, and let u' denote the same peer after the split. Further, let w denote the new peer created. It holds that k-d tree node $u.id$ is the parent of leaves $u'.id$ and $w.id$. Also, note that MIDAS ensures that $I_u = I_{u'} \uplus^{d_u} I_w$. Therefore, any point in the space I that was assigned to u is now assigned to either u' or w, but not to both. All other points remain assigned to the same peer despite the join.

In the second scenario, observe that when a physical peer joins, no changes in the k-d tree and thus in the peers' zones are made. Hence, in both scenarios, the MIDAS invariant is preserved after a physical peer joins.   □

The probabilistic nature of the join mechanism in MIDAS achieves a very important goal. It ensures that the (expected value of the) depth of the k-d tree, i.e., the maximum length of a root to leaf path, is logarithmic to the number of total k-d tree nodes (and thus of leaves and thus of peers). The following theorem proves this claim.

Theorem 1. *The expected depth of the distributed k-d tree of MIDAS when n peers join on an initially empty overlay is $O(\log n)$ with constant variance.*

Proof. Consider a MIDAS k-d tree of n peers. Since, each internal node has exactly two children (it corresponds to a split), there are $n - 1$ internal nodes. The k-d tree obtained by removing the leaves is an instance of a *random relaxed k-d tree*, as defined in [8], which is an extension of a *random k-d tree* defined in [2]. This holds because the splitting value and dimension are independently drawn from uniform distributions.

It is shown [8], [2] that the probability of constructing a k-d tree by n random insertions is the same as the probability of attaining the same tree structure by n random insertions into a binary search tree. It is generally known that, in *random binary search trees*, the expected value of a root-to-leaf path length is logarithmic to the number of nodes. However, a stronger result from [16] shows that the *maximum path length*, i.e., the depth, has expected value $O(\log n)$ and variance $O(1)$. This results carries over to the MIDAS k-d tree with n peers. □

The previous theorem is essential for establishing asymptotic bounds on the performance of MIDAS. First, it implies that the amount of information stored in each peer is logarithmic to the overlay size. Moreover, as discussed in Sections 4.1 and 4, the theorem provides bounds for the cost of query processing.

3.4 Peer Departures

When a physical peer departs, MIDAS executes the following procedure for each of the peers that it is responsible for. Two possible scenarios exist, depending on the location of the departing peer in the k-d tree.

Let y denote a peer of the departing physical peer E in the *first scenario*, which applies when the sibling of y in the k-d tree is also a leaf and thus corresponds to a peer, say u. Observe that y has a link to u, as the last entry in $y.link$ must point to u. In this scenario, when peer y departs, MIDAS adapts the k-d tree by *removing* leaves $y.id$ and $u.id$, so that their parent becomes a leaf. Peer u is properly updated so that it becomes associated with this parent. In the example of Figure 1a, assume that physical peer E, responsible for y, departs. Peer y's sibling is 000, which is a leaf and corresponds to peer u. Figure 4a shows the resulting k-d tree after E departs. Note that peer u is now responsible for a zone which is the union of y's and u's old zones.

To ensure that all necessary changes in this scenario are propagated to the network, MIDAS takes the following actions. (1) y sends to u all its tuples. (2) Peer u: (2a) drops its least significant bit from its id; (2b) decreases its depth by one; and (2c) removes the last entry from arrays $u.sdim$, $u.split$, $u.link$. (3) y notifies all its backlinks, i.e., the peers that link to y, to update their link to u instead of y. (4) u merges list $y.backlink$ with its own.

Let w be a peer of the departing physical peer C in the *second scenario*, which applies when the sibling of w in the k-d tree is not a leaf. In this case, k-d tree leaf $w.id$ cannot be removed along with its sibling. Therefore, peer w must migrate to another physical peer. Peer w chooses one of its links and asks the corresponding physical peer to assume responsibility for peer w. Ideally, the physical peer that has the lightest load

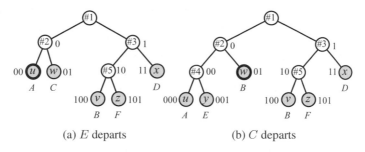

(a) E departs   (b) C departs

Fig. 4. The two scenarios for peer departures

is selected[1] (see also Section 3.6). Note that the backlinks of w must be notified about the address of the new peer responsible for w. In the example of Figure 1a, assume that physical peer C departs. C is responsible for w, whose sibling 00 in the k-d tree is not a leaf. Therefore, w contacts its link v so that its physical peer B assumes responsibility for w. Figure 4b shows the resulting k-d tree after C departs.

The following lemma shows that departures in MIDAS are *safe*, i.e., Lemma 1 continues to hold.

Lemma 3. *After a physical peer departs, the MIDAS invariant holds.*

Proof. Assume that the MIDAS invariant initially holds. Note that a physical peer departure is treated as multiple departures of all peers that it controls.

In the first scenario, a peer departure causes the removal of two k-d tree leaves. Let w be the departing peer and u be the peer responsible for the sibling of $w.id$ in the tree. Further, let u' denote peer u after the departure. Observe that the zone of u' must correspond to some old peer that split along dimension $d_{u'}$, which implies that $I_{u'} = I_u \uplus^{d_{u'}} I_w$. Therefore, any point in the space I that was assigned to either u or w is now assigned to u. All other points remain assigned to the same peer despite the departure.

In the second scenario, observe that when a peer departs, no changes in the k-d tree and thus in the peers' zones are made. Hence, in both scenarios, the MIDAS invariant is preserved after a physical peer departs.   □

3.5 Peer Failures

In a dynamic environment, it is common for peers to fail. MIDAS employs mechanisms that ensure that the distributed index continues to function. Consider that a physical peer fails; the following procedure applies for each peer under the responsibility of the failing peer. MIDAS addresses two orthogonal issues when a peer w fails: (1) another physical peer must take over w, and (2) the key-value pairs stored in w must be retrieved.

Regarding the first, note that all peers connected to w will learn that it failed; this happens because a peer periodically pings its neighbors. Each of the peers responsible

---

[1] Peers periodically inform their backlinks about their load.

for one of w's backlinks knows w's zone (i.e., the boundaries of the region for which w is responsible), but only one must take over w. This raises a distributed agreement problem common in other works; for example, in CAN, the backlinks of w would follow a protocol so that the one with the smallest zone takes over w's zone. However, communication among the peers is not necessary in MIDAS. If w's sibling in the tree is a peer (i.e., a leaf), say u, then the physical peer responsible for u will take over w. If that is not the case, the peer with the smallest id, among w's backlinks will take over w.

Regarding the second issue, note that w (or any peer for that matter) is not the owner of the data it stores. Therefore, it is the responsibility of the owner to ensure that its data exist in the distributed index. This is addressed in all distributed indices in a similar manner. Each tuple is associated with a time-to-live (TTL) parameter. The owner periodically (before the TTL expires) re-inserts the tuples in the index. Therefore, the lost key-value pairs of w will eventually be restored. To increase fault tolerance, distributed indices typically employ replication mechanisms. MIDAS is compatible with them as explained in Section 3.6.

3.6   Load Balancing and Fault Tolerance

Balancing the *load*, i.e, the amount of work, among peers is an important issue in distributed indices. MIDAS can use standard techniques. For example, one could apply the task-load balancing mechanism of Chord [19]. That is, given M physical peers, we introduce $N \gg M$ peers. Then each physical peer is assigned a set of peers so that the combined task-load *per physical peer* is uniform.

To enhance *fault tolerance*, MIDAS can utilize standard replication schemes. For example, consider the multiple reality paradigm, where each reality corresponds to an instance of the domain space indexed by a separate distributed k-d tree. Each data tuple has a replica in every reality. A physical peer contains (at least) one peer in each reality. When initiating a query, a physical peer picks randomly a reality to pose the query to; note that this also results in better load distribution. Then, in case of peer failures and before the key-value pairs are refreshed (see Section 3.5), the physical peer can pose the query to another reality.

4   Query Processing on MIDAS

This section details how MIDAS processes multi-attribute queries. In particular, Section 4.1 discusses point queries, while Section 4.2 deals with range queries.

4.1   Point Queries

The distributed k-d tree of MIDAS allows for efficient hierarchical routing. We show that a peer can process a *point query*, i.e., reach the peer responsible for a given point in the space I, in number of hops that is, in expectation, logarithmic to the total number of peers in MIDAS.

Algorithm 1 details how point queries are answered in MIDAS. Assume that peer u receives a point query message for point q. If its zone contains q, it returns the answer (the value associated with the key q) to the issuer, say w, of the query (lines 1–3). Otherwise, u needs to find the most relevant peer to forward the request to. The most relevant peer is the one that resides in the same maximal sibling subtree with the q. Therefore, peer u examines its local knowledge of the k-d tree (i.e., the $sdim$ and $split$ arrays) and determines the maximal sibling subtree that q falls in (lines 4–10). The query is then forwarded to the link corresponding to that subtree (line 7).

Algorithm 1. u.Point (q, w)
Peer u processes a point query for q issued by w

```
 1: if u.IsLocal(q) then
 2:     u.Send_to (w, u.Get_val (q))
 3:     return
 4: end if
 5: for j ← 0 to u.depth do
 6:     d ← u.sdim[j]
 7:     if (u.id[j] = 0 and q[d] ≥ u.split[j]) or (u.id[j] = 1 and q[d] < u.split[j]) then
 8:         u.link[j].Point (q, w)
 9:         return
10:     end if
11: end for
```

To illustrate the previous procedure, consider a query for point q issued by peer u. Figure 5a draws q and the local knowledge about the space I at peer u. Observe that q falls outside u's zone. Peer u thus forwards the query to its link z within the shaded area since it contains q (1st hop). Next, peer z processes the query. Point q is inside the shaded area of Figure 5a, which depicts z's local k-d tree knowledge. Subsequently, z forwards the query to its link x within that area (2nd hop). Finally, peer x responds to the issuing peer u, as point q falls inside its zone.

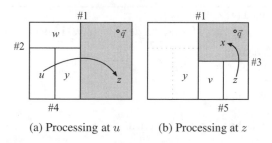

(a) Processing at u (b) Processing at z

Fig. 5. Point query example for q

Lemma 4. *The expected number of hops in a point query is $O(\log n)$.*

Proof. We first show that the number of hops required is at the worst case equal to the depth of the k-d tree.

Assume that the requested point is q. Consider a peer u that executes Algorithm 1. u determines the maximal sibling subtree that q resides in, and let k be the depth of its root. Then, u forwards the request to peer v in that subtree. We argue that v will determine a maximal sibling subtree at depth $\ell > k$. Observe that u and q fall in the same subtree rooted at depth k. Therefore, all subtrees rooted at depths higher than k that contain u will also contain q. The argument holds because all maximal sibling subtrees of u rooted at depths higher than k cannot contain q. The above argument implies that point queries will be forwarded to subtrees of successively higher depth, until the leaf which has q is reached; such a leaf exists because $q \in I$. Therefore, the number of hops is at the worst case equal to the depth of the k-d tree.

From Theorem 1, we have that the expected depth of the k-d tree is $O(\log n)$, which concludes the proof.   □

4.2   Range Queries

A range query specifies a rectangular area Q in the space, defined by a lower ℓ and a higher point h, and requests all tuples that fall in Q. Instead of locating the peer responsible for a corner of the area, e.g., ℓ, and then visit all relevant neighboring peers, MIDAS utilizes the distributed k-d tree to identify in parallel multiple peers whose zone overlaps with Q. The range partitioning idea is similar to the shower algorithm [7], which however applies only for single-dimensional data.

Algorithm 2 details the actions taken by a peer u upon receipt of a range query for area $Q = [\ell, h]$ issued by w. First, u identifies all its tuples inside Q, if any, and sends them to the issuer w (lines 1–3). Then, u examines all its maximal sibling subtrees by scanning arrays $sdim$ and $split$ (lines 4–15). If the area of a subtree overlaps Q (lines 6 and 10), peer u constructs the intersection of this area and Q (lines 7–8 and 11–12). Then, u forwards a request for this intersection to its link (lines 9 and 13). Lines 6–9 (resp. 10–14) apply when u is in the left (resp. right) subtree rooted at depth j.

Algorithm 2. u.Range (ℓ, h, w)
Peer u processes a range query for rectangle $Q = [\ell, h]$ issued by w.

1: **if** u.Overlaps (ℓ, h) **then**
2:      u.Send_to $(w, u$.Get_vals $(\ell, h))$
3: **end if**
4: **for** $j \leftarrow 0$ **to** $u.depth$ **do**
5:      $d \leftarrow u.sdim[j]$
6:      **if** $u.id[j] = 0$ and $u.split[j] < h[d]$ **then**
7:          $\ell' \leftarrow \ell$
8:          $\ell'[d] \leftarrow u.split[j]$
9:          $u.link[j]$.Range (ℓ', h, w)
10:     **else if** $u.id[j] = 1$ and $u.split[j] > \ell[d]$ **then**
11:         $h' \leftarrow h$
12:         $h'[d] \leftarrow u.split[j]$
13:         $u.link[j]$.Range (ℓ, h', w)
14:     **end if**
15: **end for**

Figure 6 illustrates an example of a range query issued by peer u. Initially, u executes Algorithm 2 for the range depicted as a bold line rectangle in Figure 6a. Peer u retrieves the tuples inside its zone that are within the range; these tuples reside in the non-shaded region of the range in Figure 6a. Then, u constructs the shaded regions, shown in Figure 6a, as the intersections of the range with the area corresponding to its maximal sibling subtrees. For each of these shaded regions, u forms a new query and sends it to the appropriate link (1st hop); the messages are depicted as arrows in Figure 6. For instance, peer z, which is u' link in the maximal sibling subtree rooted at depth one, receives a query about the light shaded area.

Peers w, y and z receive a query from u. The range for w and y falls completely within their zone. Therefore, they process them locally and do not send any other message. Figure 6b illustrates query processing at peer z, where the requested range is drawn as a bold line rectangle. Observe that this range does not overlap with z's zone; therefore, z has no tuple that satisfies the query. Then, z constructs the intersections of the range with the areas in its maximal sibling subtrees. Observe that the range does not overlap with the maximal sibling subtree rooted at depth one; hence, no peer receives a duplicate request. Peer z sends a query message to its links v and x with the shaded regions of Figure 6b (2nd hop). Finally, peers v and x process the queries locally as the requested ranges have no overlap with their zones.

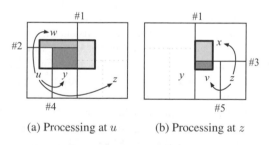

(a) Processing at u (b) Processing at z

Fig. 6. Range query example

As explained in the example of Figure 6, the query is answered in two hops. In the first, u reaches y, w, and z, and in the second, z reaches v and x. The following lemma shows that the expected number of hops is logarithmic to the number of peers.

Lemma 5. *The expected number of hops for processing a range query is $O(\log n)$.*

Proof. We show that the number of hops required is at the worst case equal to the depth of the k-d tree.

Consider that virtual peer v receives from u a query for range $Q = [\ell, h]$; assume that v is the link at depth k in the $link$ array of u. Due to its construction (the result of an intersection operation), range Q is completely contained within the subtree that contains v rooted at depth k. Therefore, Q cannot intersect with any maximal sibling subtree of v at depth lower than k. As a result, if v forwards a range query, it will be to links at depths strictly higher than k.

In the worst case, a message will be forwarded to as many virtual peers as the depth of the k-d tree. The fact that the expected depth is $O(\log n)$ (from Theorem 1) completes the proof. $\qquad\qquad\qquad\qquad\qquad\qquad\qquad\qquad\qquad\qquad\qquad\qquad\qquad\qquad$ $\square$

5   Experimental Evaluation

In order to assess our methods and validate our analytical results, we simulate a dynamic environment and study each query type. We implement two methods from the literature, CAN [15] and the VBI-tree [11], to serve as competitors to MIDAS.

5.1   Setting

Network. We simulate a dynamic topology to capture arbitrary peer joins, and departures, by implementing two distinct stages. In the *increasing stage*, peers continuously join the network, while no peer departs. It starts from an overlay of 1,000 peers and ends when 100K peers are available. On the other hand, in the *decreasing stage*, peers continuously leave the network, while no new peer joins. This stage starts from an overlay of 100K peers and ends when only 1,000 peers are left. When depicting the effect of these stages in the figures, the solid (resp. dashed) line represents the increasing (resp. decreasing) stage.

Data and Queries. We use both a *real* as well as *synthetic* datasets of varying dimensionality. The real dataset, denoted as *NE*, consists of spatial (2D) points representing 125K postal addresses in three metropolitan areas (New York, Philadelphia and Boston). The synthetic datasets contain 1M tuples uniformly and independently distributed in the domain space. The dimensionality is varied from 2 up to 19.

For each point query, we choose uniformly and independently a random location in the domain space. For each range query, we also choose a random location, while the length of the rectangular sides are selected so that the query returns approximately 50 tuples. In all figures, the reported values are the averages of executing 50K queries over 10 distinct overlay topologies.

Performance Metrics. We employ several metrics to evaluate MIDAS against other methods. The basic metric indicative of query performance is *latency*, which is defined as the maximum distance (in terms of hops) from the issuing peer to any peer reached during query processing. Clearly, lower values suggest faster response.

Distributed query processing imposes a task load on multiple peers. Two metrics quantify this load. *Precision* is defined as the ratio of the number of peers that contribute to the answer over the total number of peers reached during processing of a query; the optimal precision value is 1. *Congestion* is defined as the average number of queries processed at any node, when n uniformly random queries are issued (n is the number of peers in the network); lower values suggest lower average task load.

The next metrics are independent of the query evaluation process. Note that two types of information are stored locally in each peer. The first is overhead information (e.g., links, zone description, etc.), which is measured by the average *state* a peer must maintain. As the number of peers increase, this is an important measure of scalability. The

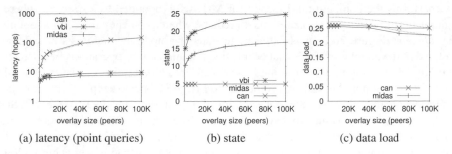

(a) latency (point queries)   (b) state   (c) data load

Fig. 7. Latency, state cardinality a peer maintains, and data load for the NE dataset

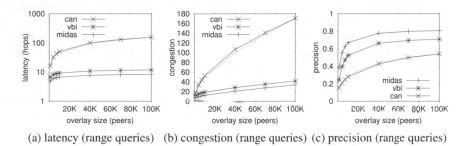

(a) latency (range queries)   (b) congestion (range queries) (c) precision (range queries)

Fig. 8. Latency, congestion, and precision for the NE dataset

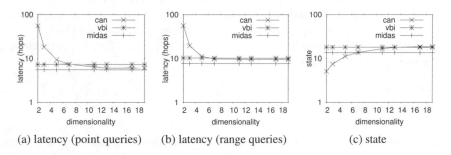

(a) latency (point queries)   (b) latency (range queries)   (c) state

Fig. 9. Latency, cardinality of the state a peer maintains for multidimensional dataset

second type is the number of tuples stored. As this depends on the dataset distribution and the network topology, imbalances can occur. *Data load* measures the percentage of the total data in the network, that resides in the top $Q\%$ most loaded peers. We show measurements for $Q = 10\%$, where the optimal fairness value is 0.1.

5.2   Results

This section presents the findings of our experimental study. Figure 7a illustrates query performance aspects for point queries for spatial workloads, where MIDAS clearly outperforms the competition. Latency for MIDAS is bounded by $O(\log n)$, where n stands for the overlay size, as Lemma 4 predicts; whereas, latency for CAN is bounded by $O(\sqrt[d]{n})$. Figure 7b depicts the average state of a node maintains, as it is directly related to the amount of traffic that occurs due to maintenance operations like detecting failures, preserving updated routing tables each time a peer joins or leaves. Note that state is increased in the VBI-tree compared to MIDAS as peers keep information about the peers on their path from the root and their siblings at the same depth of the tree. State in MIDAS has a logarithmic behavior in terms of the overlay size, as Theorem 1 states. In low dimensionality settings, such as the spatial dataset *NE*, CAN peers maintain very few links to others. Moreover, Figure 7c presents the data load. CAN achieves lower data load mainly due to its joining protocol. In particular, it chooses to halve a peer's area for a newcomer, instead of splitting its data load like in MIDAS. As a result, CAN becomes vulnerable to data skew. Note that in all methods, data load slightly decreases as the number of peers increases.

We next discuss the case of range queries. Latency shows similar logarithmic behavior with range queries (Figure 8a) as Lemma 5 predicts. Figure 8b illustrates that the congestion in MIDAS has logarithmic behavior in terms of the overlay size, and is significantly lower than its competitors.

In Figure 8c precision improves with overlay size. The reason is that peers become responsible for smaller areas, while the queries have fixed size, in other words less irrelevant peers are reached. Precision is worse for VBI-tree and CAN compared to MIDAS because of the longer routes required to reach relevant peers (see latency).

We finally discuss synthetic datasets of varying dimensionality. As Figures 9a and 9b, MIDAS is largely unaffected by dimensionality and is asymptotically better than both the VBI-tree and CAN. The expected latency for CAN is bounded by $O(\sqrt[d]{n})$, and thus, decreases as dimensionality increases. However, this comes at an extra cost. As Figure 9c shows, the expected cardinality of a node's routing table increases with dimensionality and it is confirmed to be in $O(d)$, whereas MIDAS' peer state is in $O(\log n)$ regardless dimensionality degree.

6   Conclusion

In this work we have presented MIDAS, a pure multidimensional indexing scheme for large-scale decentralized networks, which significantly differs from other popular overlays which are either single dimensional or implement a space filling curve. Our peer-to-peer variant offers the possibility of combining DHT with hierarchical space partitioning schemes, avoiding order-preserving hashing and space-filling curves. Yet, it outperforms other multidimensional structures in terms of scalability. MIDAS allows for multidimensional queries and offers guarantees concerning all operations. In particular, updates, points and range queries are resolved in $O(\log n)$ hops. Most importantly, each peer in MIDAS maintains $O(\log n)$ state only. All things considered, MIDAS constitutes an extremely attractive solution when it comes to high dimensional datasets that provides a rich and wide functionality. Finally, interesting results arose from this work. The curse of dimensionality has no impact on query performance and maintenance costs, while MIDAS achieves high levels of fairness.

References

1. Aberer, K., Cudré-Mauroux, P., Datta, A., Despotovic, Z., Hauswirth, M., Punceva, M., Schmidt, R.: P-grid: a self-organizing structured p2p system. SIGMOD Record 32(3), 29–33 (2003)
2. Bentley, J.L.: Multidimensional binary search trees used for associative searching. Commun. ACM 18(9), 509–517 (1975)
3. Bentley, J.L.: K-d trees for semidynamic point sets. In: Symposium on Computational Geometry, pp. 187–197 (1990)
4. Bharambe, A.R., Agrawal, M., Seshan, S.: Mercury: supporting scalable multi-attribute range queries. In: SIGCOMM, pp. 353–366 (2004)
5. Blanas, S., Samoladas, V.: Contention-based performance evaluation of multidimensional range search in p2p networks. In: InfoScale 2007, pp. 1–8 (2007)
6. Cai, M., Frank, M.R., Chen, J., Szekely, P.A.: Maan: A multi-attribute addressable network for grid information services. J. Grid Comp. 2(1), 3–14 (2004)
7. Datta, A., Hauswirth, M., John, R., Schmidt, R., Aberer, K.: Range queries in trie-structured overlays. In: P2P Computing, pp. 57–66 (2005)
8. Duch, A., Estivill-Castro, V., Martínez, C.: Randomized k-dimensional binary search trees. In: Chwa, K.-Y., Ibarra, O.H. (eds.) ISAAC 1998. LNCS, vol. 1533, pp. 199–208. Springer, Heidelberg (1998)
9. Ganesan, P., Yang, B., Garcia-Molina, H.: One torus to rule them all. Multidimensional queries in p2p systems. In: WebDB, pp. 19–24 (2004)
10. Jagadish, H.V., Ooi, B.C., Vu, Q.H.: Baton: A balanced tree structure for peer-to-peer networks. In: VLDB, pp. 661–672 (2005)
11. Jagadish, H.V., Ooi, B.C., Vu, Q.H., Zhang, R., Zhou, A.: Vbi-tree: A peer-to-peer framework for supporting multi-dimensional indexing schemes. In: ICDE, p. 34 (2006)
12. Karger, D., Lehman, E., Leighton, T., Panigrahy, R., Levine, M., Lewin, D.: Consistent hashing and random trees: Distributed caching protocols for relieving hot spots on the world wide web. In: ACM Symp. on Theory of Comp., pp. 654–663 (1997)
13. Maymounkov, P., Mazières, D.: Kademlia: A peer-to-peer information system based on the xor metric. In: Druschel, P., Kaashoek, M.F., Rowstron, A. (eds.) IPTPS 2002. LNCS, vol. 2429, pp. 53–65. Springer, Heidelberg (2002)
14. Plaxton, C.G., Rajaraman, R., Richa, A.W.: Accessing nearby copies of replicated objects in a distributed environment. Theory Comput. Syst. 32(3), 241–280 (1999)
15. Ratnasamy, S., Francis, P., Handley, M., Karp, R., Schenker, S.: A scalable contentaddressable network. In: SIGCOMM 2001, pp. 161–172 (2001)
16. Reed, B.A.: The height of a random binary search tree. Journal of the ACM 50(3), 306–332 (2003)
17. Rowstron, A., Druschel, P.: Pastry: Scalable, decentralized object location, and routing for large-scale peer-to-peer systems. In: Liu, H. (ed.) Middleware 2001. LNCS, vol. 2218, pp. 329–350. Springer, Heidelberg (2001)
18. Shu, Y., Ooi, B.C., Tan, K.-L., Zhou, A.: Supporting multi-dimensional range queries in peer-to-peer systems. In: Peer-to-Peer Computing, pp. 173–180 (2005)
19. Stoica, I., Morris, R., Liben-Nowell, D., Karger, D.R., Kaashoek, M.F., Dabek, F., Balakrishnan, H.: Chord: a scalable p2p lookup protocol for internet applications. IEEE/ACM Trans. Netw. 11(1), 17–32 (2003)
20. Zhao, B., Kubiatowicz, J., Joseph, A.D.: Tapestry: a resilient global-scale overlay for service deployment. IEEE Journal on Selected Areas in Comm. 22(1), 41–53 (2004)

Thread-Level Parallel Indexing of Update Intensive Moving-Object Workloads

Darius Šidlauskas[1,*], Kenneth A. Ross[2,**],
Christian S. Jensen[3], and Simonas Šaltenis[1]

[1] Aalborg University
{darius,simas}@cs.aau.dk
[2] Columbia University
kar@cs.columbia.edu
[3] Aarhus University
csj@cs.au.dk

Abstract. Modern processors consist of multiple cores that each support parallel processing by multiple physical threads, and they offer ample main-memory storage. This paper studies the use of such processors for the processing of update-intensive moving-object workloads that contain very frequent updates as well as contain queries.

The non-trivial challenge addressed is that of avoiding contention between long-running queries and frequent updates. Specifically, the paper proposes a grid-based indexing technique. A static grid indexes a near up-to-date snapshot of the data to support queries, while a live grid supports updates. An efficient cloning technique that exploits the memcpy system call is used to maintain the static grid.

An empirical study conducted with three modern processors finds that very frequent cloning, on the order of tens of milliseconds, is feasible, that the proposal scales linearly with the number of hardware threads, and that it significantly outperforms the previous state-of-the-art approach in terms of update throughput and query freshness.

1 Introduction

As wireless networks and sensing devices continue to proliferate, databases with frequently updated data will gain in prominence. A prominent instance of this scenario occurs when a database is maintained of the locations of moving objects carrying on-line devices with GPS receivers. In this setting, an update usually concerns only a single object, while a query may concern a single object (object query) or multiple objects (e.g., range, kNN, or aggregate query).

For example, consider a population of 10 million vehicles that send location updates to a central server and that subscribe to services that are enabled by queries against the vehicle location database. Assuming that a high location

* Supported by the Danish Research Council.
** Supported by NSF grants IIS-0915956 and IIS-1049898.

D. Pfoser et al. (Eds.): SSTD 2011, LNCS 6849, pp. 186–204, 2011.

accuracy of 10 meters is required and that the vehicles move with an average speed of 10 m/s, the resulting workload will contain some 10 million updates/s.

With a few notable exceptions [6, 19], past research on update performance in spatio-temporal indexing assumes that the data is disk resident [3, 11, 16, 18]. This paper's focus is on supporting extreme workloads assuming main-memory resident data using modern processors that consist of multiple cores, each with multiple hardware threads. With main memories reaching terabytes, this assumed setting is realistic. The challenge is to exploit this available parallelism, e.g., by avoiding serialization bottlenecks.

Past studies [4] suggest that standard index concurrency control mechanisms do not scale in such parallel settings; performance already degrades when the number of threads exceeds two. This is due to interference between update and query operations. An alternative strategy is to isolate the conflicting operations by allowing them to operate on different copies of data [6]. Queries are carried out on a near up-to-date, read-only snapshot of the data, whereas updates are applied to a write-only data structure. This strategy improves performance at the cost of slightly outdated query results.

Our proposal, called TwinGrid, adopts this general strategy. It takes an existing index structure [19] that couples a spatial grid index with a secondary object-id based index as its starting point. TwinGrid is equipped with a second grid index. Updates are directed to the first grid, while queries are served by the second grid. A copying mechanism is proposed that supports the efficient copying of an entire memory-resident data structure, thus enabling the second grid to be a near up-to-date copy of the first, in turn resulting in near up-to-date query results.

The copying mechanism uses the `memcpy` system call. Figure 1 shows the results of profiling the performance of `memcpy` on three processors. For example, the Nehalem processor (an 8-core Intel Xeon X5550) can copy 8.5 GB/s using at least 8 of the 16 available hardware threads. As copying involves reading and writing this copying uses about half of Nehalem's maximum memory bandwidth of 32 GB/s. Continuing our example from above with 10 million objects, if an item occupies 20 bytes, the index occupies 200 MB and can be copied in 24 ms. In contrast, the previous state-of-the-art proposal, called MOVIES, is capable of building its index in about 3 seconds on a 4-core AMD Opteron processor [6].

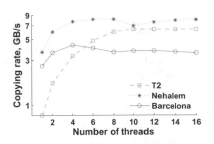

Fig. 1. `memcpy` Performance

The paper reports on an extensive empirical study of TwinGrid and a multi-threaded version of MOVIES. The study uses three different, modern processors that provide from 8 to 64 hardware threads. The results show that TwinGrid's performance scales linearly with the number of threads on all three processors and that TwinGrid is able to support up to 10 million updates per second with query results being outdated by only tens of milliseconds in realistic settings. Additionally, the study shows that TwinGrid outperforms MOVIES substantially in terms of both update throughput and query freshness.

The rest of the paper is structured as follows. The next section briefly covers preliminaries and related work. Section 3 details the design of TwinGrid and the ideas underlying its design, and it compares TwinGrid with the previous state-of-the-art approach. Section 4 reports the results of a thorough empirical study, and Section 5 concludes.

2   Preliminaries

2.1   Problem Setting

A setting is assumed where a collection of objects move within a two-dimensional space, regularly sending updates with their current locations to a server. Each object has a unique identifier termed *oid*.

To take into account future developments, we assume the objects acquire their locations using a Global Navigation Satellite System (e.g., GPS or Galileo) that offers an accuracy on the order of a few meters. An update of an object *oid* is a three-tuple (oid, x, y). The object's old location is deleted, and its new location (x, y) is inserted.

We consider the problem of efficiently supporting workloads consisting of such updates as well as spatial queries, such as range and kNN queries, against the current locations of the objects. In particular, we assume a large population of objects where each object issues updates frequently.

2.2   Related Work

Spatial indexes were initially invented primarily with efficient querying in mind and were perfectly fine for static data (e.g., [2, 7]). However, new applications such as location-based services exhibit workloads that not only contain queries, but also contain frequent updates, studies now also consider update-intensive workloads [3, 11, 16, 18]. While data has traditionally been assumed to be stored on disk, the needs for high performance and increases in main memory sizes call for techniques that assume main-memory resident data [3, 6, 10, 19, 8].

This paper considers the natural next step in this evolution, namely that of exploiting the parallelism that is increasingly becoming available in modern processors. We are not aware of any spatio-temporal indexing proposals that aim to exploit this parallelism. To the best of our knowledge, the proposal most closely related to ours is MOVIES [6], which we describe and parallelize in Section 3.4 and empirically compare with TwinGrid in Section 4.4.

Since TwinGrid is based on data snapshots, an interesting and performance critical question is how to create snapshots. One approach is to use the idea of differential files [14], where changes to the snapshot are accumulated separately from the snapshot and where the two are merged regularly. MOVIES follows this approach: logged updates and the previous snapshot are used to bulk-load a new snapshot. We show that our technique outperforms this approach.

Another snapshot creation approach is based on the idea of copy-on-write file systems [15]. A snapshot is made by maintaining an additional set of references to the elements in the main data structure. Multiple copies of the same element then represent the element at different times. Rastogi et al. [13] apply this idea in the T-tree. New versions of tree nodes are created when updates are about to modify them. The merging is relatively cheap as only pointers have to be swapped. However, the high cost of creating versions renders this technique useful only for workloads with query intensive workloads with at most 4% updates (according to results obtained with a SPARCstation 20 with two 200 MHz processors and 256 MB of main memory [13]).

2.3   Multi-threaded Processing

Modern processors, known as chip multi-processors or CMPs, have multiple cores and multiple hardware threads per core, where all hardware threads share memory and (parts of) the cache hierarchy. Although low inter-thread communication and synchronization latencies on the same die enable fine-grained parallelization, which was previously unprofitable (on symmetric multi-processors or SMPs), efficient utilization of such hardware remains a non-trivial task, especially for the type of workload we consider.

In single-threaded processing, updates and queries are executed in their order of arrival. At any point in time, either a query or an update is being processed. This guarantees consistent results, as all operations are carried out in isolation. However, updates that arrive during the processing of a query must be buffered. Thus, an update can be delayed by the time it takes to process a query. This may delay an update very substantially.

With multi-threaded processing, special care must be taken to guarantee consistent results. Often full serializability (as in database transaction processing) is not required. Rather, queries must be *correct* in the sense that they must reflect a state of the database that was correct at some point in time. This setting renders it challenging to manage the possible interference between rapid updates and relatively long-running queries.

To address this issue, a variety of concurrency control schemes can be used. However, queries must wait until all necessary exclusive latches are released by updaters so that a consistent state can be read; and updates must wait for queries holding shared latches. Long-running queries have the potential to slow down updates. Moreover, extra care must be taken to avoid deadlocks and phantom problems. Deadlocks are especially common when updates and queries acquire locks in opposite order (e.g., see bottom-up updates and top-down queries in Section 3). Phantoms can, for instance, arise when queries obtain locks on the fly and a later update alters a data item, before a query gets the chance to lock it, in such a way that it no longer satisfies the query condition. The traditional solution to phantoms is to use predicate locks or table locks, which have the effect of delaying even more updates.

3   Parallel Workload Processing

This section describes the TwinGrid indexing technique. First, we describe an existing update-efficient index structure for a single-threaded environment. Second, we extended this to efficiently support multi-threaded processing. Then, we compare our approach with the preexisting state-of-the-art proposal that also uses snapshots.

3.1   Single-Threaded Processing

TwinGrid extends an existing memory-resident grid-based index structure, called u-Grid, that offers high performance for a traffic monitoring application [19].

Structure. The structure and components of u-Grid are shown in Figure 2 and explained next. Two important design decisions underlie u-Grid. First, it was decided to use a fixed and uniform grid [1], the equally-sized square cells of which are oriented rectilinearly and partition a predefined region to be monitored. Objects with coordinates within the boundaries of a cell belong to that cell. This reduces index maintenance costs, as no grid refinement or re-balancing are needed (in contrast to adaptive grids or hierarchical space partitioning methods, e.g., [12,7]).

Second, a secondary index is employed that enables updates in a bottom-up fashion [11]. The objects are indexed on their *oid* so that the index provides direct access to an object's data in the spatial index.

The grid directory is stored as a 2-dimensional array, where each element of the array corresponds to a grid cell. Each cell stores a pointer to a linked list of buckets. Instead of storing identifiers of (or pointers to) actual data, the objects themselves are stored in the buckets to improve retrieval performance. The data to be processed during updates and queries is loaded in units of blocks (cache lines) into the CPU cache. When compared to using a linked list of objects, the use of large buckets increases data access

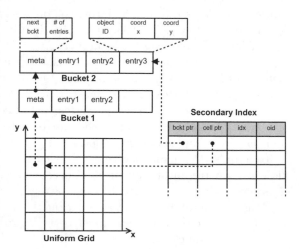

Fig. 2. Basic Indexing Scheme

locality and enables more effective prefetching by modern CPUs. Each bucket has object data and metadata fields. The metadata field contains a pointer to the next bucket and the current number of objects in the bucket.

Update. Updates are categorized as *local* or *non-local*. An update of an object determines the object's old and new cells. If the two are the same cell, the update is local, and it involves only updating the x and y coordinates. A non-local update involves deletion of the object from its old cell and insertion of it into the new cell.

The objects in a bucket are unordered. However, the insertion and deletion algorithms ensure that all but the first bucket of a grid cell are full. A new object is inserted at the end of the first bucket. In case it is full, a new bucket is allocated, and the necessary pointers are updated so that this bucket becomes the first. The deletion algorithm moves the last object of the first bucket into the place of the object to be deleted. If the first bucket becomes empty, it is removed and the next bucket becomes the first or the grid cell becomes empty and stores a null pointer. Also, since the secondary index knows where exactly an object is located (cell pointer, *cell_ptr*, bucket pointer, *bckt_ptr*, and entry index within bucket, *idx*, in Figure 2), there is no need for cell/bucket scanning during updates. This is particularly desirable for dense cells that contain many buckets.

To process an update, a secondary index entry is retrieved based on an incoming object's *oid*, and a new cell is determined according to the new x and y coordinates. The new cell is compared with the object's old cell, referenced by the pointer *cell_ptr*, to check whether the update is local or non-local. Then, *bckt_ptr* and *idx* are used to compute the direct address of the entry to be updated. If the update is non-local, pointer *cell_ptr* is also used during the deletion to move the last object from the first bucket of the old grid cell into the place of the deleted object. Therefore, a non-local update also requires modifying the affected values in *cell_ptr*, *bckt_ptr*, and *idx*.

Querying. Object queries are used in updates and involve only consulting the secondary index with *oid*.

A range query is defined by a rectangle given by two corner points $(x_{q_{min}}, y_{q_{min}})$ and $(x_{q_{max}}, y_{q_{max}})$. The algorithm first partitions the cells covered by the range query into fully covered cells and partially covered cells. The objects from the fully covered cells are put into the result list by reading the corresponding buckets. The buckets from the partially covered cells are scanned, each object being checked individually to determine whether it is within the query range.

Other types of queries can easily be supported (e.g., kNN queries [19]).

3.2   TwinGrid

As mentioned earlier, one strategy that avoids interference between updates and queries in parallel processing is to service queries using a separate, read-only snapshot of the data that does not require locking. Then updates to the latest data can progress in parallel with the queries. Based on this, TwinGrid extends the above indexing approach by maintaining two copies of the spatial index. One copy, *reader-store*, is read-only and is used to service (long-running) queries. The other, *writer-store*, is used to service updates and object queries. The writer-store represents the most up-to-date copy of the data as incoming updates are always applied to it as they are received.

To ensure near up-to-date query results, the read-only snapshot is refreshed regularly, by copying the writer-store over the reader-store. Specifically, the entire memory-resident writer-store is copied using the `memcpy` system call. Before copying, updating is paused momentarily so that the writer-store achieves a consistent state.

An important consideration when performing this copying is to ensure that pointers are handled appropriately. For instance, if a bucket list with pointers is simply copied, the pointers in the newly copied bucket still reference the original bucket list, yielding a snapshot that is not a faithful copy of the data.

To overcome this pointer problem, TwinGrid uses what we call *container-based memory allocation* in which data items are allocated within *containers*. All data structures are allocated within a single container, and all pointers in a container are interpreted as offsets from the base address of that container. When data is copied, it is copied to a new container, thus preserving the relative positioning within the container. This scheme thus enables the faithful copying of pointer-based structures. TwinGrid allocates all writer-store components within one container and copies this to the reader-store's container during refreshing.

With container-based memory allocation, it is necessary to always preallocate a large enough amount of memory to a container so that the container will not run out of space. To embrace data growth, we set a minimum free space requirement within each container. Whenever this requirement is violated, two larger containers are allocated (before refreshing) and replicated with two copies of the writer-store container. One container becomes the new (and bigger) reader-store, while the other becomes the new writer-store. The old containers are discarded. We note that in the workloads we are considering, the modifications are predominantly updates rather than insertions or deletions.

Example. To make the presentation more concrete, we consider the following example. Figure 3 shows how the uniform grid partitions the monitored region into nine cells, assigning each moving object ($o0$, $o1$, ..., $o9$) to some particular cell (numbered from 0 to 8). Figure 4 depicts the resulting TwinGrid main-memory structure. The components reader-store and writer-store are identical and represent two copies of the data. Each store is stored within a container, and each container includes the corresponding grid directory and array of buckets populated with all the object data.

Bucket positions are determined based on offsets from a container base address. Value -1 is equivalent to the `null` pointer. The cells numbered 0, 1, 3, and 8 are empty; thus, the grid directory in reader-store (also in writer-store) contains -1 in the corresponding entries. A non-empty cell contains the index of the first bucket belonging to that cell. For instance, cell 6 contains 0, and the bucket at offset 0 contains one object, which is object $o1$.

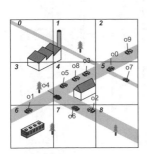

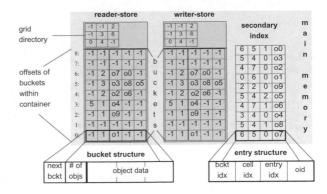

Fig. 3. Monitored Area   **Fig. 4.** TwinGrid Main-Memory Components

Consider multi-bucket cell 4 with four objects that contains 3 as the offset for the first bucket. Bucket 3 contains one object ($o4$), and the field for the next bucket indicates a bucket at offset 5. Bucket 5 contains the remaining three objects ($o3$, $o8$, and $o5$). This shows how only the first bucket may be non-full, so that scanning in the insertion and deletion algorithms is avoided.

Simple oid-based lookups (during update or object query) are processed efficiently on the current data, and the snapshot is targeted at long running queries involving many objects for which the secondary index is not helpful. Therefore, the secondary index is never copied and not stored in the container. Nevertheless, the actual object locations are described as offsets within the writer-store container too—see the entry structure of the secondary index in Figure 4.

3.3   Multi-threaded Processing

Before the writer-store can be copied to the reader-store, it must be in a consistent state. This is achieved by interrupting the processing of updates (Section 4.2 shows that the interruption is at the order of tens of milliseconds). Therefore, TwinGrid progresses by switching between two phases: update processing and cloning. This introduces a new parameter, the *cloning period* (cp), that defines how often the reader-store is refreshed. From a query perspective, the parameter defines the maximum staleness of a query result.

Figure 5 depicts the thread-level synchronization in TwinGrid. A main thread starts by creating a number of worker threads. The workers start processing incoming messages, updates and queries, while the main thread goes to sleep. After cp time units have passed, the main thread wakes up, sends a message to the workers that cloning is required, and blocks until cloning is done. When the workers notice a cloning request, they do not take on new messages, but finish their ongoing processing and go to a synchronization point (barrier). The cloning starts when all worker threads are at the synchronization point.

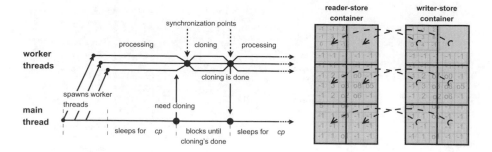

Fig. 5. Thread-Level Synchronization   **Fig. 6.** Copying in Parallel

When cloning starts, all threads take part in copying the writer-store container over the reader-store container. Each worker-thread gets assigned a memory range. Figure 6 shows such an example with 6 thread copiers. After the cloning is done (and all threads have arrived at the next synchronization point), one worker-thread (the last to finish its copying) signals the main thread that cloning is done. Again, all workers start workload processing while the main thread goes to sleep for another cloning period.

Listing 1. Thread-safe Update in TwinGrid

```
1  bool UPDATE( Obj u ) {
2    bool done = false;
3    SIEntry obj = SI.lookup(u.oid);
4    lock(obj);
5    int old_cell = obj.cell_idx;
6    int new_cell = CELLIDX(u.x, u.y);
7    if ( new_cell == old_cell ) {
8      /* Local update */
9      obj.x = u.x; obj.y = u.y;
10     done = true;
11   } else {
12     /* Non-local update */
13     LOCKCELLS(old_cell, new_cell);
14     Obj lst_obj = LASTOBJ(old_cell);
15     if ( trylock(lst_obj) ) {
16       DELETE(obj); //using lst_obj;
17       unlock(lst_obj);
18       INSERT(u, new_cell);
19       done = true;
20     }
21     UNLOCKCELLS(old_cell, new_cell);
22   }
23   unlock(obj);
24   return done;
25 }
```

TwinGrid enables multiple threads to serve queries without interfering with concurrently updating threads. However, the arrangement with multiple updates operating on the same writer-store requires careful attention.

Listing 1 shows pseudocode for the thread-safe update algorithm. Since local updates read and write single objects, they are relatively easy to perform. An updating thread retrieves the corresponding object from a secondary index (line 3), obtains a lock on it (line 4), and determines the update type (lines 5–7). If it is local, the old object coordinates are simply updated with the new ones (line 9). Single object updates are short, and due to application semantics, the server rarely, if ever, encounters concurrent update messages operating on the same object.

In case of a non-local update, the object has to be removed from its old cell and inserted into the new cell. When multiple non-local updaters operate within the same cell(s), incorrect behavior can result. For example, multiple updaters can try to move their objects into the same new cell. This implies that they all try to write to the same (last) position of the first bucket. With unlucky timing, the threads can overwrite each others insertions. Therefore, each non-local updater, in addition to object locks, obtains locks on the two cells involved (line 13). The locks are obtained in increasing order of cell number so that deadlocks are avoided. This guarantees that only one thread will operate on the two participating cells.

Next, the call LASTOBJ retrieves the last object of the first bucket in the given cell (line 14), which will be used to overwrite (delete) the old object's position. Since this object might already be locked by other updaters, we do trylock, which returns immediately with false if the object is already locked. If the lock was successfully obtained, the old object is deleted (line 16) and the just acquired lock is released (line 17). After object insertion into the new cell (lines 18), the remaining locks are released and the update completes. Note that in case we failed to lock lst_obj, all locks are released and false is returned (the update has to be retried[1]). This eliminates (very rare) deadlocks that occur when another non-local updater happens to lock lst_obj already in line 4 and waits until the same cell(s) will be unlocked by this updater.

The incoming update and query messages are placed in a queue-like data structure and are processed by worker-threads in a FIFO manner. To guarantee that the same message is not processed by more than one thread, we use the atomic compare-and-swap instruction available on most commodity hardware. In particular, each thread reads and atomically increments a counter that takes as its value the number of the message in the queue to be processed next. This mechanism causes threads to serialize their execution, thus reducing parallelism, if the queue is accessed more frequently than the atomic operation can be processed. To reduce the effects of such contention, TwinGrid uses multiple queues. Each thread then reads messages only from one queue that is assigned to it. To avoid load imbalances among the threads, the queues are filled in a round-robin fashion. Section 4.3 studies the effect the number of queues can have on TwinGrid performance.

The use of multiple queues may result in the queries and updates in a workload being processed in an order slightly different from the submission order. This is acceptable in the intended applications because some randomization in the arrival order can already occur due to differences in transmission delays, which have relatively larger effects than those caused by the use of multiple queues. To ensure that an updater sees its own updates, all updates and queries from a single data producer can be directed to the same queue. Only if there is a small number of updaters relative to the number of threads, this may result

---

[1] Similarly to reading retries in concurrency control of main-memory R-trees [8].

in load imbalances. In the target application, where updaters are vehicles, load imbalance is unlikely.

Note that the choice of copying the writer-store over the reader store can be relaxed: multiple historical snapshots can be accumulated by copying the writer-store to a newly allocated container. Old snapshots are kept as long as there are queries using them. This arrangement ensures that cloning is not de-layed by very-long running queries. This is so because cloning can be started without waiting for threads processing very-long running queries to finish; only the fast object queries need to finish before cloning can start. Specifically, with a 200MB data structure and 20GB of main memory, one can keep 100 snap-shots in main memory. At 24 ms between snapshots, these cover 2.4 s, thus enabling the processing of queries that run very long for the setting considered on memory-resident data. The maximum range query processing time in our considered workloads is on the order of ten milliseconds (see Section 4).

3.4 Comparison with MOVIES

As MOVIES [6] is the only proposal known to us that uses the snapshot-based strategy for the processing of moving-object workloads, we compare with MOVIES in detail.

Conceptual Comparison. MOVIES is based on frequently building short-lived throwaway indexes. In particular, during a time interval, called the *frame time* T_i, a read-only index, I_i, is used to answer queries, while all incoming updates are collected in a dedicated update buffer, B_i. In addition, an index I_{i+1} is being built during T_i based on the updates collected in the update buffer B_{i-1} during T_{i-1}. If the buffer B_{i-1} does not contain updates for some objects (i.e., objects not updated during T_{i-1}), index I_i is consulted for the missing data in order to build I_{i+1}.

As soon as I_{i+1} is built, the update buffer B_{i-1} and the old index I_i are destroyed, and a new time interval T_{i+1} is started, where index I_{i+1} is the new read-only index that is used to answer queries, while all incoming updates are collected in a new buffer, B_{i+1}, etc. Update buffers can be organized using an array or hash table structure, while linearized kd-tries [17] are used for indexing.

The minimum T_i duration depends on how fast the read-only index can be built using some bulk-loading technique ($T_i > T_{bulk}$). The query staleness is $2 \times T_i$ because queries issued in time frame T_{i+1} are processed using index I_{i+1} that was built in T_i using update buffer B_{i-1}. In TwinGrid, query staleness depends on the chosen cloning period because as soon as cloning is done, queries are serviced using the fresh snapshot. Moreover, it can support some queries with current data by scheduling cloning at the moment the query starts. The penalty is additional cloning time, though.

Bulk-loading is beneficial only when substantial amounts of updates have been accumulated; otherwise, an update-in-place approach is cheaper. Also it is expen-sive to do single-object look-ups in the old index to retrieve un-updated objects. MOVIES assumes that each object sends an update at least every $t_{\Delta max}$ time units, which is less than a second in the reported empirical study [6]. Since work-loads may contain many static objects (e.g., parked vehicles), this is a notable limitation. TwinGrid does not suffer from this problem. However, TwinGrid updates are applied to the index structure and are thus more expensive than MOVIES updates, which are simply buffered.

To perform updates efficiently, each write-only buffer pre-allocates enough memory to store all indexed objects. In addition, two read-only indexes are maintained in parallel. As a result, the memory requirement is four times the size of the database. This is twice that of TwinGrid.

MOVIES Implementation. To compare empirically with MOVIES, we modify the original implementation slightly.

In a *machine-level* parallelization experiment, the data is horizontally partitioned by *oid* among machines that then work independently [6]. Assuming the partitioning does not cause load imbalances[2], one can expect a linear scale-up with the number of machines employed. The same partitioning scheme can be applied to TwinGrid. However, TwinGrid is built to exploit *chip-level* parallelism. To obtain a fair and meaningful comparison with MOVIES, we use a multi-threaded variant of MOVIES.

The MOVIES implementation compresses index and object data into 64-bit values. An object identifier occupies 27 bits, a velocity vector occupies 5 bits, leaving 16 bits per dimension (a discrete space of $2^{16} \times 2^{16}$ is assumed). For example, indexing the area of Germany yields a resolution of 26.3 meters per point (index granularity). Since current devices can support accuracies of 1 meter and we do not consider time-parametrized (predictive) queries, our MOVIES implementation uses the speed vector bits to extend the discrete space (to increase index granularity) up to $2^{19} \times 2^{19}$ (one extra bit is borrowed from the *oid* representation).

Therefore, single object updates operate on 64-bit values and can be performed atomically[3]. In addition to multiple queries operating on the same read-only index, this enables us to run multiple updates on the same write-only buffer without introducing any locking scheme. Note that the same limitation (fitting data into a 64-bit value) can be applied to TwinGrid. This would improve performance further, as local update would be atomic and require no locking. Currently, all three attributes (*oid*, x, and y) are 32-bit integers, which we feel offer greater versatility.

4   Experimental Study

4.1   Setting

We study performance on three modern platforms: a dual quad-core AMD Opteron 2350 (Barcelona) with 8 hardware threads, a dual quad-core Intel Xeon X5550 (Nehalem) with 16 hardware threads, and an 8-core Sun Niagara 2 (T2)

---

[2] Empirical evidence suggests this is a strong assumption [18].

[3] In a 64-bit system, 64-bit reads and writes are atomic.

Table 1. Machines Used

	Sun UltraSPARC-T2	Dual Intel Xeon 5550	Dual AMD Opteron 2350
Chips×cores×threads	1×8×8	2×4×2	2×4×1
Clock rate (GHz)	1.2	2.67	2.0
RAM (GB)	32	24	16
L1 data cache (KB)	8 (core)	32 (core)	64 (core)
L2 (unified) cache (KB)	4k (chip)	256 (core)	512 (core)
L3 (unified) cache (MB)	none	8 (chip)	2 (chip)

with 64 hardware threads; Table 1 offers additional detail. Caches are shared by all threads in a core or an entire chip.

The indexes were studied in a range of experiments with different workloads generated using the COST benchmark [9] so as to stress test the indexes under the controlled and varying conditions.

We use a default workload that is created to represent a realistic scenario for the intended applica-

Table 2. Parameters and Their Values

Parameter	Values
objects, ×10^6	5, **10**, 15, 20
density, objects/km²	**100**
speed$_i$, km/h	**20, 30, 40, 50, 60, 90**
nodes	**500**
query selectivity, km²	1, **2**, 4, 8, 16, 32, 64
update/query ratio	**1000:1**

tions described earlier. Table 2 shows the parameters and their values; the values in bold denote the default values used. Specifically, 10M objects are moving in a square region between 500 nodes in a complete graph. Each edge simulates a two-way road. The area is chosen so that the number of objects per km² is ca. 100, which corresponds to the average number of vehicles per km² in Germany. Objects start at a random position and are assigned at random one of the six maximum speeds, $speed_i$. The update/query ratio is fixed at 1000:1[4], and the range query size is 2 km².

4.2 Optimal Index Parameters

Determining the Cloning Period. An important TwinGrid parameter is the *cloning period* that defines the cloning frequency and thus the query freshness. Frequent cloning results in a high number of interruptions of actual workload processing, as worker-threads are frequently suspended in order to refresh reader-store. On the other hand, frequent cloning also results in low query staleness, as queries operate on almost up-to-date states of the reader-store.

In an experiment, we vary the cloning period from 16 ms to 0.5 s and measure the update throughput (updates/s). The results, shown in Figure 7, confirm our speculations: With very frequent cloning, throughput deteriorates.

When cloning occurs, the CPU caches at all levels are flushed. Therefore, many cache misses initially occur when workload processing resumes. With

---

[4] We did not observe any new findings when the update/query ratio was varied.

longer cloning periods, workload processing enjoys warm caches longer. Experiments (not included) using hardware performance counters confirmed this observation.

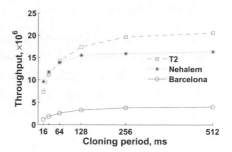

The experiment also shows that the throughput stabilizes on all platforms as the cloning period increases. For example, with a cloning period longer than 128 ms, the time Nehalem spends on cloning becomes less than 10% of the entire processing time. Larger cloning periods thus do not offer significant performance gains.

Fig. 7. Cloning Period

With cloning periods below 50 ms, Nehalem, with 16 threads, outperforms T2, with 64 threads. This is expected because it is more expensive to frequently suspend and resume many relatively slow threads than it is to do so for few and faster threads.

In Section 4.4, we shall see that MOVIES is unable to deliver a query staleness of below 0.5 ε. To obtain fair throughput comparisons, we thus conservatively set the cloning period in TwinGrid at 0.5 s in the following experiments. Even with this frequent cloning, the time spent on cloning is negligible. For example, on Nehalem TwinGrid does a 10M-object clone in circa 0.02 s (Section 4.4), which results in 144 s per hour (just 4%).

Determining the Grid Cell Size. To obtain the optimal performance for both updates and queries, we exercised TwinGrid while varying the grid cell size from 400 to 3200 meters. Since the same performance trends are observed on all machines, Figure 8 shows results just for T2. Larger cells favor updates, as fewer of them results in expensive non-local updates, while smaller cells favor queries [19]. Nevertheless, only extreme values have a significant impact on performance. Therefore, TwinGrid is configured with a grid cell size of 800 m in the subsequent experiments. The corresponding update and query throughputs are marked with circles in Figure 8.

The bucket size was tuned in a similar manner.

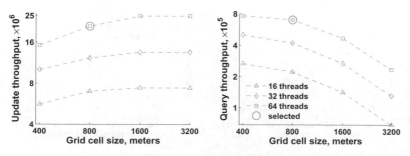

Fig. 8. Grid Cell Size (T2)

4.3   TwinGrid Scalability

To understand the multi-threaded capabilities of TwinGrid, we vary the number of worker-threads spawned by the main thread.

General Scalability. The results in Figure 9 show that TwinGrid generally scales near-linearly with the number of hardware threads on all platforms. As the number of threads used exceeds the number of hardware threads, throughput either stabilizes or deteriorates slightly.

On Nehalem, throughput first increases rapidly, then slows down after 8 threads are used. This is because the second thread on a core only helps in using pipeline slots that are unused by the first thread, which may be only 10% of all slots. Also up to 8 threads need not share L2 caches. Barcelona has one thread per core, and so we do not observe any changes in performance increase until all available hardware threads are exhausted. T2 has a two-level cache hierarchy, and its L2 cache is shared among all 8 cores. Thus, threads need to share the L2 cache regardless of whether they run on the same core or not. Interestingly, T2, with the lowest clock rate but the most hardware threads, eventually outperforms its two competitors and achieves a throughput of more than 21M update/s.

Contention Detection. As described in Section 3.3, to reduce contention among worker-threads when retrieving messages, multiple message queues are used. Figure 10 shows the quite dramatic effect on T2 of the number of queues used. Using one queue yields a sharp performance drop when more than 22 worker threads are used. Using two queues delays the drop to 44 threads, and using four queues completely eliminates contention. On Nehalem, two queues are enough, and using two queues on Barcelona slightly improves performance.

To verify that the performance drops are due to contention on the counter variable, we apply techniques that detect thread-level contention within parallel database operations [5]. Specifically, we use an alternative to the compare-and-swap intrinsic that also measures the contention during the atomic operation. This leads to the finding that contention increases ten-fold at the point where the performance drop occurs.

We also find that the used atomic operation has an execution time of circa 114 nanoseconds on T2, which implies a maximum input reading rate of 8.7M

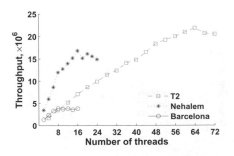

Fig. 9. TwinGrid Scalability

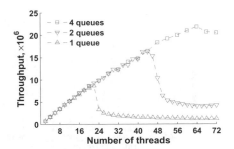

Fig. 10. Contention on T2

messages/s. This is indeed close to the maximum update throughput reached on T2 using one queue in Figure 10.

4.4   TwinGrid vs. MOVIES

In the comparisons that follow, TwinGrid and MOVIES are configured to use the maximum number of hardware threads available on the three different machines.

Update Throughput. Figure 11(b) shows the throughput when varying the number of indexed objects from 5M to 20M. TwinGrid outperforms MOVIES on all three platforms, by a factor ranging from 1.2 to 10. TwinGrid's throughput decreases slightly as more objects are being indexed. This is reasonable, as updates operate on an increasingly large index, yielding fewer opportunities for cache-sharing. The copying time during cloning also increases, as more objects need to be copied. For MOVIES, increases in the number of objects have little or no effect. This is because the increased update loads translate into increased query staleness. That is, the cost of accumulating incoming updates remains nearly the same (fast, constant-time hash table inserts), while the simultaneous building of a larger index takes more time.

Interestingly, TwinGrid achieves the best update throughput on the machine with the most hardware threads available (T2) and the worst throughput on the machine with the least hardware threads (Barcelona). In contrast, MOVIES achieves the best performance on the fastest machine (Nehalem) and the worst performance on the slowest machine (T2). This suggests that MOVIES does not make efficient use of parallelism.

Query Staleness. Both indexes report only near up-to-date query results. Figure 11(c) depicts the range query staleness for the same experiment as above. TwinGrid's staleness is slightly above 0.5 s, which corresponds to the cloning period plus the time needed for all worker threads to arrive at the synchronization points (see Section 3.3) and the time needed to do the actual copying. Since the copying time is short comparing to cloning period, the query staleness increases barely with an increasing number of objects. We also note that query staleness can be set to be several times lower than in the figure while maintaining similar throughput (recall Figure 7).

The staleness shown for MOVIES are the minimum ones possible, as we advance to the next time frame as soon as a new read-only index is built (i.e., $T_i = T_{bulk}$). The staleness increases markedly with the index size. On T2, the staleness is about an order of magnitude higher than that of TwinGrid.

Figure 11(d) compares the copying time in TwinGrid with the bulk-loading time in MOVIES, showing that copying is beyond an order of magnitude faster on all three machines.

Update Locality. Recall that in TwinGrid, a local update stays within a single cell, while a non-local update involves two cells and additional locking (see Sections 3.1 and 3.3).

To observe the effect of update locality, we vary the percentage of local updates from 4% to 92%. As can be seen in Figure 11(e), TwinGrid benefits from locality.

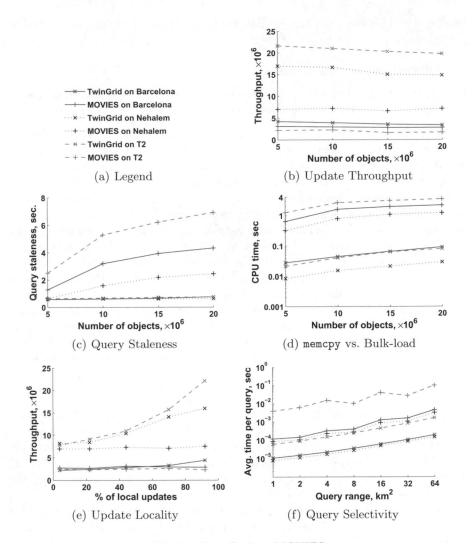

Fig. 11. TwinGrid vs. MOVIES

In contrast, updates in MOVIES simply involves accumulating the updates in an update buffer, which is unaffected by locality.

When less than half of all updates are local, MOVIES performs better than TwinGrid on Barcelona, while TwinGrid always dominates on the two other machines. When nearly all updates are local, TwinGrid's average update cost is comparable to that of MOVIES. However, since most of MOVIES' processing goes to the building of the next index, update throughput is several times better in TwinGrid, where all CPU resources are used for workload processing.

Query Performance. Since object queries are used in updates, object query performance is similar to that of update throughput as reported in Figure 11(b).

Figure 11(f) depicts range query performance of both indexes when varying the query size from 1 km^2 to 64 km^2. Both indexes are affected by the increased query size that implies that more objects need to be inspected.

MOVIES indexes z-values of two-dimensional object locations. Although the z-transformation exhibits good proximity-preservation, the resulting range queries are inherently tighter in TwinGrid's two-dimensional grid. On average, range queries are an order of magnitude faster in TwinGrid.

5 Conclusions

Motivated by the increasing chip-level parallelism in processors and the lack of studies on how to exploit this parallelism for efficient processing of update-intensive workloads, this paper presents a new spatial indexing scheme called TwinGrid that separates updates from queries by performing the latter on a near up-to-date data snapshot. The `memcpy` system call, which is highly optimized on most platforms, is used for refreshing the data snapshot. TwinGrid avoids conflicts between reads and write as well as other forms of contention.

An empirical study with three modern machines show that TwinGrid's throughput scales linearly with the number of hardware threads on the machines. TwinGrid is capable of processing some 10M updates/s while ensuring that the snapshots used for queries are stale by only about 100 ms on a single multi-threaded machine. Moreover, TwinGrid outperforms the previous state-of-the-art approach in terms of update throughput and query staleness.

References

1. Akman, V., Franklin, W.R., Kankanhalli, M., Narayanaswami, C.: Geometric computing and the uniform grid data technique. Computer Aided Design 21(7), 410–420 (1989)
2. Beckmann, N., Kriegel, H.-P., Schneider, R., Seeger, B.: The R*-tree: an efficient and robust access method for points and rectangles. In: SIGMOD, pp. 322–331 (1990)
3. Biveinis, L., Šaltenis, S., Jensen, C.S.: Main-memory operation buffering for efficient R-tree update. In: VLDB, pp. 591–602 (2007)
4. Chen, S., Jensen, C.S., Lin, D.: A benchmark for evaluating moving object indexes. PVLDB 1(2), 1574–1585 (2008)
5. Cieslewicz, J., Ross, K.A., Satsumi, K., Ye, Y.: Automatic contention detection and amelioration for data-intensive operations. In: SIGMOD, pp. 483–494 (2010)
6. Dittrich, J., Blunschi, L., Salles, M.A.V.: Indexing moving objects using short-lived throwaway indexes. In: Mamoulis, N., Seidl, T., Pedersen, T.B., Torp, K., Assent, I. (eds.) SSTD 2009. LNCS, vol. 5644, pp. 189–207. Springer, Heidelberg (2009)
7. Finkel, R.A., Bentley, J.L.: Quad trees: a data structure for retrieval on composite keys. Acta Informatica 4(1), 1–9 (1974)
8. Hwang, S., Kwon, K., Cha, S., Lee, B.: Performance evaluation of main-memory R-tree variants. In: Hadzilacos, T., Manolopoulos, Y., Roddick, J., Theodoridis, Y. (eds.) SSTD 2003. LNCS, vol. 2750, Springer, Heidelberg (2003)

9. Jensen, C., Tiesyte, D., Tradisauskas, N.: The COST Benchmark-Comparison and evaluation of spatio-temporal indexes. Database Systems for Advanced Applications, 125–140 (2006)
10. Kim, K., Cha, S.K., Kwon, K.: Optimizing multidimensional index trees for main memory access. In: SIGMOD, pp. 139–150 (2001)
11. Lee, M.L., Hsu, W., Jensen, C.S., Cui, B., Teo, K.L.: Supporting frequent updates in R-trees: a bottom-up approach. In: VLDB, pp. 608–619 (2003)
12. Nievergelt, J., Hinterberger, H., Sevcik, K.C.: The grid file: An adaptable, symmetric multikey file structure. ACM TODS 9(1), 38–71 (1984)
13. Rastogi, R., Seshadri, S., Bohannon, P., Leinbaugh, D.W., Silberschatz, A., Sudarshan, S.: Logical and physical versioning in main memory databases. In: VLDB, pp. 86–95 (1997)
14. Severance, D.G., Lohman, G.M.: Differential files: their application to the maintenance of large databases. ACM TODS 1(3), 256–267 (1976)
15. Smith, J.M., Maguire, G.Q.: Effects of copy-on-write memory management on the response time of UNIX fork operations, vol. 1, pp. 255–278. University of California Press, Berkeley (1988)
16. Song, M., Kitagawa, H.: Managing frequent updates in R-trees for update-intensive applications. IEEE TKDE 21(11), 1573–1589 (2009)
17. Tropf, H., Herzog, H.: Multidimensional range search in dynamically balanced trees. Angewandte Informatik 23(2), 71–77 (1981)
18. Tzoumas, K., Yiu, M.L., Jensen, C.S.: Workload-aware indexing of continuously moving objects. PVLDB 1(2), 1186–1197 (2009)
19. Šidlauskas, D., Šaltenis, S., Christiansen, C.W., Johansen, J.M., Šaulys, D.: Trees or grids? Indexing moving objects in main memory. In: GIS, pp. 236–245 (2009)

Efficient Processing of Top-k Spatial Keyword Queries

João B. Rocha-Junior*, Orestis Gkorgkas, Simon Jonassen, and Kjetil Nørvåg

Department of Computer and Information Science,
Norwegian University of Science and Technology (NTNU),
Trondheim, Norway
{joao,orestis,simonj,noervaag}@idi.ntnu.no

Abstract. Given a spatial location and a set of keywords, a top-k spatial keyword query returns the k best spatio-textual objects ranked according to their proximity to the query location and relevance to the query keywords. There are many applications handling huge amounts of geo-tagged data, such as Twitter and Flickr, that can benefit from this query. Unfortunately, the state-of-the-art approaches require non-negligible processing cost that incurs in long response time. In this paper, we propose a novel index to improve the performance of top-k spatial keyword queries named *Spatial Inverted Index* (S2I). Our index maps each distinct term to a set of objects containing the term. The objects are stored differently according to the document frequency of the term and can be retrieved efficiently in decreasing order of keyword relevance and spatial proximity. Moreover, we present algorithms that exploit S2I to process top-k spatial keyword queries efficiently. Finally, we show through extensive experiments that our approach outperforms the state-of-the-art approaches in terms of update and query cost.

1 Introduction

Given a location and a set of keywords, a top-k spatial keyword query returns a ranked set of the k best spatio-textual objects taking into account both 1) the spatial distance between the *objects* (spatio-textual objects) and the query location, and 2) the relevance of the text describing the objects to the query keywords. There are several applications that can benefit from top-k spatial keyword queries such as finding the tweets sent from a given location (Twitter) or finding images near by a given location whose annotation is similar to the query keywords (Flickr). There are also other applications for GPS-enabled mobile phones that can benefit from such queries.

For example, Fig. 1 shows a spatial area containing objects p (bars and pubs) with their respective textual description. Consider a tourist in São Paulo with a GPS mobile phone that wants to find a bar playing *samba* near her current location q. The tourist poses a top-3 spatial keyword query on her mobile phone

---

* On leave from the State University of Feira de Santana (UEFS).

D. Pfoser et al. (Eds.): SSTD 2011, LNCS 6849, pp. 205–222, 2011.
© Springer-Verlag Berlin Heidelberg 2011

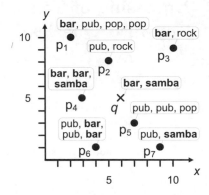

Fig. 1. Example of top-k spatial keyword query

with the keywords *bar* and *samba* (the query location q is automatically sent by the mobile phone). The top-1 result is p_4 because its description is similar to the query keywords, and it is close to the query location q. The top-2 result is p_6 that is nearer to q than p_7 and has a better textual relevance to the query keywords than p_7. Here, we are assuming for simplicity that documents with higher numbers of occurrences of query keywords are more textually relevant. Later, we will drop this assumption and present a more advanced model. Consequently, the top-3 results are $\langle p_4, p_6, p_7 \rangle$.

Top-k spatial keyword queries are intuitive and constitute a useful tool for many applications. However, processing top-k spatial keyword queries efficiently is complex and requires a hybrid index combining information retrieval and spatial indexes. The state-of-the-art approaches proposed by Cong *et al.* [4] and Li *et al.* [11] employ a hybrid index that augments the nodes of an R-tree with inverted indexes. The inverted index at each node refers to a pseudo-document that represents all the objects under the node. Therefore, in order to verify if a node is relevant for a set of query keywords, the current approaches access the inverted index at each node to evaluate the similarity between the query keywords and the pseudo-document associated with the node. This process incurs in non-negligible processing cost that results in long response time.

In this paper, we propose a novel method for processing top-k spatial keyword queries more efficiently. Instead of employing a single R-tree embedded with inverted indexes, we propose a new index named *Spatial Inverted Index* (S2I) that maps each keyword (term) to a distinct *aggregated R-tree* (*aR-tree*) [14] that stores the objects with the given term. In fact, we employ an aR-tree only when the number of objects exceeds a given threshold. As long as the threshold is not exceeded, the objects are stored in a file, one block per term. However, for ease of presentation, let us assume that we employ an aR-tree for each term. The aR-tree stores the latitude and longitude of the objects, and maintains an aggregated value that represents the maximum term impact (normalized weight) of the objects under the node. Consequently, it is possible to retrieve the best objects ranked in terms of both spatial relevance and keyword relevance efficiently and incrementally. For processing a top-k spatial keyword query with a single keyword, only few nodes of a single aR-tree are accessed. For queries with more than one keyword, we employ an efficient algorithm that aggregates the partial-scores on keyword relevance of the objects to obtain the k best results efficiently. In summary, the main contributions of this paper are:

– We present S2I, an index that maps each term in the vocabulary into a distinct aR-tree or block that stores all objects with the given term.
– We propose efficient algorithms that exploit the S2I in order to process top-k spatial keyword queries efficiently.
– Finally, we show through an extensive experimental evaluation that our approach outperforms the state-of-the-art algorithms in terms of update time, I/O cost, and response time.

The rest of this paper is organized as follows. Sect. 2 gives an overview of the related work. Sect. 3 poses the problem statement. In Sect. 4, we describe S2I. In Sect. 5 and 6, we present the algorithms for processing top-k spatial keyword queries. Finally, the experimental evaluation is presented in Sect. 7 and the paper is conclude in Sect. 8.

2   Related Work

Initially, the research on spatial keyword queries focused on improving the performance of spatial queries in search engines. Zhou *et al.* [17] did a relevant work combining inverted indexes [18] and R-trees [2], and propose three approaches: 1) indexing the data in both R-trees and inverted indexes, 2) creating an R-tree for each term, and 3) integrating keywords in the intermediary nodes of an R-tree. They found out that the second approach achieved better performance. However, they did not consider objects with a precise location (latitude and longitude), and did not provide support for top-k spatial keyword queries. Chen *et al.* [3] also had an information retrieval perspective on their work and did not provide support for exact query location of objects. In their approach, inverted indexes and R-trees are accessed separately in two different stages.

With the popularization of GPS-enabled devices, the research focused on searching for objects in a specific location. Hariharan *et al.* [7] proposed augmenting the nodes of an R-tree with keywords extracted from the objects in the sub-tree of the node. These keywords are then indexed in a structure similar to an inverted index for fast retrieval. Their approach supports conjunctive query in a given region of space. It is not clear, however, how their solution can be extended to support top-k spatial keyword queries. Later, Ian de Felipe *et al.* [5] proposed a data structure that integrates signature files and R-trees. The main idea was indexing the spatial objects in an R-tree employing a signature on the nodes to indicate the presence of a given keyword in the sub-tree of the node. Consequently, at query processing time, the nodes that cannot contribute with the query keywords can be pruned. The main problem of this approach is the limitation to Boolean queries and to a small number of keywords per document.

To the best of our knowledge, there are two previous approaches that support top-k spatial keyword queries. They were developed concurrently by Cong *et al.* [4] and Li *et al.* [11]. Both approaches augment the nodes of an R-tree with a document vector that represents all documents in the sub-tree of the node. For all terms present in the objects in the sub-tree of the node, the vector stores the maximum impact of the term (normalized weight). Consequently, the vector allows computing an upper bound for the textual score (textual relevance) that can be achieved visiting a given node. Hence, it is possible to rank the nodes according to textual relevance and spatial relevance, and decide which nodes should be accessed first to compute the top-k results.

The work of Cong *et al.* goes beyond the work of Li *et al.* by incorporating document similarity to build a more advanced R-tree namely DIR-tree. DIR-tree groups, in the same node, objects that are near each other in terms of spatial distance, and whose textual description are also similar. Furthermore, instead of comparing vectors at query time, DIR-tree employs an inverted index associated with each node that permits to retrieve the children of the node that can contribute with a given query keyword efficiently. Only the posting lists associated with the query keywords are accessed. Cong *et al.* also propose clustering the nodes of DIR-tree (CDIR-tree) to further improve the query processing performance. The main idea is grouping related entries (objects, in case of leaf-nodes) and employing a pseudo-document to represent each group. Hence, more precise bounds can be estimated at query time, consequently, improving the query processing performance. However, it is not clear if the improvement achieved at query processing time compensates the additional cost required for clustering the nodes (pre-processing), and the extra storage space demanded by CDIR-tree. Moreover, keeping a CDIR-tree updated is more complex. For this reason, we decided to compare our approach against the DIR-tree proposed by Cong *et al.*, and we consider this approach as the state-of-the-art.

3   Problem Statement

Let P be a dataset with $|P|$ spatio-textual objects $p = \langle p.id, p.l, p.d \rangle$, where $p.id$ is the identification of p, $p.l$ is the spatial location (latitude and longitude) of p, and $p.d$ is the textual document describing p (e.g., menu of a restaurant). Let $q = \langle q.l, q.d, q.k \rangle$ be a top-k spatial keyword query, where $q.l$ is the query location (latitude and longitude), $q.d$ is the set of query keywords, and $q.k$ is the number of expected results. A query q returns $q.k$ spatio-textual objects $\{p_1, p_2, \cdots, p_{q.k}\}$ from P with the highest scores $\tau(p, q)$, $\tau(p_1, q) \geq \tau(p_2, q) \geq \cdots \geq \tau(p_k, q)$. Furthermore, a spatio-textual object p is part of the result set R of q, if and only if exists at least one term $t \in q.d$ that is also in $p.d$ ($p \in R \Leftrightarrow \exists t \in q.d : t \in p.d$). The *score* of p for a given query q is defined in the following equation:

$$\tau(p, q) = \alpha \cdot \delta(p.l, q.l) + (1 - \alpha) \cdot \theta(p.d, q.d) \tag{1}$$

where $\delta(p.l, q.l)$ is the spatial proximity between the query location $q.l$ and the object location $p.l$, and $\theta(q.d, p.d)$ is the textual relevance of $p.d$ according to

$q.d$. Both measures return values within the range $[0, 1]$. The *query preference parameter* $\alpha \in (0, 1)$ defines the importance of one measure over the other. For example, $\alpha = 0.5$ means that spatial proximity and textual relevance are equally important. In the following, we define the measures more precisely.

Spatial proximity (δ). The spatial proximity is defined in the following equation:

$$\delta(p.l, q.l) = 1 - \frac{d(p.l, q.l)}{d_{max}} \tag{2}$$

where $d(p.l, q.l)$ is the Euclidean distance between $p.l$ and $q.l$, and d_{max} is the largest Euclidean distance that any two points in the space may have. The maximum distance may be obtained, for example, by getting the largest diagonal of the Euclidean space of the application.

Textual relevance (θ). There are several similarity measures that can be used to evaluate the textual relevance between the query keywords $q.d$ and the text document $p.d$ [13]. In this paper, we adopt the well-known *cosine* similarity between the vectors composed by the weights of the terms in $q.d$ and $p.d$:

$$\theta(p.d, q.d) = \frac{\sum_{t \in q.d} w_{t,p.d} \cdot w_{t,q.d}}{\sqrt{\sum_{t \in p.d}(w_{t,p.d})^2 \cdot \sum_{t \in q.d}(w_{t,q.d})^2}} \tag{3}$$

In order to compute the cosine, we adopt the approach employed by Zobel and Moffat [18]. Therefore, the weight $w_{t,p.d}$ is computed as $w_{t,p.d} = 1 + ln(f_{t,p.d})$, where $f_{t,p.d}$ is the number of occurrences (frequency) of t in $p.d$; and the weight $w_{t,q.d}$ is obtained from the following formula $w_{t,q.d} = ln(1 + \frac{|P|}{df_t})$, where $|P|$ is the total number of documents in the collection. The *document frequency* df_t of a term t gives the number of documents in P that contains t. The higher the cosine value, the higher the textual relevance. The textual relevance is a value within the range $[0, 1]$ (property of cosine).

We also define the *impact* $\lambda_{t,d}$ of a term t in a document d, where d represents the description of an object $p.d$ or the query keywords $q.d$. The impact $\lambda_{t,d}$ is the normalized weight of the term in the document $[1, 16]$, $\lambda_{t,d} = \frac{w_{t,d}}{\sqrt{\sum_{t \in d}(w_{t,d})^2}}$. The impact takes into account the length of the document and can be used to compare the relevance of two different documents according to a term t present in both documents. Consequently, the textual relevance $\theta(p.d, q.d)$ can be rewritten in terms of the impact $[16]$, $\theta(p.d, q.d) = \sum_{t \in q.d} \lambda_{t,q.d} \cdot \lambda_{t,p.d}$.

Other types of spatial proximity and textual relevance measures such as Okapi BM25 [13] can be supported by our framework. The focus of this paper is, however, on the efficiency of top-k spatial keyword queries. In the following, we present the S2I (Sect. 4) and describe the algorithms to process top-k spatial keyword queries efficiently (Sect. 5 and 6).

4   Spatial Inverted Index

The S2I was designed taking in account the following observations. First, terms with different document frequency should be stored differently. It is well-known

term	id	df_t	type	ptr	storage
bar	t_1	4	tree	↪	aR^{t_1}
pop	t_2	2	block	↪	$\langle p_1, p_5 \rangle$
pub	t_3	5	tree	↪	aR^{t_3}
rock	t_4	2	block	↪	$\langle p_2, p_3 \rangle$
samba	t_5	2	block	↪	$\langle p_4, p_7 \rangle$

(a) S2I.

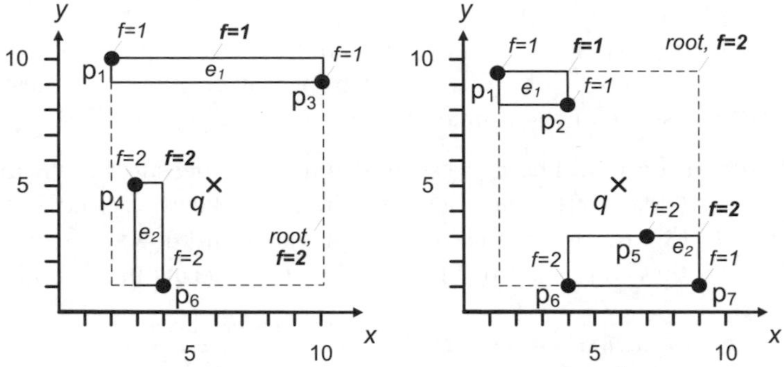

(b) aR^{t_1}.　　　　　(c) aR^{t_3}.

Fig. 2. Spatial Inverted index and the aR-tree of terms t_1 and t_3

that the document frequency of terms in a corpus follows the Zipf's law, which means that there are a small number of terms that occur frequently, while most terms occur infrequently [10]. Current approaches that support top-k spatial keyword queries ignore this property and store frequent and infrequent terms in the same way. For example, the impact of a term that occurs in a single object has to be replicated in several nodes of the DIR-tree to indicate the path in the tree for the given object. Second, good support for distribution is important for scaling applications that can benefit from top-k spatial keyword queries. Third, keeping the index update is important. The current approaches require accessing one or several inverted files to perform a single update, which has a significant cost. Finally, response time is critical for several applications. Our approach accesses less disk pages and can perform queries more efficiently.

The S2I maps each term t to an aggregated R-tree (aR-tree) or to a block that stores the spatio-textual objects p that contain t. The most frequent terms are stored in aR-trees, one tree per term. The less frequent terms are stored in blocks in a file, one block per term. Similarly to a traditional inverted index, the S2I maps terms to objects that contain the term. However, we employ two different data structures, one for less frequent terms and another for more frequent terms that can be accessed in decreasing order of keyword relevance and spatial proximity efficiently.

The S2I consists of three components: vocabulary, blocks, and trees.

- **Vocabulary.** The vocabulary stores, for each distinct term, the number of objects in which the term appears (df_t), a flag indicating the type of storage used by the term (block or tree), and a pointer to a block or aR-tree that stores the objects containing the given term.
- **Blocks.** Each block stores a set of objects, the size of a block is an application parameter. For each object, we store the object identification $p.id$, the object location $p.l$, and the impact of term t in $p.d$ $(\lambda_{t,p.d})$. The objects stored in a block are not ordered.
- **Trees.** The aggregated R-tree [14] of a term aR^t follows the same structure of a traditional R-tree. A leaf-node stores information about the data objects: $p.id, p.l$, and $\lambda_{t,p.d}$. An intermediary-node stores for each entry (child node) a *minimum bounding rectangle* (MBR) that encloses the spatial location of all objects in the sub-tree. Differently from an R-tree, the nodes of an aR-tree store also an aggregated non-spatial value among the objects in its sub-tree. In our case, the aggregated value is the maximum impact of the term t among the objects in the sub-tree of the node. Hence, we can access the objects in aR^t in decreasing order of term relevance and spatial proximity.

Example 1. Fig. 2 presents the S2I created from the objects depicted in Fig. 1. In order to simplify the presentation, we assume in all examples that the impact of a term in a document $(\lambda_{t,d})$ is defined by the number of occurrences of the term in a document. We also assume that $\alpha = 0.5$ in all examples. We drop these assumptions in the experimental evaluation. The less frequent terms are stored in a block, while the more frequent terms are stored in an aR-tree, see Fig. 2(a). The aR-tree of terms t_1 and t_3 is depicted in Fig. 2(b) and 2(c), respectively. The root of R^{t_1} contains two entries e_1 and e_2. The entry e_1 contains the spatio-textual objects $\{p_1, p_3\}$, while e_2 contains $\{p_4, p_6\}$. The objects p_1 and p_3 have one occurrence of t_1 $(f = 1)$, while p_4 and p_6 have two occurrences of t_1 $(f = 2)$. The number of occurrences of a term in bold present the maximum number of occurrences of a term among the entries of each node.

The S2I has several good properties. First, terms with different document frequency are stored differently. Second, S2I has good support for distribution. S2I employs a different data structure for each term and can benefit from techniques used by traditional distributed search engines such as term partitioning [18]. Third, S2I provides good support for updates. Although one tree or block has to be accessed for each term in an object, the operations executed at each tree or block can be performed efficiently. Furthermore, the average number of distinct terms per document is small in most of the applications that are target of this query, as it can be seen in Table 2 (Section 7). Finally, S2I allows efficient query execution. For queries with a single keyword, only one small tree (in general) or a block needs to be accessed. For queries with several keywords only few nodes of a set of small trees or blocks are accessed. No access to external inverted indexes is required. In the following, we present the algorithm to process single-keyword (Sect. 5) and multiple-keyword queries (Sect. 6) employing S2I.

Algorithm 1. *SKA(MaxHeap H^t, Query q)*

1: **INPUT:** MaxHeap H^t with entries e in decreasing order of score $\tau(e,q)$ and the query q.
2: **OUTPUT:** The next object p with highest score $\tau(p,q)$.
3: Entry $e \leftarrow H^t.pop()$ *//e is the entry with highest score $\tau(e,q)$ in H^t*
4: **while** e **is not** an object **do**
5:    **if** e is an intermediary-node **then**
6:       **for each** node $n \in e$ **do**
7:          $H^t.insert(n, \tau(n,q))$
8:       **end for**
9:    **else**   *//e is a leaf-node*
10:       **for each** spatio-textual object $p \in e$ **do**
11:          $H^t.insert(p, \tau(p,q))$
12:       **end for**
13:    **end if**
14:    $e \leftarrow H^t.pop()$
15: **end while**
16: **return** e

---

5   Single-Keyword Queries

Top-k spatial keyword queries with a single keyword t can be efficiently processed using the S2I, since only one single block or tree containing the objects with the term t is accessed. If the objects are stored in a block, the query processing steps are: 1) retrieve all objects p in the block, 2) insert the objects in a heap in decreasing order of $\tau(p,q)$, and 3) report the top-k best objects. If the objects are stored in an aR-tree, we employ an incremental algorithm (Algorithm 1) to retrieve the objects in decreasing order of $\tau(p,q)$.

The SKA (Single Keyword Algorithm, Algorithm 1) visits entries e (nodes or objects) in an aR-tree in decreasing order of $\tau(e,q)$. The score of a node n, $\tau(n,q)$, and the score of an object p, $\tau(p,q)$, are computed in a similar way. The spatial proximity $\delta(n,q)$ is obtained computing the minimum Euclidean distance between q and any border of the MBR of n. On the other hand, the textual relevance $\theta(n,q)$ is obtained multiplying the impact of t in the query $\lambda_{t,q}$ by the impact of t in the node $\lambda_{t,n}$ that is the maximum impact among any entry e in the sub-tree of n. Consequently, the score $\tau(n,q)$ is an upper bound score for any entry e in the sub-tree of n, since any entry e in the sub-tree of n has a distance to q longer or equal $d(q,n)$, $d(q,n) \leq d(q,e)$; and the term impact of n is higher or equal the term impact of any entry e, $\lambda_{t,n} \geq \lambda_{t,e}$.

SKA receives as parameter a priority queue (MaxHeap) H^t and a query q. The heap stores the entries e of R^t in decreasing order of score $\tau(e,q)$. Initially, the heap has the root of R^t. The entry e (line 3) is the entry with highest score in H^t. If e is an intermediary-node (line 5), the algorithm computes the score of all nodes n children of e and insert n in H^t in decreasing order of $\tau(n,q)$ (lines 6-8). If e is a leaf-node (line 9), the algorithm computes the score of all objects p children of e and insert p in H^t in decreasing order of score (lines 10-12).

This process repeats until an object p achieves the top of the heap (line 4). This means that there is no other object in H^t with higher score than p. Therefore, p can be reported incrementally (line 16). The algorithm keeps the state of the variables for future access.

Example 2. Assume that we want to obtain the top-1 object among the objects depicted in Fig. 1 according to the query keyword $q.d = \{bar\}$ and the query location $q.l = (6, 5)$. The query processing starts accessing the S2I and acquiring the information that objects with the term "bar" (t_1) are stored in aR^{t_1}, Fig. 2(b). The SKA algorithm starts with the root of aR^{t_1} in H^{t_1} and inserts the entries e_1 and e_2 in the heap, $H^{t_1} = \langle e_2, e_1 \rangle$. The entry e_2 is in the top of H^{t_1} because it has a better score, $\tau(e_2, q) > \tau(e_1, q)$. The score of e_2 is higher because it has a higher term frequency ($f_{e_2, t_1} = 2$) and is nearer to q. The algorithm continues removing e_2 from the heap and inserting the objects p_4 and p_6 into $H^{t_1} = \langle p_4, p_6, e_1 \rangle$. In the next iteration, p_4 reaches the top of the heap H^{t_1} and is returned as top-1.

The problem of retrieving the objects in increasing order of score is similar to the problem of retrieving the nearest neighbors incrementally [8]. In the following, we present the algorithm to process multiple keyword queries.

6 Multiple-Keyword Queries

We divide this section in two parts. First, we define *partial-score* that is the score of an object according to a single term in the query, Sect. 6.1. Second, we present the algorithm to aggregate the objects retrieved in terms of partial-score to compute the top-k results, Sect. 6.2.

6.1 Partial-Score on Keyword

Processing multiple-keyword queries in the S2I requires aggregating objects from different *sources* S_i (aR-trees or blocks), where i refers to a distinct term $t_i \in q.d$. The naïve way is to retrieve objects p from each source S_i in decreasing order of score $\tau(p, q)$ (Equation 1) replacing $q.d$ by a query term $t \in q.d$, $\tau(p, \{t\})$. The final score is obtained adding the scores retrieved for each term. However, $\tau(p, q) \neq \sum_{t \in q.d} \tau(p, \{t\})$, because the spatial proximity is replicated in the scores computed for each term $\tau(p, \{t\})$. Hence, we propose to derive a partial-score on keyword $\tau^t(p, q)$ such that the score $\tau(p, q)$ can be computed in terms of the sum of partial-scores obtained from each source, $\tau(p, q) = \sum_{t \in q.d} \tau^t(p, q)$.

In order to obtain the partial-score $\tau^t(p, q)$, we rewrite Equation 1 so that the score $\tau(p, q)$ can be obtained in terms of the *sum* of partial-scores of t:

$$\tau(p, q) = \alpha \cdot \delta(p.l, q.l) + (1 - \alpha) \cdot \theta(p.d, q.d)$$

$$= \alpha \cdot \delta(p.l, q.l) + (1 - \alpha) \cdot \sum_{t \in q.d} \lambda_{t, p.d} \cdot \lambda_{t, q.d}$$

$$= \sum_{t \in q.d} (\alpha \cdot \frac{\delta(p.l, q.l)}{|q.d|} + (1 - \alpha) \cdot \lambda_{t, p.d} \cdot \lambda_{t, q.d})$$

where $|q.d|$ is the number of distinct terms in the query. From the equation above, we define *partial-score* $\tau^t(p,q)$ of p in relation to a term t as:

$$\tau^t(p,q) = \alpha \cdot \frac{\delta(p.l, q.l)}{|q.d|} + (1 - \alpha) \cdot \lambda_{t,p.d} \cdot \lambda_{t,q.d} \ . \tag{4}$$

The partial-score reduces the weight of the spatial proximity $\delta(p.l, q.l)$ according to the number of distinct terms in the query $q.d$. Furthermore, once we have obtained an object p from a source S_i, we can also derive a lower bound partial-score $\tau^t_-(p,q)$ for p on the other sources in which p has not been seen yet, $\tau^t_-(p,q) = \alpha \cdot \frac{\delta(p.l,q.l)}{|q.d|}$. This is the lowest possible partial-score that p may have in a source where it has not been seen yet, since the proximity between p and q will not change. With the partial-scores $\tau^t(p,q)$ and $\tau^t_-(p,q)$, we can compute the lower bound and upper bound scores based on partial information about the object.

Example 3. Assume a query with two terms t_i and t_j, where the partial-score of p according to term t_i is known. The lower bond score of p, p_-, can be computed adding the partial-score of p according to t_i with the lower bound partial-score of p according to t_j, $p_- = \tau^{t_i}(p,q) + \tau^{t_j}_-(p,q)$.

In the following, we present the MKA algorithm that employs the partial-scores to compute the top-k results for queries with multiple-keywords.

6.2   Multiple Keyword Algorithm

The Multiple Keyword Algorithm (MKA) computes a top-k spatial keyword query progressively by aggregating the partial-scores of the objects retrieved for a given keyword. The objects containing a given term are retrieved in decreasing order of partial-scores from the source (aR-tree or block). In order to retrieve the objects in decreasing order of partial-score, we employ the SKA algorithm (Algorithm 1) replacing the score $\tau(p,q)$ by partial-score $\tau^t(p,q)$.

Each time an object p is retrieved from a source S_i ($S_i.next()$), we compute the lower bound score of p, update the upper bound score for any unseen object, and check if there is an object whose the lower bound score is higher or equal the upper bound of any other object. Those objects are reported progressively. We repeat this process until k objects have been found.

Algorithm 2 presents the MKA algorithm. MKA receives as parameter a top-k spatial keyword query q and reports the top-k results incrementally. MKA employ one source S_i for each distinct term $t_i \in q.d$ (line 3). Next, for each source S_i, the algorithm sets an upper bound u_i^- that maintains the highest partial-score among the objects unseen from S_i (line 6). The upper bound is updated every time that an object p is retrieved from S_i (line 10).

During each iteration (lines 7-23), MKA selects a source i (line 8), where the procedure $selectSource(q.d)$ defines the strategy to select the source. In this paper, we employ a round-robin strategy that selects a source S_i that is not empty. Other more sophisticated strategies such as keeping an indicator

---

Algorithm 2. *MKA(Query q)*

1: **INPUT:** The top-k spatial keyword query q.
2: **OUTPUT:** Progressively reports the top-k objects found.
3: $S_i \leftarrow$ *source* of the term $t_i \in q.d$
4: $C \leftarrow \emptyset$ *//List of candidate objects.*
5: $L_i \leftarrow \emptyset$ *//List of objects seen in the source S_i*
6: $u_i^- \leftarrow \infty$ *//Upper-bound score for any $p \in S_i$*
7: **while** $\exists S_i$ such that $S_i \neq \emptyset$ **do**
8:     $i \leftarrow selectSource(q.d)$
9:     $p \leftarrow S_i.next()$ *//Next object p in S_i with highest $\tau^{t_i}(p,q)$*
10:     $u_i^- \leftarrow \tau^{t_i}(p,q)$ *//Update upper bound score for source S_i*
11:     $L_i \leftarrow L_i \cup p$
12:     $p_- \leftarrow \sum_{\forall j: p \in L_j} \tau^{t_j}(p,q) + \sum_{\forall j: p \notin L_j} \tau_-^{t_j}(p,q)$ *//Update lower bound score of p*
13:     **if** $p \notin C$ **then**
14:         $C \leftarrow C \cup p$
15:     **end if**
16:     **for each** $p \in C$ **do**   *//Update upper bound score of the candidates*
17:         $p^- \leftarrow \sum_{\forall j: p \in L_j} \tau^{t_j}(p,q) + \sum_{\forall j: p \notin L_j} max(u_j^-, \tau_-^{t_j}(p,q))$
18:     **end for**
19:     **while** $\exists p \in C : p_- \geq max_{\forall o \in C, o \neq p}(o^-)$ **do**
20:         $C \leftarrow C - p$
21:         reports p as next top-k, halt if $q.k$ objects have been reported
22:     **end while**
23: **end while**
24: **if** less than $q.k$ objects have been reported **then**
25:     return the objects in C with highest lower bound score p_-
26: **end if**

---

that express the effectiveness of selecting a source [6] can also be employed. MKA continues retrieving from S_i the next object p with highest partial-score $\tau^{t_i}(p,q)$ (line 9), and updating the upper bound score u_i^- (line 10) with the new partial-score retrieved from S_i. At this point, any unseen object in S_i has a lower or equal partial-score than u_i^-. Next, MKA marks that p has been seen in S_i adding p into L_i (line 11), and updating the lower bound score for p (line 12). The lower bound score is computed by summing the partial-scores known for p with the worst possible partial-score that p may have based on the sources in which p has not been seen yet. Then, MKA checks if p is in the candidate set C, inserting p otherwise (lines 13-15). Next, MKA updates the upper bound score for any object in the candidate set C (lines 16-18). The upper bound score for a given object in a source that it has not been yet, does not decrease bellow its lower bound score on that source (line 17). The objects $p \in C$ whose lower bound p_- is higher or equal the upper bound o^- of any other object in C (lines 19-22) are reported incrementally. MKA repeats this process until k objects has been reported, or the sources are empty. If the sources are empty and less than k objects have been returned, MKA reports the objects in C with highest lower bound score (lines 24-26) as the remaining top-k.

Steps	aR^{t_1}	aR^{t_3}	Candidate set $C = \{p[p_-, p^-]\}$
	$\downarrow$	$\downarrow$	
2	$\tau^{t_1}(\mathbf{p_4}, q) \approx 1.2$	$\tau^{t_3}(\mathbf{p_5}, q) \approx 1.21$	$p_4[1.39, 2.41], p_5[1.42, 2.41]$
2	$\tau^{t_1}(\mathbf{p_6}, q) \approx 1.17$	$\tau^{t_3}(\mathbf{p_6}, q) \approx 1.17$	$p_4[1.39, \mathbf{2.37}], p_5[1.42, \mathbf{2.38}], p_6[2.34, 2.34]$
1	$\tau^{t_1}(\mathbf{p_3}, q) \approx 0.65$		$p_4[1.39, 2.37], p_5[1.42, \mathbf{1.86}], p_6[2.34, 2.34]$
1		$\tau^{t_3}(\mathbf{p_2}, q) \approx 0.68$	$p_4[1.39, \mathbf{1.88}], p_5[1.42, 1.86], p_6[2.34, 2.34]$

Fig. 3. State of some variables during the execution of MKA. The candidates p_2 and p_3 have low score and are omitted from the list of candidates.

Example 4. Assume a top-k spatial keyword query q, where $q.l = (5, 6)$, $q.d = \{bar, pub\}$, and $q.k = 1$. The MKA accesses the aR-trees aR^{t_1} (bar) and aR^{t_3} (pub) as depicted in Fig. 2. The state of some variables during the execution of MKA is shown in Fig. 3. After the second iteration (step), MKA has retrieved p_4 and p_5 from aR^{t_1} and aR^{t_3} respectively. MKA continues retrieving p_6 from both aR-trees aR^{t_1} and aR^{t_3}. Although p_6 has been found in both aR-trees, it cannot be reported as top-1 since the upper bound score of the other candidates p_4 and p_5 is still smaller than the score of p_6 (Fig. 3). However, after retrieving p_3 from aR^{t_1} and p_2 from aR^{t_3}, p_6 can be reported progressively as top-1 since its score is smaller than the upper bound score of the other candidates.

Aggregating term-scores to compute the top-k spatio-textual objects is similar to aggregating ranked inputs to compute the top-k results [9,15]. For simplicity, we omit from the description of Algorithm 2 some implementation details that permit reducing the size of the candidate set and the number of comparisons to update the upper bound score [12]. In the following, we evaluate the performance of SKA and MKA algorithms.

7   Experimental Evaluation

In this section, we compare our approach the S2I against the DIR-tree proposed by Cong *et al.* [4]. All algorithms were implemented in Java using the XXL library[1]. The nodes of the aR-trees, employed in S2I, have a block size of 4KB that is able to store between 42 and 85 entries. The blocks in the file used to store the non-frequent items has a maximum size of 4KB that permits to store a maximum of 146 entries. The intuition behind this choice is that we store objects in an aR-tree only when there are enough objects to fill more than one node of an aR-tree. Each node of a DIR-tree also has a block size of 4KB and is able to store between 46 and 92 entries. The parameter β used to balance textual similarity and spatial location during the construction of the DIR-tree was set to 0.1 as suggested by Cong *et al.* [4].

Setup. Experiments were executed on a PC with a 3GHz Dual Core AMD processor and 2GB RAM. In each experiment, we execute 100 queries to warm-

---

[1] http://dbs.mathematik.uni-marburg.de/Home/Research/Projects/XXL

Table 1. Settings used in the experiments. The default values are presented in bold.

Parameter	Values
Number of results (k)	**10**, 20, 30, 40, 50
Number of keywords	1, 2, **3**, 4, 5
Query preference rate α	0.1, **0.3**, 0.5, 0.7, 0.9
Twitter dataset	1M, **2M**, 3M, 4M
Other datasets	Data1, Wikipedia (Wiki), Flickr, OpenStreetMap (OSM)

Table 2. Characteristics of the datasets

Datasets	Tot. no. of. objects	Avg. no. of unique words per object	Tot. no. of unique words	Tot. no. of words
Twitter1	1,000,000	11.94	553,515	12,542,414
Twitter2	2,000,000	12.00	1,009,711	25,203,367
Twitter3	3,000,000	12.26	1,391,171	38,655,751
Twitter4	4,000,000	12.27	1,678,451	51,661,462
Data1	131,461	131.70	101,650	32,622,168
Wikipedia	429,790	163.65	1,871,836	109,365,635
Flickr	1,471,080	14.49	487,003	25,417,021
OpenStreetMap	2,927,886	8.76	662,334	31,526,352

up the buffers, and collect the average results of the next 800 queries. The queries are randomly generated using the same vocabulary and the same spatial area of the datasets as used by Cong *et al.* [4]. We employed a buffer whose size was fixed in 4MB for both approaches. In the experiments, we measured the total execution time (referred as response time) and the number of I/Os (page faults). All charts are plotted using a logarithmic scale on the y-axis. The main parameters and values used through the experiments are presented in Table 1.

Datasets. Table 2 shows the characteristics of the datasets used in the experiments. We employed four Twitter datasets of 1M, 2M, 3M, and 4M objects each, where each object is composed by a Twitter message (tweet) and a random location where latitude and longitude are within the range [0,100]. In order to create these datasets, we used the first 10 million non-empty tweets from the Stanford Twitter dataset[2]. The Data1 dataset was created combining texts from 20 Newsgroups dataset[3] and locations from LA streets[4]. This dataset is similar to the Data1 dataset used by Cong *et al.* [4]. However, instead of selecting only 5 groups, we employed all documents in the 20 Newsgroups dataset. We also conducted experiments on real datasets: Wikipedia, Flickr, and OpenStreetMap. The Wikipedia dataset is composed by Wikipedia articles with a spatial location. The Flickr dataset contains objects referring to photos taken in the area

---

[2] http://snap.stanford.edu/data/twitter7.html
[3] http://people.csail.mit.edu/jrennie/20Newsgroups
[4] http://barcelona.research.yahoo.net/websmapm/datasets/uk2007

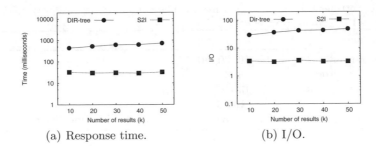

(a) Response time.   (b) I/O.

Fig. 4. Response time and I/O varying the number of results (k)

of London. Each object is composed by a spatial location and a text describing the photo (title and description). Finally, the OpenStreetMap[5] dataset contains objects downloaded from OpenStreetMap covering the entire planet.

7.1   Query Processing Performance

In this section, we evaluate the query processing performance of the S2I and DIR-tree for different setups. In all experiments, we employ the default settings (Table 1) to study the impact on I/O and response time, while varying a single parameter.

Varying the number of results (k). Fig. 4 plots the response time and I/O of the S2I and DIR-tree, while varying the number of results k. The response time achieved using S2I is one order of magnitude better than using DIR-tree (Fig. 4(a)). Furthermore, the advantage of S2I increases when the number of results increases. The main reason for this is that S2I accesses less disk pages to process a query (Fig. 4(b). In order to process a query employing DIR-tree, the inverted files at each node are accessed to obtain the posting lists of each distinct keyword in the query. For example, a query with 3 distinct keywords is performed in two steps: first, the postings lists of each keyword is retrieved in order to identify the entries of the node that can contribute to the query results, then the relevant entries are visited in decreasing order of score. Although the size of the posting lists are small, since they are bounded by the maximum capacity of a node, the process to perform such queries on inverted indexes incurs in non-negligible cost.

Varying the number of keywords. Fig. 5 depicts the response time and I/O, while varying the number of query keywords. Again, the response time (Fig. 5(a)) and I/O (Fig. 5(b)) achieved by using the S2I are one order of magnitude better. Single-keyword queries are processed efficiently employing the SKA algorithm. Hence, few pages are accessed during the query processing. As expected, the larger the number of keywords, the higher the I/O required to process the query.

---

[5] http://www.openstreetmap.org

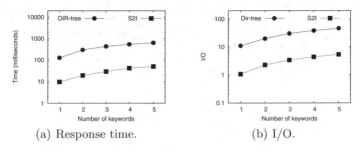

(a) Response time.   (b) I/O.

Fig. 5. Response time and I/O varying the number of keywords

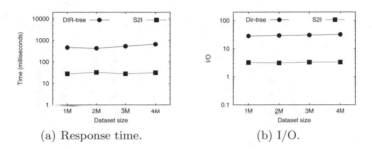

(a) Response time.   (b) I/O.

Fig. 6. Response time and I/O varying the cardinality of the Twitter dataset

However, the advantage of S2I over DIR-tree remains constant, which demonstrates the efficiency of the MKA algorithm.

Varying the cardinality. In Fig. 6, we evaluate the impact of increasing the dataset size (cardinality) on response time and I/O. Again, the response time achieved by using the S2I is around one order magnitude better than S2I, see Fig. 6(a). Moreover, the advantage increases when the dataset increases. The same behavior is noted in the number of I/Os required, see Fig. 6(b).

Varying the query preference parameter (α). In Fig. 7, we study the impact of α (Equation 1) on response time and number of I/Os. The performance of the S2I increases for higher values of the query preference parameter α (Fig. 7(a)), which means that S2I can terminate earlier if the preference parameter gives more weight to proximity over text relevance. The same behavior repeats in Fig. 7(b) that plots I/O. The query preference parameter does not present an impact on the performance of DIR-tree. The main reason for this is that the Twitter dataset has a large vocabulary and the objects are uniformly spread in the spatial space, which reduces the capacity of DIR-tree to put objects with similar content and similar location in the same node.

Varying the datasets. In Fig. 8, we present response time and I/O for different datasets. The S2I presents better response time (Fig. 8(a)) and I/O (Fig. 8(b)) for all datasets. The response time is influenced by the size of the dataset. The

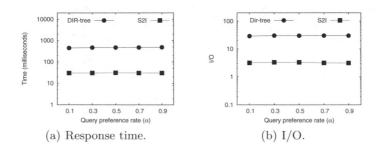

(a) Response time.   (b) I/O.

Fig. 7. Response time and I/O varying the query preference rate (α)

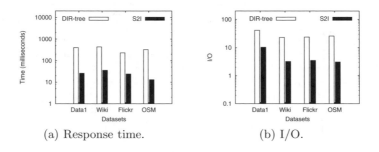

(a) Response time.   (b) I/O.

Fig. 8. Response time and I/O for different datasets

Wikipedia dataset is one of the largest dataset evaluated, since the text associated with each object has many keywords. On the other hand, Flickr and OpenStreetMap datasets have similar number of distinct terms per object, which is, in general, small. Consequently, the performance on those datasets is similar. Finally, Data1 presents the highest I/O. This happens because the total number of unique words in Data1 is small compared to its dataset size. Data1 is created combining 131,461 locations with only 20 thousand documents to create a dataset with 131,461 spatio-textual objects. Hence, there are many objects with the same textual description.

7.2 Maintenance and Space Requirements

In this section, we evaluate the maintenance cost and space requirements in both the DIR-tree and S2I.

Maintenance. We evaluate the cost of insertions, which are more frequent than updates and deletions. In order to obtain the insertion cost, we inserted 100 objects in S2I and DIR-tree and collected the average time. After each insertion, we flush the data in the index structures. The results are presented in Fig. 9.

Several storage units (blocks and trees) of the S2I, one per distinct term, are accessed due to an insertion of a new object. However, the tasks executed in a block or in an aR-tree are performed efficiently since they do not require

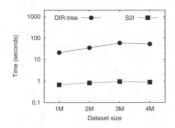

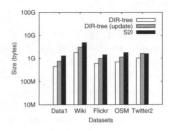

Fig. 9. Update time (seconds)   **Fig. 10.** Size of the indexes

computing the textual similarity between the new object and the objects currently stored in the index. However, the cost of inserting a new object in a DIR-tree requires comparing the similarity between the text of the new object and the text of other several objects in order to find the region of DIR-tree that better accommodates the new object. Moreover, the insertion requires updating inverted files which is also challenging and costly.

Space requirements. Fig. 10 depicts the space required for both indexes DIR-tree and S2I. The size required by S2I is larger than the size required by DIR-tree. The main reason is that S2I employs more trees that do not use the space in the nodes effectively. The figure shows also the space required by DIR-tree to execute updates. This space includes the size of the vectors of the pseudo-documents at each node. The vectors are required only during updates, since at query time DIR-tree employs only inverted files. The space required by S2I to perform search and update is the same.

8   Conclusions

In this paper, we present a new index named Spatial Inverted Index (S2I) and algorithms (SKA and MKA) to support top-k spatial keyword queries efficiently. Similar to an inverted index, S2I maps distinct terms to the set of objects that contains the term. The list of objects that contain a term are stored differently according to the document frequency of the term. If the term occurs often in the collection, the objects with the term are stored in an aggregated R-tree and can be retrieved in decreasing order of partial-score efficiently. Differently, the objects of infrequent term are stored together in a block in a file. Furthermore, we present algorithms to process single-keyword (SKA) queries and multiple-keyword (MKA) queries efficiently. Finally, we show through extensive experiments that our approach outperforms the state-of-the-art approach in terms of query and update cost.

Acknowledgments. We are thankful to Massimiliano Ruocco for providing the Flickr dataset used in the experimental evaluation.

References

1. Anh, V.N., de Kretser, O., Moffat, A.: Vector-space ranking with effective early termination. In: Proc. of ACM Special Interest Group on Information Retrieval (SIGIR), pp. 35–42 (2001)
2. Beckmann, N., Kriegel, H., Schneider, R., Seeger, B.: The R*-tree: An efficient and robust access method for points and rectangles. In: Proceedings of the ACM Int. Conf. on Management of Data (SIGMOD), pp. 322–331 (1990)
3. Chen, Y., Suel, T., Markowetz, A.: Efficient query processing in geographic web search engines. In: Proceedings of the ACM Int. Conf. on Management of Data (SIGMOD), pp. 277–288 (2006)
4. Cong, G., Jensen, C.S., Wu, D.: Efficient retrieval of the top-k most relevant spatial web objects. In: Int. Conf. on Very Large Data Bases (VLDB), pp. 337–348 (2009)
5. Felipe, I.D., Hristidis, V., Rishe, N.: Keyword search on spatial databases. In: Proceedings of Int. Conf. on Data Engineering (ICDE), pp. 656–665 (2008)
6. Güntzer, U., Balke, W., Kießling, W.: Optimizing Multi-Feature queries for image databases. In: Proceedings of the Int. Conf. on Very Large Data Bases (VLDB), pp. 419–428 (2000)
7. Hariharan, R., Hore, B., Li, C., Mehrotra, S.: Processing spatial-keyword (SK) queries in geographic information retrieval (GIR) systems. In: Proceedings of the Int. Conf. on Scientific and Statistical Database Management (SSDBM), pp. 1–10 (2007)
8. Hjaltason, G.R., Samet, H.: Distance browsing in spatial databases. ACM Transactions on Database Systems (TODS) 24(2), 265–318 (1999)
9. Ilyas, I.F., Beskales, G., Soliman, M.A.: A survey of top-k query processing techniques in relational database systems. ACM Comp. Surveys 40(4), 1–58 (2008)
10. Joachims, T.: A statistical learning model of text classification for support vector machines. In: Proc. of ACM Special Interest Group on Information Retrieval (SIGIR), pp. 128–136 (2001)
11. Li, Z., Lee, K.C., Zheng, B., Lee, W.-C., Lee, D., Wang, X.: IR-tree: An efficient index for geographic document search. Proceedings of the IEEE Transactions on Knowledge and Data Engineering (TKDE) 99(4), 585–599 (2010)
12. Mamoulis, N., Yiu, M.L., Cheng, K.H., Cheung, D.W.: Efficient top-k aggregation of ranked inputs. ACM Transactions on Database Systems (TODS) 32(3), 19 (2007)
13. Manning, C., Raghavan, P., Schütze, H.: Introduction to Information Retrieval. Cambridge University Press, Cambridge (2008)
14. Papadias, D., Kalnis, P., Zhang, J., Tao, Y.: Efficient OLAP operations in spatial data warehouses. In: Proceedings of the Int. Symposium on Advances in Spatial and Temporal Databases (SSTD), pp. 443–459 (2001)
15. Rocha-Junior, J.B., Vlachou, A., Doulkeridis, C., Nørvåg, K.: Efficient processing of top-k spatial preference queries. Proceedings of the VLDB Endowment (PVLDB) 4(2), 93–104 (2010)
16. Salton, G., Buckley, C.: Term-weighting approaches in automatic text retrieval. Information Processing and Management 24(5), 513–523 (1988)
17. Zhou, Y., Xie, X., Wang, C., Gong, Y., Ma, W.: Hybrid index structures for location-based web search. In: Proceedings of Int. Conf. on Information and Knowledge Management (CIKM), pp. 155–162 (2005)
18. Zobel, J., Moffat, A.: Inverted files for text search engines. ACM Comp. Surveys 38(2), 1–56 (2006)

Retrieving k-Nearest Neighboring Trajectories by a Set of Point Locations

Lu-An Tang[1,2], Yu Zheng[2], Xing Xie[2], Jing Yuan[3], Xiao Yu[1], and Jiawei Han[1]

[1] Computer Science Department, UIUC
[2] Microsoft Research Asia
[3] University of Science and Technology of China
{tang18,xiaoyu1,hanj}@illinois.edu
{yuzheng,xingx}@microsoft.com, yuanjing@mail.ustc.edu.cn

Abstract. The advance of object tracking technologies leads to huge volumes of spatio-temporal data accumulated in the form of location trajectories. Such data bring us new opportunities and challenges in efficient trajectory retrieval. In this paper, we study a new type of query that finds the *k Nearest Neighboring Trajectories* (*k*-NNT) with the minimum aggregated distance to a set of query points. Such queries, though have a broad range of applications like trip planning and moving object study, cannot be handled by traditional *k*-NN query processing techniques that only find the neighboring points of an object. To facilitate scalable, flexible and effective query execution, we propose a *k*-NN trajectory retrieval algorithm using a candidate-generation-and-verification strategy. The algorithm utilizes a data structure called *global heap* to retrieve candidate trajectories near each individual query point. Then, at the verification step, it refines these trajectory candidates by a lower-bound computed based on the global heap. The global heap guarantees the candidate's *completeness* (*i.e.*, all the *k*-NNTs are included), and reduces the computational overhead of candidate verification. In addition, we propose a *qualifier expectation* measure that ranks partial-matching candidate trajectories to accelerate query processing in the cases of non-uniform trajectory distributions or outlier query locations. Extensive experiments on both real and synthetic trajectory datasets demonstrate the feasibility and effectiveness of proposed methods.

1 Introduction

The technical advances in location-acquisition devices have generated a huge volume of location trajectories recording the movement of people, vehicle, animal and natural phenomena in a variety of applications, such as social networks, transportation systems and scientific studies: In Foursquare [1], the check-in sequence of a user in restaurants and shopping malls can be regarded as a location trajectory. In many GPS-trajectory-sharing websites like Geolife [17, 18, 19], people upload their travel routes for the purpose of memorizing a journey and sharing life experiences with friends. Many taxis in big cities have been embedded with GPS sensors to report their locations. Such reports formulate a large amount of trajectories being used for resource allocation, security management and traffic analysis [8]. Biologists solicit the moving

D. Pfoser et al. (Eds.): SSTD 2011, LNCS 6849, pp. 223–241, 2011.
© Springer-Verlag Berlin Heidelberg 2011

trajectories of animals like migratory birds for their research [2]. Similarly, climatologists are busy collecting the trajectories of natural phenomena such as hurricane and ocean currents [3].

In the above-mentioned applications, people usually expect to retrieve the trajectories passing a set of given point locations. For example, the social network users want to retrieve their friend's trails of visiting some scenic spots as references for trip planning. The fleet operators expect to analyze the business of their taxis traveling around several hot spots by the GPS traces. The biologists are interested in study the migration trails of birds passing some mountains, lakes and forests. In general, these applications need to efficiently query and access trajectories from large datasets residing on disks by geospatial locations. Note that, the system needs to select the top k trajectories with the minimum aggregated distance to the given locations instead of the trajectory exactly passing those locations, since in most case exact match may lead to no result or not the best results returned. This study aims to provide an efficient method to expedite a novel geospatial query, the *k-Nearest Neighboring Trajectory Query* (*k*-NNT query), in a trajectory database.

Unfortunately, the *k*-NNT query is not efficiently supported in existing systems. Most traditional *k*-NN query processing methods are designed to find point objects [11, 6, 5]. On the other hand, the traditional trajectory search techniques focus on retrieving the results with similar shapes to a sample trajectory [9, 13]. The new problem, searching top-*k* trajectories given a set of geospatial locations, poses the following challenges:

- *Huge size*: Many databases contain large volumes of trajectories. For example, the T-drive system [8] collects the trajectories from over 33,000 taxis for 3 months. The total length of the trajectories is more than 400 million kilometers and the total number of GPS points reaches 790 million. The huge I/O overhead is the major cost in query processing.

- *Distance computation*: The distance computation in *k*-NNT query is more complex than traditional spatial queries. To compute the aggregated distance from a trajectory to a set of query points, the system has to check all the member points of the trajectory, find out the closest one to each individual query point (*i.e.*, shortest matching pairs) and sum up all the matching pairs as the measure. The techniques of point *k*-NN queries, such as *best-first search* [6] and *aggregate k-NN* [5], cannot handle this problem.

- *Non-uniform distribution*: In many real applications, the distributions of trajectories are highly skewed, *e.g.*, taxi trajectories are much denser in downtown than suburban areas. In addition, query points are given by users in an ad-hoc manner and some of them may be far from all the trajectories.

In this study, we propose a robust, systematic and efficient approach to process *k*-NNT queries in the trajectory database. The system employs a data structure called *global heap* to generate candidate trajectories by only accessing a small part of the data and verifies the candidates with the lower-bound derived from global heap. To handle the skewed trajectory data and outlier query locations, a *qualifier expectation* measure is designed to rank the candidates and accelerate query processing.

The rest of the paper is organized as follows. Section 2 provides the background and problem definition, Section 3 describes detailed query processing framework, and

Section 4 introduces the qualifier expectation-based algorithm. Section 5 evaluates the approaches by extensive experiments on both real and synthetic datasets. Section 6 discusses related studies. Finally, Section 7 concludes the paper.

2 Problem Formulation

The trajectory data are collected in the form of point sequences. Trajectory R_i can be represented as $R_i = \{p_{i,1}, p_{i,2}, \ldots p_{i,n}\}$, where $p_{i,j}$ is the j-th member point of R_i. The input of k-NNT query Q, according to applications, is specified by a set of point locations, $Q = \{q_1, q_2, \ldots q_m\}$. In the following we first define the distance measures between trajectories and query points.

Definition 1. Let trajectory $R_i = \{p_{i,1}, p_{i,2}, \ldots p_{i,n}\}$ and q be a query point. The *matching pair* of a member point $p_{i,j}$ and q is denoted as $<p_{i,j}, q>$. If $\forall p_{i,k} \neq p_{i,j}$, $dist(p_{i,j}, q) \leq dist(p_{i,k}, q)$, $<p_{i,j}, q>$ is the *shortest matching pair* of R_i and q.

Definition 2. Let trajectory $R_i = \{p_{i,1}, p_{i,2}, \ldots, p_{i,n}\}$ and query $Q = \{q_1, q_2, \ldots, q_m\}$. The distance between R_i and a query point q is the distance of the shortest matching pair $<p_{i,j}, q>$, the *aggregated distance* between R_i and Q is the sum of distances of the shortest matching pairs from R_i to all query points.

$$dist(R_i, Q) = \sum_{q \in Q} dist(R_i, q) = \sum_{q \in Q} dist(p_{i,j}, q)$$

Example 1. Figure 1 shows a matching example of two trajectories and three query points. The query points q_1, q_2 and q_3 are matched with the closest points in R_1 and R_2. The nearest neighboring trajectory is selected by the aggregated distance of all the query points. Even R_2 is more distant to q_1 and q_3, its aggregated distance is still smaller than R_1. So R_2 should be returned as the query result.

With the distance measures, now we formally describe the task of k-NNT query.

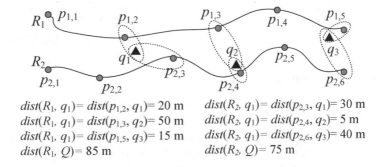

$dist(R_1, q_1)= dist(p_{1,2}, q_1)= 20$ m $dist(R_2, q_1)= dist(p_{2,3}, q_1)= 30$ m
$dist(R_1, q_1)= dist(p_{1,3}, q_2)= 50$ m $dist(R_2, q_1)= dist(p_{2,4}, q_2)= 5$ m
$dist(R_1, q_1)= dist(p_{1,5}, q_3)= 15$ m $dist(R_2, q_1)= dist(p_{2,6}, q_3)= 40$ m
$dist(R_1, Q)= 85$ m $dist(R_2, Q)= 75$ m

Fig. 1. Aggregated Distance Example

Task Definition: Given the trajectory dataset, D, and a set of query points, Q, the k-NNT query retrieves k trajectories K from D, $K = \{R_1, R_2, ..., R_k\}$ that for $\forall R_i \in K, \forall R_j \in D - K, \ dist(R_i, Q) \leq dist(R_j, Q)$.

A direct way to process k-NNT query is to scan the whole trajectory dataset, compute all the shortest matching pairs for each trajectory, and then compare their aggregated distances. The I/O overhead is very high since all the trajectories are retrieved.

A candidate-generation-and-verification framework was first proposed by Fagin *et al.* for processing top k aggregation queries for distributed systems and middleware [12]. The Fagin's algorithm generates a candidate set by searching the objects listed at the top in each individual site, and then carries out the detailed verification only on such candidates. The general idea can be employed to solve this problem, since the k-NNTs usually have member points close to some query locations. The system should search for candidate trajectories around each query point, and then verify them for final results. However, since Fagin's algorithm is not designed for spatial query processing, the candidate search operations on individual sites are carried out in parallel (*i.e.*, the system retrieves the top member from each site, then gets the second member of each of the sites, and so on). In the scenarios of k-NNT query, the situations around each query points are different. It is inefficient to search in such parallel manner. Thus the key problem becomes how to coordinate the candidate searching processes. Meanwhile, the algorithm must guarantee that all the k-NNTs are included in the candidate set (*i.e.*, *completeness*). Section 3 will discuss those issues in detail. Table 1 lists the notations used in the following sections.

Table 1. A List of Notations

Notation	Explanation	Notation	Explanation
D	the trajectory dataset	K	the k-NNT result set
R	a trajectory	$p_{i,j}, p_i$	member points of traj.
Q	the k-NNT query	q, q_j	k-NNT query points
H	the individual heap list	h_i	an individual heap
G	the global heap	N	an R-tree node
C	the candidate set	δ	the pruning threshold
μ	the full-matching ratio	k	a constant given by user

3   Query Processing

3.1   Candidate Generation

The first task of candidate generation is to retrieve the neighboring points around query locations. In this study, we utilize the *best-first* strategy to search for k-NN points [6]. The best-first strategy traverses R-tree index from root node and always visits the node with the least distance to the query point, until it reaches the leaf node and returns it as the result. Based on the best-first search strategy, we construct a data structure of *individual heap* to search k-NN points.

Definition 3. Let q_i be a query point. The *individual heap* h_i is a minimum heap whose elements are the matching pairs of trajectory member point and q_i. The matching pairs are sorted by their distances to q_i.

The individual heap takes q_i as the input and visits R-tree nodes with the shortest distance. If the R-tree node is an inner node, the heap retrieves all its children node and keeps to traverse the tree; if the R-tree node is leaf node (*i.e.*, a trajectory's member point p_j), the heap composes a matching pair of p_j with q_i.

There are two advantages of individual heap: (1) It employs the best-first strategy to traverse R-tree and achieves optimal I/O efficiency [6]. (2) The individual heap does not need to know k in advance, it can pop out the matching pairs incrementally.

For a query $Q = \{q_1, q_2, ..., q_m\}$, the system constructs a heap list $H = \{h_1, h_2, ..., h_m\}$ and finds out the shortest matching pairs from the m heaps. Since there are multiple heaps, the key problem is to coordinate the searching processes of individual heaps. To this end, we introduce the *global heap*.

Definition 4. Let k-NNT query $Q = \{q_1, q_2, ..., q_m\}$ and individual heap list $H = \{h_1, h_2, ..., h_m\}$. The *global heap* G consists of m matching pairs, $G = \{<p_1, q_1>, <p_2, q_2>, ..., <p_m, q_m>\}$, where $<p_i, q_i>$ is popped from the individual heap h_i. G is a minimum heap that sorts the matching pairs by their distances.

The global heap has two operations, *pop* and *retrieve*. The pop operation simply outputs the shortest matching pair of G. The retrieve operation is carried out immediately after popping a pair $<p_i, q_i>$. The global heap retrieves another matching pair $<p_i', q_i>$ from the corresponding individual heap h_i. In this way, there are always m matching pairs in G.

The popped matching pairs are kept in a candidate set. In the beginning, the candidate trajectories only have a few matching pairs, we call them *partial-matching candidates*. When the global heap pops out more matching pairs, several trajectories will eventually complete all the matching pairs for query Q, they are called *full-matching candidates*. In Figure 2, trajectory R_1 is a full-matching candidate with all shortest matching pairs, and R_2 and R_4 are partial-matching candidates since they miss the pairs of several query points. One may notice that, not all the matching pairs popped out by global heap G are added to the candidate set. For example, the current top element of G is $<p_{1,4}, q_1>$, and there is already a shortest matching pair $<p_{1,2}, q_1>$ in candidate R_1. Since the individual heap h_1 reports the k-NN points in incremental manner, the oldest pair $< p_{1,2}, q_1>$ is guaranteed to be the shortest one from R_1 to q_1. The new pair $<p_{1,4}, q_1>$ is then a *useless pair*. It should be thrown away.

The last issue of candidate generation is the stop criterion. The algorithm should not stop unless the candidate set has already contained all the k-NNTs.

Property 1. If a candidate set has at least k full-matching candidates whose shortest matching pairs are all popped from the global heap, then the candidate set is complete.

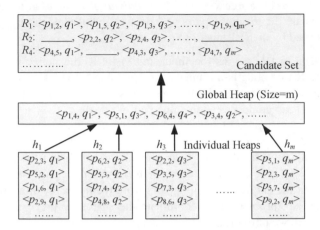

Fig. 2. Data Structures for Candidate Generation

Proof: A candidate set C is complete if it contains all the k-NNTs. That is, for any trajectory $R_i \notin C$, we need to prove that R_i cannot be a k-NNT.

For $\forall q \in Q$, we denote the shortest matching pair from R_i to q as $<p_i, q>$. Since $R_i \notin C$, $<p_i, q>$ has not been popped out yet. The global heap G pops out matching pairs in the order of increasing distance, the distance of $<p_i, q>$ is large than or equal to current matching pair $<p_G, q>$ in G, that is: $dist(p_G, q) \leq dist(p_i, q)$.

Let $R_j \in C$ be a full-matching trajectory candidate, whose matching pair for q is $<p_j, q>$. Since $<p_j, q>$ is already popped from G, its distance is less than $<p_G, q>$: $dist(p_j, q) < dist(p_G, q)$. Then $dist(p_j, q) < dist(p_i, q)$. Hence we have:

$$dist(R_j, Q) = \sum_{q \in Q} dist(p_j, q) < \sum_{q \in Q} dist(p_i, q) = dist(R_i, Q)$$

There are at least k full-matching trajectories in C, their aggregated distances are all smaller than R_i, so R_i is not possible to be a k-NNT, candidate set C is complete. ■

Based on Property 1, we develop the algorithm of k-NNT candidate generation. The algorithm first constructs the individual heaps and initializes the global heap and candidate set (Lines 1--3). Each individual heap pops a shortest matching pair to the global heap (Lines 4--5). In this way the global heap has m matching pairs. Then the candidate generation process begins. Once the global heap is full with m pairs, it pops out the shortest one. The system checks whether the candidate set already contains an old pair with the same trajectory and query point. If there is no such pair, the new popped pair is a shortest matching pair, it is then added to the candidate set (Lines 7--9). After that, the global heap retrieves another pair from the corresponding individual heap (Line 10). This process stops when there are k full-matching trajectories in the candidate set (Line 11).

Algorithm 1. k-NNT Candidate Generation
Input: Trajectory dataset D, Query Q, k
Output: k-NNT Candidate Set C
1.　　**for** each $q_i \in Q$
2.　　　　construct the individual heap h_i on D;
3.　　initialize the global heap G and candidate set C;
4.　　**for** each individual heap h_i
5.　　　　pop a matching pair and push it to G;
6.　　**repeat**
7.　　　　pop the shortest pair $<p_j, q_j>$ from G;
8.　　　　**if** $<p_j, q_j>$ is a shortest matching pair, **then**
9.　　　　　　add $<p_j, q_j>$ to C;
10.　　　pop a matching pair from h_j and push it to G;
11.　　**until** C contains k full-matching candidates
12.　　**return** C;

Example 2. Figure 3 shows an example of the candidate generation algorithm. Suppose k is set to 1. The algorithm first constructs the global heap with matching pairs $<p_{1,4}, q_2>$, $<p_{1,6}, q_3>$ and $<p_{1,2}, q_1>$. In the first iteration the pair $<p_{1,4}, q_2>$ is popped to the candidate list, candidate R_1 is generated. Meanwhile the global heap retrieves the another pair $<p_{5,5}, q_2>$ from q_2's individual heap. In the next three iterations, the global heap pops matching pairs $<p_{1,6}, q_3>$, $<p_{5,5}, q_2>$ and $<p_{4,5}, q_3>$ and generates two partial-matching candidates R_4 and R_5. At the 5th iteration, $<p_{1,2}, q_1>$ is popped out and a full-matching candidate R_1 is generated. The algorithm then stops and outputs the candidate set for further verification.

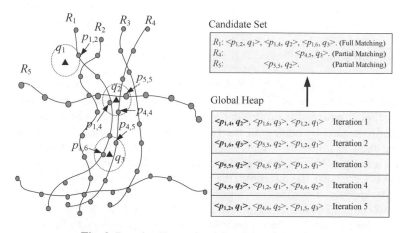

Fig. 3. Running Example of Candidate Generation

3.2　Candidate Verification

After generating the candidates, the system needs to verify them to select k-NNTs. For the partial-matching candidates, the algorithm has to make up their missing pairs.

The computational and I/O costs are high for this task. Suppose the candidate set's size is l (with k full-matching candidates), each trajectory contains average n member points and the number of query points is m. In the worst case, each partial-matching candidate only has one matching pair, the system has to carry out $(l - k) * n * (m - 1)$ times of distance calculation to make up the missing pairs. And it needs to access the $l - k$ partial-matching trajectories. Fortunately the global heap provides a short-cut to enhance the efficiency of candidate verification.

Property 2. Let Q be the query, G be the global heap, R be a partial-matching candidate trajectory and $Q_R \subset Q$ be the subset of query points that are contained in R's matching pairs. Then $LB(R, Q)$, as defined in the following equation, is a lower-bound of R's aggregated distance.

$$LB(R, Q) = \sum_{q_i \in Q_R} dist(p_R, q_i) + \sum_{q_j \in Q - Q_R} dist(p_G, q_j)$$

where $<p_R, q_i>$ is a matching pair in R and $<p_G, q_j>$ is a matching pair in G.

Proof: For $\forall q_j \in Q - Q_R$, we denote the shortest matching pair from R to q_j as $<p_R, q_j>$. $<p_R, q_j>$ is a missing pair that has not been popped out from the global heap G yet. Since G is a minimum heap and pops out matching pairs with increasing distance, $dist(p_R, q_j) \geq dist(p_G, q_j)$. So we have:

$$dist(R, Q) = \sum_{q_i \in Q} dist(p_R, q_i) = \sum_{q_i \in Q_R} dist(p_R, q_i) + \sum_{q_j \in Q - Q_R} dist(p_R, q_j)$$

$$\geq \sum_{q_i \in Q_R} dist(p_R, q_i) + \sum_{q_j \in Q - Q_R} dist(p_G, q_j) = LB(R, Q)$$

Hence $LB(R,Q)$ is a lower-bound of R's aggregated distance.   ∎

Note that, for $\forall q_i \in Q_R$ and $\forall q_j \in Q - Q_R$, the distances of $<p_R, q_i>$ and $<p_G, q_j>$ have been already computed. Thus $LB(R, Q)$ is calculated without any I/O overhead.

Algorithm 2 outlines the processing of candidate verification based on Property 2. The algorithm starts by adding all the full-matching trajectories to the result set and obtaining the k-th trajectory's aggregated distance as the pruning threshold (Lines 1--4). Then it computes the lower-bound for each partial-matching candidate. If the lower-bound is larger than the threshold, the trajectory is pruned without further computation (Lines 6--7). Otherwise, the algorithm has to access the trajectory's member points and compute its aggregated distance (Lines 8--9). If the system finds a trajectory with a shorter distance than threshold, it adds the trajectory to the result set and updates the threshold (Lines 9--12). After processing all the partial-candidates, the algorithm outputs the top k trajectories in the result set as k-NNTs (Line 13). This pruning strategy is especially powerful if there are many partial-matching candidates with only one or two matching pairs. The more matching pairs a candidate lacks, the lager lower-bound it will have, and the higher probability it will be pruned.

Algorithm 2. k-NNT Candidate Verification

Input: k-NNT candidate set C, global heap G, query Q.
Output: k-NNTs.
1. initialize result set K;
2. add all full-matching candidate of C to K;
3. sort K in the order of increasing distance;
4. threshold $\delta \leftarrow k$-th trajectory's aggregated distance in K;
5. **for** each partial-matching candidate R in C
6.   compute $LB(R,Q)$;
7.   **if** $LB(R,Q) \geq \delta$ **then continue**;
8.   **else**
9.   compute $dist(R,Q)$;
10.   **if** $dist(R,Q) < \delta$ **then**
11.   add R to K;
12.   sort K and update δ;
13. **return** the top k trajectories in K;

Example 3. Figure 4 shows the process to verify the candidates from Example 2. There are one full-matching candidate, R_1, and two partial-matching candidates R_4 and R_5. The algorithm first calculates $LB(R_4, Q)$ and $LB(R_5, Q)$. The calculations are carried out based on the results in the global heap and candidate set. Since $LB(R_4, Q)$ is larger than the threshold, R_4 is pruned directly. The system only accesses the member points of R_5 for further computation. Finally, R_1 is returned as the result.

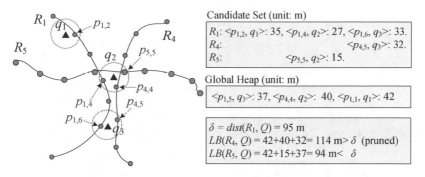

Fig. 4. Running Example of Candidate Verification

4   Qualifier Expectation

In k-NNT query processing, there are two steps that involve I/O overheads: (1) In candidate generation, the individual heaps traverse the R-tree to pop out matching pairs; (2) In candidate verification, if the lower-bound of a partial-matching candidate is less than the threshold, the system needs to access the trajectory's member points for distance computation. The key to reduce I/O costs is to generate *tight* candidate set, *i.e.*, the number of the candidates should be as small as possible. A tight candidate set costs less time to generate and is easier to be verified since the number of partial-matching trajectories is also smaller.

The major function of global heap is to raise the candidate set's tightness in the premise of guaranteeing completeness. The global heap controls searching processes of different individual heaps, restricts the search regions as equal-radius circles, as illustrated in Figure 5. When the trajectories and query points are all uniformly distributed, the global heap can pop out a similar number of matching pairs to different locations. In this way, the k full-matching candidates are soon found and the candidate generation algorithm stops early. However, many real trajectory datasets are skewed: the taxi trajectories are dense in the downtown areas, the animal movements are concentrated around water/food sources. In addition, the query points are ad-hoc. It is possible that a user may provide an *outlier location* that is distant from all the trajectories. In such cases, the matching pairs of outlier locations are much longer, they could be stuck in the global heap and significantly delay query processing.

Example 4. Figure 5 shows an example of outlier location. The query point q_3 is an outlier since it is far from all the trajectories. The distance of its shortest matching pair $<p_{2,7}, q_3>$ is much larger than the pairs from other individual heaps. This pair cannot be popped out from the global heap and no full-matching candidate is found. The global heap has to increase the search radius and keep on popping useless pairs. Finally the algorithm ends with a large candidate set. And the system has to cost even more time in the verification step. The query efficiency is affected seriously due to a single outlier location.

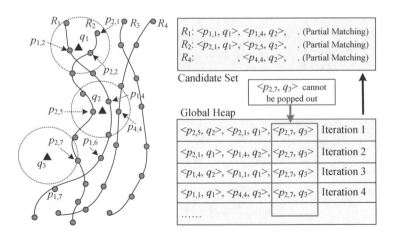

Fig. 5. Example of Outlier Point

In the cases of outlier locations, the cost is high to wait global heap to pop k full-matching candidates. *Then can we compute them directly?* The system can retrieve some partial-matching trajectories and make up their missing pairs. The key point is to guarantee the completeness of generated candidates.

Property 3. If a candidate set has at least k full-matching candidates, and their aggregated distances are smaller than the accumulated distance of all the matching pairs in the global heap, then the candidate set is complete.

Proof: A candidate set C is complete if it contains all the k-NNTs. In another word, for $\forall R_i \notin C$, we need to prove that R_i cannot be a k-NNT.

For $\forall q \in Q$, we denote the shortest matching pair from R_i to q as $<p_i, q>$ and the current matching pair in G as $<p_G, q>$. According to the proof of Property 1, we have $dist(p_G, q) \leq dist(p_i, q)$, then

$$Sum(G) = \sum_{q \in Q} dist(p_G, q) \leq \sum_{q \in Q} dist(p_i, q) = dist(R_i, Q)$$

Let R_j denote the full-matching candidate that, $R_j \in C$ and $dist(R_j, Q) < Sum(G)$. Then $dist(R_j, Q) < dist(R_i, Q)$. There are at least k such full-matching candidates in C, then R_i is not possible to be a k-NNT, candidate set C is complete. ∎

Based on Property 3, if the candidate set contains k full-matching candidates with smaller distances than the distance sum of global heap's matching pairs, the candidate generation can end safely. We call such full-matching candidates as *qualifiers*. Since the number of partial-matching candidates is usually much larger than k, the system needs to select out the ones that are most likely to become qualifiers to make up. A measure is thus required to represent the expectation of a partial-candidate to be a qualifier. To reveal the essential factors of such measure, let us investigate the following example.

Example 5. Figure 6 lists out the matching pairs in a candidate set. There are three partial-matching candidates R_1, R_2 and R_4 with current aggregated distances as 70m, 80m and 70m. q_3 is an outlier location, the distance of its matching pair is much larger than others. Now if the system has to select a candidate and makes it up, which one is most likely to be a qualifier?

Candidate Set (Unit: m)

R_1: $<p_{1,2}, q_1>$, 10; $<p_{1,4}, q_2>$, 20;	$;<p_{1,6}, q_4>$, 40. [Agg. : 70]	
R_2: $\quad$; $<p_{2,5}, q_2>$, 45;	$;<p_{2,8}, q_4>$, 35. [Agg. : 80]	
R_4: $\quad$; $<p_{4,4}, q_2>$, 40;	$;<p_{4,5}, q_4>$, 30. [Agg. : 70]	

Global Heap (Accumulated distance: 340 m)

$<p_{2,6}, q_2>$, 50; $<p_{2,1}, q_1>$, 65; $<p_{6,3}, q_4>$, 75; $<p_{1,7}, q_3>$, 150.

Fig. 6. Partial-Matching Candidates

Intuitively, we prefer R_4 to R_2, they have the same number of matching pairs but R_4's aggregated distance is smaller. Furthermore, R_1 is better than R_4, their aggregated distances are the same but R_4 has one more missing matching pair than R_1. To become a qualifier, R_4 needs to make up two matching pairs but R_1 only needs one.

From the above example, we can find out that a candidate's *qualifier expectation* is determined by two factors: the number of missing pairs and the advantage of existing matching pairs over the corresponding ones in global heap.

Let R be a partial-matching candidate trajectory and $Q_R \subset Q$ be the subset of query points that are contained in R's matching pairs. The *qualifier expectation* of R is given in the following equation:

$$Expect(R) = \frac{\sum_{q \in Q_R}(dist(p_G, q) - dist(p_R, q))}{|Q - Q_R|}$$

The qualifier expectation actually denotes an upper-bound of the average distance that R's missing pairs could be larger than the corresponding ones in global heap. The larger this value is, the more likely that R will be a qualifier. Note that, the computation of qualifier expectation can be done with the current matching pairs in the candidate set and global heap, and no more I/O access is needed.

Algorithm 3. Qualifier Expectation-based Generation

Input: Trajectory dataset D, Query Q, Full-matching ratio μ
Output: k-NNT Candidate Set C

1. (Lines 1-10 are the same as Algorithm 1) ......
11.   **while** (|full-matching candidate| / |C|< μ)
12.   compute partial-matching candidate's expectation;
13.   retrieve the candidate R with highest expectation;
14.   make up the matching pairs for R;
15.   **until** C contains k qualifiers;
16.   **return** C;

With the help of qualifier expectation, we can improve the candidate generation algorithm as shown in Algorithm 3. The first few candidate generating steps are the same as Algorithm 1 (Lines 1--9). The difference is at Line 10, Algorithm 3 controls the size of full-matching candidates by a ratio parameter μ. Each time the global heap pops out a matching pair to candidate set, the proportion of full-matching candidate is compared with μ. If a full-matching candidate needs to be generated, the algorithm first calculates the qualifier expectations of all partial-matching candidates and picks the one with the highest expectation for making up. (Lines 11--13). The algorithm stops if there are k qualifiers (Lines 14--15).

Example 6. Figure 7 illustrates the qualifier expectation-based method. Suppose k is set as 1 and the full-matching ratio μ is 0.33. At the 5th iteration, the candidate set size is 3 and a full-matching candidate should be generated. The algorithm calculates the qualifier expectations of the three partial-matching candidates. R_1 is the one with the highest expectation. The algorithm then retrieves R_1's member points and makes up the missing pairs. Since $dist(R_1, Q)$ is less than the global heap's accumulated distance, R_1 is a qualifier. The candidate generation ends and R_1, R_2 and R_4 are returned as candidates.

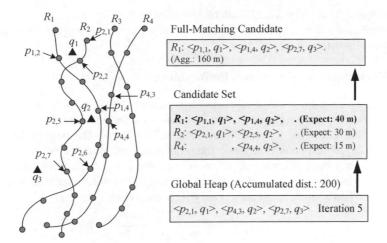

Fig. 7. Qualifier Expectation-based Method

5   Performance Evaluation

5.1   Experiment Settings

Datasets: We conduct extensive experiments to evaluate the proposed methods, using both real-world and synthetic trajectory datasets. The real datasets D_3 is retrieved from the Microsoft GeoLife and T-Drive projects [8, 17, 18, 7]. The trajectories are generated from GPS devices with sampling rate from 5 seconds to 10 minutes. Meanwhile, to test the algorithm's scalability, we also generate two synthetic datasets, being comprised of both uniform and skewed trajectory distributions, with a size more than 2 GB.

Environments: The experiments are conducted on a PC with Intel 7500 Dual CPU 2.20G Hz and 3.00 GB RAM. The operating system is Windows 7 Enterprise. All the algorithms are implemented in Java on Eclipse 3.3.1 platform with JDK 1.5.0. The parameter settings are listed in Table 2.

Table 2. Experimental Settings

Dataset	Type	Traj. #	Total Points	File Size
Syn 1 (D_1)	syn., uniform	40,000	$4.0*10^7$	2.0 GB
Syn 2 (D_2)	syn., skewed	40,000	$4.0*10^7$	2.0 GB
Real (D_3)	taxi	12,643	$1.1*10^6$	54 M

The value of k: 4 – 20, default 20

The query size $|Q|$ (number of query points): 2 –10, default 10

The full matching ratio μ: 20% – 100%, default 40%

Competitors: The proposed *Global Heap-based algorithm* (GH) and *Qualifier Expectation-based method* (QE) are compared with *Fagin's Algorithm* (FA) [12] and *Threshold Algorithm* (TA) [14].

5.2   Evaluations on Algorithm's Performance

We first evaluate the algorithm using uniform dataset D_1. We start the experiments by tuning different k value. Figure 8 shows the query time and accessed R-tree nodes. Note that the y-axes are in logarithmic scale. GH achieves the best performance in both time and I/O efficiency. Because the trajectories are uniformly distributed in D_1, the global heap does a good job to coordinate the candidate search around query points. No matching pair is stuck in the global heap. It is thus unnecessary to directly make up the partial-matching candidates, which involves higher cost.

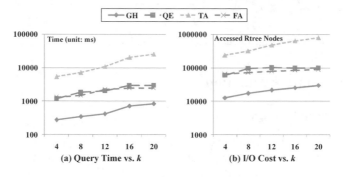

Fig. 8. Performances vs. k on D_1

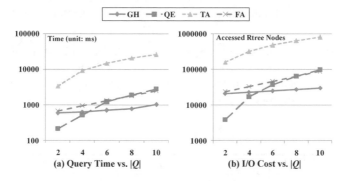

Fig. 9. Performances vs. $|Q|$ on D_1

Figure 9 illustrates the influences of query size $|Q|$ in the experiments. When the query size is small, QE has better performance than GH, because in such cases, the partial-matching candidates have higher probability to become qualifiers. Hence the candidate generation ends earlier. When the query size grows larger, GH outperforms other algorithms with the power of coordinated candidate search. It is also more robust than other competitors.

The second part of our experiments is carried out on skewed dataset D_2. Based on default settings, we evaluate the time and I/O costs of the four algorithms with different values of k and $|Q|$. From Figures 10 and 11 one can clearly see the problem of outlier locations on skewed datasets. The best algorithm on D_1, GH, has degenerated to two orders of magnitude slower. FA also has the same problem. Fortunately, QE still achieves a steady performance, it can process the k-NNT query in 10 seconds even with the largest k and $|Q|$ ($k = 20$, $|Q| = 10$).

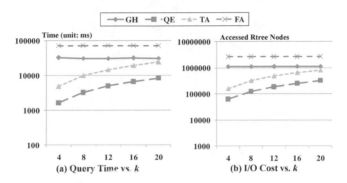

Fig. 10. Performances vs. k on D_2

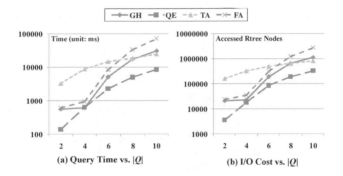

Fig. 11. Performances vs. $|Q|$ on D_2

An interesting observation is that, in Figure 10, GH's time and I/O costs are almost not influenced by the number of k, but the costs increase rapidly with $|Q|$ in Figure 11. This phenomenon can be illustrated by the mechanism of global heap. In the case of outlier point locations, the global heap is difficult to pop out the first matching pair, since such query is distant from all the trajectories. The global heap has to wait for a long time before the first full-matching candidate is generated. Once the global heap pops out the first pair of an outlier point location, it may quickly pop out more matching pairs of that outlier, because the search region has already been enlarged to reach the dense areas of trajectories. As an illustration, please go back to Figure 5, in which GH uses four more iterations to pop out $<p_{2,7}, q_3>$. But once this pair is popped out, other pairs such as $<p_{1,7}, q_3>$ will soon be popped. Hence if GH generates the first full candidate, it can quickly generate more to form a complete candidate set.

We also conduct experiments on the real dataset D_3. In the experiments of tuning k value, QE is the winner, as shown in Figure 12. Comparing to Figures 8 and 10, the algorithm performances on D_3 are more to close to the ones on skewed dataset D_2, since the distribution of real trajectories is more likely to be skewed.

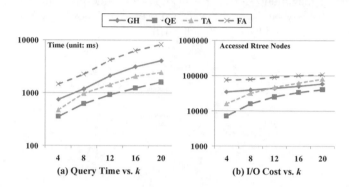

Fig. 12. Performances vs. k on D_3

In Figure 13, GH has better performance than QE when the query size is less than 6. As query size grows, GH's performance degenerates rapidly and QE will be the most efficient algorithm. With more query points, the probability of outlier location becomes higher. GH suffers from such a problem, but QE is relatively robust.

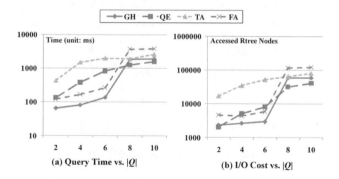

Fig. 13. Performances vs. $|Q|$ on D_3

Finally, we tune the full-matching ratio μ used in QE. The system starts by setting μ as 20% and gradually increases the parameter. The results are recorded in Figure 14. Overall, QE's performance is much better on the real dataset since the data size is much smaller. From the figures one can learn that, when μ is smaller than 40%, the qualifiers are not enough to stop the candidate generation process, but if μ is larger than 60%, the system has to make up too many partial-matching candidates that are unlikely to be qualifiers. QE achieves the best efficiency when μ ranges from 40% to 60%.

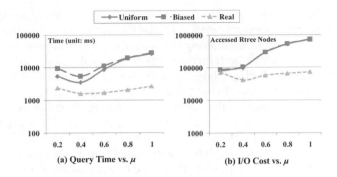

Fig. 14. Performances vs. μ

5.3 Discussions

In summary, GH achieves the best efficiency when the trajectories are uniformly distributed, QE is more suitable for the skewed dataset. Although there is no overall winner on real dataset, we suggest QE as the best choice since it is much robust than GH. In addition, QE is better to handle the complex queries with more location points.

When processing k-NNT query, we set the upper-bounds of k as 20 and |Q| as 10. Since most users are only interested in the first few results, and it is impractical for them to enter tens of query points as input. From the trends of performances of QE and GH, we can see that the algorithms work without problem with even larger k and |Q|. In the experiments, datasets are indexed by R-trees. However, the proposed techniques are flexible to higher dimensions or alternative indexes, such as R* Trees and A-trees. The methods can also be extended to road network. We only need to make minor changes by adjusting the distance computation of individual heap for road network distances.

In the experiments, we use the GPS datasets with high sampling frequency. It is possible that the raw trajectory data collected from other tracking devices are not as ideal as expected. The trajectories could be sparse due to device limitations or users' turning off tracking sensors. On the other hand, if a very long trajectory traverses the entire region, it has higher probability to be selected as k-NNT. In such cases, the trajectories should be preprocessed with similar sampling frequency and length.

6   Related Work

A number of algorithms were proposed to process k-NN queries for point objects. Roussopoulos et al. propose a *depth-first* algorithm for k-NN query in 2D Euclidean space [11]. Hjaltason et al. improve the algorithm with a *best-first* search strategy [6]. Papadias et al. propose the concept of *Aggregate k-NN* (ANN) query, for the first tiem, the problem is extended to multiple query points [5]. Those methods search k-NN as points, while the objects in k-NNT query are trajectories. The distance measures are different. It is difficult to use them directly for k-NNT query processing.

There are many studies about *searching trajectories by sample* based on different similarity measures. Some representative works are: Chen *et al.* propose the *ERP distance* to support local time shifting [9], they show that ERP delivers superb pruning power and search time performance; Sherkat *et al.* develop a *uDAE-based MBR approximation* approach [13]. Those methods define the similarity functions on the shape of trajectories, but they do not consider the spatial properties.

There are several studies in the category of searching trajectory by point locations. As a pioneering work, Frentzos *et al.* propose the concepts of *Moving Object Query* (MOQ) to find the trajectories near a single query point [4]. The moving object query can be seen as a special case of k-NNT query with a single query point.

More recently, Chen *et al.* propose the *k-Best Connected Trajectory* (k-BCT) query [20]. For the first time, they connect the trajectory query to multiple query locations, with both spatial distance and order constraints. The biggest difference between k-BCT and k-NNT is the distance measure. In the k-BCT query, the similarity function between a trajectory R and query locations Q is defined by an exponential function, where $Sim(R, Q) = \sum_{q \in Q} e^{-dist(R,q)}$. With the exponential function, the k-BCT query assigns a larger contribution to closer matched query points and trajectories than far away ones. However, the exponential function of k-BCT query may not be robust to the distance unit.

For a query with m point locations, the FA algorithm generates candidates in a parallel manner around each point location [12]. Without the coordination of global heap, the FA algorithm costs more time on candidate search and also more time to prune the candidates since the lower-bound cannot be computed to help processing. Fagin *et al.* propose another Threshold Algorithm (TA) for top k aggregation query in distributed systems and middleware [14]. For any generated candidates, TA computes their aggregated distances immediately. Indeed it can be seen as a special case of qualifier expectation-based method when the full candidate ratio μ is set to 100%.

7   Conclusions and Future Work

In this study we present the *k-Nearest Neighboring Trajectory* (k-NNT) query to retrieve top k trajectories with the minimum aggregated distance to a set of point locations. This k-NNT query will facilitate a board range of applications, such as travel recommendation, traffic analysis and biological research. We propose a global heap-based method, which coordinates candidate generation, guarantees the completeness of candidates and offers a lower-bound for candidate verification. Meanwhile, by leveraging the proposed measure of qualifier expectation, our method handles the trajectory dataset with skewed distribution and outlier query locations, which are practical situations we need to face in the real world. We evaluate our methods using both real-world and synthetic trajectory datasets, and compare our methods with the state-of-the-art algorithms. The results demonstrate the feasibility and effectiveness of our proposed methods.

This paper is the first step of our trajectory search study. We plan to evaluate the k-NNT queries with different constraints, *e.g.*, temporal constraints and traffic congestions. We are also interested in applying them to advanced spatial applications such as driving route recommendation and traffic management.

Acknowledgements. The work was supported in part by U.S. NSF grants IIS-0905215, CNS-0931975, CCF-0905014, IIS-1017362, the U.S. Army Research Laboratory under Cooperative Agreement No. W911NF-09-2-0053 (NS-CTA). The views and conclusions contained in this document are those of the authors and should not be interpreted as representing the official policies, either expressed or implied, of the Army Research Laboratory or the U.S. Government. The U.S. Government is authorized to reproduce and distribute reprints for Government purposes notwithstanding any copyright notation here on.

References

[1] http://foursquare.com/
[2] http://www.movebank.org
[3] http://weather.unisys.com/hurricane/atlantic
[4] Frentzos, E., Gratsias, K., Pelekis, N., Theodoridis, Y.: Nearest Neighbor Search on Moving Object Trajectories. In: Anshelevich, E., Egenhofer, M.J., Hwang, J. (eds.) SSTD 2005. LNCS, vol. 3633, pp. 328–345. Springer, Heidelberg (2005)
[5] Papadias, D., Tao, Y., Mouratidis, K., Hui, K.: Aggregate Nearest Neighbor Queries in Spatial Databases. ACM TODS 30(2), 529 576
[6] Hjaltason, G.R., Samet, H.: Distance browsing in spatial databases. ACM TODS 24(2), 265–318 (1999)
[7] GeoLife GPS Trajectories Datasets. Released at,
     http://research.microsoft.com/en-us/downloads/b16d359d-d164-469e-9fd4-daa38f2b2e13/default.aspx
[8] Yuan, J., Zheng, Y., Zhang, C., Xie, W., Xie, X., Sun, G., Huang, Y.: T-Drive: Driving Directions Based on Taxi Trajectories. In: ACM SIGSPATIAL GIS (2010)
[9] Chen, L., Ng, R.: On the marriage of lp-norms and edit distance. In: VLDB (2004)
[10] Tang, L., Yu, X., Kim, S., Han, J., et al.: Tru-Alarm: Trustworthiness Analysis of Sensor Networks in Cyber-Physical Systems. In: ICDM (2010)
[11] Roussopoulos, N., Kelley, S., Vincent, F.: Nearest neighbor queries. In: SIGMOD (1995)
[12] Fagin, R.: Combining fuzzy information from multiple systems. J. Comput. System Sci. 58, 83–89 (1999)
[13] Sherkat, R., Rafiei, D.: On efficiently searching trajectories and archival data for historical similarities. In: PVLDB (2008)
[14] Fagin, R., Lotem, A.: Optimal aggregation algorithms for middleware. In: PODS (2001)
[15] Zheng, V.W., Zheng, Y., Xie, X., Yang, Q.: Collaborative location and activity recommendations with GPS history data. In: WWW (2010)
[16] Zheng, Y., Zhang, L., Xie, X., Ma, W.Y.: Mining interesting locations and travel sequences from GPS trajectories. In: WWW (2009)
[17] Zheng, Y., Wang, L., Xie, X., Ma, W.Y.: GeoLife: Managing and understanding your past life over maps. In: MDM (2008)
[18] Zheng, Y., Xie, X., Ma, W.Y.: GeoLife: A Collaborative Social Networking Service among User, location and trajectory. IEEE Data Engineering Bulletin 33(2), 32–40
[19] Zheng, Y., Chen, Y., Xie, X., Ma, W.Y.: GeoLife2.0: A Location-Based Social Networking Service. In: MDM (2009)
[20] Chen, Z., Shen, H.T., Zhou, X., Zheng, Y., Xie, X.: Searching trajectories by locations: an efficiency study. In: SIGMOD (2010)

Towards Reducing Taxicab Cruising Time Using Spatio-Temporal Profitability Maps⋆

Jason W. Powell[1], Yan Huang[1], Favyen Bastani[1], and Minhe Ji[2]

[1] University of North Texas
{jason.powell,huangyan}@unt.edu, FavyenBastani@my.unt.edu
[2] East China Normal University
mhji@geo.ecnu.edu.cn

Abstract. Taxicab service plays a vital role in public transportation by offering passengers quick personalized destination service in a semi-private and secure manner. Taxicabs cruise the road network looking for a fare at designated taxi stands or alongside the streets. However, this service is often inefficient due to a low ratio of *live miles* (miles with a fare) to *cruising miles* (miles without a fare). The unpredictable nature of passengers and destinations make efficient systematic routing a challenge. With higher fuel costs and decreasing budgets, pressure mounts on taxicab drivers who directly derive their income from fares and spend anywhere from 35-60 percent of their time cruising the road network for these fares. Therefore, the goal of this paper is to reduce the number of cruising miles while increasing the number of live miles, thus increasing profitability, without systematic routing. This paper presents a simple yet practical method for reducing cruising miles by suggesting profitable locations to taxicab drivers. The concept uses the same principle that a taxicab driver uses: follow your experience. In our approach, historical data serves as experience and a derived Spatio-Temporal Profitability (STP) map guides cruising taxicabs. We claim that the STP map is useful in guiding for better profitability and validate this by showing a positive correlation between the cruising profitability score based on the STP map and the actual profitability of the taxicab drivers. Experiments using a large Shanghai taxi GPS data set demonstrate the effectiveness of the proposed method.

Keywords: Profitability, Spatial, Temporal, Spatio-temporal, Taxi, Taxicabs.

1 Introduction

Taxicab service plays a vital role in public transportation by offering passengers quick personalized destination service in a semi-private and secure manner. A

⋆ This work was partially supported by the National Science Foundation under Grant No. IIS-1017926.

D. Pfoser et al. (Eds.): SSTD 2011, LNCS 6849, pp. 242–260, 2011.

2006 study reported that 241 million people rode New York City Yellow Medallion taxicabs and taxis performed approximately 470,000 trips per day, generating \$1.82 billion in revenue. This accounted for 11% of total passengers, an estimated 30% of total public transportation fares, and yielded average driver income per shift of \$158 dollars [1]. Taxicab drivers earn this by cruising the road network looking for a passenger at designated taxi stands or alongside the streets. However, this service is often inefficient from expensive vehicles with low capacity utilization, high fuel costs, heavily congested traffic, and a low ratio of *live miles* (miles with a fare) to *cruising miles* (miles without a fare).

With higher fuel costs and decreasing budgets, pressure mounts on taxicab drivers who directly derive their income from fares yet spend anywhere from 35-60 percent of their time cruising the road network for fares [1]. The unpredictable nature of passengers and destinations make efficient systematic routing a challenge. Therefore, the goal is simultaneously reducing cruising miles while increasing live miles, thus increasing profitability, without systematic routing.

This paper presents a simple yet practical method for suggesting profitable locations that enable taxicab drivers to reduce cruising miles. The concept uses the same principle that a taxicab driver uses: follow your experience. We propose a framework to guide taxi drivers in locating fares. Specifically, this paper makes three contributions. First, the proposed framework uses historical GPS data to model the potential profitability of locations given the current location and time of a taxi driver. This model considers the main factors contributing to the profitability: time and the profit loss associated with reaching a location. Second, this framework makes personalized suggestions to a taxi driver based on location and time. This avoids the problem of communicating the same information to all drivers, which may result in non-equilibrium in supply and demand. Third, we demonstrate the effectiveness of the proposed framework using a large dataset of Shanghai taxicab GPS traces and use correlation to compare the suggested locations with actual driver behavior.

2   Related Work

Taxicab service falls into two general categories and research follows this, occasionally attempting to bridge them. The first category is *dispatching* where companies dispatch taxicabs to customer requested specific locations. A request may be short-term (e.g., a customer requests a taxi for pickup within the next 20 minutes) or long-term (e.g., arrangements come hours or days in advance). Logic dictates that the farther in advance the request, the easier it is to plan efficient taxi service because routing algorithms already exist (mostly based on Dijkstra's work); the shorter the request time, the more challenging the routing problem. The second category is *cruising*. The taxicab driver cruises the road network looking for a fare at designated taxi stands or alongside the streets, using experience as a guide. This leads to an inefficient system where taxi drivers spend significant time without a fare and often serve hot spots, leading to a supply and demand imbalance. Since cruising is a profit loss, this paper will refer to non-live miles as a *cruising trip* and live miles as a *live trip*. The following highlights some recent research in this area.

Yamamoto et al. propose a fuzzy clustering based dynamic routing algorithm in [2]. Using a taxicab driver's daily logs, the algorithm creates an optimal route solution based on passenger frequency on links (i.e., paths). The routes, not intended to be used directly by the taxi driver, are shared among taxis through mutual exchanges (i.e., path sharing) that assigns the most efficient path to a taxi as they cruise. This potentially reduces the competition for a potential fare, excessive supply to popular areas, and traffic congestion while increasing profitability. Similarly, Li et al. present an algorithm using taxi GPS traces to create a usage based road segment hierarchy from the frequency of taxis traversing a road segment [3]. This hierarchy inherently captures the taxi driver experience and is usable in route planning. In these two examples, the focus is on routing but trip profitability—a key factor in the driver's decision—is not explicitly addressed.

Another example of taxicab routing is T-Drive, developed by Yuan et al. to determine the fastest route to a destination at a given departure time [16]. T-Drive uses historical GPS trajectories to create a time-dependent landmark graph in which the nodes are road segments frequently traversed by taxis and a variance-entropy-based clustering approach determines the travel time distribution between two landmarks for a given period. A novel routing algorithm then uses this graph to find the fastest practical route in two stages. The first stage, *rough routing*, searches the graph for the fastest route for a sequence of landmarks; the second stage, *refined routing*, creates the real network route using the rough route. Similar to the previous example, this system does inherently capture taxi driver experience and suggests faster routes than alternative methods; however, this method does not suggest profitable locations for taxicabs.

A thesis by Han Wang proposes a methodology for combining short-term and long-term dispatching [4]. If a customer's starting and ending locations follow the path of a taxi as it heads to a different dispatch call, the taxi can pick up the fare. This allows a reduction in cruising and an increase in profitability. The catch is that it may not be common for passenger routes to align exactly. Therefore, Wang proposes the Shift Match Algorithms that suggest drivers and/or customers to adjust locations, creating a reasonable short delay in service but an improvement overall. In this study, the cruise trips are different from those in the aforementioned cruising category because they result from dispatching, not from intent to cruise. This method is practical for dispatching but not for general cruising.

Another approach, given by Cheng et al., focuses on customer queuing at taxi stands and taxis switching between serving stands and cruising [5]. Phithakkit-nukoon et al. developed an inference engine with error based learning to predict vacant taxis [6] while Hong-Cheng et al. studied travel time variability on driver route choices in Shanghai taxi service [7]. Additional research covers a variety of issues from demand versus supply to pricing issues [8,9,10,11,12]; however, these studies do not consider location profitability, which is inherent to the driver's decision.

The work most similar to ours is by Ge et al., who provide a novel technique in extracting energy-efficient transportation patterns from taxi trajectory traces and a mobile recommender system for taxis [17]. The technique extracts a group of successful taxi drivers and clusters their pick-up points into centroids with an assigned probability of successful pick-up. The resulting centroids become the basis for pick-up probability routes that the system distributes among taxis to improve overall business success. The major contribution is how the system evaluates candidate routes using a monotonic Potential Travel Distance (PTD) function that their novel route-recommendation algorithm exploits to prune the search space. They also provide the SkyRoute algorithm that reduces the computational costs associated with skyline routes, which dominate the candidate route set. This recommender system potentially improves success by using probabilities; however, probabilities can be misleading in relation to the profitability since high probabilities do not necessarily translate into highly profitable live trips. In addition, the algorithm clusters locations using fix periods regardless of when the taxicab actually arrives at a location. Furthermore, our framework suggests a customized map of locations based on the taxicab's current location to eliminate route creation cost and taxi-route assignment distribution issues; however, their algorithms could enhance our framework by suggesting customized paths for the taxi driver using a time series of STP maps.

3   Methodology

The taxicab driver is not concerned with finding profitable locations during a live trip. Once the live trip is complete, assuming there is not a new passenger available at the location, the driver must decide where to go. They may stay in that general vicinity for a time in hopes of a passenger or, more likely, head to another location based on experience. At this moment, the driver considers two variables: profitable locations and reasonable driving distances. However, a driver might be unaware of both variables. For example, a driver may be unreasonably far from a highly profitable airport but unsure of closer profitable locations less often visited. Given a map identifying these locations, the driver can make an informed decision quickly and reduce cruising time.

Figure 1 summarizes the proposed methodology for identifying these locations. When a taxicab begins a cruising trip, the current location and time are parameters for querying a historical database that serves as driver experiences. The experience information coincides with locations and becomes a location-based profitability score. The process assembles these scores into an STP map that suggests potentially profitable locations to the taxicab driver. By following the suggestions, the driver can reduce cruising time thus increase profitability.

STP map generation occurs when a taxicab is ready for a new fare, i.e. when the driver begins a cruising trip. At this moment, and based on the current location, the map encompasses a region of interest within a reasonable driving

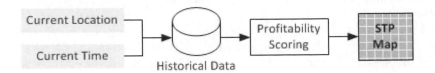

Fig. 1. The STP map generation process. The current location and time are parameters for retrieving historical data that becomes profitability scores in an STP map.

distance and uses historical data to determine the profitability of locations within this region. The map is personalized to the driver since each driver is at a different location. This mechanism can prevent sending the same information to multiple drivers, which could result in localized competition and a non-equilibrium state. It is possible for multiple drivers to receive the same STP map if their closeness is within error bounds of the distance calculations, but this occurs infrequently.

This region can be large enough to encompass all the historical data, but since the taxicab moves spatially and temporally, it is not necessary to model the entire region. The first step in generating this map is to define a sub-region M around the taxicab's current location in region R such that $M \subseteq R$. In other words, region M is for short-term planning, the taxicab driver's inherent process—the driver moves towards locations of high live trip probability and profitability. Figure 2 demonstrates this concept. At time t_1, the taxi driver drops off the passenger at the end of a live trip and receives STP map $M1$ of the surrounding location. The driver chooses a profitable location within the region, moves to that location, and picks up a passenger. This new live trip continues until t_2 when the passenger is dropped off and the driver receives a new STP map. This process continues until the taxi goes out of service. With this knowledge, the driver can reduce overall cruising time.

This method defines locations within M and determines a profitability score for each location. The simplest implementation is to divide M into a grid of equally sized cells such that $M = \{x_1, x_2, ..., x_n\}$, with the taxicab located at center cell x_c (see Figure 3). The grid granularity is important. The cell sizes should be large enough to represent a small immediate serviceable area and to provide enough meaningful historical information to determine potential profitability. For instance, if the cell of interest x_i has little or no associated historical information, but the cells surrounding it do, it may be beneficial to increase the granularity. On the other hand, it should also not be too large as to become meaningless and distorted in terms of profitability. For instance, a cell the size of a square kilometer may be unrepresentative of a location.

As mentioned previously, the historical data determines the cell profitability since it captures the taxi drivers' experiences in terms of trips. The natural inclination is to use the count of live trips originating from the cell as the profitability indicator; however, this can be misleading since it does not consider the probability of getting a live trip and because a trip fare calculation, which de-

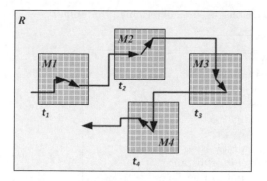

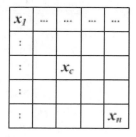

Fig. 2. A taxicab moving through region R. At each time t_i, the driver receives a new STP map M customized to that location and time with the taxicab located at the center.

Fig. 3. An example sub-region M composed of n cells with the taxi located at x_c. Each x_i represents a location with a profitability score.

termines profitability, uses distance and time, implying that the average of some distance to time ratio for a location's trips may be more appropriate. This ratio still would not capture true profitability because trip fares are not a direct ratio of distant to time. It is common practice to charge a given rate per distant unit while the taxi is moving and a different rate when idling. Therefore, a location's profitability is a factor of trip counts (cruising and live), trip distances, and trip times (idling and moving).

The formula for calculating the fare for live trip j, $F(j)$, at starting location x_i is the amount of idle time $t_i(j)$ charged at rate r_i, plus the distanced traveled $d_l(j)$ charged at r_l[1] (see Eq. 1). If unknown, one can estimate idle time t_i using the total trip time t and the taxicab speed s as $t - d_l/s$. The cost of reaching the starting point of trip j is the distance traveled between x_c and x_i, $D(x_c, x_i)$, times some proportion θ of r_l since the rate charged has the cost factored into it. Therefore the profitability of trip j with the taxi currently located at x_c, $P(x_c, j)$, is $F(j)$ minus the cost associated with reaching x_i from x_c (see Eq. 2). The profitability of location x_i with respect to current location x_c, $P(x_c, x_i)$, is the sum of all historical live trip fares from that location divided by the total count of trips, both live n_l and cruising n_c, minus the cost between x_c and x_i (see Eq. 3). Eq. 1 is actually a simplification of the fare pricing structure which can vary for different cities, different locations within the city, and different times of day. The fare price may be fixed; for example, the price from JFK airport in New York is fixed, or more commonly, there is a fixed charged for a given distance, a different charge per distance unit for a bounded additional distance, and a third charge after exceeding a given distance. As an example, Table 2 gives the Shanghai taxi price structure.

$$F(j) = (t_i(j) * r_i) + (d_l(j) * r_l) \tag{1}$$

---

[1] Note that r_l is some proportion of r_i.

Table 1. A summary of the variables used to determine profitability

Variable	Description
d_l	Total distance of a live trip
$D(x_1, x_2)$	Distance between locations x_1 and x_2
$F(j)$	The fare of trip j
n_c	Number of cruise trips
n_l	Number of live trips
$P(x_c, j)$	Profitability of trip j
$P(x_c, x_i)$	Profitability of an STP map location
r_l	Charge rate per unit of distance
r_i	Charge rate per unit of idle time
s	Taxicab speed
t	Total time of a trip
t_c	Time of a cruise trip
t_i	Time of taxi idling
t_l	Time of a live trip
x_c	The taxicab location, which is the center of M
x_i	Location of interest within M
θ	Proportion relating unit cost
ϵ	Adjustment for low trip counts

$$P(x_c, j) = F(j) - (D(x_c, x_i) * r_l * \theta) \tag{2}$$

$$P(x_c, x_i) = (\sum_{j=1}^{n_l} F(j))/(n_l + n_c) - (D(x_c, x_i) * r_l * \theta) \tag{3}$$

Table 2. The Shanghai taxicab service price structure [15]. Taxi drivers charge a flat rate of 12 Yuan for the first 3 kilometers plus a charge for each additional kilometer.

Trip Description	5am - 11pm	11pm - 5am
0 to 3 km	12.00 Total	16.00 Total
3 to 10 km	$12.00 + 2.40/km$	$16.00 + 3.10/km$
Over 10 km	$12.00 + 3.60/km$	$16.00 + 4.70/km$
Idling	2.40/5 minutes	5.10/5 minutes

It is difficult to use these equations without detailed historical information; however, there is an exploitable relationship. Time can represent the profitability by converting each variable to time; distance converts to time by dividing by the speed and the rate charged for idle time is proportional to the rate charge for movement time. Furthermore, the profit earned by a taxi driver is directly proportional to the ratio of live time t_l to cruising time t_c. The average of all live trip times originating from x_i, minus the cost in time to get to x_i from x_c, can represent the profitability score of cell x_i (see Eq. 4). The ϵ ensures that the profitability reflects true trip probability when trip counts are low.

$$P(x_i) = (\sum_{j=1}^{n_l} t_l)/(n_l + n_c + \epsilon) - (D(x_c, x_i) * r_l * \theta)/s \qquad (4)$$

Each of these variables is derivable from GPS records. Given a set of records with an indication of the taxicab's occupancy status, the time stamps and status can determine the live and cruising times, the GPS coordinates determine the distances, and the distances and times determine speeds.

It is not a requirement to use all the historical data from the GPS records to determine the location profitability used in the STP map; in fact, using all the data may be misleading due to the changing conditions throughout the day. For example, profitability for a specific location may be significantly different during rush hour than during night traffic. Since the taxicab driver is looking for a fare in the here and now, a small data window will better represent the driver's experience for this period. For each location, the data selected should represent what the conditions will be when the driver reaches that location. For example, if it is 1:00pm and takes 10 minutes to reach the location, the historical data should begin at 1:10pm for the location. The size of this *Delayed Experience Window* (DEW) may be fixed or variable as necessary, but the size is important. If the DEW is too large, it may include data not representative of the profitability; if too small, it may not include enough data. The following case study gives an example of the proposed methodology applied to Shanghai taxicab service.

4   Case Study – Shanghai Taxi Service

Shanghai is a large metropolitan area in eastern China with over 23 million denizens [13] and a large taxicab service industry with approximately 45 thousand taxis operated by over 150 companies [14]. To demonstrate our method, we use a collection of GPS traces for May 29, 2009. The data set contains over 48.1 million GPS records (WGS84 geodetic system) for three companies between the hours of 12am and 6pm and over 468,000 predefined live trips of 17,139 taxicabs. We divided the data into the three companies and focused on the first company, which yielded data for 7,226 taxis. The region R was limited to 31.0°-31.5° N, 121.0°-122.0° E to remove extreme outliers and limit trips to the greater metropolitan area. Furthermore, only trips greater than five minutes are included since erratic behavior occurred more often in those below that threshold. Similar erratic behavior occurred with trips above three hours, often the result of the taxi going out of service, parking, and showing minute but noticeable movement from GPS satellite drift. The three-hour threshold is partially arbitrary and partially based on the distribution of trips times. While relatively rare, there are times when taxis spend over an hour on a cruising trip, but cruising trips over three hours occur much less frequently.

These reductions left 144 thousand live trips remaining for the first company from which we constructed cruising trips. For each taxicab, we defined the cruising trips as the time and distance between the ending of one live trip and the beginning another. We assumed there is a cruising trip before the first chronological live trip if there is at least one GPS record before the live trip starting time that indicated no passenger in the vehicle. We also assumed, that the end of the taxi's last live trip indicated that the taxicab was out of service and did not incur any additional cruising trips. For example, if the taxicab first appears at 12:04am, but its first live trip is at 12:10am, 12:04-12:10am became a cruising trip. If the taxicab's last live trip ended at 3:07pm, this became the last trip considered. This resulted in 948 fewer cruise trips than live trips, but did eliminate all outlier trips outside the period.

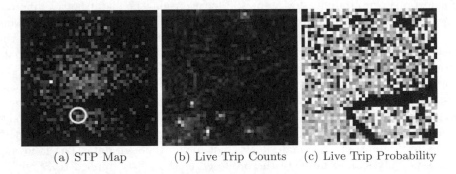

(a) STP Map   (b) Live Trip Counts   (c) Live Trip Probability

Fig. 4. Results for the downtown region at 1:00pm with a 60-minute DEW, 190.5-meter cell length, M size of 67.1 km^2, and with the taxicab in the center. The Oriental Pearl Tower is encircled. Using the live trip counts or probability could cause the taxicab driver to incur a higher cost compared to using the STP map.

For the first demonstration, the square region around the taxicab location was approximately 8.1x8.1 km^2 divided into 43x43 square cells of approximately 190.5 meters in length. The DEW is 60 minutes, starting at a time delay based on the time required for to reach the cell. We chose the taxi's current time as 1:00pm since the 1:00-3:00pm period has the largest percentage in data distribution. The time and distance required to reach the cell's center came from the L_1 Manhattan distance and average speed of 11 km/h based on instantaneous speeds recorded in the GPS data. The distances of the live trips originating from the cell is the sum of L_2 Euclidean distances from the trip's individual GPS records. For the profitability score, θ was deduced from the data to be approximately 0.333; although the exact value is unknown, it can be estimated by analyzing trip times. Figure 4 and Figure 5 display the STP maps near the downtown region and near the Shanghai Hongqiao International Airport, respectively, with the taxicab at the center of the map. The figures also include the live trip counts and probability for the areas as a comparison to the profitability. Additionally, to show that two taxicabs at the same time get two distinct STP maps, Figure 6

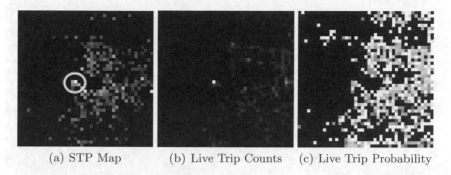

(a) STP Map   (b) Live Trip Counts   (c) Live Trip Probability

Fig. 5. Results for the Shanghai International Airport region at 1:00pm with a 60-minute DEW, 190.5-meter cell length, M size of 67.1 km^2, and with the taxicab in the center. The airport terminal is encircled. The high count of live trips from the terminal shadows the other locations, hiding other potentially profitable locations that our STP map captures.

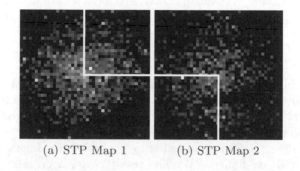

(a) STP Map 1   (b) STP Map 2

Fig. 6. Results for overlapping downtown regions at 1:00pm with a 60-minute DEW, 90.5-meter cell length, M size of 67.1 km^2, and with the taxicab in the center. The top-right quadrant of STP Map 1 overlaps the bottom-left quadrant of STP Map 2. There is a clear difference between the overlapped regions as lower profitability areas in one are often higher profitability areas in the other.

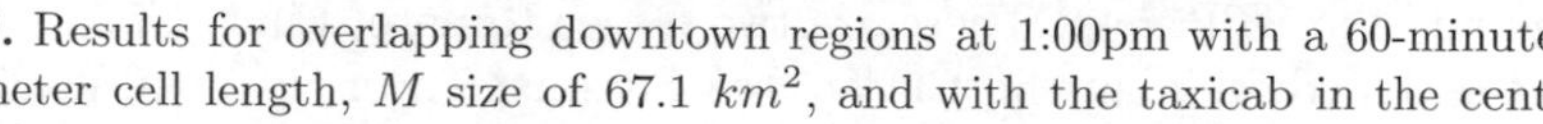

Fig. 7. Color scale from low to high values for Figures 4, 5, and 6

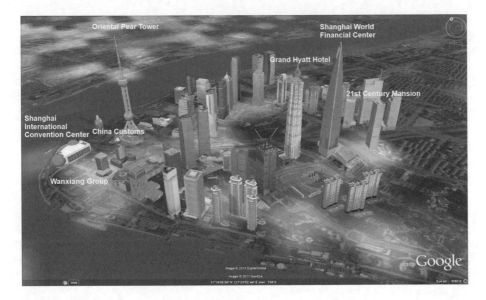

Fig. 8. Results for the downtown region STP map over laid with the Google Earth's satellite image at 1:00pm, a 60-minute DEW, 190.5 meters cell length, and M size of 67.1 km^2. Lighter areas represent higher profitability scores and often correlate with areas expected to be profitable, such as the Shanghai International Convention Center.

shows two overlapping regions in which the top-right quadrant of Figure 6(a) overlaps the bottom-left quadrant of Figure 6(b). For visualization purposes, negative profitability areas are set to zero to highlight the profitable regions, which are of interest to the taxi driver.

Figure 8 overlays the STP map in Figure 4(a) with the downtown area using Google Earth and one-hour DEW. The results show a correlation with office buildings and the STP map. There are two issues to note. First, Google Earth distorts the cell edges in an effort to stretch the image over the area, leading to potential misinterpretation, although minor. Second, the cell granularity plays an important role in the results. Near the image center, the construction area near the Grand Hyatt Hotel shows high profitability while the Grand Hyatt itself does not show as high profitability as would be expected. This is because the cell boundary between these locations is splitting the trips between them. While this is an issue, it is more typical for a group of close cells to have similar profitability scores. From the viewpoint of a taxicab driver, this is not an issue because the goal to find general locations of high profitability, not necessarily the specific 190.5 by 190.5 square meters. Figure 9 similarly shows an STP map overlaying the airport. The airport is one of the hottest locations, producing numerous profitable trips that make it a favorite location among taxi drivers. In this case, the entire terminal area has similar profitability even though the cells maybe splitting the activity among them. A graphical glitch is preventing the red cell from completely showing near the image center.

Fig. 9. Results for the Shanghai International Airport region STP map over laid with the Google Earth's satellite image at 1:00pm, a 60-minute DEW, 190.5 meters cell length, and M size of 67.1 km^2. Lighter areas represent higher profitability scores and correlates with airport terminal and surrounding area. Note that a graphical glitch is causing the red center cell to be distorted.

5   Validation

To validate this method, we must show that the STP maps correlate with actual profitability. If assumed that taxicab drivers move towards high profitable areas when cruising, then it is logical that the ending location of a cruising trip (i.e., the beginning location of a live trip) is a profitable location. If these ending locations correlate to the higher profitable areas in the STP maps generated for the taxicab throughout the day, and this correlates with known taxi profitability, then the STP map correctly suggests good locations. In other words, if the aggregate profitability scores associated with the ending locations of cruise trips throughout the day correlates to actual profitability, which can be determined by live time to total time for a taxi, then the correlation should be positive.

We selected five distinct test sets of 600 taxicabs and removed taxis with less than 19 total trips to focus on those that covered the majority of the day. This resulted in 516-539 taxis per test set. For each taxicab, we followed their path of live and cruising trips throughout the day. When a taxicab switched from live to cruising, we generated an STP map using a 15-minute DEW for a 15.8 by 15.8 km^2 area divided into 167 by 167-square cells (approximately 95 meters in length) with the taxi at the center. We summed the profitability scores at cruising trip ending locations and correlated them with the real live time to total time ratio that defines actual profitability. We then repeated the experiment, increasing the cell size and DEW while holding the region size constant.

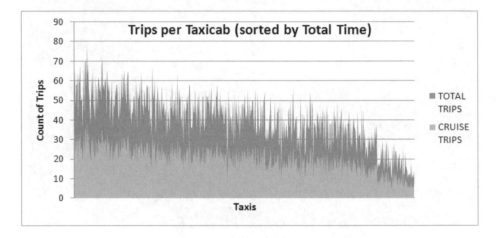

Fig. 10. The total and cruise trips for one dataset with 538 taxis, sorted by total time. Increasing the total time tends to disproportionately increase the count of cruise trips to live trips.

Figures 10, 11, and 12 visualize typical characteristics of the test sets. Figure 10 shows trip counts and Figure 11 shows trip times with taxis sorted by total time. There is a distinct group having higher total times, but this results from a larger percent of cruising time relative to the other taxis. Comparing this with Figure 12, the taxicab profits with taxis sorted by total time, reveals that the total time in service does not necessarily improve profits; in fact, it has a tendency to have the opposite effect. Figure 12 also shows that an increase in total time typically yields more trips, but does not necessarily increase overall profits.

Figure 13 displays the resulting average correlation over the datasets for three cell sizes and DEWs. The average correlations approached 0.50 with a slightly higher median. The trend in correlation clearly demonstrates the effect of cell sizes. Small sizes do not accurately represent the profitability and larger sizes tend to distort. Additionally, the DEW shows a definite trend. The more historical data, the better the correlation; however, caution should be taken. Increasing the DEW increases the amount of historical data, but may cause it to include data not representative of the current period. For example, if the DEW includes both rush hour and non-rush hour traffic, then the profitability may not reflect real profitability. In addition, if a taxi only cruises for a few minutes, the extra 50 minutes of a 60-minute DEW has less importance in making a decision. Figure

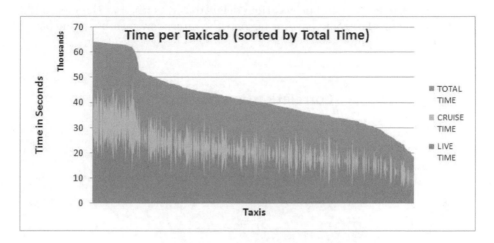

Fig. 11. The total, live, and cruise times in seconds for one dataset with 538 taxis, sorted by total time. The amount of cruising time is often greater than live time.

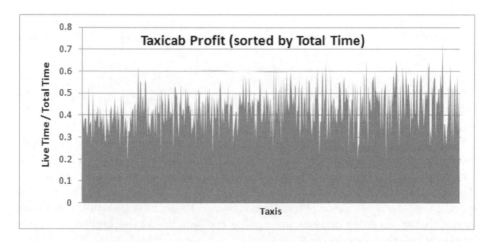

Fig. 12. The profit (live time/total time) for one dataset with 538 taxis, sorted by total time. There is a slight upward trend in profits as the total time decreases, indicating that an increase in total time does not guarantee an increase in profit.

14 confirms this hypothesis—holding the 190x190 m^2 cell size constant, the correlation increases with the increasing DEW until past the 90-minute mark. Since the DEW starts at 1:00pm but was time delayed as described in the method, it started including the traffic pattern beyond the afternoon rush hour but before the evening rush hour.

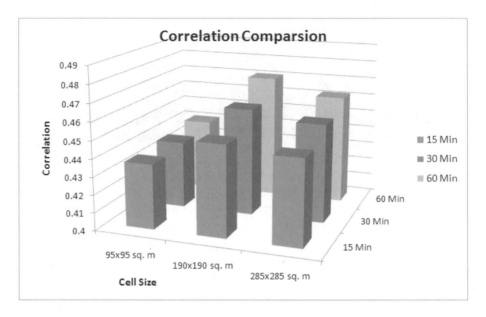

Fig. 13. Average correlation of the five test datasets for a given cell size and DEW. The cell size 190x190 m^2 produced the best overall correlation, reaching 0.51 for one of the five datasets.

The positive correlation was not as high as expected, but investigating the scatter plots revealed that there is a good correlation. As an example, Figure 15 shows the scatter plot correlation for one test set with a 30-minute DEW and a 190x190 m_2 cell size. The Hit Profit is the sum of all profit scores from cells where the taxicab ended a cruising trip and the Live Time/Total Time is the profitability of the taxicab for that day. As indicated by the trend line, the higher the taxicab profitability, the higher the Hit Profit. While the correlation for this specific set was 0.51, there is a definite upward trend in correlation among all sets with the majority of taxis are ending cruise trips in the higher profitable locations based on our method.

6   Future Work

There are several potential improvements for this method. First, we did not focus on the temporal aspect beyond shifting the DEW at a delayed time and

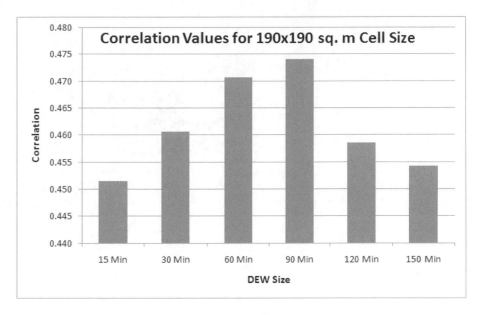

Fig. 14. Correlations values for the 190x190 m^2 cell size with varying DEW size starting at 1:00pm. A DEW size greater than 90 minutes causes a significant decrease in correlation, demonstrating it's importance.

adjusting the size. Patterns in time may affect results by allowing it to include data from two distinct periods in relation to the traffic pattern; for example, rush hour traffic data included with non-rush hour data. To a lesser extent, the cell sizes and region M may need to adjust with time as well; for example, late at night, there may be a need to increase the cell size due to lower probability of live trips and an increase in M to include more potential locations. For validation purposes, M was large enough to ensure that all cruising trips considered ended within the area with our profitability scores; otherwise, the score would be zero when in reality it should be positive or negative. The goal would be to develop a dynamic STP mapping system that adjusts each of these components given current conditions and time.

Another improvement involves the distance calculations. The L_1 Manhattan distance formula determined the distance between the current taxi location and the location of interest. While this is more realistic than using the L_2 Euclidean distance, it relies on a grid city model, which is not always applicable to all areas of the city. The live trip distances used the L_2 Euclidean distance between individual GPS records to determine total distance, which has an associated error in accuracy as well. These calculations also did not consider obstacles; for example, drivers must cross rivers at bridges or tunnels, which may add distance and time to a trip. One potential solution is to use the road network, current traffic conditions, and known obstacles to find the best path to a location and then use the path to determine profitability. An alternative is to capture

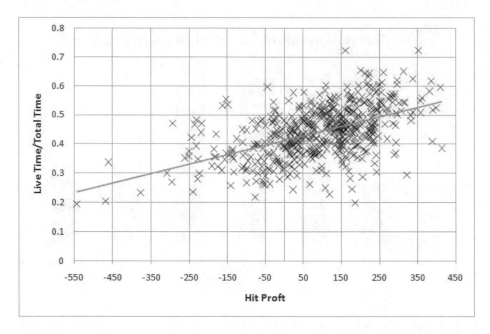

Fig. 15. An example of correlation results with an × marker indicating a taxicab. This test set used a cell size 190x190 m^2 with a 30-minute DEW. The Hit Profit is the sum of all profit scores from cells where the taxicab ended a cruising trip and the Live Time/Total Time is the profitability of the taxicab for that day. The majority of taxicabs are ending in the more profitable locations, providing a positive correlation between the STP map and actual profitability as shown by the upward trend line.

the driver's intuitive nature to find the best path or to use distances of common paths traveled by multiple taxis. A preliminary investigation into this alternative revealed that it is a possibility given enough GPS records.

Since the ultimate goal is for a taxi driver to use the STP map, the system needs to be real-time and use visuals that are easy to understand and not distractive to driving. It could also take into consideration the current traffic flow to determine a more accurate profitability score, and give higher potential profitability path suggestions leading to a profitable location. This could increase the probability of picking up a passenger before reaching the suggested location and thereby further reduce cruise time and increase profits.

7   Conclusion

The growing demand for public transportation and decreasing budgets have placed emphasis on increasing taxicab profitability. Research in this area has focused on improving service through taxi routing techniques and balancing supply and demand. Realistically, cruising taxicabs do not easily lend themselves to routing because of the nature of the service and the driver's desire for short-term profitability. Since the live and cruising times define the overall profitability, and a taxicab may spend 35-60 percent of time cruising, the goal is reducing cruise time while increasing live time. Our framework potentially improves profitability by offering location suggestions to taxicab drivers, based on profitability information using historical GPS data, which can reduce overall cruising time. The method uses spatial and temporal data to generate a location suggesting STP map at the beginning of a cruise trip based on a profitability score defined by the live time to total time profitability definition. A case study of Shanghai taxi service demonstrates our method and shows the potential for increasing profits while decreasing cruise times. The correlation results between our method and actual profitability shows a promising positive correlation and potential for future work in increasing taxicab profitability.

References

1. Schaller Consulting: The New York City Taxicab Fact Book, Schaller Consulting, Brooklyn, NY (2006), http://www.schallerconsult.com/taxi/taxifb.pdf
2. Yamamoto, K., Uesugi, K., Watanabe, T.: Adaptive Routing of Cruising Taxis by Mutual Exchange of Pathways. Knowledge-Based Intelligent Information and Engineering Systems 5178, 559–566 (2008)
3. Li, Q., Zeng, Z., Bisheng, Y., Zhang, T.: Hierarchical route planning based on taxi GPS-trajectories. In: 17th International Conference on Geoinformatics, Fairfax, pp. 1–5 (2009)
4. Wang, H.: The Strategy of Utilizing Taxi Empty Cruise Time to Solve the Short Distance Trip Problem, Masters Thesis. The University of Melbourne (2009)
5. Cheng, S., Qu, X.: A service choice model for optimizing taxi service delivery. In: 12th International IEEE Conference on Intelligent Transportation Systems, ITSC 2009, St. Louis, pp. 1–6 (2009)
6. Phithakkitnukoon, S., Veloso, M., Bento, C., Biderman, A., Ratti, C.: Taxi-aware map: identifying and predicting vacant taxis in the city. In: de Ruyter, B., Wichert, R., Keyson, D.V., Markopoulos, P., Streitz, N., Divitini, M., Georgantas, N., Mana Gomez, A. (eds.) AmI 2010. LNCS, vol. 6439, pp. 86–95. Springer, Heidelberg (2010)
7. Hong-Cheng, G., Xin, Y., Qing, W.: Investigating the effect of travel time variability on drivers' route choice decisions in Shanghai, China. Transportation Planning and Technology 33, 657–669 (2010)
8. Li, Y., Miller, M.A., Cassidy, M.J.: Improving Mobility Through Enhanced Transit Services: Transit Taxi Service for Areas with Low Passenger Demand Density. University of California, Berkeley (2009)
9. Cooper, J., Farrell, S., Simpson, P.: Identifying Demand and Optimal Location for Taxi Ranks in a Liberalized Market. Transportation Research Board 89th Annual Meeting (2010)
10. Sirisoma, R.M.N.T., Wong, S.C., Lam, W.H.K., Wang, D., Yan, H., Zhang, P.: Empirical evidence for taxi customer-search model. Transportation Research Board 88th Annual Meeting 163, 203–210 (2009)
11. Yang, H., Fung, C.S., Wong, K.I., Wong, S.C.: Nonlinear pricing of taxi services. Transportation Research Part A: Policy and Practice 44, 337–348 (2010)

12. Chintakayala, P., Maitra, B.: Modeling Generalized Cost of Travel and Its Application for Improvement of Taxies in Kolkata. Journal of Urban Planning and Development 136, 42–49 (2010)
13. Wikipedia, Shanghai — Wikipedia The Free Encyclopedia (2011), `http://en.wikipedia.org/w/index.php?title=Shanghai\&oldid=412823222` (accessed May 21, 2011)
14. TravelChinaGuide.com, Get Around Shanghai by Taxi, Shanghai Transportation (2011), `http://www.travelchinaguide.com/cityguides/shanghai/transportation/taxi.htm` (accessed February 9, 2011)
15. Shanghai Taxi Cab Rates and Companies, Kuber (2011), `http://kuber.appspot.com/taxi/rate` (accessed February 9, 2011)
16. Yuan, J., Zheng, Y., Zhang, C., Xie, W., Xie, X., Sun, G., Huang, Y.: T-drive: driving directions based on taxi trajectories. In: Proceedings of the 18th SIGSPATIAL International Conference on Advances in Geographic Information Systems, GIS 2010, San Jose, pp. 99–108 (2010)
17. Ge, Y., Xiong, H., Tuzhilin, A., Xiao, K., Gruteser, M., Pazzani, M.: An energy-efficient mobile recommender system. In: Proceedings of the 16th ACM SIGKDD international conference on Knowledge discovery and data mining, KDD 2010, Washington, DC, pp. 899–908 (2010)

Computing the Cardinal Direction Development between Moving Points in Spatio-temporal Databases

Tao Chen, Hechen Liu, and Markus Schneider*

Department of Computer and Information Science and Engineering,
University of Florida,
Gainesville, FL 32611, USA
{tachen,heliu,mschneid}@cise.ufl.edu

Abstract. In the same way as moving objects can change their location over time, the spatial relationships between them can change over time. An important class of spatial relationships are cardinal directions like *north* and *southeast*. In spatial databases and GIS, they characterize the relative directional position between *static* objects in space and are frequently used as selection and join criteria in spatial queries. Transferred to a spatiotemporal context, the simultaneous location change of different moving objects can imply a temporal evolution of their directional relationships, called *development*. In this paper, we provide an algorithmic solution for determining such a temporal development of cardinal directions between two moving points. Based on the *slice representation* of moving points, our solution consists of three phases, the *time-synchronized interval refinement phase* for synchronizing the time intervals of two moving points, the *slice unit direction evaluation phase* for computing the cardinal directions between two slice units that are defined in the same time interval from both moving points, and finally the *direction composition phase* for composing the cardinal directions computed from each slice unit pair. Finally, we show the integration of spatio-temporal cardinal directions into spatio-temporal queries as spatio-temporal directional predicates, and present a case study on the hurricane data.

1   Introduction

Objects that continuously change their positions over time, so-called *moving objects*, have recently received a lot of interest. Examples are moving points like vehicles, mobile devices, and animals, for which the time-dependent position is relevant. Temporal movements of spatial objects induce modifications of their spatial relationships over time, called *developments*. In spatial databases and

---

* This work was partially supported by the National Science Foundation under grant number NSF-IIS-0812194 and by the National Aeronautics and Space Administration (NASA) under the grant number NASA-AIST-08-0081.

D. Pfoser et al. (Eds.): SSTD 2011, LNCS 6849, pp. 261–278, 2011.

GIS, spatio-temporal queries are particularly interesting when they ask for temporal changes in the spatial relationships between moving objects. An important class of spatial relationships are cardinal directions like *north* and *southeast* that characterize the relative directional position between spatial objects. Cardinal directions between two static objects have been extensively studied and have been frequently used as selection and join criteria in spatial queries. Transferred to a spatio-temporal context, the simultaneous location change of different moving objects can imply a change of their directional relationships. For example, a fishing boat that is southwest of a storm might be north of it some time later. We call this a *cardinal direction development*. Such a development between two moving objects describes a temporally ordered sequence of cardinal directions where each cardinal direction holds for a certain time interval during their movements. A development reflects the impact of time on the directional relationships between two moving objects, and usually proceeds continuously over time if the movements of the two objects are continuous.

It is an open, interesting, and challenging problem to capture the cardinal direction development between moving objects. Consider a database containing information about weather conditions. The query whether a hurricane stayed all the time to the southeast of another hurricane, and the query whether a hurricane has ever moved to the southeast of another hurricane can be particularly interesting to hurricane researchers to understand dynamic weather movement patterns. To answer these queries with current approaches and systems, we would need to check the validity of the spatial directional predicate, e.g. *southeast*, at all time instances during the common life time of both hurricanes. However, this is not possible since the movements of the hurricanes are continuous. The fact that the traditional, static cardinal directions cannot describe continuous, time dependent relationships leads to the need for new modeling strategies.

We have proposed a modeling strategy for cardinal direction developments in our previous work, in which we have defined the development of cardinal directions over time as a sequence of temporally ordered and enduring cardinal directions. In this paper, we propose our solution from an algorithmic perspective. We base our solution on the *slice representation* of moving points, which represents the temporal development of a point with a sequence of timely ordered units called *slices*. We propose a three-phase solution for determining the developments of the directional relationships between two moving points. In a *time-synchronized interval refinement phase*, two moving points are refined by synchronizing their time intervals. As a result, each slice unit of the refined slice representation of the first moving point has a matching slice unit in the refined slice representation of the second moving point with the time interval. In the second phase, the *slice unit direction evaluation phase*, we present a strategy of computing cardinal directions between two slice units from both moving points. Finally, in the *direction composition phase*, the development of the cardinal direction is determined by composing cardinal directions computed from all slices pairs from both moving points.

Section 2 introduces the related work in the literature. In Section 3, we review our modeling stratergy for cardinal direction developments. We propose a three-phase approach to computing the developments of cardinal directions between two moving points in Section 4. Section 5 defines spatio-temporal directional predicates for integrating cardinal direction developments into spatial-temporal databases and query languages. We present a case study on the hurricane best track data collected from National Hurricane Center (NHC) in Section 6, and show how the cardinal direction developments can help hurricane researchers to identify interesting weather event patterns. In Section 7, we draw some conclusions and discuss future work.

2   Related Work

A number of spatio-temporal models have been proposed to represent and manage moving objects. Early approaches tried to extend the existing spatial data models with temporal concepts. One approach is to store the location and geometry of moving objects with discrete snapshots over time. In [1], a spatio-temporal object o is defined as a time-evolving spatial object whose evolution is represented by a set of triplets (o_{id}, s_i, t_i), where o_{id} identifies the object o and s_i is the location of o at time instant t_i. Another approach in [2] applies linear constraints for modeling spatio-temporal data. It associates the spatial features like location and geometry of a moving object with consecutive time intervals. A common drawback of the two approaches mentioned so far is that, ultimately, they are incapable of modeling continuous changes of spatial objects over time. New approaches have been proposed to support a more integrated view of space and time, and to incorporate the treatment of continuous spatial changes. In [3,4], the concept of *spatio-temporal data types* is proposed as *abstract data types* (*ADTs*) whose values can be integrated as complex entities into databases. A temporal version of an object of type α is given by a function from time to α. Spatio-temporal objects are regarded as special instances of temporal objects where α is a spatial data type like *point* or *region*. A *point* (representing an airplane, for example) that changes its location in the Euclidean plane over time is called a *moving point*. In this paper, we follow the specification of *spatio-temporal data types*, particularly the moving point data type, and take it as our basis for modeling cardinal directions.

Qualitative spatial relationships have a long tradition in GIS and spatial databases. They can be grouped into three categories: *topological, directional* and *distance*. The same classification holds for the relationships between moving objects. The distinction is that spatial relationships between moving objects can have a temporal evolution, i.e. they may change over time. So far, the focus has been mainly on spatio-temporal topological relationships like *cross* and *enter* [5,6], and spatio-temporal distance relationships like *moving towards, moving away from,* [7] and *opposite_direction* [6]. Cardinal directions in a spatio-temporal context have been largely neglected in the literature. Static cardinal directions like *north* and *northeast* represent important qualitative spatial relationships that describe relative direction positions between static spatial objects. Many models follow a *projection-based* approach, where direction relationships are defined using projection lines orthogonal to the coordinate axes [8,9]. Some models apply a *cone-based* approach that defines direction relations by using angular zones [10,11]. Others like the *Minimum Bounding Rectangle (MBR)* model [12] make use of the minimum bounding rectangles of both operand objects and apply Allen's 13 interval relations to the rectangle projections on the x- and y-axes respectively. However, all existing cardinal direction models only consider static directional relationships, and when transferred to a spatio-temporal context, none of the models is capable of modeling directional relationships that continuously change over time. In [13], an attempt has been made to model *moving spatio-temporal relationships (mst-relation)*, which includes both topological relations and directional relations. During a time interval I_k, the mst-relation between two moving objects A_i and A_j is expressed as A_i (α, β, I_k) A_j, where α is any topological relation among *Equal, Inside, Contain, Cover, Covered By, Overlap, Touch* and *Disjoint* and β is one of the 12 directional relations, *South, North, West, East, Northwest, Northeast, Southwest, Southeast, Left, Right, Below* and *Above*. Both A_i α A_j and A_i β A_j are true during the interval I_k. This model provides a way of describing the topological and directional relationships between two moving objects. However, it is not clear how the relationships are determined. There are currently no well established strategies for modeling cardinal directions between two moving objects, and it is the main goal of this paper to bridge this gap.

We have presented a modeling strategy for cardinal direction developments in [14], in which the cardinal direction development between two moving points is formally defined. In this paper, we focus on the design of algorithms for computing such a cardinal directional development.

3   A Review of the Modeling Strategy for Cardinal Direction Developments between Moving Points

The approach that is usually taken for defining cardinal directions between two static points in the Euclidean plane is to divide the plane into partitions using the two points. One popular partition method is the *projection-based* method that uses lines orthogonal to the x- and y-coordinate axes to make partitions [12,8]. The point that is used to create the partitions is called the *reference* point, and the other point is called the *target* point. The direction relation between two points is then determined by the partition that the *target* object is in, with respect to the *reference* object. Let *Points* denote the set of static point objects, and let $p, q \in Points$ be two static point objects, where p is the target point and q is the reference point. A total of 9 mutually exclusive cardinal directions are possible between p and q. Let CD denote the set of 9 cardinal directions, then $CD=\{northwest\ (NW),\ restrictednorth\ (N),\ northeast\ (NE),\ restrictedwest\ (W),\ sameposition(SP),\ restrictedeast\ (E),\ southwest\ (SW),\ restrictedsouth\ (S),\ southeast\ (SE)\}$. Let $dir(p, q)$ denote the function that returns the cardinal

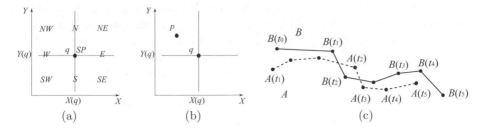

Fig. 1. Plane partitions using the reference object q (a), and an example in which p is *northwest* of q (b); the trajectories of moving points A and B in the time interval $[t_0, t_5]$ (c)

direction between two static points p and q where q is the reference point, then we have $dir(p, q) \in CD$. Figure 1a shows the partitions of the plane with respect to the reference point q, where each partition corresponds to the definition of one cardinal direction. The example in Figure 1b gives the case when p is to the *northwest* of q, i.e. $dir(p, q) = NW$.

When two points change their locations over time, the direction relation between them becomes time related, and may or may not change. First, we consider the cardinal directions at time instances. Let *time* denote the temporal data type representing time and *MPoints* denote the spatio-temporal data type that represents moving points. Figure 1c shows an example of two moving points A and B. For $A, B \in MPoints$, let $A(t)$ and $B(t)$ denote the snapshots of A and B at a time instance $t \in time$. If both A and B are defined at time t, then $A(t), B(t) \in Points$. The cardinal direction between A and B at t is therefore $dir(A(t), B(t)) \in CD$. For example, in Figure 1c, at time t_1 when A and B locate at $A(t_1)$ and $B(t_1)$, the cardinal direction between A and B at time instance t_1 is $dir(A(t_1), B(t_1)) = SW$. At the time instance t_2 when A and B move to $A(t_2)$ and $B(t_2)$, the cardinal direction between them becomes $dir(A(t_2), B(t_2)) = NE$. We propose our solution to determine what happened in between and to answer the question whether there exists a time instance t ($t_1 < t < t_2$) such that $dir(A(t), B(t)) = W$ in the following sections. This scenario shows that within a common time interval, we may get different cardinal directions at different time instances. However, the change of time does not necessarily imply the change of cardinal directions between two moving points. In Figure 1c, from time t_3 to time t_4, A moves from $A(t_3)$ to $A(t_4)$ and B moves from $B(t_3)$ to $B(t_4)$. One observation that we can make is that although the positions of A and B have changed, the cardinal direction between A and B does not change. In this case, A is always to the *southwest* of B between t_3 and t_4. In other words, the cardinal direction between two moving points holds for a certain period of time before it changes. Based on this fact, we propose our modeling strategy. To determine the cardinal directions between two moving points during their life time, we first find out the *common life time intervals* between two moving points, on which both two moving points are defined. This is necessary because only when both moving points exist, we can determine the cardinal directions between them. In this case, the common life time interval between A and B in Figure 1c is $[t_1, t_5]$, and during the time interval $[t_0, t_1]$, cardinal directions between A and B cannot be determined. Then, each common life interval is split into a list of smaller sub-intervals such that during each sub-interval the cardinal direction between two moving points does not change Further, on adjacent sub-intervals, different cardinal directions hold. Finally, we compose all cardinal directions determined on the common life time intervals of two moving points, and define it as the development of cardinal directions between the two moving points. Let $DEV(A, B)$ denote the function that computes the cardinal direction developments between two moving points A and B. Then we define $DEV(A, B)$ as $DEV(A, B) = d_1 \triangleright d_2 \triangleright ... \triangleright d_n$, where $d_i \in CD$ or $d_i = \bot$ ($1 \leq i \leq n$ and $\bot$ means undefined). Further, we restrain the transition between two adjacent cardinal directions to follow a so-called *state transition diagram*. The formal definitions and the detailed explanations can be found in [14].

4   Computing Developments between Moving Points

The concept we have introduced in the previous section serves as a specification for describing the changing cardinal directions between two moving points. However, issues like how to find common life time intervals and how to split them are left open. In this section, we overcome the issues from an algorithmic perspective. We first introduce the underlying data structure, called *slice representation*, for representing moving points. Then we propose a three phase strategy including the *time-synchronized interval refinement phase*, the *slice unit direction evaluation phase*, and *the direction composition phase*.

4.1   The Slice Representation for Moving Points

Since we take the specification of the moving point data type in [3,4] as our basis, we first review the representation of the moving point data type. According to the definition, the moving point date type describes the temporal development of a complex point object which may be a point cloud. However, we here only consider the simple moving point that involves exactly one single point. A *slice representation* technique is employed to represent a moving point object. The basic idea is to decompose its temporal development into fragments called "slices", where within each slice this development is described by a simple linear function. A slice of a single moving point is called a *upoint*, which is a pair of values (*interval, unit-function*). The *interval* value defines the time interval for which the unit is valid; the *unit-function* value contains a record (x_0, x_1, y_0, y_1) of coefficients representing the linear function $f(t)=(x_0 + x_1 t, y_0 + y_1 t)$, where t is a time variable. Such functions describe a linearly moving point. The time intervals of any two distinct slice units are disjoint; hence units can be totally ordered by time. More formally, let A be a single moving point representation, *interval = time × time, real4 = real × real × real × real*, and *upoint = interval × real4*. Then A can be represented as an array of slice units ordered by time, that is, $A = \langle (I_1, c_1), (I_2, c_2), ..., (I_n, c_n) \rangle$ where for $1 \leq i \leq n$ holds that

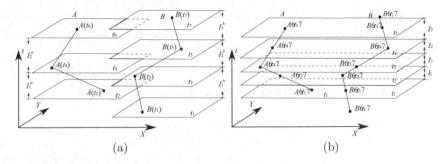

Fig. 2. An example of the slice representations of two single moving points A and B (a), and the time-synchronized slice representation of two moving points A and B (b)

$I_i \in interval$ and $c_i \in real4$ contains the coefficients of a linear unit function f_i. Further, we require that $I_i < I_j$ holds for $1 \le i < j \le n$.

Figure 2 shows the slice representations of two single moving points A and B. In this example, t_i $(1 \le i \le 7)$ is a time instance and for $1 \le i < j \le 7$, $t_i < t_j$. The moving point A is decomposed into two slices with intervals $I_1^A = [t_2, t_4]$ and $I_2^A = [t_4, t_6]$. Let the function f_1^A with its coefficients c_1^A and the function f_2^A with its coefficients c_2^A describe the movement of A in the intervals I_1^A and I_2^A respectively. Then A is represented as $A = \langle (I_1^A, c_1^A), (I_2^A, c_2^A) \rangle$. The snapshots $f_1^A(t_2)$ and $f_1^A(t_4)$ of the moving point A at the times t_2 and t_4 are the start and end points of the first slice, and $f_2^A(t_4)$ and $f_2^A(t_6)$ are the start and end points of the second slice. Similarly, the moving point B can be represented as $B = \langle (I_1^B, c_1^B), (I_2^B, c_2^B), (I_3^B, c_3^B) \rangle$ where c_1^B, c_2^B, and c_3^B contain the coefficients of the three linear functions f_1^B, f_2^B, and f_3^B that describe the linear movement of B in its three slice units. If the function f_i^A or f_i^B that in a slice unit maps a time instant t to a point value in A or B is not important, we allow the notations $A(t)$ and $B(t)$ respectively to retrieve the location of a moving point A or B at the time instant t.

Further, we introduce a few basic operations for retrieving information from the slice representation, which will be used by our algorithm later for computing cardinal directions between moving points.

The first set of operations is provided for manipulating moving points. The *get_first_slice* operation retrieves the first slice unit in a slice sequence of a moving point, and sets the current position to 1. The *get_next_slice* operation returns the next slice unit of the current position in the sequence and increments the current position. The predicate *end_of_sequence* yields *true* if the current position exceeds the end of the slice sequence. The operation *create_new* creates an empty *MPoint* object with an empty slice sequence. Finally, the operation *add_slice* adds a slice unit to the end of the slice sequence of a moving point.

The second set of operations is provided for accessing elements in a slice unit. The operation *get_interval* returns the time interval of a slice unit. The operation *get_unit_function* returns a record that represents the linear function of a slice unit. The *create_slice* operation creates a slice unit based on the provided time interval and the linear function.

Based on the slice representation and the basic operations, we are now ready to describe our strategy for computing the cardinal directions between two moving points.

4.2   The Time-Synchronized Interval Refinement Phase

Since a slice is the smallest unit in the slice representation of moving points, we first consider the problem of computing cardinal directions between two moving point slices. According to our definitions in [14] the cardinal directions only make sense when the same time intervals are considered for both moving points. However, matching, i.e., equal, slice intervals can usually not be found in both moving points. For example, in Figure 2, the slice interval $I_1^A = [t_2, t_4]$ of A does not match any of the slice intervals of B. Although the slice interval $I_1^B = [t_1, t_3]$ of B overlaps with I_1^A, it also covers a sub-interval $[t_1, t_2]$ that is not part of I_1^A, which makes the two slices defined in I_1^A and I_1^B incomparable. Thus, in order to compute the cardinal directions between two moving point slices, a *time-synchronized interval refinement* for both moving points is necessary.

We introduce a linear algorithm *interval_sync* for synchronizing the intervals of both moving points. The input of the algorithm consists of two slice sequences $mp1$ and $mp2$ that represent the two original moving points, and two empty lists $nmp1$ and $nmp2$ that are used to store the two new interval refined moving points. The algorithm performs a parallel scan of the two original slice sequences, and computes the intersections between the time intervals from two moving points. Once an interval intersection is captured, two new slices associated with the interval intersection are created for both moving points and are added to the new slice sequences of the two moving points. Let $I = [t_1, t_2]$ and $I' = [t_1', t_2']$ denote two time intervals, and let *lower_than* denote the predicate that checks the relationship between two intervals. Then we have $lower\_than(I, I') = true$ if and only if $t_2 < t_2'$. Further, let *intersection* denote the function that computes the intersection of two time intervals, which returns $\emptyset$ if no intersection exists. We present the corresponding algorithm *interval_sync* in Figure 3.

As a result of the algorithm, we obtain two new slice sequences for the two moving points in which both operand objects are synchronized in the sense that for each unit in the first moving point there exists a matching unit in the second moving point with the same unit interval and vice versa. For example, after the time-synchronized interval refinement, the two slice representations of the moving points A and B in Figure 2 become $A = \langle (I_1, c_1^A), (I_2, c_1^A), (I_3, c_2^A), (I_4, c_2^A) \rangle$ and $B = \langle (I_1, c_1^B), (I_2, c_2^B), (I_3, c_2^B), (I_4, c_3^B) \rangle$, where the c_i^A with $i \in \{1, 2\}$ contain the coefficients of the linear unit functions f_i^A, the c_i^B with $i \in \{1, 2, 3\}$ contain the coefficients of the linear unit functions f_i^B, and $I_1 = intersection(I_1^A, I_1^B) = [t_2, t_3]$, $I_2 = intersection(I_1^A, I_2^B) = [t_3, t_4]$, $I_3 = intersection(I_2^A, I_2^B) = [t_4, t_5]$, and $I_4 = intersection(I_2^A, I_3^B) = [t_5, t_6]$.

Now we analyze the complexity of the algorithm for function *interval_sync*. Assume that the first moving point mp_1 is composed of m slices, and the second moving point mp_2 is composed of n slices. Since a parallel scan of the slice

method *interval_sync* (*mp1, mp2,*	1 **method** *compute_dir_dev*(*sl1, sl2*)
nmp1, nmp2)	2 *dev_list* ← empty list
s1 ← *get_first_slice*(*mp1*)	3 *s1* ← *get_first_slice*(*sl1*)
s2 ← *get_first_slice*(*mp2*)	4 *s2* ← *get_first_slice*(*sl2*)
while not *end_of_sequence*(*mp1*)	5 *slice_dir_list* ← *compute_slice_dir*(*s1,s2*)
and not *end_of_sequence*(*mp2*) **do**	6 *append*(*dev_list, slice_dir_list*)
i1 ← *get_interval*(*s1*)	7 **while not** *end_of_sequence*(*sl1*)
i2 ← *get_interval*(*s2*)	8 **and not** *end_of_sequence*(*sl2*) **do**
i ← *intersection*(*i1, i2*)	9 (*b, e*) ← *get_interval*(*s1*)
if *i* ≠ ∅ **then**	10 *s1* ← *get_next_slice*(*sl1*)
f1 ← *get_unit_function*(*s1*)	11 *s2* ← *get_next_slice*(*sl2*)
f2 ← *get_unit_function*(*s2*)	12 (*b_new, e_new*) ← *get_interval*(*s1*)
ns1 ← *create_slice*(*i, f1*)	13 **if** *e* < *b_new* **then**
ns2 ← *create_slice*(*i, f2*)	14 *append*(*dev_list,* ⟨⊥⟩)
add_slice(*nmp1, ns1*)	15 **endif**
add_slice(*nmp2, ns2*)	16 *slice_dir_list* ← *compute_slice_dir*(*s1,s2*)
endif	17 *last_dir* ← *get_last_in_list*(*dev_list*)
if *lower_than*(*i1, i2*) **then**	18 *new_dir* ← *get_first_in_list*(*slice_dir_list*)
s1 ← *get_next_slice*(*mp1*)	19 **if** *last_dir* = *new_dir* **then**
else	20 *remove_first*(*slice_dir_list*)
s2 ← *get_next_slice*(*mp2*)	21 **endif**
endif	22 *append*(*dev_list, slice_dir_list*)
endwhile	23 **endwhile**
end	24 **return** *dev_list*
	25 **end**

Fig. 3. The algorithm *interval_sync* that computes the time-synchronized interval refinement for two moving points, and the algorithm *compute_dir_dev* that computes the cardinal direction development for two moving points

sequences from two moving points is performed, the complexity is therefore $O(m + n)$ and the result contains at most $(m + n)$ intervals.

4.3   The Slice Unit Direction Evaluation Phase

From the first phase, the *time-synchronized interval refinement* phase, we obtain two refined slice sequences of both moving points that contain the same number of slice units with synchronized time intervals. In the second phase, we propose a solution for computing the cardinal directions between any pair of time-synchronized slice units.

We adopt a two-step approach to computing the cardinal directions between two slice units. The first step is to construct a mapping and apply it to both slice units so that one of the slice units is mapped to a slice unit that consists of a point that does not change its location. We prove that the mapping is a *cardinal direction preserving mapping* that does not change the cardinal direction relationships between the two slice units. The second step is to determine the cardinal directions between the two mapped slice units.

The difficulty of computing cardinal directions between two slice units comes from the fact that the positions of the moving points change continuously in both slice units. A much simpler scenario is that only one slice unit consists of a moving point, whereas the other slice unit involves no movement. In this simpler case, the cardinal directions can be easily determined. Thus, the goal is to find a mapping that maps two slice units su_a and su_b to two new slice units su_a' and su_b' that satisfy the following two conditions: (i) su_a' and su_b' have the same cardinal directions as units su_a and su_b, that is, the mapping is a *cardinal direction preserving mapping*; (ii) either su_a' or su_b' does not involve movement.

In order to find such a mapping for two slice units, we first introduce a simple *cardinal direction preserving mapping* for static points. Let p and q denote two points with coordinates (x_p, y_p) and (x_q, y_q). Let $X(r)$ and $Y(r)$ denote the functions that return the x-coordinate and y-coordinate of a point r. We establish a simple translation mapping $M(r) = (X(r) - x_0, Y(r) - y_0)$, where x_0 and y_0 are two constant values. We show that the cardinal direction between p and q is preserved by applying such a mapping.

Lemma 1. *Given* $p = (x_p, y_p)$, $q = (x_q, y_q)$, *the mapping* $M(r)=(X(r) - x_0,$ $Y(r) - y_0)$, *where* r *is a point and* x_0 *and* y_0 *are two constant values, and* $p' = M(p)$ *and* $q' = M(q)$, *we have* $dir(p, q) = dir(p', q')$

Proof. According to the definition in Section 3, the cardinal direction $dir(p, q)$ between two points p and q is based on the value of $X(p) - X(q)$ and the value of $Y(p) - Y(q)$. Since we have $X(p') - X(q') = X(p) - X(q)$ and $Y(p') - Y(q') = Y(p) - Y(q)$, we obtain $dir(p, q) = dir(p', q')$.   □

In Figure 4(a), two points p and q are mapped to p' and q', and the cardinal direction is preserved after the mapping, i.e., $dir(p, q) = dir(p', q') = NW$.

Now we are ready to define a *cardinal direction preserving mapping* for two slice units. Let su^A and su^B denote two slice units (*upoint* values) from the time-synchronized moving points A and B where $su^A = (I, c^A)$ and $su^B = (I, c^B)$ with $I \in interval$ and $c^A, c^B \in real4$. Let f^A and f^B be the two corresponding linear unit functions with the coefficients from c^A and c^B respectively. We establish the following mapping M for a unit function $f \in \{f^A, f^B\}$: $M(f) = f - f^B$.

We show in Lemma 2 that by mapping the unit functions of the slices su_A and su_B to two new unit functions, the cardinal directions between the slice units su_A and su_B are still preserved.

Lemma 2. *Let* $su^A = (I, c^A) \in$ upoint *and* $su^B = (I, c^B) \in$ upoint, *and let* f^A *and* f^B *be the corresponding linear unit functions with the coefficients from* $c^A = (x_0^A, x_1^A, y_0^A, y_1^A)$ *and* $c^B = (x_0^B, x_1^B, y_0^B, y_1^B)$ *respectively. We consider the mapping* $M(f) = f - f^B$, *where* f *is a linear unit function, and the translated upoint values* $su_t^A = (I, c_t^A)$ *and* $su_t^B = (I, c_t^B)$ *where* c_t^A *and* c_t^B *contain the coefficients of* $M(f^A)$ *and* $M(f^B)$ *respectively. Then, the cardinal directions between the slice units* su_t^A *and* su_t^B *are the same as the cardinal directions between slice units* su^A *and* su^B.

Proof. Let $I = [t_1, t_2]$, $f^A(t) = (x_0^A + x_1^A t, y_0^A + y_1^A t)$, and $f^B(t) = (x_0^B + x_1^B t, y_0^B + y_1^B t)$. Then we have $M(f^A) = (x_0^A - x_0^B + (x_1^A - x_1^B)t, y_0^A - y_0^B +$

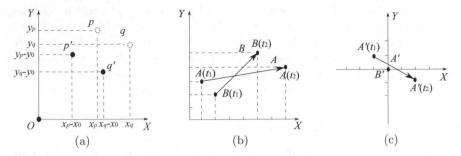

Fig. 4. A simple cardinal direction preserving mapping from p, q to p',q' (a); the cardinal direction preserving mapping from slice unit A, B (b) to A',B' (c)

$(y_1^A - y_1^B)t)$ and $M(f^B) = (0,0)$. Assume there exists a time t_0 $(t_1 \leq t_0 \leq t_2)$ such that $dir(f^A(t_0), f^B(t_0)) \neq dir(M(f^A)(t_0), M(f^B)(t_0))$. Let $x^B = x_0^B + x_1^B t_0$ and $y^B = y_0^B + y_1^B t_0$ denote two constant values. Since $M(f^A)(t_0) = (x_0^A - x_0^B + (x_1^A - x_1^B)t_0, y_0^A - y_0^B + (y_1^A - y_1^B)t_0)$ and $M(f^B)(t_0) = (0,0)$, we have $M(f^A)(t_0) = (X(f^A(t_0)) - x^B, Y(f^A(t_0)) - y^B)$ and $M(f^B)(t_0) = (X(f^B(t_0)) - x^B, Y(f^B(t_0)) - y^B)$. This matches the *cardinal direction preserving mapping* function $M(r) = (X(r) - x_0, Y(r) - y_0)$. Thus, the assumption $dir(f^A(t_0), f^B(t_0)) \neq dir(M(f^A)(t_0), M(f^B)(t_0))$ contradicts to Lemma 1.   □

After applying the *cardinal direction preserving mapping* $M(f)$ to both unit functions f^A and f^B, we now obtain two new unit functions f_a' and f_b' as follows:

$$g^A(t) = M(f^A)(t) = (x_0^A - x_0^B + (x_1^A - x_1^B)t, y_0^A - y_0^B + (y_1^A - y_1^B)t)$$
$$g^B(t) = M(f^B)(t) = (0,0)$$

The unit function g^A describes a linear movement in the unit interval, while the unit function g^B describes a static point that holds its position during the entire unit interval. In other words, su^A is mapped to a new slice unit su_t^A which has a linear movement, and su^B is mapped to a new slice unit su_t^B that has no movement during the unit interval. Figure 4 shows an example of mapping the slice units A and B to slice units A' and B'. In this example, $A = [I, c^A]$ and $B = [I, c^B]$ where $I = [1,2]$, c^A and c^B contain the coefficients of the two unit functions f^A and f^B respectively, $f^A(t) = (-5 + 6t, 2 + t)$ and $f^B(t) = (-1 + 3t, -1 + 3t)$. Thus, $A(t_1) = f^A(1) = (1,3)$, $A(t_2) = f^A(2) = (7,4)$, $B(t_1) = f^B(1) = (2,2)$, and $B(t_2) = f^B(2) = (5,5)$. After applying the mapping, we obtain $g^A(t) = (-4 + 3t, 3 - 2t)$ and $g^B(t) = (0,0)$. Thus, $A'(t_1) = g^A(1) = (-1,1)$, $A'(t_2) = g^A(2) = (2,-1)$, and $B'(t_1) = B'(t_2) = (0,0)$.

So far, we have managed to reduce the problem of computing the cardinal directions between two moving slice units to the problem of computing the cardinal directions between one moving slice unit and one static slice unit. The second step is to compute the cardinal directions between su_t^A and su_t^B.

Since su_t^B is located constantly at $(0,0)$ during the time interval and since the trajectory of su_t^A is a linear function with respect to time t, we apply the projection based approach (Section 3) to determining the cardinal directions.

The idea is to take su_t^B as the reference point and to create partitions by using the x- and y-coordinate axes. Then we project the slice unit su_t^A to the xy-plane, and the cardinal directions are determined by the partitions that its trajectory intersects. Finally, the cardinal directions are ordered according to the time when they occurred and are stored into a list. For example, the cardinal directions between A' and B' in Figure 4b are NW, N, NE, E, and SE.

4.4   The Direction Composition Phase

Finally, in the direction composition phase, we iterate through all slice units, compose all cardinal directions that have been detected in slice units, and form a complete cardinal direction list in the temporal order. Further, we remove duplicates between consecutive cardinal directions.

We introduce the linear algorithm *compute_dir_dev* in Figure 3 for computing the final cardinal direction development (line 24) between two synchronized moving points. The input of the algorithm consists of two lists of slices $sl1$ and $sl2$ (line 1) that stem from the *time-synchronized interval refinement phase*. Since the two slice lists are guaranteed to have the same length, the algorithm takes a slice from each list (lines 3, 4, 10 and 11), determines the cardinal directions for each pair of slices (lines 5 and 16), which have the same unit interval, and traverses both lists in parallel (lines 7 and 8). For two consecutive pairs of slices, we have to check whether the slice intervals are adjacent (lines 9, 12, and 13). If this is not the case, we add the list with the single element $\perp$ to the global list *dev_list* in order to indicate that the cardinal direction development is undefined between two consecutive slice intervals (lines 13 to 15).

For each pair of slices, the function *compute_slice_dir* determines their cardinal directions according to the strategy discussed in Section 4.3 (lines 5 and 16). We maintain a list *slice_dir_list* to keep these newly computed cardinal directions from the current slice pair and compare its first cardinal direction with the last cardinal direction that has been computed from the last slice pair and is stored in the global list *dev_list* (lines 17 to 19). If both cardinal directions are the same, the first cardinal direction from the list *slice_dir_list* is removed in order to avoid duplicates (lines 19 to 21). The newly computed cardinal directions in the list *slice_dir_list* are added to the global list *dev_list* (lines 6 and 22).

The algorithm *compute_dir_dev* deploys a number of auxiliary list functions. The function *get_first_in_list* returns the first element in a list. The function *get_last_in_list* returns the last element in a list. The function *append* adds a list given as its second argument to the end of another list given as its first argument. The function *remove_first* removes the first element from a list.

Now we analyze the complexity of the algorithm for function *compute_dir_dev*. Assume that the first moving point mp_1 consists of m slices, and the second moving point mp_2 consists of n slices. The inputs of the function *compute_dir_dev* are two lists of slices generated from the time-synchronized interval refinement phase, thus each list contains at most $m+n$ slices. The function *compute_dir_dev* iterate through all slices in both list and compose the cardinal directions computed. So the time complexity is $O(m + n)$.

5   Defining Spatial-temporal Direction Predicates within Databases

In this section, we discuss how cardinal direction developments can be integrated into spatio-temporal databases and query languages. This requires the formal definition of cardinal direction developments as binary predicates since it will make the query processing easier when using pre-defined predicates as selection conditions. In the following part, we define some important predicates which will be sufficient for most queries on cardinal direction developments between moving objects.

First of all, we give the definition of *existential direction predicates*. This type of predicates finds out whether a specific cardinal direction existed during the evolution of moving objects. For example, a query like "Find all ships that appeared north of ship Fantasy" belongs to this category. It requires a predicate named *exists_north* as a selection condition of a join. This predicate can be defined as follows,

Definition 1. *Given two moving points $A, B \in MPoints$, their cardinal direction development $DEV(A, B) = d_1 \triangleright d_2 \triangleright \ldots \triangleright d_n$ with $n \in \mathbb{N}$ and $d_i \in CD$ or $d_i = \bot$ for all $1 \le i \le n$. Then we define the* existential direction predicate exists_north *as*

$$exists\_north(A, B) = true \overset{\text{def}}{\Leftrightarrow} \exists 1 \le i \le n : d_i = N$$

Definition 1 indicates that the predicate *exists_north* is true if the direction *north* exists in the sequence of the cardinal direction development. It can help us define the above query. Assume that we have the following relation schema for ships

```
ships(id:integer, name:string, route:mpoint)
```

The query can be expressed using an SQL-like query language as follows:

```
SELECT s1.name FROM ships s1, ships s2
WHERE  s2.name = 'Fantasy' AND exists_north(s1.route, s2.route);
```

The other existential cardinal direction predicates *exists_south*, *exists_east*, *exists_west*, *exists_sameposition*, *exists_northeast*, *exists_southeast*, *exists_northwest*, and *exists_southwest* are defined in a similar way.

Another important category of predicates expresses that one moving object keeps the same direction with respect to another moving object. For example, assume that there is a group of ships traveling from north to south and each ship follows the ship in front of the group. Now the leader of the group wants to know which ships belong to the group. The problem is to find out which ships are keeping a northern position with respect to the leading ship.

Definition 2. *Given two moving points $A, B \in MPoints$. The predicate* keeps_north *is defined as*

$$
\begin{aligned}
keeps\_north(A, B) = \ &exists\_north(A, B) &&\land \ \neg exists\_south(A, B) \\
&\land \ \neg exists\_southeast(A, B) &&\land \ \neg exists\_east(A, B) \\
&\land \ \neg exists\_sameposition(A, B) &&\land \ \neg exists\_northwest(A, B) \\
&\land \ \neg exists\_northeast(A, B) &&\land \ \neg exists\_southwest(A, B) \\
&\land \ \neg exists\_west(A, B)
\end{aligned}
$$

Definition 2 shows that the relationship *keeps_north* between two moving objects implies that the only existential direction predicate in the cardinal direction development of these moving objects is *exists_north* without any other existential direction predicates. In other words, we have $DEV(A, B) = N$.

We consider the above example and assume that the identifier of the leader ship is 1001. Then the query "Find all ships keeping a position north of the leader ship 1001" can be expressed as

```
SELECT s1.id FROM ships s1, ships s2
WHERE  s2.id = '1001' AND keeps_north(s1.route, s2.route);
```

The other predicates that express that one moving object remains in the same direction with respect to another moving object are *keeps_south*, *keeps_east*, *keeps_west*, *keeps_sameposition*, *keeps_northeast*, *keeps_southeast*, *keeps_northwest*, and *keeps_southwest*.

Another useful predicate checks for the transition between two cardinal directions in a cardinal direction development. The transition can be either a direct change or an indirect change through a set of intermediate directions. We name this predicate as *from_to*. For example, the query "Find all ships that have traveled from the south to the north of the ship Fantasy" can be answered by using this predicate.

Definition 3. *Given two moving points $A, B \in MPoints$, their cardinal direction development* $\mathrm{DEV}(A, B) = d_1 \triangleright d_2 \triangleright \ldots \triangleright d_n$ *such that $d_i \in CD$ or $d_i = \bot$ for all $1 \leq i \leq n$, and two cardinal directions $d', d'' \in CD$. We define the predicate* from_to *as follows:*

$$
from\_to(A, B, d', d'') = true \stackrel{\mathrm{def}}{\Leftrightarrow} d' \neq \bot \ \land \ d'' \neq \bot \ \land \\
\exists 1 \leq i < j \leq n : d_i = d' \ \land \ d_j = d''
$$

We formulate the above query as follows:

```
SELECT s1.id FROM ships s1, ships s2
WHERE  s2.name = 'Fantasy' AND
       from_to(s1.route, s2.route, 'S', 'N');
```

Finally, we define the predicate *cross_north* which checks whether a moving point traverses a large extent of the region in the north of another moving point.

Definition 4. *Given two moving points $A, B \in MPoints$ and their cardinal direction development* $\mathrm{DEV}(A, B) = d_1 \triangleright d_2 \triangleright \ldots \triangleright d_n$ *such that $d_i \in CD$ or $d_i = \bot$ for all $1 \leq i \leq n$. We define the predicate* crosses_north *as follows:*

$$crosses\_north(A, B) = true \stackrel{\text{def}}{\Leftrightarrow} n \geq 3 \; \wedge \; \exists 2 \leq i \leq n - 1 :$$
$$(d_{i-1} = \text{NW} \; \wedge \; d_i = N \; \wedge \; d_{i+1} = \text{NE}) \; \vee$$
$$(d_{i-1} = \text{NE} \; \wedge \; d_i = N \; \wedge \; d_{i+1} = \text{NW})$$

The query "Find all the ships that have crossed the north of ship Fantasy" can be expressed as follows:

```
SELECT s1.id FROM ships s1, ships s2
WHERE  s2.name = 'Fantasy' AND crosses_north(s1.route, s2.route);
```

The other predicates $cross\_south$, $cross\_east$, and $cross\_west$ can be defined in a similar way.

6   Case Study: Cardinal Direction Development in Hurricane Research

In this section, we apply our strategy to a real world application, and show how the evaluation of cardinal direction development can help with the hurricane research.

We have integrated the directional predicates into a moving object database (MOD) developed for the NASA workforce. The moving object database is a full-fledged database with additional support for spatial and spatiotemporal data in its data model and query language. It maintains tropical cyclone and hurricane data provided by public sources, and the weather data derived from the NASA mission sensor measurements. It also provides functionality in terms of spatiotemporal operations and predicates that can be deployed by decision makers and scientists in ad-hoc queries. By enabling the capability of evaluating cardinal direction developments among hurricanes, the scientists can have a better understanding of dynamic patterns on weather events. We establish our experiments on the historical hurricane data collected from National Hurricane Center (NHC). The original data is available on the web site of NHC [15]. The sensors collect six data points per day for a specific hurricane, i.e., at 00:00, 06:00, 12:00 and 18:00 UTC time. The data collected are the hurricane locations in terms of longitudes and latitudes, time, and other thematic data like wind speed and category. We load these data points into moving point types, and represent the trajectory of each hurricane as a moving point in MOD. In this paper, we present a case study on all hurricanes in year 2005 on the Atlantic Ocean. The following table is created in the database:

```
test_moving(id:integer, name:string, track:mpoint)
```

In the schema $test\_moving$, $name$ is the attribute that stores hurricane names and $track$ is a moving point type attribute that stores the trajectory of hurricanes. A total of 28 hurricanes that have been active on the Atlantic Ocean in the year 2005 are loaded in the data table. Due to the space limit, we evaluate the following two types of directional queries: the cardinal direction development query and the top-k query.

Fig. 5. The trajectories of hurricanes PHILIPPE and RITA

First, consider the query: "Find the cardinal direction development between PHILIPPE and RITA.", we can post the following SQL query:

```
SELECT m1.name, m2.name, mdir(m1.track, m2.track),
FROM   test_moving m1, test_moving m2
WHERE  m1.name = 'PHILIPPE' AND m2.name = 'RITA';
```

The function *mdir* is a user defined function registered at the database end that computes the cardinal direction developments between two moving points. A string representation is returned as the result. In this case, we obtain the following result:

```
NAME      NAME   MDIR(M1.TRACK,M2.TRACK)
--------  -----  -------------------------------
PHILIPPE  RITA   ->undefined[2005091712,2005091800)
                 ->NW[2005091800,2005092212)
                 ->W[2005092212,2005092212]
                 ->SW(2005092212,2005092402)
                 ->W[2005092402,2005092402]
                 ->NW(2005092402,2005092406)
                 ->undefined[2005092406,2005092606)
```

The result is a list of timely ordered cardinal directions. In the time interval [2005-09-17 12:00:00,2005-09-18 00:00:00), RITA is not evolved yet, thus the cardinal direction is *undefined*. When RITA is "born", it starts from the northwest of PHILIPPE, moves to the north of PHILIPPE. Then it crosses the west of PHILIPPE and moves to the southwest of PHILIPPE on date 2005-09-22. In the following two days, it moves back to the northwest of PHILIPPE. The visualization of the two hurricane is shown in Figure 5. The result shows an interesting movement pattern between the two hurricanes, which may suggest the hurricane researchers to investigate the correlations in terms of wind speed, air pressure, and ocean currents during a certain time interval between the two hurricanes.

Another type of query that is intersecting to hurricane researchers is the top-k query. Here, the top-k evaluates the lasting time of cardinal directions between two hurricanes. Thus, given two hurricanes, we are able to find the top-k cardinal directions between them. Let us consider the query: "find top 2 cardinal directions between MARIA with other hurricane tracks". We can formulate the SQL query as follows:

```
SELECT m1.name, m2.name, topKDir(m1.track,m2.track,3)
FROM   test_moving m1, test_moving m2
WHERE  m1.name='MARIA' AND m1.name<>m2.name
AND    topKDir(m1.track,m2.track,2) <> ' '
```

The function $topKDir(m1.track, m2.track, 2)$ returns the top 2 cardinal directions (excluding the undefined direction) between two moving points that last the longest, and it returns empty string if there does not exist defined cardinal directions between them. We get the following result:

```
NAME    NAME      TOPKDIR(M1.TRACK,M2.TRACK,2)
------  --------  ---------------------------
MARIA   LEE       NW NE
MARIA   NATE      SW
MARIA   OPHELIA   SW
```

The result shows that the top two cardinal directions lasting the longest between MARIA and LEE are NW and NE. NATE and OPHELIA are always to the SW of MARIA. From this result, we can observe that during the life time of MARIA, two hurricanes spent most of their time moving in the southwest of MARIA and one hurricane spent most of its time in the northwest of MARIA. No hurricanes exists in the other directions like SE or NE of MARIA. This observation may raise the intersects of hurricane researchers to investigate the causes and the facts that lead to the pattern, or to make conclusions from this pattern.

7   Conclusions and Future Work

In this paper, we present a three-phase solution for computing the cardinal directions between two moving points from an algorithmic perspective. We show the mapping of cardinal direction developments between moving points into spatio-temporal directional predicates and the integration of these predicates into the spatio-temporal query language of a moving objects database. We present a case study on the hurricane data to show a real world application for the cardinal direction development. In the future, we will implement a comprehensive set of predicates for querying cardinal direction development. We will also extend our concept to more complex moving objects like moving regions and moving lines.

References

1. Theodoridis, Y., Sellis, T.K., Papadopoulos, A., Manolopoulos, Y.: Specifications for Efficient Indexing in Spatiotemporal Databases. In: 10th Int. Conf. on Scientific and Statistical Database Management (SSDBM), pp. 123–132 (1998)

2. Grumbach, S., Rigaux, P., Segoufin, L.: Spatio-temporal Data Handling with Constraints. GeoInformatica, 95–115 (2001)
3. Erwig, M., Güting, R.H., Schneider, M., Vazirgiannis, M.: Spatio-temporal Data Types: an Approach To Modeling and Querying Moving Objects in Databases. GeoInformatica 3(3), 269–296 (1999)
4. Forlizzi, L., Guting, R., Nardelli, E., Schneider, M.: A Data Model and Data Structures for Moving Objects Databases. In: ACM SIGMOD Int. Conf. on Management of Data, pp. 319–330 (2000)
5. Erwig, M., Schneider, M.: Spatio-temporal Predicates. IEEE Trans. on Knowledge and Data Engineering (TKDE) 14(4), 881–901 (2002)
6. Su, J., Xu, H., Ibarra, O.H.: Moving Objects: Logical Relationships and Queries. In: 7th Int. Symp. on Spatial and Temporal Databases (SSTD), pp. 3–19 (2001)
7. de Weghe, N.V., Bogaert, P., Delafontaine, M., Temmerman, L.D., Neutens, T., Maeyer, P.D., Witlox, F.: How To Handle Incomplete Knowledge Concerning Moving Objects. Behaviour Monitoring and Interpretation, 91–101 (2007)
8. Frank, A.: Qualitative Spatial Reasoning: Cardinal Directions As an Example. International Journal of Geographical Information Science 10(3), 269–290 (1996)
9. Skiadopoulos, S., Koubarakis, M.: Composing Cardinal Direction Relations. Artificial Intelligence 152, 143–171 (2004)
10. Haar, R.: Computational Models of Spatial Relations. Technical Report: TR-478, (MSC-72-03610) (1976)
11. Skiadopoulos, S., Sarkas, N., Sellis, T., Koubarakis, M.: A Family of Directional Relation Models for Extended Objects. IEEE Trans. on Knowledge and Data Engineering (TKDE) 19 (2007)
12. Papadias, D., Theodoridis, Y., Sellis, T.: The Retrieval of Direction Relations Using R-trees. In: Karagiannis, D. (ed.) DEXA 1994. LNCS, vol. 856, pp. 173–182. Springer, Heidelberg (1994)
13. Li, J.Z., Ozsu, M.T., Tamer, M., Szafron, D., Ddi, S.G.: Modeling of Moving Objects in a Video Database. In: IEEE International Conference on Multimedia Computing and Systems, pp. 336–343 (1997)
14. Chen, T., Liu, H., Schneider, M.: Evaluation of Cardinal Direction Developments between Moving Points. In: ACM Symp. on Geographic Information Systems (ACM GIS), pp. 430–433 (2010)
15. NHC Archive of Hurricane Seasons, http://www.nhc.noaa.gov/pastall.shtml

Continuous Probabilistic Count Queries in Wireless Sensor Networks*

Anna Follmann[1], Mario A. Nascimento[2], Andreas Züfle[1], Matthias Renz[1], Peer Kröger[1], and Hans-Peter Kriegel[1]

[1] Department of Computer Science, Ludwig-Maximilians-Universität, Germany
follmann@cip.ifi.lmu.de, {zuefle,kroegerp,renz,kriegel}@dbs.ifi.lmu.de
[2] University of Alberta, Canada
mn@cs.ualberta.ca

Abstract. Count queries in wireless sensor networks (WSNs) report the number of sensor nodes whose measured values satisfy a given predicate. However, measurements in WSNs are typically imprecise due, for instance, to limited accuracy of the sensor hardware. In this context, we present four algorithms for computing continuous *probabilistic* count queries on a WSN, i.e., given a query Q we compute a probability distribution over the number of sensors satisfying Q's predicate. These algorithms aim at maximizing the lifetime of the sensors by minimizing the communication overhead and data processing cost. Our performance evaluation shows that by using a distributed and incremental approach we are able to reduce the number of message transfers within the WSN by up to a factor of 5 when compared to a straightforward centralized algorithm.

1 Introduction

A wireless sensor network (WSN) is usually defined as a set of spatially distributed autonomous sensors that cooperatively monitor physical or environmental conditions in an area of interest. A single sensor node consists of one (or more) sensor(s), a microprocessor, a small amount of memory, a radio transceiver and a battery. However, measurements by nodes in WSNs are typically imprecise, be it because of the sensor's hardware or because of fluctuations in the environment itself [1]. Processing uncertain data sets leads to a variety of novel problems and demands more complex query algorithms. Typically, queries on uncertain data known as probabilistic queries involve computation of probability distributions over possible answers. In addition, if we want to process probabilistic queries in wireless sensor networks we have to consider the general characteristics and limitations of such networks. Even if sensors are gaining in computing ability they remain constrained by limited batteries. With communication being the primary drain on power it is crucial that we reduce transmissions to extend the lifetime of the network [2].

* Research partially supported by NSERC (Canada) and DAAD (Germany).

D. Pfoser et al. (Eds.): SSTD 2011, LNCS 6849, pp. 279–296, 2011.
© Springer-Verlag Berlin Heidelberg 2011

Table 1. Running example of probabilistic sensor readings

Sensor-ID	Location	Timestamp	Probability that temperature exceeds $27°$
s_1	Room 101	11:40	0.2
s_2	Room 102	11:40	0.8
s_3	Room 103	11:40	0.7
s_4	Room 203	11:40	0.4
s_1	Room 101	11:50	0.0
s_2	Room 102	11:50	0.8
s_3	Room 103	11:50	1.0
s_4	Room 203	11:50	0.4

This paper addresses count queries in WSNs. Such type of queries are very useful for many applications, for example, if we want to control the climate of a building complex using a number of sensors monitoring the current temperature at different locations within the building. We could, for example, turn on the heating if *exactly k*, *at most k* or *at least k* sensors measure a temperature below (or above) a specific threshold temperature. Traditional count queries simply count the number of sensors that satisfy the given query predicate and return the value of the counter. In fact, in-network aggregation can be applied easily to optimize the query's energy cost [3]. However, adding uncertainty raises new issues for the processing of the query itself as well as for the aggregation strategies that can be applied. Instead of simply counting sensors fulfilling the query predicate, now, for each sensor, we have to consider the probability that it satisfies the query predicate. Thus, instead of counting the number of sensors that satisfy the query, we now have a probability value that a specific number of sensors satisfies the query. It can intuitively be understood that the result of such a count query is in fact a probability distribution.

Consider the example in Table 1 which we will use as a running example throughout the paper. Let $S = \{s_1, s_2, s_3, s_4\}$ be a WSN with four sensors monitoring a building. Sensors s_1, s_2 and s_3 are installed on the first floor and s_4 is placed on the second floor. They measure the temperature and send their data periodically (for example every 10 min.). Each sensor reading contains a given timestamp as well as the probability that it satisfies a query Q, e.g. *"Temperature exceeds $27°$"*. Table 1 shows example readings of the four sensors in such an application. Examples for probabilistic count queries on this table could be "What is the probability that exactly (at least/at most) two sensors satisfy the query?" As our main contribution in this paper we investigate four algorithms to answer these types of queries within WSNs with the ultimate goal of minimizing the energy cost of processing such queries. We show that using an incremental and distributed algorithm we can reduce, by up to 80%, the number of messages transferred within the WSN.

The remainder of this paper is organized as follows. Next, we briefly discuss related work. A formal description of our data model and query computation is presented in Section 3. In Section 4 we develop four algorithms for solving the stated problem of continuous probabilistic count queries in WSNs, and experimentally evaluate them in Section 5. Our main findings and directions for further work are discussed in Section 6.

2   Related Work

Due to the steady increase in the number of application domains where uncertain data arises naturally, such as data integration, information extraction, sensor networks, persuasive computing etc., modelling and querying probabilistic, uncertain, incomplete, and/or fuzzy data in database systems is a fast growing research direction. Previous work has spanned a range of issues from theoretical development of data models and data languages [4], to practical implementation issues such as indexing techniques, e.g. [5,6] and probabilistic similarity query techniques [7]. In this paper we adopt the uncertainty model proposed in [8], where uncertainty is attached to tuples instead of individual attributes.

In [9] Ross et. al. addressed the problem of answering probabilistic count queries. However, they proposed a solution that requires the individual consideration of each possible world. This implies a computational cost that is exponential in the number of sensor nodes. The problem of answering probabilistic count queries is related to the problem of answering probabilistic top-k queries, e.g., [10,11,12,13,7]. In order to determine the rank of an uncertain tuple t, the number of tuples which have a score higher than t needs to be counted. In this work, we will generalize efficient techniques used to solve the probabilistic ranking of uncertain objects in the context of databases in order to apply these techniques to answer probabilistic count queries on sensor networks where usually completely different parameters need to be optimized.

WSNs are studied in their various aspects with work ranging from optimization [14], over practical implementation issues [15], to experimental analysis [16]. However most previous work considers data to be certain. To the best of our knowledge Wang et al. are the only ones who studied the field of probabilistic queries in a distributed system such as a WSN [17]. They address the problem of answering probabilistic top-k queries in WSN which is related but different to our problem. In addition to previous approaches, we introduce update strategies to handle the continuous stream of data that is produced by the sensor network.

3   Background

3.1   Probabilistic Data Model

Given a WSN $S = \{s_1, s_2, \ldots, s_n\}$ composed by a set of n sensors yielding n sensor values[1] at a given time t. For a given query Q, each sensor s_i has a probability $P_{s_{i,t}}^{Q}$ of satisfying Q's predicate at time t. Consider the temperature monitoring application of our first example. Each tuple in Table 1 consists of a

---

[1] In the remainder we use the term sensor to refer to the corresponding sensor value.

Table 2. Possible Worlds of Table 1

Possible World $W_{k,j}$	Probability $P(W_{k,j})$
$W_{4,1} = \{s_1, s_2, s_3, s_4\}$	0.2*0.8*0.7*0.4 = 0.0448
$W_{3,1} = \{s_1, s_2, s_3\}$	0.2*0.8*0.7*(1-0.4) = 0.0672
$W_{3,2} = \{s_1, s_2, s_4\}$	0.2*0.8*(1-0.7)*0.4 = 0.0192
...	...
$W_{1,1} = \{s_1\}$	0.2*(1-0.8)*(1-0.7)*(1-0.4) = 0.0072
$W_{1,2} = \{s_2\}$	(1-0.2)*0.8*(1-0.7)*(1-0.4) = 0.1152
$W_{1,3} = \{s_3\}$	(1-0.2)*(1-0.8)*0.7*(1-0.4) = 0.0672
$W_{1,4} = \{s_4\}$	(1-0.2)*(1-0.8)*(1-0.7)*0.4 = 0.0192
$W_{0,1} = \emptyset$	(1-0.2)*(1-0.8)*(1-0.7)*(1-0.4) = 0.0288

sensor id, a location reading, a time stamp t, a temperature value and a probability value $P^Q_{s_{i,t}}$ indicating the likelihood that sensor s_i satisfies the predicate of a given query Q at time point t^2. Thereby we assume data vectors of two sensors $s_x \neq s_y$ to be mutually independent, i.e. each of the probability values assigned to the sensors is an independent Bernoulli random variable with $P(X_i = 1) = 1 - P(X_i = 0) = P^Q_{s_{i,t}}$.

For solving probabilistic queries on our uncertain sensor network model, we apply the possible worlds semantics model which was originally proposed by Kripke [18] for modal logics and is commonly used for representing knowledge with uncertainties. However, there have been different adaptations of the model for probabilistic databases [19], [4], [8]. Here, we use the model as proposed in [8], specifically, a possible world is a set of sensors satisfying Q associated with the probability that this world is true at a certain time t. In particular, we define $W_{k,j}$ as the jth world where exactly k sensors satisfy Q at time t. The probability $P(W_{k,j})$ of a possible world $W_{k,j}$ at time t is computed by multiplying $P^Q_{s_{i,t}}$ for each sensor $s_i \in W_{k,j}$. For our four entries in the Table 1 there are $2^4 = 16$ possible worlds in total. Table 2 displays a few of those possible worlds and their respective probabilities.

3.2   Probabilistic Count Query

Given the WSN with the set of uncertain sensors S and any query Q as described above and a count parameter k. The problem to be solved for a probabilistic count query is to compute the probability $P^t(k, S, Q)$ that exactly k sensors in the network S satisfy Q at time point t. We call $P(k, S, Q)$ *probabilistic count*. Instead of a single probabilistic count, a probabilistic count query returns the probability distribution of $P^t(k, S, Q)$ over k $(0 \leq k \leq |S|)$ called *count histogram*. To lighten the notation, we use P_{s_j} equivalent for $P^Q_{s_{j,t}}$ and $P(k, S, Q)$ equivalent for $P^t(k, S, Q)$, i.e., all probabilities and data are considered with respect to the same point in time.

---

^{2} The predicate of Q is irrelevant for our observations. In the following we assume a given Q as a base query on top of which we can later build our actual probabilistic count queries.

The distribution of the count probabilities depends on the probability values P_{s_i}. However, as we shall see in the discussion that follows, the computation of the distribution does not depend on the actual probability values.

A naive method to answer a probabilistic count query is to first enumerate all possible worlds of S and then sum up the probabilities of those worlds where exactly k tuples appear, that is:

$$P(k, S, Q) = \sum_j P(W_{k,j}) \tag{1}$$

Considering Table 1 and a query Q at $t = 11{:}40$, we can intuitively compute the probabilities of all possible worlds and then sum up the possible worlds that contain the same amount of sensors by using Equation 1. Figure 1 shows the resulting count histogram. Since the number of possible worlds is exponential in the number of sensors this naive method is not very efficient.

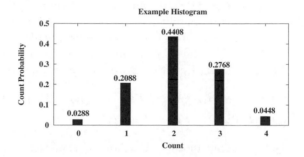

Fig. 1. Probability Count Histogram for Example 1

In the following, we show how to efficiently compute a probabilistic count query in general, before we address the problem of answering *continuous* probabilistic count queries in *wireless sensor networks* in Section 4. In [1], Hua et al. have proposed an efficient algorithm based on the *Poisson Binomial Recurrence* to avoid searching all possible worlds in a different setting. In fact, the problem description and definitions given there can be adjusted to our problem.

3.3   Poisson Binomial Recurrence

Let S be a set of sensors and the order of the sensors $s_j \in S$ is irrelevant. Every sensor s_j satisfies Q with a probability P_{s_j}. The probabilistic count $P(k, S, Q)$ is the probability that k $(0 \le k \le |S|)$ sensors in S satisfy Q's predicate. Moreover, the subset probability $P(k, S_j, Q)$ is the probability that k sensors in the subset $S_j = S \setminus s_j$ fulfill Q. Then, the probabilistic count $P(k, S, Q)$ depends only on the $(k - 1)$ other tuples in the subset S_j. That is, k sensors in S satisfy Q only when s_j satisfies Q and at the same time $k - 1$ sensors in S_j satisfy Q or when s_j does not satisfy Q and at the same time k sensors in S_j satisfy Q. If we define base cases $P(0, \emptyset, Q) = 1$, $P(j, \emptyset, Q) = 0$ for $(0 \leq j \leq |S|)$ and $P(-1, S, Q) = 0$ for any set of sensors, then

$$P(j, S, Q) = P(j - 1, S_j, Q)P_{s_j} + P(j, S_j, Q)(1 - P_{s_j}) \tag{2}$$

Equation 2 can be iteratively applied to compute the result of a probabilistic count query. In fact, in [20] (Sec. 1.7) this is called the *Poisson Binomial Recurrence*, and it is shown that: $P(0, S, Q) = P(0, S_j, Q)(1 - P_{s_j}) = \prod_{j=1}^{k}(1 - P_{s_j})$, and $P(j, S, Q) = P(j - 1, S_j, Q)P_{s_j} + P(j, S_j, Q)(1 - P_{s_j})$ for $0 < j \leq |S|$.

In each iteration we can omit the computation of any $P(j, S, Q)$ where $j \geq k$, since we are not interested in any counts other than k and do not need them to process the query. In total, for each $0 \leq j < k$ and each of the n sensors $s_i \in S$, $P(j, S, Q)$ has to be computed resulting in $O(k \cdot n)$ time complexity.

Example 1. Consider the sensor readings of our example application in Section 1 with a query Q at time $t = 11 : 40$ and $k = 2$. Assuming that all tuples satisfy Q's predicate with the probabilities listed in Table 1, we have $P_{s_1} = 0.2$, $P_{s_2} = 0.8$, $P_{s_3} = 0.7$ and $P_{s_4} = 0.4$. Note that the order in which we process the sensor readings is irrelevant. For the sake of simplicity, we process the table line by line starting at the top. Using Equation 2 we have $S = \{s_1\}$, $S_1 = \emptyset$ and

$$P(0, S, Q) = P(-1, S_1, Q)P_{s_1} + P(0, S_1, Q)(1 - P_{s_1}) = 0.8$$
$$P(1, S, Q) = P(0, S_1, Q)P_{s_1} + P(1, S_1, Q)(1 - P_{s_1}) = 0.2$$

Next we process s_2 by adding another iteration of Equation (2). With $S = \{s_1, s_2\}$ and $S_2 = \{s_1\}$ we obtain the following:

$$P(0, S, Q) = P(-1, S_2, Q)P_{s_2} + P(0, S_2, Q)(1 - P_{s_2}) = 0.16,$$
$$P(1, S, Q) = P(0, S_2, Q)P_{s_2} + P(1, S_2, Q)(1 - P_{s_2}) = 0.68,$$
$$P(2, S, Q) = P(1, S_2, Q)P_{s_2} + P(2, S_2, Q)(1 - P_{s_2}) = 0.16.$$

With s_3 and $S = \{s_1, s_2, s_3\}$ we have: $P(0, S, Q) = 0.048$, $P(1, S, Q) = 0.316$ and $P(2, S, Q) = 0.524$. Since $k = 2$ we can stop at that point and do not need to compute $P(3, S, Q)$. Finally, we add s_4 ($S = \{s_1, s_2, s_3, s_4\}$): $P(0, S, Q) = 0.0288$, $P(1, S, Q) = 0.2088$ and $P(2, S, Q) = 0.4408$, and we can return the result of our query: $P(2, S, Q) = 0.4408$.

As illustrated in Example 1, Equation 2 can be used to efficiently compute the probabilistic count $P(k, S, Q)$ that k sensors satisfy a query Q. Hence, we can easily compute the probability, that *at most* or *at least* k sensors satisfy Q's predicate. To compute the probability that *at most* k sensors satisfy a query Q, we intuitively sum up all probabilistic counts $P(j, S, Q)$ with $0 \leq j \leq k$, i.e.,

$$P_-(k, S, Q) = \sum_{j=0}^{k} P(j, S, Q) \tag{3}$$

To compute the probability that *at least* k sensors satisfy Q it is useful to know that the values of a complete count histogram sum up to 1. Hence, we can compute the probability of *at least* k sensors satisfying Q by using Equation 3 as follows:

$$P_+(k, S, Q) = 1 - \sum_{j=0}^{k-1} P(j, S, Q) \tag{4}$$

3.4   Allowing Certainty

Until now, we assumed all sensors to be uncertain at all times $(0 < P_{s_j} < 1)$. But certain circumstances or queries call for the existence of certain values. On one hand, sensors can die or wake up from one round to another, messages can get lost, or there can be restrictions to the query, i.e., we only want to query sensors in a bounded area. Thus, particular sensors do not participate in a query at all, and we can safely set their probability P_{s_j} to *zero*. On the other hand there could be reasons to set the probability P_{s_j} to *one*, e.g., if a sensor's samplings are very stable over a sufficient period of time. In the following, we explore the effects of $P_{s_j} = 0$ and $P_{s_j} = 1$.

Effect of $P_{s_j} = 0$. Reconsider Example 1. We now add a fifth sensor s_5 with $P_{s_5} = 0$ and observe the effect of a *zero probability*. After processing the four sensors in Example 1 we had $P(0, S_4, Q) = 0.048$, $P(1, S_4, Q) = 0.316$ and $P(2, S_4, Q) = 0.524$. We now incorporate P_{s_5} and compute the probability that exactly two sensors satisfy the query predicate (c.f. Equation 2): $P(2, S, Q) = P(1, S_4, Q)P_{s_4} + P(2, S_4, Q)(1 - P_{s_4}) = 0.316 \cdot 0 + 0.524 \cdot 1 = 0.524 = P(2, S_4, Q)$. The incorporation of a value $P_{s_j} = 0$ does not affect $P(j, S, Q)$ $(0 \le j \le k)$ at all. The following lemma formalizes this observation.

Lemma 1. *Let* $0 \le k \le n, P_{s_j} = 0$. *It holds that*

$$\forall k : P(k, S, Q) = P(k, S \setminus s_j, Q).$$

Proof. Using Equation 2 we obtain:

$$P(k, S, Q) = P(k - 1, S \setminus s_j, Q) \cdot 0 + P(k, S \setminus s_j, Q)(1 - 0) = P(k, S \setminus s_j, Q).$$

Thus, Lemma 1 allows us to ignore any sensor s_j with $P_{s_j} = 0$ in the computation of the Poisson binomial recurrence.

Effect of $P_{s_j} = 1$. Again, we use the example of the previous section, but now assuming $P_{s_5} = 1$. The probability that exactly two sensors satisfy the query predicate is derived by using Equation 2: $P(2, S, Q) = P(1, S_4, Q)P_{s_4} + P(2, S_4, Q)(1 - P_{s_4}) = 0.316 \cdot 1 + 0.524 \cdot 0 = 0.316 = P(1, S_4, Q)$.

The incorporation of a value $P_{s_j} = 1$ shifts all values in the Poisson Binomial Recurrence to the right, formalized by the following lemma:

Lemma 2. *Let $0 \le k \le n, P_{s_j} = 1$. It holds that*

$$\forall k : P(k, S, Q) = P(k - 1, S \setminus s_j, Q).$$

Proof. Using Equation 2 we obtain:

$$P(k, S, Q) = P(k - 1, S \setminus s_j, Q) \cdot 1 + P(k, S \setminus s_j, Q)(1 - 1) = P(k - 1, S \setminus s_j, Q).$$

Lemma 2 allows us to avoid iterations of the Poisson binomial recurrence for each sensor s_j with $P_{s_j} = 1$. Instead we can use a counter which is incremented by one for each such s_j, thus, counting the number of positions that the Poisson Binomial Recurrence has to be shifted to the right.

In summary, Lemma 1 and Lemma 2 allow us to handle zero and one values in a very efficient way.

4   Probabilistic Count Queries in Wireless Sensor Networks

As mentioned in Section 1, it is crucial for applications on WSNs to reduce energy cost, and this is mainly achieved through reducing CPU costs and communication. In the previous section showed how to reduce the CPU costs for computing the count distribution. In this section, we focus on reducing the communication costs. For that purpose, we must consider the typical underlying characteristics of WSNs such as network topology, routing and scheduling. We will propose four algorithms which solve the problem of answering continuous count queries in a WSN. Thus, we now take the local distribution of data as well as the temporal dimension into consideration and reinstate the notation $P^t(k, S, Q)$. We assume that the nodes in S are connected together via a logical tree where the sink node (or base-station) is the tree's root. The choice of the tree's topology does matter, but is outside the scope of this paper. For the sake of simplicity, we assume it to be a hop-based shortest path tree commonly used in other works, e.g., [3].

4.1   A Centralized Algorithm

In the centralized approach all probability values are sent to the root node at every round without previous processing. Thus, the sink node can be seen as a central database that receives and temporarily stores the readings of all sensor nodes within the network and that centrally processes queries on the WSN. The probabilistic count histogram can then be easily computed by using Equation 2.

While leaf nodes send only one value to their parent node, intermediate nodes send their own value plus all values received from their child nodes. Thus, the payload size of the packages that are sent within the network increases as we get closer to the root node. With $0 \le P_{s_j} \le 1$ for all sensors and unlimited payload size for any message sent within the network, $n - 1$ messages are sent to the root in every round. However, in reality, the number of messages is likely to be much larger, depending on the topology and the fixed payload size.

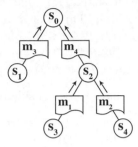

Fig. 2. Example Network Structure

With the knowledge that we gained in Section 3.4, aggregation strategies can be applied whenever a sensor has a *zero probability* or a *one probability*. According to Lemma 1, all sensors s_j with $P_{s_j} = 0$ can safely be ignored in the computation. For sensors s_j with $P_{s_j} = 1$, Lemma 2 applies and we use a counter variable C_1^t that denotes the number of such tuples. Thus, Equation 2 is only required for sensors s_j for which $0 < P_{s_j} < 1$, allowing us to have shorter messages, thus saving energy in packet transmission.

Example 2. Consider the sensors in Example 1 having a network structure as shown in Figure 2. Sensors s_i are depicted as circles, m_i denotes a message sent from one node to the other with i indicating the ordering among messages. In the first round ($t = 11{:}40$) all sensors report probability values. At time $t = 11{:}50$, s_1 has *zero probability* and therefore no message is sent at all. Whereas sensor s_3 is certain to satisfy Q with probability $P_{s_3} = 1$. Instead of sending a value, s_3 increments its counter ($C_1^t = 1$) and sends the counter only. Sensor s_4 sends its value as usual and sensor s_2 forwards the two probability values as well as the counter to the sink. With two iterations of Equation 2 we obtain for $t = 11{:}50$: $P^t(0, S, Q) = 0.12$, $P^t(1, S, Q) = 0.56$, $P^t(2, S, Q) = 0.32$ and $C_1^t = 1$.

4.2   A Centralized Incremental Algorithm

The naive solution works well if the probability values P_{s_j} change often from one round t to the next round $t + 1$. However, this is expensive in terms of messages sent when only a few values change. In some cases it is more efficient to send all probability values and process the query as proposed in Section 4.1 only in the initial computation. In all subsequent rounds, only sensors that have changed their probability will send an update. The sink node will then compute the new count probability incrementally as described next.

We can update $P^t(k, S, Q)$ using the results of previous iterations. Let s_x be a sensor and let $P_{s_x,t}^Q$ and $P_{s_x,t+1}^Q$ denote the previous and new probability that s_x satisfies Q, respectively. Our update algorithm has two phases, summarized next:

- **Phase 1:** We remove the effect that $P^Q_{s_x,t}$ had on the previous probabilistic counts $P^t(j, S, Q)$, $0 \le j \le k$. This yields an intermediate result of the probabilistic counts $\hat{P}^t(j, S, Q)$, $0 \le j \le k$.
- **Phase 2:** We incorporate the new probability $P^Q_{s_x,t+1}$ by adding it to the temporary probabilistic counts $\hat{P}^t(j, S, Q)$, $0 \le j \le k$ using Equation (2).

In Phase 1, the following cases have to be considered:

Case 1: $P^Q_{s_x,t} = 0$. In this case, nothing has to be done to remove the effect of $P^Q_{s_x,t} = 0$ and $\hat{P}^t(j, S, Q) = P^t(j, S, Q)$.

Case 2: $P^Q_{s_x,t} = 1$. When $P^Q_{s_x,t} = 1$ we must decrement the counter C^t_1 by one. Thus, $\hat{P}^t(j, S, Q) = P^t(j, S, Q)$ and $\hat{C}^t_1 = C^t_1 - 1$.

Case 3: $0 < P^Q_{s_x,t} < 1$. To remove the effect of any probability $P^Q_{s_x,t}$ from all $P^t(j, S, Q)$, $(0 \le j \le k)$ we look at its incorporation via Equation 2:
$$P^t(j, S, Q) = \hat{P}^t(j-1, S_x, Q)P^Q_{s_x,t} + \hat{P}^t(j, S_x, Q)(1 - P^Q_{s_x,t}).$$

We can remove the effect of $P^Q_{s_x,t}$ by resolving Equation 2 as follows:

$$\hat{P}^t(j, S, Q) = \frac{P^t(j, S, Q) - \hat{P}^t(j-1, S, Q) \cdot P^Q_{s_x,t}}{1 - P^Q_{s_x,t}} \qquad (5)$$

Since any $P^t(-1, S, Q) = 0$ we have:

$$\hat{P}^t(0, S, Q) = \frac{P^t(0, S, Q)}{1 - P^Q_{s_x,t}} \qquad (6)$$

for $j = 0$ and can step by step compute $\hat{P}^t(j, S, Q)$ by using $\hat{P}^t(j-1, S, Q)$ and Equation (5) for any $0 < j \le k$.

In Phase 2 we have to consider the same cases as in Phase 1:

Case 1: $P^Q_{s_x,t+1} = 0$ has no influece on the result at time $t + 1$ and $P^{t+1}(j, S, Q) = \hat{P}^t(j, S, Q)$.

Case 2: $P^Q_{s_x,t+1} = 1$. When $P^Q_{s_x,t} = 1$ we must increment the counter C^t_1 by one. Thus, $\hat{P}^t(j, S, Q) = P^t(j, S, Q)$ and $\hat{C}^t_1 = C^t_1 + 1$.

Case 3: $0 < P^Q_{s_x,t+1} < 1$. We can incorporate the new probability $P^Q_{s_x,t+1}$ by an additional iteration of Equation (2):
$$P^{t+1}(j, S, Q) = \hat{P}^t(j-1, S, Q)P^Q_{s_x,t+1} + \hat{P}^t(j, S, Q)(1 - P^Q_{s_x,t+1}).$$

This means that, whenever a sensor sends an update, it has to send both its previous probability and its new probability. Thus, in the worst case we send twice the number of values compared to the centralized algorithm. On the other hand, the less updates are sent, the better the results for the incremental approach.

Regarding the computational complexity, the following holds for both, Phase 1 and Phase 2: Case 1 and 2 have a cost of $O(1)$ since either nothing has to be done, or C^t_1 has to be incremented or decremented. Case 3 has a total cost of $O(k)$ leading to a total execution time of $O(k)$ per old-new value pair in the root node.

Example 3. Reconsider Example 1 where at time $t = 11{:}40$ $P^Q_{s_1,t} = 0.2$, $P^Q_{s_2,t} = 0.8$, $P^Q_{s_3,t} = 0.7$ and $P^Q_{s_4,t} = 0.4$. At time $t + 1 = 11{:}50$ only s_1 and s_3 change their values: $P^Q_{s_1,t+1} = 0.0$ and $P^Q_{s_3,t+1} = 1.0$.

We start with $P^t(0, S, Q) = 0.0288$, $P^t(1, S, Q) = 0.2088$, $P^t(2, S, Q) = 0.4408$ and $C^t_1 = 0$ and remove the effect of $P^Q_{s_1,t}$ by using Equations 5 and 6:

$$\hat{P}^t(0, S, Q) = \frac{P^t(0, S, Q)}{1 - P^Q_{s_1,t}} = 0.036$$

$$\hat{P}^t(1, S, Q) = \frac{P^t(1, S, Q) - \hat{P}^t(0, S, Q) \cdot P^Q_{s_1,t}}{1 - P^Q_{s_1,t}} = 0.252$$

$$\hat{P}^t(2, S, Q) = \frac{P^t(2, S, Q) - \hat{P}^t(0, S, Q) \cdot P^Q_{s_1,t}}{1 - P^Q_{s_1,t}} = 0.488$$

Next we incorporate the new probability of s_1 but notice that $P^Q_{s_1,t+1} = 0$, so Phase 2 can be skipped. We go on with removing the effect of $P^Q_{s_3,t}$ and obtain:

$$\hat{P}^t(0, S, Q) = \frac{P^t(0, S, Q)}{1 - P^Q_{s_1,t}} = 0.12$$

$$\hat{P}^t(1, S, Q) = \frac{P^t(1, S, Q) - \hat{P}^t(0, S, Q) \cdot P^Q_{s_1,t}}{1 - P^Q_{s_1,t}} = 0.56$$

$$\hat{P}^t(2, S, Q) = \frac{P^t(2, S, Q) - \hat{P}^t(0, S, Q) \cdot P^Q_{s_1,t}}{1 - P^Q_{s_1,t}} = 0.32$$

Since $P^Q_{s_3,t+1} = 1$ we only need to increment the counter $C^t_1 = 1$.

4.3   An In-Network Algorithm

In both previous algorithms the sink computes the histogram in a centralized manner. Despite the counter, there is no aggregation strategy applied. But we can benefit from computing the histogram at intermediate nodes as we send the values up to the root node. On the one hand, we can decrease the number of messages sent. On the other hand, intermediate count histograms could be used to query subtrees or apply early stopping conditions if a subtree satisfies the query [1].

Like in the centralized algorithm every sensor sends its value in every round. The idea is that every intermediate node s_j computes the probability histogram of its subtree by pairwise multiplying the probability histograms of its child nodes and its own probability $P^Q_{s_j}$ on the fly. *Zero probabilities* and *one probabilities* are processed as usual. As we only need the first k count probabilities to answer

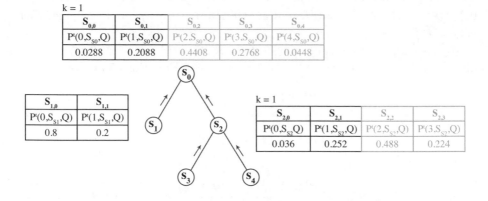

Fig. 3. Multiplication of two Count Histograms

a query, it is sufficient to compute only the first k count probabilities and then forward them. Hence, a maximum number of $k + 1$ values per message is sent at every hop up to the sink and, compared to the centralized approach, we can gain as the maximum payload size gets smaller.

Each probability P_{s_i} can be modelled as a minimal probabilistic count histogram with $P^t(0, S, Q) = 1 - P_{s_i}$ and $P^t(1, S, Q) = P_{s_i}$. To obtain the count histogram in an intermediate sensor node, we can use Equation 2 as long as all child nodes are leaf nodes. The more general task of merging two count histograms where $k > 1$ can be solved by multiplying them *component-wise*.

Consider Figure 3 where s_1, s_2 are intermediate nodes and send their count histograms to root node s_0. We can model their probabilistic distribution (shown in the tables besides the nodes) as polynomials, e.g., for node s_1 we would have $S_1(x) = S_{1,0} + S_{1,1}x + S_{1,2}x^2 + \cdots + S_{1,n}x^n$, where each $S_{1,j} = P^t(j, S, Q)$ and a similarly defined polynomial $S_2(x)$ for node s_2. Merging the probabilistic counts of those two nodes into the root node s_0 becomes a matter of simply computing:

$$S_0(x) = S_1(x) \times S_2(x) = S_{0,0} + S_{0,1}x + S_{0,2}x^2 + \cdots + S_{0,n+m}x^{2n+m},$$

from where we can obtain each probabilistic count at s_0, i.e., $P^t(j, S, Q) = S_{0,j}$

In every node, we can stop the computation at k, because we only need the first k probabilistic counts to answer our queries. It suffices to implement a simple (component-wise) polynomial multiplication algorithm taking $O(k^2)$ time.

Example 4. Consider the wireless sensor network in Table 1 at time $t = 11{:}40$ and the topology of Figure 2. Let $k = 1$. Sensors s_1, s_3 and s_4 are leaf nodes and send their values to their parent node. s_2 is an intermediate node and computes the intermediate probabilistic counts for $k = 0$ and $k = 1$. This means s_2 needs to merge $S_3(x) = 0.3 + 0.7x$ and $S_4(x) = 0.6 + 0.4x$ by multiplying them pairwise and then incorporate its own probability value $P_{s_2} = 0.8$. Again the order is irrelevant. Here we only have two polynomials $S_3(x)$ and $S_4(x)$ resulting in $S_3(x) \times S_4(x) = 0.18 + 0.54x$. Next, we incorporate P_{s_2} and get our final intermediate histogram with $S_2(x) = 0.036 + 0.252x$. Sensor s_2 forwards the probabilistic counts to the root. The sink finally multiplies $S_1(x) = 0.8 + 0.2x$ and $S_2(x) = 0.036 + 0.252x$ to compute the probabilistic counts, i.e., the coefficients of resulting polynomial, of the whole tree $S_0(x) = 0.0288 + 0.2088x$.

4.4   An Incremental In-Network Algorithm

Next we present an algorithm that brings together both, the in-network aggregation of Section 4.3 and the incremental update strategy of Section 4.2. To enable updates of intermediate count probabilities, every intermediate node has to save its current count histogram as well as the count histograms of its child nodes along with their associated id as unique key. In the first round $t = 0$ we process the query as proposed in Section 4.3 but store all required values in intermediate sensors. In all subsequent rounds only updates are reported and processed as described in the following. We start with the updating process for child count histograms.

There are two types of updates: either a number of old-new value pairs, or a whole updated function. Whenever an intermediate node receives an update, it either updates the child polynomial as described in Section 4.2 or it replaces the whole function with the new function. When $\hat{P}^t(0, S, Q) = 1$ and $\hat{P}^t(j, S, Q) = 0$ for all $(0 \leq j \leq k)$ after removing an old probability value and at the same time no new probability value is sent (sensor either changed its probability to zero or one), we remove the whole count histogram[3]. After processing the updates of the child nodes the intermediate node needs to update its own polynomial. Again we need to consider two cases: We can either merge all updated child polynomials analog to Section 4.2 or use the old-new value pairs of all child nodes to compute the new probabilistic count function. Obviously, we have to multiply all child polynomials as soon as the histogram of a child node was replaced by a new one.

For the number of messages the following applies: If we start with sending updates in the manner of the incremental centralized algorithm and continue sending only old-new value pairs, the number of messages equals the number of messages sent in approach 4.2. Thus, the number of values increases as we come closer to the root note. However, we want to make sure that analog to the in-network approach, the maximum number of values per message is fixed to $k + 1$. This means we send the whole polynomial as soon as the number of old-new value pairs exceeds $\frac{k}{2}$.

With this strategy, we reduce the number of messages if only few sensors report an update. In addition, we ensure that in the worst case (all sensors update) the number of messages does not surpass the number of messages sent in the in-network approach.

Example 5. We process the initialization analog to Example 4 but store all relevant data structures. At time $t + 1 = 11:50$ only s_1 and s_3 change their values: $P^Q_{s_1, t+1} = 0.0$ and $P^Q_{s_3, t+1} = 1.0$. Sensor s_3 sends an update message with its

---

[3] This usually happens only when leaf nodes are updated.

unique key, old value and in this case no new value but its incremented counter $C_1^t = 1$ to sensor s_2. Sensor s_2 uses the key to find the matching count histogram and updates it by using Equations 5 and 6: $\hat{P}^t(0, S, Q) = \frac{P^t(0,S,Q)}{1-P_{s_1,t}^Q} = 1$ and $\hat{P}^t(1, S, Q) = \frac{P^t(1,S,Q)-\hat{P}^t(0,S,Q)\cdot P_{s_1,t}^Q}{1-P_{s_1,t}^Q} = 0$.

Since all values but the first one of s_3's count histogram are 0 and the $\hat{P}^t(1, S, Q)$ is 1, we delete its entry in s_2 and increment the counter of s_2 by the value of the sent counter. With s_4 and s_2 keeping their values we only need to update the resulting count histogram in s_2 to complete the update. As no other child node sent an update with a completely new count histogram, we also use Equations 5 and 6 to update the resulting probabilistic counts: $\hat{P}^t(0, S, Q) = \frac{P^t(0,S,Q)}{1-P_{s_1,t}^Q} = 0.12$ and $\hat{P}^t(1, S, Q) = \frac{P^t(1,S,Q)-\hat{P}^t(0,S,Q)\cdot P_{s_1,t}^Q}{1-P_{s_1,t}^Q} = 0.56$.

As the number of updated values exceeds the threshold value $(1 > \frac{1}{2})$, we forward the whole polynomial to the intermediate node. The update of sensor s_1 is processed analog to the update of s_3 in s_2 and the entry of s_1 is deleted in s_0. Now, sensor s_2 sends its update with the entire count histogram. Sensor s_0 replaces the histogram and increments its counter. To update the histogram in s_0, we now need to merge the child polynomials as described in Section 4.3. But as there is only one entry, we do not need to perform a polynomial multiplication and have our final result.

5   Performance Evaluation

We performed our experiments by varying five parameters: the number of sensors within the network (n), the percentage of uncertain sensors (γ), the probability that a sensors probability value changes from one round to the next round (δ), the probabilistic count (k), and the message size measured in bytes (m). Table 3 shows the values used for those parameters.

The positions of the sensors were randomly chosen within a 100m × 100m area and each sensor node was assumed to have a fixed wireless radio range of 30m. As mentioned in Section 3, the actual distribution of the probability values is not relevant for the query computation in terms of messages sent – hence not a relevant parameter for our performance evaluation. But for the sake of completeness the probability values follow a normal distribution N(0.5,0.5). Results are based on an average of 10 simulation runs whereas each run consists of 100 time stamps. All generated instances of the WSNs used a hop-wise shortest-path tree as the routing topology. We assume in all experiments that messages are delivered using a multi-hop setup. Since the query is only sent once from the root to all child nodes and will be amortized over time, we only measure nodes-to-root messages.

Every coefficient was taken into account with 8 bytes, counters as well as ids were taken into account with 4 bytes each. For the sake of simplicity, we assumed data packages with a header size of 0 bytes.

Table 3. Parameter Values Used in the Performance Evaluation (Default Values Printed in Bold Face)

Parameter	Values
n (Number of Sensors)	100, 500, **1000**, 2500
γ (Ratio of Uncertain Sensors)	25%, 50%, **100%**
δ (Probability of Change)	25%, **50%**, 75%, 100%
k (Number of Coefficients)	1, 5, **10**, 25
m (Message Size in bytes)	64, **128**, 256

In the following figures the arithmetic mean is plotted with upper (lower) error bar denoting the overall best (worst) performance. We abbreviate the centralized algorithm with "Central", the centralized incremental algorithm with "IncCentral", the in-network algorithm with "InNet" and the incremental in-network algorithm with "IncInNet".

5.1   Experiments and Results

The results of our experiments are summarized in Figures 4 to 8. In each experiment all parameters but the one in focus are fixed to their default value.

The foremost trend that we can see in Figure 4 is that IncInNet consistently sends less messages than all others. Both in-network algorithms further improve as the network grows bigger. Particularly in the case of IncInNet the improvement is significant. The reason for this is that the number of values sent per message is restricted to $k + 1$ which is 11 with k set to a default of 10. Thus, with a default maximum payload size of 128 bytes there is no need to send more than one message per sensor. This also explains, why for the InNet approach best case and worst case coincide with the arithmetic mean (no error bars). Since for the InNet algorithm every sensor sends exactly one message per round to its parent node, independent of their respective topology, always $n - 1$ messages are sent[4]. As expected, when n increases, the costs for all algorithms rise as well, however, both in-network algorithms grow slower than the centralized algorithms. Overall, InNet and IncInNet offers better scalability with IncInNet being the overall best solution saving up to 80% of the communication cost compared to Central.

Figure 5 illustrates how a counter as introduced in Section 4.1 affects the performance of the algorithms. Note that for our experiments, the number of certain sensors was equally split into *one* and *zero probabilities*. A counter is taken into account with constantly 4 bytes per message. The larger the number of uncertain sensors the better perform the in-network algorithms. Since in every round different nodes have zero probabilities the error bars become wider.

Varying δ (Figure 6) creates a scenario that allows observing how the dynamics of the observed probabilities affect the algorithms' performance. Since Central and InNet do not mind updates but always compute the count probabilities from

---

[4] This explanation also applies to the plots that follow.

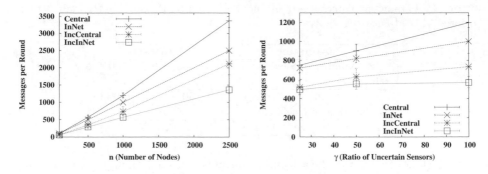

Fig. 4. Effect of n Fig. 5. Effect of γ

scratch, varying δ does not affect them at all. However, for IncCentral and IncIn-
Net naturally applies that the more dynamic the observed values, the more updates will be required. In essence, increasing δ creates more communication traffic
in the tree. It is interesting to note that Central outperforms IncCentral for high
values ($\delta \geq 75\%$) while IncInNet and InNets performance results even up.

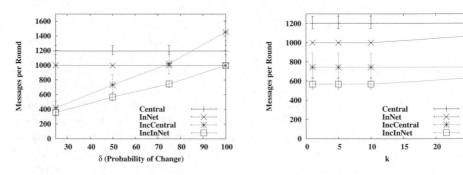

Fig. 6. Effect of δ Fig. 7. Effect of k

Figure 7 shows the results of the experiments conducted with varying k. As
expected, k only has an influence on InNet and IncInNet. As k increases, the
performance of both algorithms slowly decreases. Since we chose k relatively
small, varying k does not seem to have a significant effect on the performance at
all. Nonetheless, in the extreme case the centralized algorithms outperform the
in-network algorithms for $k = n$. This is because InNet as well as IncInNet send
$k + 1$ values resulting in $n + 1$ values (IncInNet additionally also sends a value
to identify the sender).

Following [16] we chose a (payload) message size of 128 bytes. For the following
experiment, m varies within a range of 64 bytes and 256 bytes. Obviously, the
smaller the size of a single message, the more messages need to be sent. Figure 8
illustrates the result of our experiments. For Central and IncCentral we therefore
observe a steady decreasing number of sent messages. InNet and IncInNet also

record a higher number of messages for m =64 bytes. However, the number of messages is constant for m =128 bytes and m =256 bytes, indicating that the total number of bytes never exceeds 128 bytes. Thus, InNet sends $n-1$ messages in total. As mentioned in Section 4.1 with unlimited payload size this also applies to Central. In general, with increasing payload size Central draws near to InNet and IncCentral draws near to IncInNet until they finally concide ($m = \infty$).

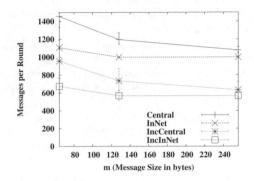

Fig. 8. Effect of m

6   Conclusions

In this paper, we studied the problem of answering continuous probabilistic count queries in wireless sensor networks. After formalizing the problem, we first discussed how to reduce the CPU overhead for processing the query and we then proposed four different algorithms aiming at minimizing the communication cost. All algorithms were examined empirically in a performance evaluation. The results show that the incremental in-network algorithm is the overall best solution, in particular when we have a small message size, a small probability of change, a high number of nodes, a high ratio of uncertain nodes and a small count.

To the best of our knowledge, this paper is the first one that addresses continuous probabilistic count queries in wireless sensor networks. In this paper we assumed probabilities of two sensors to be mutually independent. One future goal should be finding solutions to cope with more complex uncertainty models where readings of two different sensors can be correlated. Another interesting project would be the transfer of our algorithms to different topologies. By simulating our algorithm on different tree structures we could gain further insight into the strengths and weaknesses of the algorithms. Finally, intermediate count histograms could be used to query subtrees or apply early stopping conditions if a subtree satisfies the query. Future work could address those topics and thereby further reduce transmission costs.

References

1. Hua, M., et al.: Ranking queries on uncertain data: a probabilistic threshold approach. In: Proc. of ACM SIGMOD, pp. 673–686 (2008)
2. Schurgers, C., et al.: Optimizing sensor networks in the energy-latency-density design space. IEEE TMC 1, 70–80 (2002)
3. Madden, S., et al.: Tag: a tiny aggregation service for ad-hoc sensor networks. SIGOPS Operating Systems Review 36, 131–146 (2002)
4. Dalvi, N., Suciu, D.: Efficient query evaluation on probabilistic databases. The VLDB Journal 16, 523–544 (2007)
5. Cheng, R., et al.: Efficient indexing methods for probabilistic threshold queries over uncertain data. In: Proc. of VDLB, pp. 876–887 (2004)
6. Kriegel, H.P., et al.: Probabilistic similarity join on uncertain data. In: Li Lee, M., Tan, K.-L., Wuwongse, V. (eds.) DASFAA 2006. LNCS, vol. 3882, pp. 295–309. Springer, Heidelberg (2006)
7. Bernecker, T., et al.: Scalable probabilistic similarity ranking in uncertain databases. IEEE TKDE 22(9), 1234–1246 (2010)
8. Sarma, A., et al.: Working models for uncertain data. In: Proc. of IEEE ICDE, pp. 7–7 (2006)
9. Ross, R., Subrahmanian, V.S., Grant, J.: Aggregate operators in probabilistic databases. J. ACM 52, 54–101 (2005)
10. Soliman, M.A., Ilyas, I.F., Chang, K.C.C.: Top-k query processing in uncertain databases. In: Proc. of IEEE ICDE, pp. 896–905 (2007)
11. Yi, K., et al.: Efficient processing of top-k queries in uncertain databases. In: Proc. of IEEE ICDE, pp. 1406–1408 (2008)
12. Cormode, G., Li, F., Yi, K.: Semantics of ranking queries for probabilistic data and expected ranks. In: Proc. of IEEE ICDE, pp. 305–316 (2009)
13. Li, J., Saha, B., Deshpande, A.: A unified approach to ranking in probabilistic databases. Proc. of VLDB 2, 502–513 (2009)
14. Malhotra, B., Nascimento, M.A., Nikolaidis, I.: Exact top-k queries in wireless sensor networks. IEEE TKDE (2010) (to appear)
15. Ye, W., Heidemann, J., Estrin, D.: An energy-efficient mac protocol for wireless sensor networks. In: Proc. of IEEE INFOCOM, pp. 1567–1576 (2002)
16. Pinedo-Frausto, E., Garcia-Macias, J.: An experimental analysis of zigbee networks. In: 33rd IEEE Conference on Local Computer Networks, LCN 2008, pp. 723–729 (2008)
17. Wang, S., Wang, G., Gao, X., Tan, Z.: Frequent items computation over uncertain wireless sensor network. In: Proc. of ICHIS, pp. 223–228 (2009)
18. Kripke, S.A.: Semantical analysis of modal logic i normal modal propositional calculi. Mathematical Logic Quaterly 9, 67–96 (1963)
19. Antova, L., Koch, C., Olteanu, D.: 10 worlds and beyond: efficient representation and processing of incomplete information. The VLDB Journal 18, 1021–1040 (2009)
20. Lange, K.: Numerical analysis for statisticians. Springer, Heidelberg (1999)

Geodetic Point-In-Polygon Query Processing in Oracle Spatial

Ying Hu, Siva Ravada, and Richard Anderson

One Oracle Drive, Nashua, NH, 03062, USA
{Ying.Hu,Siva.Ravada,Richard.Anderson}@oracle.com

Abstract. As Global Positioning Systems (GPSs) are increasingly ubiquitous, spatial database systems are also encountering increasing use of location or point data, which is often expressed in geodetic coordinates: longitude and latitude. A simple but very important question regarding this data is whether the locations lie within a given region. This is normally called the point-in-polygon (PIP) problem. Within the Geodetic space, PIP queries have additional challenges that are not present in the Cartesian space. In this paper, we discuss several techniques implemented in Oracle Spatial to speed up geodetic PIP query processing. Our experiments utilizing real-world data sets demonstrate the PIP query performance can be significantly improved using these new techniques.

Keywords: Point-In-Polygon, Geodetic Data, Spatial Databases, R-tree.

1 Introduction

To better approximate the Earth as an ellipsoid, many spatial database systems, including IBM Informix Geodetic Datablade [3], Microsoft SQL server 2008 [8], Oracle Spatial [9], and PostGIS [12], currently support geography or geodetic data types. When compared with projected geometry data types that treat the Earth as a flat surface, geography or geodetic data types are useful to represent the curved surface of the Earth, and they can give more accurate results for a large area such as North America, Europe or the entire world as a single entity. Geodetic coordinates are commonly expressed in the form of longitude and latitude. The latest version of the World Geodetic System (WGS 84) standard is widely used in many applications. For example, WGS 84 is the reference coordinate system used by the Global Positioning System (GPS).

As GPS is now widely prevalent, many spatial database systems see increasing use of point data as expressed in the longitude and latitude format. For example, an insurance company stores the location of its insured properties in its database, and needs to quickly determine which properties might be affected by a natural disaster such as a hurricane. Normally, the area affected by the disaster is specified as a polygon. Another example is a truck company that can track the location of its trucks periodically, store the truck location as point data in its database, and determine how many of them lie inside a given region, such as New York City. Thus, the point-in-polygon (PIP)

D. Pfoser et al. (Eds.): SSTD 2011, LNCS 6849, pp. 297–312, 2011.

problem not only is one of fundamental computational geometry problems [10], but also arises naturally in many geospatial or Location-Based Services (LBS) applications [7]. Furthermore, because PIP in geodetic space has additional challenges that are not present in the Cartesian space, new techniques are needed to meet these challenges. This paper reports our experience of implementing several techniques in Oracle Spatial to speed up geodetic PIP query processing.

The rest of this paper is organized as follows. Section 2 reviews prior art. Section 3 discusses new techniques used in Oracle Spatial to speed up geodetic PIP queries. Section 4 presents results of an experimental study using real-world data sets. Section 5 concludes the paper.

2 Prior Art

A well-known solution to the PIP problem [7, 10] is to draw a ray or line segment from the candidate point (e.g. the location of a property or a truck) to a point known to be outside the query polygon, and count the number of intersections between this ray and the edges of the query polygon (e.g. for either a natural disaster or New York City). If the number of intersections is odd, the point is inside; otherwise, it is outside. In a two-dimensional flat plane, the ray is normally either horizontal or vertical to simplify computations. When geodetic coordinates are used, the ray can be obtained by selecting a point that is outside the polygon and connecting the two points along a geodesic[1]. Thus, the ray is not necessarily horizontal or vertical for the geodetic case.

The ray-crossing algorithm described above works fine when there are a small number of query points. However, two factors can cause performance problems for this ray-crossing algorithm: (i) number of query points; and (ii) number of vertices in the query polygon. When there are millions of query points, spatial indexes have to be built to reduce the search scope. This is the two-step query processing: (1) the filter step is to use spatial indexes and returns a candidate set; (2) the refinement step is a test on exact geometries in the candidate set [2]. For example, the ray-crossing algorithm can be used in the refinement step to determine if a point is inside a polygon. In commercial database systems, two classes of spatial indexes—Quad-tree indexes and R-tree indexes are supported. Microsoft SQL server supports multiple-level Quad-tree indexes [4], and Oracle Spatial supports both Quad-tree indexes and R-tree indexes [6]. Because Quad-tree indexes need more fine-tuning to set an appropriate tessellation level, and they can be used to index only 2D non-geodetic geometries in Oracle Spatial, we will focus on the use of R-tree indexes in this paper. To use R-tree indexes to manage geodetic geometries, Oracle Spatial converts geodetic coordinates to 3D Earth-centered coordinates, and builds 3D minimal bounding boxes (MBBs) on them. Consequently, while 2D geodetic geometries have two dimensions (longitude, latitude), R-tree indexes built on them are 3D (Geocentric 3D) in Oracle Spatial.

To further reduce the candidate set returned from the filter step, previous work [5, 1] discusses how interior approximations for a non-geodetic query polygon help achieve the goal: a non-geodetic query polygon is tessellated, and if a point or an

---

[1] A geodesic is the shortest path between two points on the ellipsoid.

MBR is inside an interior tile, this point or MBR is also inside the query polygon. For geodetic geometries, Voronoi tessellations [3] can be used to speed up geodetic computations. However, users have to carefully select appropriate Voronoi tessellations for their data sets, as some Voronoi tessellations are good for some data sets, yet degrade query performance on other data sets. To help users avoid the difficulties of choosing right tessellations, this paper reports new techniques, based on an in-memory R-tree structure, which are designed and implemented in Oracle Spatial. This in-memory R-tree structure not only requires no tuning from users, but also significantly improves the performance of many geodetic queries, including geodetic PIP queries that are the focus of this paper.

3 Geodetic PIP Query Processing

In the refinement step of spatial index-based query processing, the problem with the ray-crossing algorithm is that for each candidate point, every line segment in a query polygon has to be checked once. If the number of vertices in the query polygons is large, this can lead to performance degradation. If we build a hierarchy structure (e.g. a tree structure) on the query polygon, we can quickly determine which line segments are intersected. We choose an in-memory R-tree structure, because it is versatile to approximate different shaped geometries. For example, this in-memory R-tree structure not only can handle 3D MBBs for geodetic geometries, which is the focus of our paper, but also can handle 2D MBRs for non-geodetic geometries, which we will discuss in the Appendix.

Fig. 1 shows how a line segment on the boundary of Great Britain is used to build a 3D MBB when Great Britain is used as a query polygon. Since there are 53 line segments in the simplified version of Great Britain, 53 MBBs are obtained, and an in-memory R-tree structure can be built on top of these 53 MBBs. Note that because the in-memory R-tree structure is built once per query polygon, the cost of building it is amortized over a large number of data points. Furthermore, this new in-memory R-tree structure is used not only in the refinement step, but also in the filter step, as discussed in Section 3.1.

3.1 New R-Tree Index Query Processing

In the original two-step R-tree index query processing method [2], the filter step only uses MBRs (or MBBs in 3D[2]) to quickly return a candidate set. In other words, both query polygons and data geometries are approximated by MBBs, and R-tree index nodes are built on top of their children (MBBs). Therefore, only MBBs are used in the filter step to determine the candidate set. However, for a large point data set and a relatively large query polygon, we find many index nodes that can be either inside or outside the query polygon. If we can use the above in-memory R-tree structure to

---

[2] In the rest of this paper except the Appendix, we will only consider MBBs because Oracle Spatial geodetic R-tree indexes are 3D.

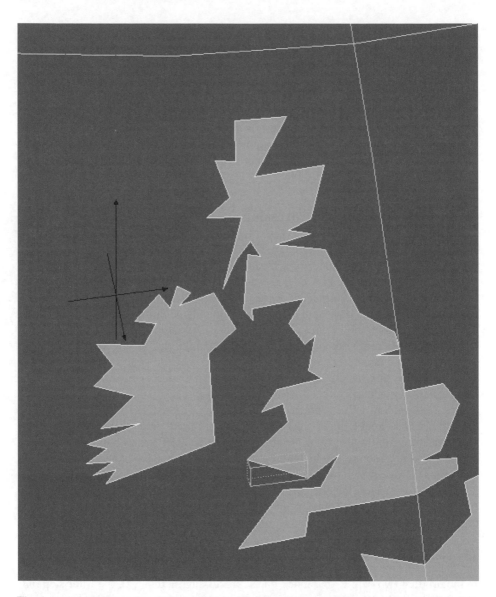

Fig. 1. A simplified version of Great Britain: The pink line is the Prime Meridian and the white line has a latitude 60° north. Three axes in black have their origins at the center of the Earth. The 3D MBB (in green) is obtained from a line segment of Great Britain. Some green lines are dotted because they are under the Earth's surface.

quickly determine the topological relationship between these index nodes and the query polygon in the filter step, we don't need to wait until the refinement step to determine if a point is inside a polygon. For example, if we know that an index node is inside (or outside) a query polygon, all its descendant data points are inside (or outside) a query polygon. However, because in an index node, its MBB can cover at most 8 regions in the Earth, we have to make sure all these regions are either inside or outside the query polygon (See Fig. 2). Note that this is different from the 2D Cartesian case, where an index node corresponds an MBR, i.e. only one region. We will discuss the geodetic case in more detail in Section 3.2. In summary, the query polygon is used to check if R-tree index nodes are inside the polygon as early as possible.

The new R-tree index based PIP query processing is described in the following pseudo code. Note that it is tailored to make PIP query processing clearer. For example, other topological relationship determination (such as TOUCH, etc) [11] and other data geometries (e.g. lines, etc) are purposely not described. As R-tree index leaf nodes already contain the information about data points, we just use these index leaf nodes to determine if they are inside a query polygon or not. Therefore, although the filter step and the refinement step are well separated in theory, the two steps are seamlessly integrated in Oracle Spatial to improve query performances.

```
Algorithm PIP R-tree Index Query Processing
1.   build an In-memory R-tree structure (IR) of the
     query polygon (Q)
2.   push the R-tree index root into stack
3.   while (stack is not empty)
4.   {
5.     pop an R-tree index entry (RE) from a stack;
6.     if (RE is not from an R-tree index leaf node)
7.     {
8.        if (RE's inclusion flag is set)
9.        {
10.         for each child entry of RE,
11.         {
12.           set its inclusion flag;
13.           push it into stack;
14.         }
15.       }
16.       else
17.       {
18.         determine the topological relationship
            between RE and Q using IR;
19.         if (RE is inside Q)
20.         {
21.           for each child entry of RE,
22.           {
23.             set its inclusion flag;
24.             push it into stack;
25.           }
26.         }
27.         else if (RE intersects with Q)
28.           for each child entry of RE,
29.             push it into stack;
30.         else /* RE is outside Q */
31.           continue;
32.       }
33.     }
34.     else  /* an entry in a leaf node */
35.     {
```

```
36.        if (RE's inclusion flag is set)
37.           put this data point into the result set;
38.        else
39.        {
40.           determine the topological relationship
              between RE and Q using IR;
41.           if (RE is inside Q)
42.              put this data point into the result set;
43.           else
44.              continue;
45.        }
46.     }
47. }
```

Note that the above algorithm works for both geodetic and non-geodetic geometries. We will discuss the non-geodetic case in the Appendix.

3.2 Determining Topological Relationship between R-Tree Index Entry and Query Polygon

To determine the topological relationship between an R-tree index entry (MBB) and a query polygon, we have a two-step procedure:

(1) Use the MBB to search the in-memory R-tree structure that is built for the query polygon, and check if there is any intersection between the MBB and the line segments of the query polygon. The search can be simply fulfilled using a normal in-memory R-tree search algorithm. Note that for a leaf entry in the in-memory R-tree structure, we not only use its MBB, but also use the line segment inside its MBB. For example, in Fig. 1 it is possible that the MBB is intersected while the line segment inside the MBB is not intersected at all. If there is no intersection, go to the next step.

(2) Compute how many regions this MBB can cover and, for each region, select an arbitrary point and check if the selected point is inside the query polygon. If all points are inside the query polygon, we are certain that any regions covered by this MBB are also inside the query polygon. If all points are outside the query polygon, we are certain that any regions covered by this MBB are also outside the query polygon. Otherwise, this MBB intersects with the query polygon.

Note that step (2) is executed only if there is no intersection from step (1). Because there is no intersection from step (1), we are certain that the points covered by the MBB are either inside or outside the query polygon. No points that are covered by the MBB can be on the query polygon. Otherwise, there must be one intersection between the MBB and the line segments of the query polygon. Furthermore, if a point in a region is inside or outside the query polygon, all points in this region will be inside or outside the query polygon. If some points in this region are inside the query polygon while others in this region are outside the query polygon, there must be a point in this region that is on the query polygon because this point has to cross the boundary of the query polygon. This is in contradiction with the fact that step (2) is executed only if there is no intersection from step (1). Therefore, we can select an arbitrary point from each region, check if the selected point is inside the query polygon, and determine if the whole region is inside the query polygon. But it is possible that one region is completely inside the query window while another region is completely outside the query window, if the two regions are not connected. Thus, this case can pass step (1), and it is determined in step (2) that this MBB intersects with the query polygon.

We will discuss in Section 3.3 how to use the in-memory R-tree structure to determine if a point is inside a query polygon, while in this section we focus on the question: "Why can an MBB cover at most 8 regions?" For example, Fig. 2 shows that an MBB (in green) covers 4 regions on the Western Hemisphere and we assume there are another 4 similar regions on the Eastern Hemisphere. Note that in Fig. 2 we only draw the portion of the MBB that is above the Earth's surface, and the portion under the Earth's surface is not shown. For example, the portion of the top straight line that connects point A (-120°, 30°) and point B (-60°, 30°) is not drawn because it is under the Earth's surface. If we have the Northern Hemisphere as a query polygon, the MBB in Fig. 2 does not intersect with any line segments of the query polygon, i.e. the Equator. So this MBB passed step (1). In step (2), we select one point from each of 8 regions. It is easy to see there are 4 points inside the Northern Hemisphere, and another 4 points outside the Northern Hemisphere. So this MBB intersects with the Northern Hemisphere.

Although an MBB can cover at most 8 regions on the Earth like the MBB in Fig. 2, most MBBs typically cover only 1 region on the Earth. To determine how many regions are covered by an MBB, we can label 8 corner points in a way as shown in Fig. 3. Since we can associate a region with each corner point, we call them region-0 through region-7. Thus, the problem is reduced to determining how many regions are connected. For example, if only region-0 and region-1 are connected, and other regions are not connected, there will be 7 regions. We can run breadth-first search (BFS) twice: (a) Forward BFS: {0} -> {0, 1, 2, 4} -> {0, 1, 2, 4, 3, 5, 6} -> {0, 1, 2, 4, 3, 5, 6, 7} and (b) Reverse BFS: {7} -> {7, 6, 5, 3} -> {7, 6, 5, 3, 4, 2, 1} -> {7, 6, 5, 3, 4, 2, 1, 0}. In the Forward BFS, if region-0 and region-1 are connected, we set region-1 to "0". Thus if region-7 is connected to region-3, then connected to region-1, and region-0, region-7 is also set to "0". So once the Forward BFS is done, every region is set to the lowest value from the Forward BFS. The Reverse BFS deals with the case where region-4 is connected to region-0, because they are connected by region-1 and region-5, not because of the direct link between region-0 and region-4. So once the Reverse BFS is done, every region is set to the lowest value from all possible connections. Finally, we just count different values from the 8 regions, and obtain how many regions are covered by this MBB.

3.3 Geodetic Point-In-Polygon Methods

For geodetic coordinates, as we already discussed in Section 2, a ray or line segment could be obtained by selecting another point that is outside the polygon, and connecting the two points along a geodesic. Note that we cannot simply use horizontal or vertical lines, such as lines of latitude or longitude. For example, assume that we have the tropics as a query polygon and we would like to determine if a point on the Equator is inside the tropics. If we draw a horizontal ray, i.e. the Equator, this ray will not intersect with any line segments of the tropics (i.e. the Tropic of Cancer or

Fig. 2. An MBB (in green) covers 4 regions (in pink) on the Western Hemisphere. Assume there are another 4 similar regions on the Eastern Hemisphere. If we have the Northern Hemisphere as a query polygon, the MBB does not intersect with the Equator, i.e. the boundary of the query polygon. However, there are some points (such as points A and B) inside the Northern Hemisphere and other points (such as points C and D) outside the Northern Hemisphere.

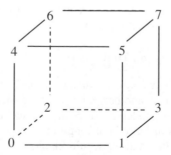

Fig. 3. 8 corner points are labeled in an MBB

the Tropic of Capricorn). Therefore, we have to use a different method. For example, we can use the first point of the query polygon, draw a line segment by connecting it with the point in question along a geodesic, use this line segment to search the in-memory R-tree structure we already built for the query polygon to see how many intersections occur between this line segment and line segments of the query polygon, and determine if the point in question is inside the query polygon. This is similar to the ray-crossing algorithm; i.e., one intersection means one boundary crossing. However, we still have to determine how the line segment ends with the first point of the query polygon. In other words, we need to determine whether this line segment gets to the first point of the query polygon from outside the query polygon, or inside the query polygon. This can be easily accomplished, because in Oracle Spatial, coordinates are defined sequentially around the polygon (counterclockwise for an exterior polygon ring, clockwise for an interior polygon ring). So the sequential order can be used to determine if a point is inside or outside a query polygon.

However, if the first point of the query polygon is very far from the point in question, the line segment connecting them along a geodesic can be very long so that it can have many intersections with line segments of the query polygon and 3D MBBs on top of these line segments. To avoid this problem, we can also use the R-tree Nearest Neighbor (NN) search method to get a point on the query polygon that is the closest to the point in question. And we get an open line segment that connects the point in question and its closest point on the query polygon (along a geodesic). Because they are the nearest neighbors, there is no intersection between this open line segment and line segments of the query polygon. We just need to determine whether this open line segment is inside the query polygon or not. This is also done in the same way as how we determine the line segment gets to the first point of the query polygon (i.e. either from inside the query polygon, or from outside the query polygon). If the open line segment is inside (or outside) the query polygon, the point in question is also inside (or outside) the query polygon. In summary, the NN-based PIP method can be proved in the following theorem.

Theorem 1. If a point is inside (or outside) a polygon, the open line segment that connects the point and its closest point on the polygon (along a geodesic for the geodetic case) is also inside (or outside) a polygon.

Proof Sketch: Assume that a point (P1) is inside a polygon and an open line segment connects this point (P1) and its closest point (P2) on the polygon (along a geodesic for the geodetic case). If there is a point (P3) in the open line segment that is outside the polygon, the open line segment must intersect with the polygon somewhere (say point (P4)) between the two points (P1) and (P3). It is obvious that P4 is on the polygon, and P4 is closer to P1 than P2. This is in contradiction with the fact that P2 is P1's closest point on the polygon. Therefore, every point on the open line segment must be inside the polygon. And the same technique can be applied for the case that a point is outside a polygon.

Note the above theorem also works for the non-geodetic case, if we ignore the geodesic and simply connect the point and its closest point on the polygon along a straight line, which is the shortest path in 2D Cartesian space. We will discuss the non-geodetic case in the Appendix.

3.4 Memory Usage

Oracle Spatial is a spatial database system not only applicable for a small group of users, but also for a large number of enterprise users. To make it highly scalable, each user query needs to consume as little memory as possible. To limit the memory usage associated with the in-memory R-tree structure, we group several line segments into a single entry to construct the in-memory R-tree structure. For example, assume that a query polygon has 50,000 line segments. We can take 10 continuous line segments, and put them into one entry. Thus, there will be only 5,000 entries vs. 50,000 entries originally. Since the number of entries is reduced, the total memory usage for the in-memory R-tree structure is also reduced by almost 90%. In Section 4, we show that the performance does not degrade much when we set smaller limits of entries.

4   Experiments

Three real-world data sets are used for our experimental study. The first data set is from an energy company; it has about 39,000 locations for its properties in the United States, and a large region about 4 times as big as Texas is used as the query polygon. Note that this region is a multi-polygon with many voids and it has 197,146 line segments. The second real-world data set is from a transportation system; it has 3 million locations and 30 regions around world, such as "Canada/Mountain", "America/Honolulu", and "Germany". The 30 regions have an average of approximately 59,000 line segments, and "Norway" has the most line segments (343,395). For each region we run a PIP query to determine how many locations are in this region, and we report the total execution time. The third real-world data set is the US Business Area (ABI) data set consisting over 10 million locations. For this data set, we use two sets of query polygons: a) 50 US states, and b) 1061 local regions, such as "Manhattan NY", and report the total execution time for each of the two sets of query polygons. The 50 US states have an average of 1,755 line segments, and "Alaska" has the most line segments (18,603). The 1061 local regions have an average of 520 line segments, and "Long Island" has the most line segments (6,915). The areas of the 1061 local regions range from 5.83 km^2 to 78,200 km^2, with an average of 4,330 km^2. In Section 4.1, we present performance results of the three data sets under different optimization configurations, in Section 4.2, we show that performance results with lower limits of entries in the in-memory R-tree structure (such as 4,096 entries) are almost as good as those without a limit, and in Section 4.3, we compare two geodetic PIP methods, which use the in-memory R-tree.

4.1   Compare Geodetic PIP with Different Configurations

In this subsection, we compare results from the following configurations: (A) the filter step uses MBBs only and the refinement step uses the original ray-crossing algorithm. (B) The filter step uses MBBs only and the refinement step uses the new geodetic PIP algorithm based on the in-memory R-tree structure, which is discussed in Section 3.3. (C) Both the refinement step and the filter step use the in-memory R-tree structure. Note that in Configuration (C), the in-memory R-tree structure is used for not only index leaf entries, but also other index entries, as shown in the pseudo code of Section 3.1.

Table 1. Geodetic PIP query execution times under different optimization configurations.

	The first data set	The second data set	The third data set with US states	The third data set with local regions
Configuration (A)	204s	20883s	2559s	11685s
Configuration (B)	10.8s	349s	277s	282s
Configuration (C)	1.4s	88s	52s	81s

Table 1 shows query execution times from three data sets. It is clear that when we replace the ray-crossing algorithm with the new algorithm based on the in-memory R-tree structure in the refinement step, query performance improves about 20 (204/10.8) times as fast as before in the first data set, 60 (20883/349) times as before in the second data set, 9.2 (2559/277) times as before in the third data set with US states as query polygons, and 41 (11685/282) times as before in the third data set with local regions as query polygons. And when we also use the in-memory R-tree structure in the filter step, query performance improves further: 7.7 (10.8/1.4) times in the first data, 4.0 (349/88) times in the second data set, 5.3 (277/52) times in the third data set with US states as query polygons, and 3.5 (282/81) times in the third data set with local regions.

4.2   Different Limits of Entries in In-Memory R-Tree Structure

Table 2 shows the query execution times when running the in-memory R-tree structure based algorithm with different limits of entries in the in-memory R-tree structure. It is clear that the results with lower limits (such as 4,096 entries) are almost as good as those without a limit. Note that when the limit of entries is 8,192, the query execution time for the first data set is actually better than that without a limit. The reason is that the first data set is small, and the benefit of building a big in-memory R-tree structure is offset by the cost of building it. For instance, if there's no maximum number of entries, it takes 0.44s to build an in-memory R-tree structure for the query polygon in the first data set, which has 197,146 entries because of 197,146 line segments. If the limit of entries is set to 8,192, it takes only 0.07s to build an in-memory R-tree structure. Furthermore, when using lower limits, the memory usage is also reduced significantly. For example, if there is no entry limit, a memory of 28M bytes is used for building the in-memory R-tree structure in the first data set and a max memory of 48M bytes is used for building the in-memory R-tree structure (for "Norway") in the second data set. If the entry limit is 8192, both data sets will use only 1.15M bytes to build their in-memory R-tree structures. So in practice, a lower limit (4096) is used by default because of benefits of lower memory usage.

Table 2. Geodetic PIP query execution times with different limits of entries.

Limit of entries	The first data set	The second data set	The third data set with US states	The third data set with local regions
32	28s	543s	170s	162s
64	18s	237s	124s	127s
128	10s	154s	90s	99s
256	5.7s	126s	70s	89s
512	3.85s	106s	61s	83s
1024	2.26s	95s	56s	82s
2048	1.81s	90s	55s	82s
4096	1.4s	88s	53s	82s
8192	1.22s	88s	52s	81s
NONE	1.4s	88s	52s	81s

4.3 Compare NN-Based PIP and Ray-Based PIP

To compare the NN-based geodetic PIP method with other geodetic PIP methods, we implement a simpler version of the ray-based geodetic PIP method: we assume the North Pole is an exterior point to query polygons and connect the North Pole and the point in question along a geodesic to construct a ray, and use this ray to search the in-memory R-tree to obtain the number of intersections between ray and the line segments of the query polygon. Note that this simple implementation is just used to compare the performances of different PIP methods, because we cannot always guarantee that the North Pole is an exterior point to query polygons. But since our query polygons in our previous experiments do not contain the North Pole, the simple implementation works for them. We use the default limit of entries (4096) to run this experiment. To make them clearer, we put the results for the default limit of entries (4096) from Table 2 into Table 3, and list them as "NN-Based". "Ray-based" is for results from the simple implementation of the Ray-based PIP method. It is clear that the NN-based geodetic PIP method is slightly better than the ray-based PIP method.

Table 3. Geodetic PIP query execution times with NN-based and ray-based PIP methods

	The first data set	The second data set	The third data set with US states	The third data set with local regions
NN-Based	1.4s	88s	53s	82s
Ray-Based	1.5s	90s	66s	96s

5 Conclusions

This paper presents new techniques implemented in Oracle Spatial to speed up geodetic point-in-polygon query processing. These new techniques not only require no user tuning, but also improve geodetic point-in-polygon query performances significantly.

Acknowledgments. We thank Chuck Murray for reviewing this paper.

References

1. Badawy, W.M., Aref, W.G.: On Local Heuristics to Speed Up Polygon-Polygon Intersection Tests. In: ACM-GIS 1999, pp. 97–102 (1999)
2. Brinkhoff, T., Horn, H., Kriegel, H.-P., Schneider, R.: A Storage and Access Architecture for Efficient Query Processing in Spatial Database Systems. In: Abel, D.J., Ooi, B.-C. (eds.) SSD 1993. LNCS, vol. 692, pp. 357–376. Springer, Heidelberg (1993)
3. IBM Informix, IBM Informix Geodetic DataBlade Module User's Guide Version 3.12 (2007)
4. Fang, Y., Friedman, M., Nair, G., Rys, M., Schmid, A.-E.: Spatial indexing in Microsoft SQL server 2008. In: SIGMOD Conference 2008, pp. 1207–1216 (2008)
5. Kothuri, R.K., Ravada, S.: Efficient Processing of Large Spatial Queries Using Interior Approximations. In: Jensen, C.S., Schneider, M., Seeger, B., Tsotras, V.J. (eds.) SSTD 2001. LNCS, vol. 2121, pp. 404–424. Springer, Heidelberg (2001)
6. Kothuri, R.K., Ravada, S., Abugov, D.: Quadtree and R-tree Indexes in Oracle Spatial: A Comparison Using GIS Data. In: SIGMOD Conference 2002, pp. 546–557 (2002)
7. Longley, P.A., Goodchild, M.F., Maguire, D.J., Rhind, D.W.: Geographic Information Systems and Science. John Wiley & Sons Ltd, West Sussex (2005)
8. Microsoft SQL Server, Spatial Indexing Overview, http://technet.microsoft.com/en-us/library/bb964712.aspx
9. Oracle Spatial, Oracle® Spatial Developer's Guide 11g Release 2 (11.2), Part Number E11830-07 (2010)
10. O'Rourke, J.: Computational Geometry in C. Cambridge University Press, Cambridge (1998)
11. Papadias, D., Theodoridis, Y., Sellis, T.K., Egenhofer, M.J.: Topological Relations in the World of Minimum Bounding Rectangles: A Study with R-trees. In: SIGMOD Conference 1995, pp. 92–103 (1995)
12. PostGIS Geography, http://postgis.refractions.net/docs/ch04.html#PostGIS_Geography

Appendix: Non-geodetic PIP Query Processing

In this Appendix, we discuss how our new techniques can be applied to non-geodetic geometries in a similar but simpler way to speed up non-geodetic PIP query processing in Oracle Spatial.

A. Algorithms

Fig. 4 shows that there are 14 line segments in a query polygon (in blue). Each line segment is used to get its MBR, which corresponds to an entry in a leaf node. An in-memory R-tree can be built on the 14 MBRs: in Fig. 4, 14 MBRs are with dashed lines, and 4 MBRs (leaf nodes) are with dotted lines. (Some lines are both dashed and dotted.) However, the root MBR is not shown, in order to make the 4 MBRs or leaf nodes clearer. Again, note that because the in-memory R-tree is built once per query polygon, the cost of building it is amortized over a large number of data points.

As we already discussed in Section 3.1, an R-tree index is built on a large point data set to speed up PIP query processing. The pseudo code in Section 3.1 works for both geodetic geometries and non-geodetic geometries. In the Appendix, we just focus on how to determine the topological relationship between an R-tree index entry and a query polygon. For example, Fig. 4 shows that a green MBR (A), which is from a disk-based R-tree index, is inside the query polygon. It is obvious that A's descendant nodes (MBRs and points in index leaf nodes) will be inside the query polygon.

To decide if an MBR is inside a query polygon, we use a two-step procedure, which is similar to but simpler than that described in Section 3.2:

(1) Search the in-memory R-tree to check if there is any intersection between the MBR and line segments of the query polygon. Note that although the MBR (A) intersects with the entry MBR (C) in Fig. 4, it does not intersect with any line segments in the query polygon. If there is any intersection, child MBRs of the MBR (A) will be fetched from disk or buffer cache for further processing recursively. Otherwise, go to the next step.

(2) Choose the right-top corner point of the MBR (A) and draw a horizontal line segment (B, shown in red) from this point to cross the maximum X value of the query polygon. Then this line segment (B) can be used to search this in-memory R-tree again and get how many line segments of the query polygon intersect with it. As the ray-crossing algorithm is already discussed, if the number of intersections is odd we are sure that the corner point is inside the query polygon, and thus the whole MBR and its descendants are also completely inside the query polygon. Otherwise, the MBR and its descendants can be skipped.

Note that in step (2), for an MBR we choose only one point (i.e. the right-top corner point of the MBR), because in 2D Cartesian space, all points in an MBR belong to a single region. This is different from the geodetic case where at most 8 points have to be selected from an MBB to determine if the MBB is inside the query polygon. And in step (2), we can also use the R-tree Nearest Neighbor (NN) search method to find out the point on the query polygon that is the closest to the right-top corner point of the MBR (A), and use Theorem 1 to determine if the right-top corner point of the MBR (A) is inside the query polygon. We find that using the ray or the line segment (such as the horizontal line segment (B)) to search the in-memory R-tree is better than the R-tree NN search method. This is different from the geodetic case, because the ray in the non-geodetic case is simply a degenerated MBR (i.e. without area) and can be directly used to search the in-memory R-tree.

The above two steps can also be applied to a point. If we ignore the case of point-on-polygon, we may skip the first step and use only the second step. In Oracle Spatial, the case of point-on-polygon is supported, so that we use the same steps to process a point. In other words, the first step can decide if a point is on, or touches, a polygon.

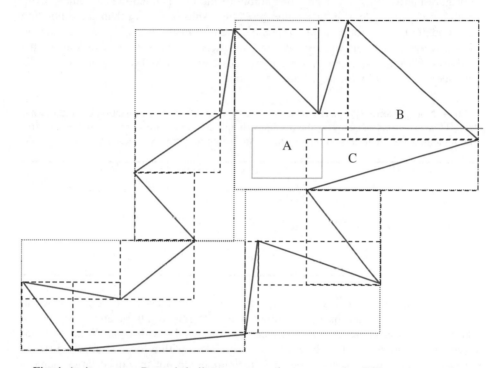

Fig. 4. An in-memory R-tree is built on a query polygon to speed up PIP query processing

B. Experiments

We use the same real-world data sets as described in Section 4 and run the Mercator projection to transform them to 2D projected (or non-geodetic) geometries. Since we have already seen in Section 4.1 that Configuration (A), where the filter step uses MBBs only and the refinement step uses the original ray-crossing algorithm, is significantly slower than other configurations, we replace Configuration (A) with Configuration (A'), where interior tile approximations are used in both the filter and refinement steps, and if in the refinement step interior tile approximations cannot determine a candidate geometry, then the original ray-crossing algorithm is used, as described in [5]. Configurations (B) and (C) are similar to those in Section 4.1 and the default PIP method is to use a ray to search the in-memory R-tree, as described in step (2) of Section A of this Appendix. We also have the R-tree NN search method to replace the ray-based R-tree search method in step (2) of Section A of this Appendix, and run the same queries under Configurations (B) and (C). These results of using the R-tree NN search method are put into the same column with parentheses.

Table 4 shows query execution times from three non-geodetic data sets. It is clear that our new techniques also work well for non-geodetic geometries, and they are significantly faster than interior tile approximations [5]. Although geodetic computation is more complex than non-geodetic computation, the geodetic results under Configuration (C) in Table 1 are comparable to the non-geodetic results under Configuration (C) in Table 4. In fact, some geodetic results are better than corresponding non-geodetic results, for example, the experiment with the second geodetic data set takes 88s while that with the second non-geodetic takes 190s. One of reasons is that MBBs in geodetic R-tree indexes are more clustered than MBRs in non-geodetic R-tree indexes, especially for large query polygons.

Table 4. Non-geodetic PIP query execution times under different optimization configurations. Under Configurations (B) and (C), the results without parentheses are from the ray-based R-tree search method and the results with parentheses are from the R-tree NN search method.

	The first data set	The second data set	The third data set with US states	The third data set with local regions
Configuration (A')	71s	7778s	935s	2158s
Configuration (B)	2.98s (13.41s)	226s (504s)	122s (333s)	130s (333s)
Configuration (C)	2.44s (2.55s)	190s (190s)	58s (62s)	64s (75s)

Table 5 shows the non-geodetic query execution times when running the in-memory R-tree structure based algorithm with different limits of entries in the in-memory R-tree structure. Again, it is clear that the results with lower limits (such as 4,096 entries) are as good as those without a limit. So with a lower limit of entries in the in-memory R-tree structure, our new techniques do not introduce large memory overhead for both geodetic and non-geodetic cases, and meanwhile they improve geodetic and non-geodetic PIP query performance significantly.

Table 5. Non-geodetic PIP query execution times with different limits of entries. The results without parentheses are from the ray-based R-tree search method and the results with parentheses are from the R-tree NN search method.

Limit of entries	The first data set	The second data set	The third data set with US states	The third data set with local regions
32	8.43s (16.79s)	240s (298s)	92s (120s)	80s (108s)
64	5.05s (10.49s)	213s (258s)	76s (97s)	69s (92s)
128	3.12s (6.96s)	201s (216s)	68s (80s)	68s (81s)
256	2.50s (4.50s)	195s (200s)	62s (71s)	64s (77s)
512	2.16s (3.20s)	191s (191s)	60s (66s)	64s (76s)
1024	2.09s (2.72s)	191s (191s)	59s (64s)	64s (76s)
2048	2.08s (2.42s)	191s (191s)	59s (63s)	64s (76s)
4096	2.03s (2.22s)	190s (190s)	58s (62s)	64s (75s)
8192	2.02s (2.16s)	190s (190s)	58s (62s)	64s (75s)
NONE	2.44s (2.55s)	190s (190s)	58s (62s)	64s (75s)

MSSQ: Manhattan Spatial Skyline Queries*

Wanbin Son, Seung-won Hwang, and Hee-Kap Ahn

Pohang University of Science and Technology, Korea
{mnbiny,swhwang,heekap}@postech.ac.kr

Abstract. Skyline queries have gained attention lately for supporting effective retrieval over massive spatial data. While efficient algorithms have been studied for spatial skyline queries using Euclidean distance, or, L_2 norm, these algorithms are (1) still quite computationally intensive and (2) unaware of the road constraints. Our goal is to develop a more efficient algorithm for L_1 norm, also known as Manhattan distance, which closely reflects road network distance for metro areas with well-connected road networks. Towards this goal, we present a simple and efficient algorithm which, given a set P of data points and a set Q of query points in the plane, returns the set of spatial skyline points in just $O(|P|\log|P|)$ time, assuming that $|Q| \leq |P|$. This is significantly lower in complexity than the best known method. In addition to efficiency and applicability, our proposed algorithm has another desirable property of independent computation and extensibility to L_∞ norm, which naturally invites parallelism and widens applicability. Our extensive empirical results suggest that our algorithm outperforms the state-of-the-art approaches by orders of magnitude.

1 Introduction

Skyline queries have gained attention [1,2,3,4,5] because of their ability to retrieve "desirable" objects that are no worse than any other object in the database. Recently, these queries have been applied to spatial data, as we illustrate with the example below:

Example 1. Consider a hotel search scenario for an SSTD conference trip to Minneapolis, where the user marks two locations of interest, e.g., the conference venue and an airport, as Figure 1 (a) illustrates. Given these two query locations, one option is to identify hotels that are close to both locations. When considering Euclidean distance, we can say that hotel H5, located in the middle of the two query points is more desirable than H4, i.e., H5 "dominates" H4. The goal of a spatial skyline is to narrow down the choice of hotels to a few desirable hotels that are not dominated by any other objects, i.e., no other object is closer to all the given query points simultaneously.

However, as Figure 1 (b) shows, considering these query and data points on the map, Euclidean distance, quantifying the length of the line segment between H5 and query points, does not consider the road constraints and thus severely underestimates the actual distance. For well-connected metro areas like Minneapolis, Manhattan distance,

---

* Work by Son and Ahn was supported by National IT Industry Promotion Agency (NIPA) under the program of Software Engineering Technologies Development and Experts Education. Work by Hwang was supported by Microsoft Research Asia.

D. Pfoser et al. (Eds.): SSTD 2011, LNCS 6849, pp. 313–329, 2011.

or L_1 norm, would be more reliable. Going back to Figure 1 (a), we can now assume the dotted lines represent the underlying road network and revisit the problem to identify desirable objects with respect to L_1 norm. In this new problem, H4 and H5 are equally desirable, as both are three blocks away from the conference venue and two blocks from the airport.

This problem has been actively studied for Euclidean distance [6,7,8,9] and the most efficient algorithm known so far has the complexity of $O(|P|(|S|\log|\mathcal{CH}(Q)|+\log|P|))$ [8,9] for the given set P of data points and set Q of query points in the plane. We denote set spatial skyline points as S, and the *convex hull* of Q as $\mathcal{CH}(Q)$.

One may think the above existing algorithms for L_2 could run faster for L_1, as there are such cases for other spatial algorithms. However, this is not the case if we follow the approach in [8,9]. It is not difficult to see that all the properties of skyline points for the Euclidean distance metric in [8,9] also hold for L_1 (or L_∞), based on which we can compute a "subset" of the spatial skylines by constructing

(a) the rectilinear convex hull of the queries (all data points lying in the convex hull are skylines), and
(b) the rectilinear Voronoi diagram of data points (if a Voronoi cell intersects the convex hull, its site is a skyline),

The above procedure takes $O(|Q|\log|Q|)$ time and $O(|P|\log|P|)$ time, respectively. However, Figure 2 shows that there are still some skyline points not belonging to the two cases above. For example, p_2 is a skyline, because none of the other points dominates it. Moreover, as its Voronoi cell (gray region) does not intersect $\mathcal{CH}(Q)$, it does not belong to the cases (a) or (b).

The above example shows that, we need not only to maintain the set of skyline points for cases (a) and (b), but also check whether the remaining data points are skylines or not. This takes $O(|P||S|\log|\mathcal{CH}(\mathcal{Q})|)$ time, which is exactly the same as the total time complexity required for L_2 norm.

In a clear contrast, we develop a simple and efficient algorithm that computes skylines in just $O(|P|\log|P|)$ time for L_1 metric, assuming $|Q| \leq |P|$. Our extensive

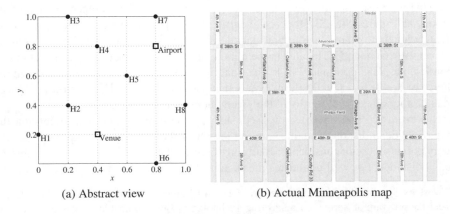

(a) Abstract view   (b) Actual Minneapolis map

Fig. 1. Hotel search scenario

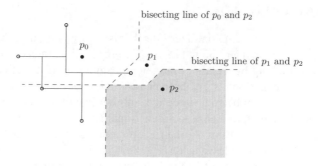

Fig. 2. The point p_2 is a skyline, which satisfies neither case (a) nor (b)

empirical results suggest that our algorithm outperforms the state-of-the-art algorithms in spatial and general skyline problems significantly, especially for large queries and high dimensional data. Our contributions can be summarized as follows:

- We study the Manhattan Spatial Skyline Queries (MSSQ) problem, which can apply to spatial queries in well-connected metro areas. We show a straight extension of the existing algorithm using L_2 metric is inefficient for our problem, then present a simple and efficient algorithm that computes skylines in just $O(|P| \log |P|)$ time.
- We show that our algorithm, by computing each skyline independently, can easily be parallelized. Our algorithm also straightforwardly extends for chebyshev distance, or L_∞ norm, used for spatial logistics in warehouses.
- We extensively evaluate our framework using synthetic data and show that our algorithms are faster by orders of magnitude than the current-state-of-the-art approaches.

2   Related Work

This section provides a brief survey of work related to (1) skyline query processing, and (2) spatial query processing.

2.1   Skyline Computation

Since early work on skyline queries [1,2] studied the maximal vector problem and applied to database applications, many algorithms and structures have been proposed. State-of-the-art approaches include using bitmaps or B+-trees [3], extending nearest neighbor algorithm [10], or developing branch and bound skyline (BBS) algorithm [4], sort-filter-skyline (SFS) algorithm leveraging pre-sorted lists [5], and linear elimination-sort for skyline (LESS) algorithm [11] for efficient skyline query processing. While the above efforts aims at enhancing efficiency, another line of work focuses on enhancing the quality of results [12,13,14], by narrowing down skyline results, often too large when dimensionality is high, using properties such as skyline frequency, k-dominant skylines, and k-representative skylines.

However, both lines of work address general skyline problems, and thus do not exploit spatial properties in spatial skyline problems.

2.2   Spatial Query Processing

The most extensively studied spatial query mechanism is ranking neighboring objects by the distance to a single query point [15,16,17]. For multiple query points, Papadias et al. [18] studied ranking by a class of monotone "aggregation" functions of the distances from multiple query points. As these nearest neighbor queries require a distance function, which is often cumbersome to define, another line of research studied skyline query semantics which do not require such functions.

For a spatial skyline query using L_2 metric, efficient algorithms have been proposed [8,9] and the best known complexity is $O(|P|(|S|\log|\mathcal{CH}(Q)| + \log|P|))$. To the best of our knowledge, our work is the first for L_1 metric and is significantly more efficient with complexity $O(|P|\log|P|)$. Meanwhile, extending existing L_2 algorithms for L_1 metric cannot reduce their complexity, as discussed in Section 1.

3   Problem Definition

In the spatial skyline query problem, we are given two point sets: a set P of data points and a set Q of query points in the plane. Distance function $d(p,q)$ returns the L_1 distance between a pair of points p and q, that is, the sum of the absolute differences of their coordinates.

Given this distance metric, our goal is to find the set of spatial skyline points. Our definitions are consistent with prior literatures [8], as we restate below.

Definition 1. *We say that p_1 spatially dominates p_2 if and only if $d(p_1, q) \leq d(p_2, q)$ for every $q \in Q$, and $d(p_1, q') < d(p_2, q')$ for some $q' \in Q$.*

Definition 2. *A point $p \in P$ is a* spatial skyline point *with respect to Q if and only if p is not spatially dominated by any other point of P.*

4   Observation

The basic idea of an algorithm for this problem is the following. To determine whether $p \in P$ is a skyline or not, existing approach under the Euclidean distance metric is to perform dominance tests with the current skylines (which we later discuss in details, denoted as baseline algorithm *PSQ*, in Section 6).

Under L_1 distance metric, we use a different approach in which we check the existence of a point that dominates p. To do this, we introduce another definition (below) on spatial dominance between two points which is equivalent to Definition 1. We denote by $C(p,q)$ the L_1 disk (its closure) centered at q with radius $d(p,q)$.

Definition 3. *We say that p_1 spatially dominates p_2 if and only if p_1 is always contained in $C(p_2, q)$ for every $q \in Q$, and is contained in the interior of $C(p_2, q')$ for some $q' \in Q$.*

Based on this new definition above, a trivial approach would be, for each data point p, to compute L_1 disks for every $q \in Q$, and check whether there is any data point satisfying the definition. However, this already takes $O(|Q|)$ time only for computing L_1 disks. Instead, we use some geometric properties of the L_1 disks $C(p, q)$ and compute the common intersection of L_1 disks in $O(\log |Q|)$ time for each data point p, and perform the dominance test efficiently. We denote by $R(p)$ the common intersection of $C(p, q)$ for every $q \in Q$. Note that p itself is always contained in $R(p)$ (in fact p is on the boundary of $R(p)$).

By Definitions 2 and 3, we have the following three cases for data points contained in $R(p)$:

(a) There is no data point in $R(p)$, other than p, or
(b) There is some data point p' in the interior of $R(p)$, or
(c) There is some data point p' in $R(p)$, other than p, but no data point in the interior of $R(p)$.

Case (a) obviously implies that p is a skyline point. For case (b), p' dominates p, and therefore p is not a skyline point. For case (c), if p' is contained in the interior of some L_1 disk $C(p, q)$ for a $q \in Q$, then p' dominates p, and therefore p is not a skyline point. If every data point in $R(p)$ lies on the boundary of $C(p, q)$ for all $q \in Q$, p is a skyline point.

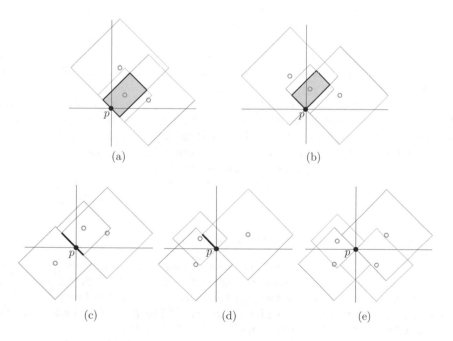

Fig. 3. The cases of common intersection $R(p)$ (gray) in the plane

5   Algorithm

In this section, we show how to handle each of the three cases efficiently so as to achieve an $O(\log |P|)$ time algorithm for determining whether a data point is a skyline point or not.

5.1   Data Structure

We first introduce data structures we build on P or Q, to support "range counting query" and "segment dragging query" efficiently. These two queries are the building blocks of our proposed algorithm.

Range counting is a fundamental problem in computational geometry and spatial databases. Given a set points P, we want to preprocess P such that, for the given query range R, it returns the number of points of P lying in R efficiently. Among the specific results, we implement the range counting structure proposed in [19], building a balanced binary tree for one dimension and storing the information for the other dimension as well. This structure on n points in the plane can be constructed in $O(n \log n)$ time, and the structure answers a range counting query in $O(\log n)$ time. We build this structure for both P and Q, which we denote as rCountP and rCountQ respectively. Note, rCountP can be built once offline, while rCountQ needs to be built at query time. (However, this cost is negligible, as we will empirically report in Section 7).

Segment dragging query, informally speaking, is to determine the next point "hit", by the given query line (or, segment) $\overline{st}$, when it is "dragged" along two rays. Given a set P of points and three orientations θ, ϕ_s, and ϕ_t, we want to preprocess P such that for any segment $\overline{st}$ parallel to θ, it returns the point hit by the segment $\overline{s't'}$ efficiently where s' slides along the ray from s in direction ϕ_s and t' slides along the ray from t in direction ϕ_t and $\overline{s't'}$ is parallel to θ. There are two types of queries, *parallel* and *dragging out of a corner*. When two rays are parallel, therefore ϕ_s and ϕ_t are parallel, such queries belong to *parallel* type. When the initial query segment $\overline{st}$ is a point, such queries belong to *dragging out of a corner* type. From [20,21], it was shown that, one can preprocess a set P of n points into a data structure of $O(n)$ size in $O(n \log n)$ time that answers a segment dragging query of "parallel" or "dragging out of a corner" type in $O(\log n)$ time. We build this structure on Q, which we denote as sDragQ.

5.2   Computing the Common Intersection $R(p)$

Because each $C(p,q)$ is a "diamond" (or, more formally, a square rotated by 45 degree) in the xy plane, $R(p)$ is obviously a rectangle with sides parallel to the lines $y = x$ or $y = -x$ (See Figure 3 (a).) Therefore, $R(p)$ is determined by at most four query points: when $R(p)$ is a point, it is the common intersection of four L_1 disks. Otherwise, it is determined by at most three query points.

Lemma 1. *There are at most four query points that determine the sides of $R(p)$. We can identify them in $O(\log |Q|)$ time after $O(|Q| \log |Q|)$ time preprocessing.*

Proof. We first show that the sides of $R(p)$ are determined by at most four query points.

Consider the subdivision of the plane into regions (quadrants) defined by the vertical line through p and the horizontal line through p. Then at least one of four quadrants contains some query points in its closure unless $Q = \phi$. Without loss of generality, we assume that the top right quadrant contains some query points. That is, the set $Q_1 = \{q \mid q.x \geq p.x, q.y \geq p.y, q \in Q\}$ is not empty. We also denote by Q_2, Q_3, and Q_4 the set of query points in the top left, the bottom left, and the bottom right quadrants, respectively. Note that a data point lying on the border of two quadrants belongs to both sets.

Consider the case that $Q_1 = Q$. Then the bottom left side of $R(p)$ is determined by p. The other three sides are determined by the three query points: the one with the smallest x-coordinate determines the bottom right side, the one with the smallest y-coordinate determines the top left side, and the one with the smallest L_1 distance from p determines the top right side of $R(p)$. Figure 3 (a) shows the case.

Consider now the case that $Q_2 \neq \phi$ and $Q = Q_1 \cup Q_2$ as shown in Figure 3 (b). In this case, p is the bottom corner of $R(p)$, and therefore the bottom left and the bottom right sides are determined by p. The top right side is determined either by the query point in Q_1 with the smallest L_1 distance from p or by the query point in Q_2 with the smallest y-coordinate. The top left side is determined by one of two such query points as above, after switching the role of Q_1 and Q_2.

If $Q_i \neq \phi$ for all $i = 1, 2, 3, 4$, $R(p)$ is a point that coincides with p as shown in Figure 3 (e).

Otherwise, $R(p)$ is just a line segment. If $Q_3 \neq \phi$ and $Q = Q_1 \cup Q_3$, then $C(p, q) \cap C(p, q')$ for any $q \in Q_1$ and $q' \in Q_3$ is a line segment. Figure 3 (c) illustrates the case. The lower endpoint of the segment is determined either by the query point in Q_1 with the smallest x-coordinate or by the query point in Q_3 with the largest y-coordinate. If two of Q_2, Q_3, and Q_4 are not empty, $R(p)$ is a line segment whose one endpoint is p. Figure 3 (d) shows the case that both Q_2 and Q_3 are not empty. In this case, the lower endpoint of $R(p)$ is p and the upper endpoint is determined by one of three query points: the query point in Q_1 with the smallest y-coordinate or the query point in Q_3 with the largest x-coordinate or the query point in Q_2 with the smallest L_1 distance from p. For the cases in which Q_2 and Q_4 are not empty, or that Q_3 and Q_4 are not empty, one endpoint of $R(p)$ is p and the other endpoint is determined by one of three such extreme query points.

For a data point p, we perform four range counting queries on rCountQ, using each region (or quadrant) subdivided by the vertical and horizontal lines through p as a query. Based on the results of the range counting queries, we determine the case it belongs to and identify query points that determine the sides of $R(p)$. Once we construct the segment dragging query structure sDragQ, we can find these query points in at most four segment dragging queries as follows. For the query point in Q_1 that has the smallest x-coordinate (or y-coordinate), we use "dragging by parallel tracks" with the vertical (or horizontal) line through p as in Figure 4 (a) (or (b)). For the query point in Q_1 that has the smallest L_1 distance from p, we use "dragging out of a corner" with p as in Figure 4 (c). The extreme query points in Q_2, Q_3, and Q_4 can also be found similarly by using the same segment dragging queries. □

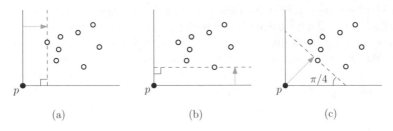

Fig. 4. Segment dragging queries to find (a) the query point with the smallest x-coordinate, (b) the query point with the smallest y-coordinate, and (c) the query point with the smallest L_1 distance from p in Q_1

Once we identity four query points that determine the sides of $R(p)$, we perform a counting query on rCountP using range $R(p)$ as a query. If there is only one data point p in $R(p)$, it belongs to the case (a) in Section 4 and p is a skyline point. Otherwise, we perform a range counting query with the interior of $R(p)$ to check additional eight counting queries on P, whether there is any data point in the interior of $R(p)$. If this is the case (case (b) in Section 4), p is not a skyline point. If there is no data point in the interior of $R(p)$, this case corresponds to case (c) in Section 4.

5.3   For Data Points on the Boundary of $R(p)$

We will show how to handle the case (c) in Section 4. For this case, we should check whether p' lies on the boundary of $C(p, q)$ for every $q \in Q$ or not. We will show that p' lies on the boundary of $C(p, q)$ for every $q \in Q$ if and only if all points in Q lie on the specific regions related to p'. We can check whether all points in Q lie on these specific regions in $O(\log |Q|)$ time, after $O(|Q| \log |Q|)$ time preprocessing. Without loss of generality, we assume that Q_1 is not empty, that is, p lies on the bottom left side of $R(p)$ as in Section 5.2.

Recall that $C(p, q)$ for every $q \in Q$ contains $R(p)$ by definition. We denote by ℓ_v and ℓ_h the vertical line through the bottom corner of $R(p)$ and the horizontal line through the left corner of $R(p)$, respectively. We denote by ℓ_s the line consisting of points at equidistance from the bottom left side and the top right side of $R(p)$ (See Figure 5 (a)).

Lemma 2. *Assume that the interior of $R(p)$ is not empty and p lies in the interior of the bottom left side of $R(p)$. A data point $p'(\neq p)$ lies on the boundary of $C(p, q)$ for every $q \in Q$ if and only if*

   (i) p' lies on the bottom left side of $R(p)$ and all query points lie above or on ℓ_h and ℓ_s, and lie on the right of or on ℓ_v (Figure 5 (a)),

  (ii) p' lies on the top left side of $R(p)$ and all query points lie on ℓ_h but lie above or on ℓ_s (Figure 5 (b)),

 (iii) p' lies on the bottom right side of $R(p)$ and all query points lie on ℓ_v but lie above or on ℓ_s, or

  (iv) p' lies on the top right side of $R(p)$ and all data points lies on ℓ_s but lie above or on ℓ_h and on the right of or on ℓ_v (Figure 5 (c)).

Proof. As it is straightforward to see that the necessary condition holds, we only prove the sufficiency condition. Since the interior of $R(p)$ is not empty and p lies in the interior of the bottom left side of $R(p)$, p always lies on the bottom left side of $C(p, q)$ for every $q \in Q$. If there is a query point q' below ℓ_h, the left corner of $C(p, q')$ lies on the interior of the bottom left side of $R(p)$ and $R(p)$ is not contained in $C(p, q')$, which contradicts the definition of $R(p)$. We can show that all query points lie on or in the right of ℓ_v analogously. For any query point q' below ℓ_s, $C(p, q')$ does not contain the top right side of $R(p)$, which again contradicts the definition of $R(p)$.

Consider case (i) in which there is a data point p' on the bottom left side of $R(p)$. Then p' lies on the bottom left side of $C(p, q)$ for every $q \in Q$. Since the bottom left side of $R(p)$ is the common intersection of the bottom left sides of all $C(p, q)$ for every $q \in Q$, p' does not impose any additional constraint on the locations of query points.

Consider case (ii) in which there is a data point p' on the top left side of $R(p)$. Then p' lies on the top left side of $C(p, q)$ for every $q \in Q$. Therefore the only additional constraint is that all query points lie on ℓ_h. Case (iii) can be shown analogously.

Consider case (iv) in which there is a data point p' on the top right side of $R(p)$. Then p' lies on the top right side of $C(p, q)$ for every $q \in Q$. Therefore the only additional constraint is that all query points lie on ℓ_s. $\square$

Note that when p' lies on a corner of $R(p)$, we consider it contained on both sides of $R(p)$ sharing the corner. Therefore the lemma holds if every query point satisfies

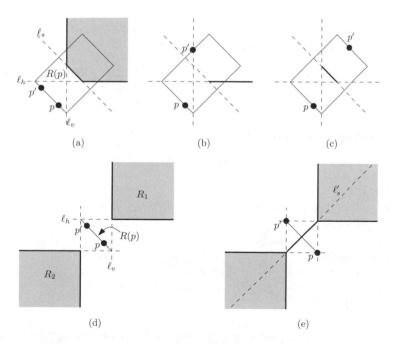

Fig. 5. A data point $p'(\neq p)$ lies on the boundary of $C(p, q)$ for every $q \in Q$ if and only if all query points lie in the gray region or thick line segments

one of two conditions of the sides. The lemma above implies that we can test whether $d(p',q) = d(p,q)$ for every $q \in Q$ in $O(\log |Q|)$ time by performing a few range counting queries on rCountQ.

The case in which p lies on the bottom or the left corner of $R(p)$ can be handled as follows. When $Q_2 \neq \phi$, p lies on the bottom corner of $R(p)$, and all query points in Q_2 must lie on or above the horizontal line through the right corner of $R(p)$. When $Q_4 \neq \phi$, p lies on the left corner of $R(p)$, and all query points in Q_4 must lie on or in the right of the vertical line through the top corner of $R(p)$.

It remains to show the case in which $R(p)$ degenerates to a line segment (See Figure 3 (c) and (d)). (Note that when $R(p)$ is a point, $d(p',q) = d(p,q)$ for every $q \in Q$ if and only if $p = p'$. (See Figure 3 (e)). We denote by R_1 the region bounded from below by ℓ_h and bounded from the left by ℓ_v. We denote by R_2 the region bounded from above by the horizontal line through the lower endpoint of $R(p)$ and bounded from right by the vertical line through the upper endpoint of $R(p)$. Let ℓ'_s be the line consisting of points at equidistance from p and p' (or, bisector).

Lemma 3. *Assume that $R(p)$ is a line segment. A data point $p' (\neq p)$ lies on the boundary of $C(p,q)$ for every $q \in Q$ if and only if*

 (i) *p or p' lies in the interior of the segment $R(p)$ and all query points lie in $R_1 \cup R_2$ (Figure 5 (d)), or*
(ii) *p and p' lie on the opposite endpoints of $R(p)$ and all query points lie in $R_1 \cup R_2 \cup \ell'_s$ (Figure 5 (e)).*

Again, we can test whether $d(p',q) = d(p,q)$ for every $q \in Q$ in $O(\log |Q|)$ time by performing a few range counting queries on rCountQ.

Lemma 4. *We can decide in $O(\log |Q|)$ time whether the data points on a side of $R(p)$ dominate p or not.*

5.4   Computing all the Skyline Points

The following pseudocodes summarize our algorithm.

Algorithm *MSSQ*
Input: a set P of data points (and range counting structure rCountP) and a set Q of query points
Output: the list S of all skylines
1.   initialize the list S
2.   construct range counting query structures rCountQ of Q
3.   construct a segment dragging query structure sDragQ of Q
4.   **for** $i \leftarrow 1$ **to** $|P|$
5.   **do**
6.   determine the quadrants containing query points, by querying rCountQ with quadrants of p_i /* Section 5.2 */
7.   determine the side of $R(p_i)$, by querying sDragQ with p_i /* Section 5.2 */

8. $count \leftarrow$ query rCountP with $R(p_i)$
9.   **if** $count = 1$ /* p_i is the only point in $R(p_i)$ */
10.   **then** insert p_i to S /* p_i is a skyline */
11.   **else** query rCountP with interior of $R(p_i)$
12.   **if** there is no data point in the interior of $R(p_i)$
13.   **then** query rCountP with the regions defined by sides or cor-
                         ners of $R(p_i)$ to check whether they contain data points /*
                         Section 5.3 */
14.   query rCountQ with regions defined by ℓ_h, ℓ_v, and ℓ_s (or
                         R_1, R_2, ℓ'_s) for sides and corners of $R(p_i)$ containing a data
                         point /* Section 5.3 */
15.   **if** all query points lie in the regions defined above
16.   **then** insert p_i to S /* p_i is a skyline */
17. **return** S

In Line 2 of algorithm *MSSQ*, we construct range counting query structures for Q which take $O(|Q| \log |Q|)$ time and $O(|Q|)$ space [19]. The segment dragging query structure in Line 3 can be constructed in $O(|Q| \log |Q|)$ time and $O(|Q|)$ space [20,21]. In the *for-loop*, we use four queries to rCountQ to determine the quadrants containing query points, at most four queries to sDragQ to find the query points determining the sides of $R(p_i)$, and eight queries to rCountP and at most six queries to rCountQ to determine whether any data point on the boundary of $R(p_i)$ dominates p. Each such query can be answered in logarithmic time - a query to sDragQ or rCountQ takes $O(\log |Q|)$ time, and a query to rCountP takes $O(\log |P|)$ time [19]. Therefore the *for-loop* takes $O(|P|(\log |P| + \log |Q|))$ time in total. Because we assume that $|P| \geq |Q|$, the total time complexity of algorithm *MSSQ* is $O(|P| \log |P|)$ and the space complexity is $O(|P|)$.

Theorem 1. *Given a set P of data points and a set Q of query points in the plane, the algorithm MSSQ returns the set of all skyline points in $O(|P| \log |P|)$ time.*

Our algorithm has the following two desirable properties:

– **Easily parallelizable:** As shown in algorithm *MSSQ*, each loop represents an independent computation for p_i and does not depend on other points. This property
  naturally invites loop parallelization.

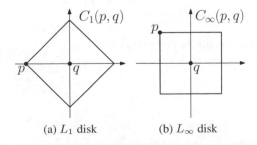

(a) L_1 disk   (b) L_∞ disk

Fig. 6. Comparing L_1 and L_∞ disks

- **Easily extensible to** L_∞: Our observations for L_1 disk $C(p,q)$ also holds for L_∞ disk, which is simply a 45 degree rotation of L_1 disk as illustrated in Figure 6. This illustrates that our algorithm can straightforwardly support L_∞ metrics, widely used for spatial logistics in warehouses, simply by rotating the input dataset by 45 degree around the origin.

6   Implementation

In our implementation of *MSSQ*, an R-tree is used to efficiently prune out the points that cannot be skyline points from P. More specifically, we first find a range bounding Q and read a constant number of points in this region from the R-tree. For each such point p, we identified the bounding box for $\cup_{i=1}^{|Q|} C(p,q_i)$. Any point outside of this bounding box can be safely pruned as it would be dominated by p. We intersect such bounding boxes and retrieve the points falling into this region, which can be efficiently supported by R-tree, as the intersected region will also be a rectangular range. We call the reduced dataset P'.

For fair comparison, we build all our baselines to use this reduced dataset P'. In this section, we discuss more on our two baselines– *PSQ* and *BBS*– representing the current-state-of-the-art for spatial and classic skyline algorithms respectively.

6.1   *PSQ*

PSQ builds upon Lemma 1 and 2 in [9] to compute skyline queries in L_1 metric space. By these two lemmas, after sorting points in P in an ascending order of distance from some query point $q \in Q$, we can compute all skyline points in $O(|P||S||Q|)$ time. Specifically, in this sorting, if some points have same distance from q, then we break the tie by the distance from the other query points in Q. After this sorting, we check each point p in the sorted order, to check whether it is a skyline or not, by testing dominance with the skylines points already found. This algorithm is essentially [8,9]. The following pseudocodes formally presents *PSQ*.

Algorithm *PSQ*
Input: P', Q
Output: S
1.   initialize the list S, array A
2. $A \leftarrow (p, d(p,q))$ for all $p \in P'$ and one query point $q \in Q$
3.   sort A by distance in ascending order
4.   **for** $i \leftarrow 1$ **to** $|P'|$
5.   **do if** $A[i]$ is not dominated by points in S
6.   **then** insert $A[i]$ to S
7.   **return** S

6.2   *BBS*

Meanwhile, BBS is a well-known algorithm for general skyline problems [4]. To apply a general skyline algorithm for spatial problems, we need to compute the distance of each point in P' from all query points, then generate $|Q|$-dimensional data, where each dimension i of point p representing the distance of p to q_i. The following pseudocodes summarize this transformation procedure.

Algorithm *BBS*
Input: P', Q
Output: S
1. initialize the list S, array A_i where i is an integer $1 \leq i \leq |Q|$
2. **for** $i \leftarrow 1$ **to** $|Q|$
3.   **do** $A_i \leftarrow d(p, q_i)$ for all $p \in P'$
4. run BBS for A to get S
5. **return** S

7 Experimental Evaluation

In this section, we outline our experimental settings, and present evaluation results to validate the efficiency and effectiveness of our framework. We compare our algorithm (*MSSQ*) with *PSQ* and *BBS*. As datasets, we use both synthetic datasets and a real dataset of points of interest (POI) in California. We carry out our experiments on Linux with Intel Q6600 CPU and 3GB memory, and the algorithms are coded in C++.

7.1 Experimental Settings

Synthetic dataset: A synthetic dataset contains up to two million uniformly distributed random locations in a 2D space. The space of the datasets is limited to the unit space, i.e., the upper and lower bound of all points are 0 and 1 for each dimension, respectively. Specifically, we use four synthetic datasets with 100 K, 500 K, 1 M, and 2 M uniformly distributed points. Data points in two dimensions are not related, i.e., they are independent, as mentioned in Table 1.

We also randomly generate queries using the parameters in Table 1. Query points are normally distributed with deviation σ to control the distribution. When σ is low, query points are clustered in a small area, and when high, they are scattered over a wide area.

Table 1. Parameters used for synthetic datasets

Parameter	Setting
Dimensionality	2
Distribution of data points	Independent
Dataset cardinality	100K, 500K, 1M, 2M
The number of points in a query	4, 8, 12, 16, 20
Standard deviation of points in a query	0.06

POI dataset: We also validate our proposed framework using a real-life dataset. In particular, we use a sampled POI dataset that has 104,770 locations in 63 different categories, as shown in Figure 10 (a).

7.2   Efficiency

We first validate the efficiency in Figure 7, by comparing the response time of the three algorithms over varying datasize $|P|$. From the figure, we can observe that our proposed algorithm is highly scalable over varying $|P|$, consistently outperforming both baselines. The performance gap only increases as $|P|$ and $|Q|$ increase. For example, when $|P| = 2M$ and $|Q| = 20$, our algorithm is up to 100 times faster than *BBS*.

Figure 8 similarly studies the effect of the query size $|Q|$. We similarly observe that our algorithm is the clear winner in all settings, outperforming *BBS* by up to 100 times, when $|P| = 2M$ and $|Q| = 20$.

For closer observation, Figure 9 shows the breakdown of our response time reported in Figure 8 (a) (i.e., when $|Q| = 4$). From this breakdown, we can observe that I/O costs (of traversing the R-tree) dominate the response time, and the second dominant factor is computation of dominance tests. The remaining cost, including that of building data structures, is left insignificant.

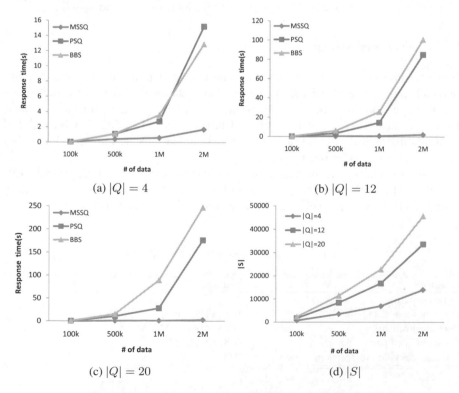

Fig. 7. Effect of the dataset cardinality for synthetic datasets

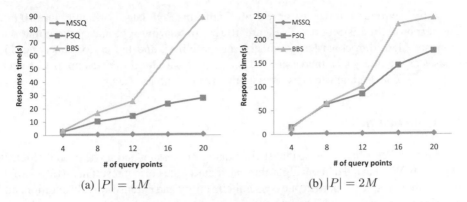

(a) $|P| = 1M$ (b) $|P| = 2M$

Fig. 8. Effect of $|Q|$ for synthetic datasets

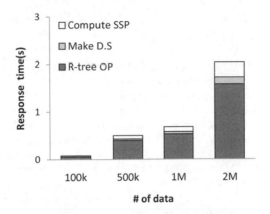

Fig. 9. Break down of the response time in Figure 8(a)

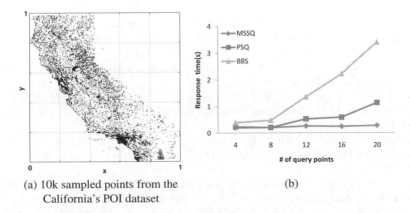

(a) 10k sampled points from the   (b)
    California's POI dataset

Fig. 10. Effect of the dataset cardinality for POI datasets

Lastly, we report our results for real-life California POI data shown in Figure 10 (a). The response time is reported in Figure 10 (b) for varying $|Q|$, and observe that our proposed algorithm consistently outperforms baselines, and the performance gap increases as the query size increases. The speedup of our algorithm was up to 12 (when $|Q| = 20$). This finding is consistent with that from synthetic data.

8 Conclusion

We have studied Manhattan spatial skyline query processing and presented an efficient algorithm. We showed that our algorithm can identify the correct result in $O(|P| \log |P|)$ time with desirable properties of easy parallelizability and extensibility. In addition, our extensive experiments validated the efficiency and effectiveness of our proposed algorithms using both synthetic and real-life data.

References

1. Kung, H.T., Luccio, F., Preparata, F.P.: On finding the maxima of a set of vectors. Journal of the Association for Computing Machinery 22(4), 469–476 (1975)
2. Börzsönyi, S., Kossmann, D., Stocker, K.: The skyline operator. In: ICDE 2001: Proc. of the 17th International Conference on Data Engineering, p. 421 (2001)
3. Tan, K., Eng, P., Ooi, B.C.: Efficient progressive skyline computation. In: VLDB 2001: Proc. of the 27th International Conference on Very Large Data Bases, pp. 301–310 (2001)
4. Papadias, D., Tao, Y., Fu, G., Seeger, B.: An optimal and progressive algorithm for skyline queries. In: SIGMOD 2003: Proc. of the 2003 ACM SIGMOD International Conference on Management of Data, pp. 467–478 (2003)
5. Chomicki, J., Godfery, P., Gryz, J., Liang, D.: Skyline with presorting. In: ICDE 2007: Proc. of the 23rd International Conference on Data Engineering (2007)
6. Sharifzadeh, M., Shahabi, C.: The spatial skyline queries. In: VLDB 2006: Proc. of the 32nd International Conference on Very Large Data Bases, pp. 751–762 (2006)
7. Sharifzadeh, M., Shahabi, C., Kazemi, L.: Processing spatial skyline queries in both vector spaces and spatial network databases. ACM Transactions on Database Systems (TODS) 34(3), 1–43 (2009)
8. Son, W., Lee, M.W., Ahn, H.K., Hwang, S.w.: Spatial skyline queries: An efficient geometric algorithm. In: Mamoulis, N., Seidl, T., Pedersen, T.B., Torp, K., Assent, I. (eds.) SSTD 2009. LNCS, vol. 5644, pp. 247–264. Springer, Heidelberg (2009)
9. Lee, M.W., Son, W., Ahn, H.K., Hwang, S.w.: Spatial skyline queries: exact and approximation algorithms. GeoInformatica, 1–33 (2010)
10. Kossmann, D., Ramsak, F., Rost, S.: Shooting stars in the sky: An online algorithm for skyline queries. In: VLDB 2002: Proc. of the 28th International Conference on Very Large Data Bases, pp. 275–286 (2002)
11. Godfrey, P., Shipley, R., Gryz, J.: Maximal vector computation in large data sets. In: VLDB 2005: Proc. of the 31st International Conference on Very Large Data Bases, pp. 229–240 (2005)
12. Chan, C.Y., Jagadish, H., Tan, K., Tung, A.K., Zhang, Z.: On high dimensional skylines. In: Ioannidis, Y., Scholl, M.H., Schmidt, J.W., Matthes, F., Hatzopoulos, M., Böhm, K., Kemper, A., Grust, T., Böhm, C. (eds.) EDBT 2006. LNCS, vol. 3896, pp. 478–495. Springer, Heidelberg (2006)

13. Chan, C.Y., Jagadish, H., Tan, K.L., Tung, A.K., Zhang, Z.: Finding k-dominant skylines in high dimensional space. In: SIGMOD 2006: Proc. of the 2006 ACM SIGMOD International Conference on Management of Data (2006)
14. Lin, X., Yuan, Y., Zhang, Q., Zhang, Y.: Selecting stars: The k most representative skyline operator. In: ICDE 2007: Proc. of the 23rd International Conference on Data Engineering, pp. 86–95 (2007)
15. Roussopoulos, N., Kelley, S., Vincent, F.: Nearest neighbor queries. In: SIGMOD 1995: Proc. of the 1995 ACM SIGMOD international conference on Management of data, pp. 71–79 (1995)
16. Berchtold, S., Böhm, C., Keim, D.A., Kriegel, H.P.: A cost model for nearest neighbor search in high-dimensional data space. In: PODS 1997: Proc. of the 16th ACM SIGACT-SIGMOD-SIGART symposium on Principles of database systems, pp. 78–86 (1997)
17. Beyer, K.S., Goldstein, J., Ramakrishnan, R., Shaft, U.: When is nearest neighbor meaningful? In: Beeri, C., Bruneman, P. (eds.) ICDT 1999. LNCS, vol. 1540, pp. 217–235. Springer, Heidelberg (1998)
18. Papadias, D., Tao, Y., Mouratidis, K., Hui, C.K.: Aggregate nearest neighbor queries in spatial databases. ACM Transactions on Database Systems 30(2), 529–576 (2005)
19. Agarwal, P., Erickson, J.: Geometric Range Searching and Its Relatives. Advances in Discrete and Computational Geometry, pp. 1–56 (1999)
20. Chazelle, B.: An algorithm for segment-dragging and its implementation. Algorithmica 3(1), 205–221 (1988)
21. Mitchell, J.: L_1 shortest paths among polygonal obstacles in the plane. Algorithmica 8(1), 55–88 (1992)

Inverse Queries for Multidimensional Spaces

Thomas Bernecker[1], Tobias Emrich[1], Hans-Peter Kriegel[1], Nikos Mamoulis[2],
Matthias Renz[1], Shiming Zhang[2], and Andreas Züfle[1]

[1] Institute for Informatics, Ludwig-Maximilians-Universität München
Oettingenstr. 67, D-80538 München, Germany
{bernecker,emrich,kriegel,renz,zuefle}@dbs.ifi.lmu.de
[2] Department of Computer Science, University of Hong Kong
Pokfulam Road, Hong Kong
{nikos,smzhang}@cs.hku.hk

Abstract. Traditional spatial queries return, for a given query object q, all database objects that satisfy a given predicate, such as epsilon range and k-nearest neighbors. This paper defines and studies *inverse* spatial queries, which, given a subset of database objects Q and a query predicate, return all objects which, if used as query objects with the predicate, contain Q in their result. We first show a straightforward solution for answering inverse spatial queries for any query predicate. Then, we propose a filter-and-refinement framework that can be used to improve efficiency. We show how to apply this framework on a variety of inverse queries, using appropriate space pruning strategies. In particular, we propose solutions for inverse epsilon range queries, inverse k-nearest neighbor queries, and inverse skyline queries. Our experiments show that our framework is significantly more efficient than naive approaches.

1 Introduction

Recently, a lot of interest has grown for *reverse* queries, which take as input an object o and find the queries which have o in their result set. A characteristic example is the reverse k-NN query [6,12], whose objective is to find the query objects (from a given data set) that have a given input object in their k-NN set. In such an operation the roles of the query and data objects are reversed; while the k-NN query finds the *data* objects which are the nearest neighbors of a given *query* object, the reverse query finds the objects which, if used as queries, return a given data object in their result. Besides k-NN search, reverse queries have also been studied for other spatial and multidimensional search problems, such as top-k search [13] and dynamic skyline [7]. Reverse queries mainly find application in data analysis tasks; e.g., given a product find the customer searches that have this product in their result. [6] outlines a wide range of such applications (including business impact analysis, referral and recommendation systems, maintenance of document repositories).

In this paper, we generalize the concept of reverse queries. We note that the current definitions take as input a *single* object. However, similarity queries such as k-NN queries and ε-range queries may in general return more than one result. Data analysts are often interested in the queries that include two or more given objects in their result.

D. Pfoser et al. (Eds.): SSTD 2011, LNCS 6849, pp. 330–347, 2011.

Such information can be meaningful in applications where only the result of a query can be (partially) observed, but the actual query object is not known. For example consider an online shop selling a variety of different products stored in a database $\mathcal{D}$. The online shop may be interested in offering a *package* of products $Q \subseteq \mathcal{D}$ for a special price. The problem at hand is to identify customers which are interested in *all* items of the package, in order to direct an advertisement to them. We assume that the preferences of registered customers are known. First, we need to define a predicate indicating whether a user is interested in a product. A customer may be interested in a product if

- the distance between the product's features and the customer's preference is less than a threshold ε;
- the product is contained in the set of his k favorite items, i.e., the k-set of product features closest to the user's preferences;
- the product is contained in the customer's dynamic skyline, i.e., there is no other product that better fits the customer's preferences in every possible way.

Therefore, we want to identify customers r, such that the query on $\mathcal{D}$ with query object r, using one of the query predicates above, contains Q in the result set. More specifically, consider a set $\mathcal{D} \in \mathbb{R}^d$ as a database of n objects and let $d(\cdot)$ denote the Euclidean distance in $\mathbb{R}^d$. Let $\mathcal{P}(q)$ be a query on $\mathcal{D}$ with predicate $\mathcal{P}$ and query object q.

Definition 1. *An inverse $\mathcal{P}$ query (IPQ) computes for a given set of query objects $Q \subseteq \mathcal{D}$ the set of points $r \in \mathbb{R}^d$ for which Q is in the $\mathcal{P}$ query result; formally:*

$$IPQ = \{r \in \mathbb{R}^d : Q \subseteq \mathcal{P}(r))\}$$

Simply speaking, the result of the *general* inverse query is the subset of the space defined by all objects r for which all Q-objects are in $\mathcal{P}(r)$. Special cases of the query are:

- The mono-chromatic inverse $\mathcal{P}$ query, for which the result set is a subset of $\mathcal{D}$.
- The bi-chromatic inverse $\mathcal{P}$ query, for which the result set is a subset of a given database $\mathcal{D}' \subseteq \mathbb{R}^d$.

In this paper, we study the inverse versions of three common query types in spatial and multimedia databases as follows.

Inverse ε-Range Query ($I\varepsilon$-RQ). The inverse ε-range query returns all objects which have a sufficiently low distance to all query objects. For a *bi-chromatic* sample application of this type of query, consider a movie database containing a large number of movie records. Each movie record contains features such as humor, suspense, romance, etc. Users of the database are represented by the same attributes, describing their preferences. We want to create a recommendation system that recommends to users movies that are sufficiently similar to their preferences (i.e., distance less than ε). Now, assume that a group of users, such as a family, want to watch a movie together; a bi-chromatic $I\varepsilon$-RQ will recommend movies which are similar to *all* members of the family. For a mono-chromatic case example, consider the set $Q = \{q_1, q_2\}$ of query objects of Figure 1(a) and the set of database points $\mathcal{D} = \{p_1, p_2, \cdots, p_6\}$. If the range ε is as illustrated

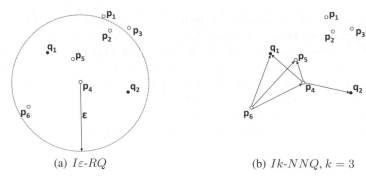

(a) $I\varepsilon$-RQ (b) Ik-$NNQ, k = 3$

Fig. 1. Examples of inverse queries

in the figure, the result of the $I\varepsilon$-$RQ(Q)$ is $\{p_2, p_4, p_5\}$ (e.g., p_1 is dropped because $d(p_1, q_2) > \varepsilon$).

Inverse k-NN Query (Ik-NNQ). The inverse k-NN query returns the objects which have all query points in their k-NN set. For example, *mono-chromatic* inverse k-NN queries can be used to aid crime detection. Assume that a set of households have been robbed in short succession and the robber must be found. Assume that the robber will only rob houses which are in his close vicinity, e.g. within the closest hundred house-holds. Under this assumption, performing an inverse 100NN query, using the set of robbed households as Q, returns the set of possible suspects. A mono-chromatic inverse 3NN query for $Q = \{q_1, q_2\}$ in Figure 1(b) returns $\{p_4\}$. p_6, for example, is dropped, as q_2 is not contained in the list of its 3 nearest neighbors.

Inverse Dynamic Skyline Query (I-DSQ). An inverse dynamic skyline query returns the objects, which have all query objects in their dynamic skyline. A sample application for the *general* inverse dynamic skyline query is a product recommendation problem: assume there is a company, e.g. a photo camera company, that provides its products via an internet portal. The company wants to recommend products to their customers by analyzing the web pages visited by them. The score function used by the customer to rate the attributes of products is unknown. However, the set of products that the cus-tomer has clicked on can be seen as samples of products that he or she is interested in, and thus, must be in the customer's dynamic skyline. The inverse dynamic skyline query can be used to narrow the space which the customers preferences are located in. Objects which have all clicked products in their dynamic skyline are likely to be inter-esting to the customer. In Figure 1, assuming that $Q = \{q_1, q_2\}$ are clicked products, I-$DSQ(Q)$ includes p_6, since both q_1 and q_2 are included in the dynamic skyline of p_6.

For simplicity, we focus on the mono-chromatic cases of the respective query types (i.e., query points and objects are taken from the same data set); however, the proposed techniques can also be applied for the bi-chromatic and the general case. For details, refer to the full version of this paper [2].

Motivation. A naive way to process any inverse spatial query is to compute the corre-sponding reverse query for each $q_i \in Q$ and then intersect these results. The problem of this method is that running a reverse query for each q_i multiplies the complexity of the reverse query by $|Q|$ both in terms of computational and I/O-cost. Objects that are not shared in two or more reverse queries in Q are unnecessarily retrieved, while objects that are shared by two or more queries are redundantly accessed multiple times. We propose a filter-refinement framework for inverse queries, which first applies a number of filters using the set of query objects Q to prune effectively objects which may not participate in the result. Afterwards, candidates are pruned by considering other database objects. Finally, during a *refinement* step, the remaining candidates are verified against the inverse query and the results are output. When applying our framework to the three inverse queries under study, filtering and refinement are sometimes integrated in the same algorithm, which performs these steps in an iterative manner. Although for $I\varepsilon$-RQ queries the application of our framework is straightforward, for Ik-NNQ and I-DSQ, we define and exploit special pruning techniques that are novel compared to the approaches used for solving the corresponding reverse queries.

Outline. The rest of the paper is organized as follows. In the next section we review previous work related to inverse query processing. Section 3 describes our framework. In Sections 4-6 we implement it on the three inverse spatial query types; we first briefly introduce the pruning strategies for the single-query-object case and then show how to apply the framework in order to handle the multi-query-object case in an efficient way. Section 7 is an experimental evaluation and Section 8 concludes the paper.

2   Related Work

The problem of supporting reverse queries efficiently, i.e. the case where Q only contains a single database object, has been studied extensively. However, none of the proposed approaches is directly extendable for the efficient support of inverse queries when $|Q| > 1$. First, there exists no related work on reverse queries for the ε-range query predicate. This is not surprising since the the reverse ϵ-range query is equal to a (normal) ε-range query. However, there exists a large body of work for reverse k-nearest neighbor (Rk-NN) queries. Self-pruning approaches like the RNN-tree [6] and the RdNN-tree [14] operate on top of a spatial index, like the R-tree. Their objective is to estimate the k-NN distance of each index entry e. If the k-NN distance of e is smaller than the distance of e to the query q, then e can be pruned. These methods suffer from the high materialization and maintenance cost of the k-NN distances.

Mutual-pruning approaches such as [10,11,12] use other points to prune a given index entry e. TPL [12] is the most general and efficient approach. It uses an R-tree to compute a nearest neighbor ranking of the query point q. The key idea is to iteratively construct Voronoi hyper-planes around q using the retrieved neighbors. TPL can be used for inverse k-NN queries where $|Q| > 1$, by simply performing a reverse k-NN query for each query point and then intersecting the results (i.e., the brute-force approach).

For reverse dynamic skyline queries, [3] proposed an efficient solution, which first performs a filter-step, pruning database objects that are globally dominated by some point in the database. For the remaining points, a window query is performed in a refinement step. In addition, [7] gave a solution for reverse dynamic skyline computation on uncertain data. None of these methods considers the case of $|Q| > 1$, which is the focus of our work.

In [13], the problem of reverse top-k queries is studied. A reverse top-k query returns, for a point q and a positive integer k, the set of linear preference functions for which q is contained in their top-k result. The authors provide an efficient solution for the 2D case and discuss its generalization to the multidimensional case, but do not consider the case where $|Q| > 1$. Although we do not study inverse top-k queries in this paper, we note that it is an interesting subject for future work.

Inverse queries are very related to group queries, i.e. similarity queries that retrieve the top-k objects according to a given similarity (distance) aggregate w.r.t. a given set of query points [9,8]. However, the problem addressed by group queries generally differs from the problem addressed in this paper. Instead of minimizing distance aggregations, here we have to find efficient methods for converging query predicate evaluations w.r.t. a set of query points. Hence, new strategies are required.

3   Inverse Query (IQ) Framework

Our solutions for the three inverse queries under study are based on a common framework consisting of the following filter-refinement pipeline:

Filter 1: Fast Query Based Validation: The first component of the framework, called *fast query based validation*, uses the set of query objects Q only to perform a quick check on whether it is possible to have any result at all. In particular, this filter verifies simple constraints that are necessary conditions for a non-empty result. For example, for the Ik-NN case, the result is empty if $|Q| > k$.

Filter 2: Query Based Pruning: *Query based pruning* again uses the query objects only to prune objects in $\mathcal{D}$ which may not participate in the result. Unlike the simple first filter, here we employ the topology of the query objects.

Filters 1 and 2 can be performed very fast because they do not involve any database object except the query objects.

Filter 3: Object Based Pruning: This filter, called *object based pruning*, is more advanced because it involves database objects additional to the query objects. The strategy is to access database objects in ascending order of their maximum distance to any query point; formally:

$$MaxDist(o, Q) = \max_{q \in Q}(d(e, q)).$$

The rationale for this access order is that, given any query object q, objects that are close to q have more pruning power, i.e., they are more likely to prune other objects w.r.t. q than objects that are more distant to q. To maximize the pruning power, we prefer to examine objects that are close to all query points first.

Note that the applicability of the filters depends on the query. *Query based pruning* is applicable if the query objects suffice to restrict the search space which holds for the inverse ε-range query and the inverse skyline query but not directly for the inverse k-NN query. In contrast, the *object based pruning* filter is applicable for queries where database objects can be used to prune other objects which for example holds for the inverse k-NN query and the inverse skyline query but not for the inverse ε-range query.

Refinement: In the final *refinement* step, the remaining candidates are verified and the *true hits* are reported as results.

4 Inverse ε-Range Query

We will start with the simpler query, the inverse ε-range query. First, consider the case of a query object q (i.e., $|Q| = 1$). In this case, the inverse ε-range query computes all objects, that have q within their ε-range sphere. Due to the symmetry of the ε-range query predicate, all objects satisfying the inverse ε-range query predicate are within the ε-range sphere of q as illustrated in Figure 2(a). In the following, we consider the general case, where $|Q| > 1$ and show how our framework can be applied.

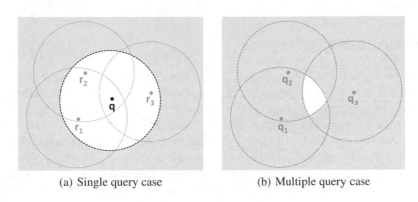

(a) Single query case   (b) Multiple query case

Fig. 2. Pruning space for $I\varepsilon\text{-}RQ$

4.1 Framework Implementation

Fast Query Based Validation: There is no possible result if there exists a pair q, q' of queries in Q, such that their ε-ranges do not intersect (i.e., $d(q, q') > 2 \cdot \varepsilon$). In this case, there can be no object r having both q and q' within its ε-range (a necessary condition for r to be in the result).

Query Based Pruning: Let $S_i^\varepsilon \subseteq \mathbb{R}^d$ be the ε-sphere around query point q_i for all $q_i \in Q$, as depicted in the example shown in Figure 2(b). Obviously, any point in the intersection region of all spheres, i.e. $\cap_{i=1..m} S_i^\varepsilon$, has all query objects $q_i \in Q$ in its ε-range. Consequently, all objects outside of this region can be pruned. However, the computation of the search region can become too expensive in an arbitrary high dimensional space; thus, we compute the intersection between rectangles that minimally bound the hyper-spheres and use it as a filter. This can be done quite efficiently even in high dimensional spaces; the resulting filter rectangle is used as a window query and all objects in it are passed to the refinement step as candidates.

Object Based Pruning: As mentioned in Section 3 this filter is not applicable for inverse ε-range queries, since objects cannot be used to prune other objects.

Refinement: In the refinement step, for all candidates we compute their distances to all query points $q \in Q$ and report only objects that are within distance ε from all query objects.

4.2   Algorithm

The implementation of our framework above can be easily converted to an algorithm, which, after applying the filter steps, performs a window query to retrieve the candidates, which are finally verified. Search can be facilitated by an R-tree that indexes $\mathcal{D}$. Starting from the root, we search the tree, using the filter rectangle. To minimize the I/O cost, for each entry P of the tree that intersects the filter rectangle, we compute its distance to all points in Q and access the corresponding subtree only if all these distances are smaller than ε.

5   Inverse k-NN Query

For inverse k-nearest neighbor queries (Ik-NNQ), we first consider the case of a single query object (i.e., $|Q| = 1$). As discussed in Section 2, this case can be processed by the bi-section-based Rk-NN approach (TPL) proposed in [12], enhanced by the rectangle-based pruning criterion proposed in [4]. The core idea of TPL is to use bi-section-hyperplanes between database objects o and the query object q in order to check which objects are closer to o than to q. Each bi-section-hyperplane divides the object space into two half-spaces, one containing q and one containing o. Any object located in the half-space containing o is closer to o than to q. The objects spanning the hyperplanes are collected in an iterative way. Each object o is then checked against the resulting half-spaces that do not contain q. As soon as o is inside more than k such half-spaces, it can be pruned. Next, we consider queries with multiple objects (i.e., $|Q| > 1$) and discuss how the framework presented in Section 3 is implemented in this case.

5.1   Framework Implementation

Fast Query Based Validation. Recall that this filter uses the set of query objects Q only, to perform a quick check on whether the result is empty. Here, we use the obvious rule that the result is empty if the number of query objects exceeds query parameter k.

Query Based Pruning. We can exploit the query objects in order to reduce the Ik-NN query to an Ik'-NN query with $k' < k$. A smaller query parameter k' allows us to terminate the query process earlier and reduce the search space. We first show how k can be reduced by means of the query objects only. The proofs for all lemmas can be found in the full version of this paper [2].

Lemma 1. *Let* $\mathcal{D} \subseteq \mathbb{R}^d$ *be a set of database objects and* $Q \subseteq \mathcal{D}$ *be a set of query objects. Let* $\mathcal{D}' = \mathcal{D} - Q$. *For each* $o \in \mathcal{D}'$, *the following statement holds:*

$$o \in Ik\text{-}NNQ(Q) \text{ in } \mathcal{D} \Rightarrow \forall q \in Q : o \in Ik'\text{-}NNQ(\{q\}) \text{ in } \mathcal{D}' \cup \{q\},$$

$$\text{where } k' = k - |Q| + 1.$$

Simply speaking, if a candidate object o is not in the Ik'-$NNQ(\{q\})$ result of some $q \in Q$ considering only the points $\mathcal{D}' \cup \{q\}$, then o cannot be in the Ik-$NNQ(Q)$ result considering all points in $\mathcal{D}$ and o can be pruned. As a consequence, Ik'-$NNQ(\{q\})$ in $\mathcal{D}' \cup \{q\}$ can be used to prune candidates for any $q \in Q$. The pruning power of Ik'-$NNQ(\{q\})$ depends on how $q \in Q$ is selected.

From Lemma 1 we can conclude the following:

Lemma 2. *Let $o \in \mathcal{D} - Q$ be a database object and $q^o_{ref} \in Q$ be a query object such that $\forall q \in Q : d(o, q^o_{ref}) \geq d(o, q)$. Then*

$$o \in Ik\text{-}NNQ(Q) \Leftrightarrow o \in Ik'\text{-}NNQ(\{q^o_{ref}\}) \text{ in } \mathcal{D}' \cup \{q\},$$

where $k' = k - |Q| + 1$.

Lemma 2 suggests that for any candidate object o in $\mathcal{D}$, we should use the farthest query point to check whether o can be pruned.

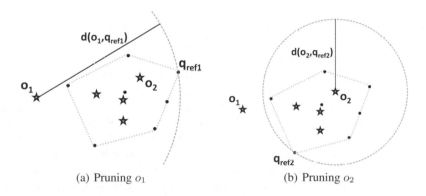

(a) Pruning o_1 (b) Pruning o_2

Fig. 3. Ik-NN pruning based on Lemma 4

Object Based Pruning. Up to now, we only used the query points in order to reduce k in the inverse k-NN query. Now, we will show how to consider database objects in order to further decrease k.

Lemma 3. *Let Q be the set of query objects and $\mathcal{H} \subseteq \mathcal{D} - Q$ be the non-query(database) objects covered by the convex hull of Q. Furthermore, let $o \in \mathcal{D}$ be a database object and $q^o_{ref} \in Q$ a query object such that $\forall q \in Q : d(o, q^o_{ref}) \geq d(o, q)$. Then for each object $p \in \mathcal{H}$ it holds that $d(o, p) \leq d(o, q^o_{ref})$.*

According to the above lemma the following statement holds:

Lemma 4. *Let Q be the set of query objects, $\mathcal{H} \subseteq \mathcal{D} - Q$ be the database (non-query) objects covered by the convex hull of Q and let $q^o_{ref} \in Q$ be a query object such that $\forall q \in Q : d(o, q^o_{ref}) \geq d(o, q)$. Then for a given database object $o \in \mathcal{D}$*

$$\forall o \in \mathcal{D} - \mathcal{H} - Q : o \in Ik\text{-}NNQ(Q) \Leftrightarrow$$

at most $k' = k - |\mathcal{H}| - |Q|$ objects $p \in \mathcal{D} - \mathcal{H}$ are closer to o than q^o_{ref}, and

$$\forall o \in \mathcal{H} : o \in Ik\text{-}NNQ(Q) \Leftrightarrow$$

at most $k' = k - |\mathcal{H}| - |Q| + 1$ objects $p \in \mathcal{D} - \mathcal{H}$ are closer to o than q^o_{ref}.

Based on Lemma 4, given the number of objects in the convex hull of Q, we can prune objects outside of the hull from Ik-NN(Q). Specifically, for an Ik-NN query we have the following pruning criterion: An object $o \in \mathcal{D}$ can be pruned, as soon as we find more than k' objects $p \in \mathcal{D} - \mathcal{H}$ outside of the convex hull of Q, that are closer to o than q^o_{ref}. Note that the parameter k' is set according to Lemma 4 and depends on whether o is in the convex hull of Q or not. Depending on the size of Q and the number of objects within the convex hull of Q, $k' = k - |\mathcal{H}| + 1$ can become negative. In this case, we can terminate query evaluation immediately, as no object can qualify the inverse query (i.e., the inverse query result is guaranteed to be empty). The case where $k' = k - |\mathcal{H}| + 1$ becomes zero is another special case, as all objects outside of $\mathcal{H}$ can be pruned. For all objects in the convex hull of Q (including all query objects) we have to check whether there are objects outside of $\mathcal{H}$ that prune them.

As an example of how Lemma 4 can be used, consider the data shown in Fig. 3 and assume that we wish to perform an inverse 10NN query using a set Q of seven query objects, shown as points in the figure; non-query database points are represented by stars. In Figure 3(a), the goal is to determine whether candidate object o_1 is a result, i.e., whether o_1 has all $q \in Q$ in its 10NN set. The query object having the largest distance to o_1 is q_{ref1}. Since o_1 is located outside of the convex hull of Q (i.e, $o \in \mathcal{D} - \mathcal{H} - Q$), the first equivalence of Lemma 4, states that o_1 is a result if at most $k' = k - |\mathcal{H}| - |Q| = 10 - 4 - 7 = -1$ objects in $\mathcal{D} - \mathcal{H} - Q$ are closer to o_1 than q_{ref1}. Thus, o_1 can be safely pruned without even considering these objects (since obviously, at least zero objects are closer to o_1 than q_{ref1}). Next, we consider object o_2 in Figure 3(b). The query object with the largest distance to o_2 is q_{ref2}. Since o_2 is inside the convex hull of Q, the second equivalence of Lemma 4 yields that o_2 is a result if at most $k' = k - |\mathcal{H}| - |Q| + 1 = 10 - 4 - 7 + 1 = 0$ objects $\mathcal{D} - \mathcal{H} - Q$ are closer to o_2 than q_{ref2}. Thus, o_2 remains a candidate until at least one object in $\mathcal{D} - \mathcal{H} - Q$ is found that is closer to o_2 than q_{ref2}.

Refinement. Each remaining candidate is checked whether it is a result of the inverse query by performing a k-NN search and verifying whether its result includes Q.

5.2   Algorithm

We now present a complete algorithm that traverses an *aggregate* R-tree (*ARTree*), which indexes $\mathcal{D}$ and computes $Ik\text{-}NNQ(Q)$ for a given set Q of query objects, using Lemma 4 to prune the search space. The entries in the tree nodes are augmented with the cardinality of objects in the corresponding sub-tree. These counts can be used to accelerate search, as we will see later.

In a nutshell, the algorithm, while traversing the tree, attempts to prune nodes based on the lemma using the information known so far about the points of $\mathcal{D}$ that are included in the convex hull (*filtering*). The objects that survive the pruning are inserted in the

---

Algorithm 1. Inverse kNNQuery

Require: $Q, k, ARTree$
 1: //*Fast Query Based Validation*
 2: **if** $|Q| > k$ **then**
 3:     return "no result" and terminate algorithm
 4: **end if**
 5: pq PriorityQueue ordered by $max_{q_i \in Q}$MinDist
 6: $pq.add(ARTree.root$ entries)
 7: $|\mathcal{H}| = 0$
 8: LIST $candidates, prunedEntries$
 9: //*Query/Object Based Pruning*
10: **while** $\neg pq.isEmpty()$ **do**
11:     $e = pq.poll()$
12:     **if** $getPruneCount(e, Q, candidates, prunedEntries, pq) > k - |\mathcal{H}| - |Q|$ **then**
13:         $prunedEntries.add(e)$
14:     **else if** $e.isLeafEntry()$ **then**
15:         $candidates.add(e)$
16:     **else**
17:         $pq.add(e.getChildren())$
18:     **end if**
19:     **if** $e \in convexHull(Q)$ **then**
20:         $|\mathcal{H}| + = e.agg\_count$
21:     **end if**
22: **end while**
23: //*Refinement Step*
24: LIST $result$
25: **for** $c \in candidates$ **do**
26:     **if** $q^o_{ref} \in knnQuery(c, k)$ **then**
27:         $result.add(c)$
28:     **end if**
29: **end for**
30: return $(result)$

---

candidates set. During the *refinement* step, for each point c in the candidates set, we run a k-NN query to verify whether c contains Q in its k-NN set.

Algorithm 1 is a pseudocode of our approach. The $ARTree$ is traversed in a best-first search manner [5], prioritizing the access of the nodes according to the maximum possible distance (in case of a non-leaf entry we use *MinDist*) of their contents to the query points Q. In specific, for each R-tree entry e we can compute, based on its MBR, the farthest possible point q^o_{ref} in Q to a point p indexed under e. Processing the entries with the smallest such distances first helps to find points in the convex hull of Q earlier, which helps making the pruning bound tighter.

Thus, initially, we set $|\mathcal{H}| = 0$, assuming that in the worst case the number of non-query points in the convex hull of Q is 0. If the object which is deheaped is inside the convex hull, we increase $|\mathcal{H}|$ by one. If a non-leaf entry is deheaped and its MBR is contained in the hull, we increase $|\mathcal{H}|$ by the number of objects in the corresponding sub-tree, as indicated by its augmented counter.

During tree traversal, the accessed tree entries could be in one of the following sets (i) the set of *candidates*, which contains objects that could possibly be results of the inverse query, (ii) the set of *pruned entries*, which contains (pruned) entries whose subtrees may not possibly contain inverse query results, and (iii) the set of entries which are currently in the priority queue. When an entry e is deheaped, the algorithm checks whether it can be pruned. For this purpose, it initializes a *prune_counter* which is a lower bound of the number of objects that are closer to every point p in e than Q's farthest point to p. For every entry e' in all three sets (candidates, pruned, and priority queue), we increase the *prune_counter* of e by the number of points in e' if the following condition holds: $\forall p \in e, \forall p' \in e' : dist(e, e') < dist(e, q^o_{ref})$. This condition can efficiently be checked [4]. An example where this condition is fulfilled is shown in Figure 4. Here the *prune_counter* of e can be increased by the number of points in e'.

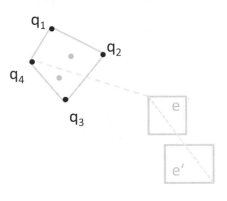

Fig. 4. Calculating the *prune_count* of e

While updating *prune_counter* for e, we check whether *prune_counter* $> k - |\mathcal{H}| - |Q|$ (*prune_counter* $> k - |\mathcal{H}| - |Q| + 1$) for entries that are entirely outside of (intersect) the convex hull. As soon as this condition is true, e can be pruned as it cannot contain objects that can participate in the inverse query result (according to Lemma 4). Considering again Figure 4 and assuming the number of points in e' to be 5, e could be pruned for $k \leq 10$ (since *prune_counter*$(5) > k(10) - |\mathcal{H}|(2) - |Q|(4)$ holds). In this case e is moved to the set of pruned entries. If e survives pruning, the node pointed to by e is visited and its entries are enheaped if e is a non-leaf entry; otherwise e is inserted in the *candidates* set. When the queue becomes empty, the filter step of the algorithm completes with a set of *candidates*. For each object c in this set, we check whether c is a result of the inverse query by performing a k-NN search and verifying whether its result includes Q. In our implementation, to make this test faster, we replace the k-NN search by an aggregate ε-range query around c, by setting $\varepsilon = d(c, q^c_{ref})$. The objective is to count whether the number of objects in the range is greater than k. In this case, we can prune c, otherwise c is a result of the inverse query. $ARTree$ is used to process the aggregate ε-range query; for every entry e included in the ε-range, we just increase the aggregate count by the augmented counter to e without having to traverse the corresponding subtree. In addition, we perform batch searching for candidates that are close to each other, in order to optimize performance. The details are skipped due to space constraints.

6   Inverse Dynamic Skyline Query

We again first discuss the case of a single query object, which corresponds to the reverse dynamic skyline query [7] and then present a solution for the more interesting case where $|Q| > 1$. Let q be the (single) query object with respect to which we want to compute the inverse dynamic skyline. Any object $o \in \mathcal{D}$ defines a pruning region, such that any object o' in this region cannot be part of the inverse query result. Formally:

Definition 2 (Pruning Region). *Let $q = (q^1, \ldots, q^d) \in Q$ be a single d-dimensional query object and $o = (o^1, \ldots, o^d) \in \mathcal{D}$ be any d-dimensional database object. Then the pruning region $PR_q(o)$ of o w.r.t. q is defined as the d-dimensional rectangle where the ith dimension of $PR_q(o)$ is given by $[\frac{q^i + o^i}{2}, +\infty]$ if $q^i \leq o^i$ and $[-\infty, \frac{q^i + o^i}{2}]$ if $q^i \geq o^i$.*

The pruning region of an object o with respect to a single query object q is illustrated by the shaded region in Figure 5(a).

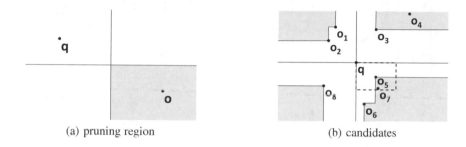

(a) pruning region   (b) candidates

Fig. 5. Single-query case

Filter step. As shown in [7], any object $p \in \mathcal{D}$ can be safely pruned if p is contained in the pruning region of some $o \in \mathcal{D}$ w.r.t. q (i.e. $p \in PR_q(o)$). Accordingly, we can use q to divide the space into 2^d partitions by splitting along each dimension at q. Let $o \in \mathcal{D}$ be an object in any partition P; o is an *I-DSQ* candidate, iff there is no other object $p \in P \subseteq \mathcal{D}$ that dominates o w.r.t. q.

Thus, we can derive all *I-DSQ* candidates as follows: First, we split the data space into the 2^d partitions at the query object q as mentioned above. Then in each partition, we compute the skyline[1], as illustrated in the example depicted in Figure 5(b). The union of the four skylines is the set of the inverse query candidates (e.g., $\{o_1, o_2, o_3, o_5, o_6, o_8\}$ in our example).

Refinement. The result of the reverse dynamic skyline query is finally obtained by verifying for each candidate c, whether there is an object in $\mathcal{D}$ which dominates q w.r.t. c. This can be done by checking whether the hypercube centered at c with extent $2 \cdot |c^i - q^i|$ at each dimension i is empty. For example, candidate o_5 in Figure 5(b) is not a result, because the corresponding box (denoted by dashed lines) contains o_7. This means that in both dimensions o_7 is closer to o_5 than q is.

---

[1] Only objects within the same partition are considered for the dominance relation.

6.1 IQ Framework Implementation

Fast Query Based Validation. Following our framework, first the set Q of query objects is used to decide whether it is possible to have any result at all. For this, we use the following lemma:

Lemma 5. *Let $q \in Q$ be any query object and let S be the set of 2^d partitions derived from dividing the object space at q along the axes into two halves in each dimension. If in each partition $r \in S$ there is at least one query object $q' \in Q$ ($q' \neq q$), then there cannot be any result.*

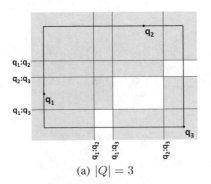
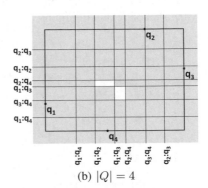

(a) $|Q| = 3$ (b) $|Q| = 4$

Fig. 6. Pruning regions of query objects

Query Based Pruning. We now propose a filter, which uses the set Q of query objects only in order to reduce the space of candidate results. We explore similar strategies as the fast query based validation. For any pair of query objects $q, q' \in Q$, we can define two pruning regions, according to Definition 2: $PR_q(q')$ and $PR_{q'}(q)$. Any object inside these regions cannot be a candidate of the inverse query result because it cannot have both q_1 and q_2 in its dynamic skyline point set. Thus, for every pair of query objects, we can determine the corresponding pruning regions and use their union to prune objects or R-tree nodes that are contained in it. Figure 6 shows examples of the pruning space for $|Q| = 3$ and $|Q| = 4$. Observe that with the increase of $|Q|$ the remaining space, which may contain candidates, becomes very limited.

The main challenge is how to encode and use the pruning space defined by Q, as it can be arbitrarily complex in the multidimensional space. As for the Ik-NNQ case, our approach is not to explicitly compute and store the pruning space, but to check on-demand whether each object (or R-tree MBR) can be pruned by one or more query pairs. This has a complexity of $O(|Q|^2)$ checks per object. In the full version of the paper [2], we show how to reduce this complexity for the special 2D case. The techniques shown there can also be used in higher dimensional spaces, with lower pruning effect.

Object Based Pruning. For any candidate object o that is not pruned during the query-based filter step, we need to check if there exists any other database object o' which dominates some $q \in Q$ with respect to o. If we can find such an o', then o cannot have q in its dynamic skyline and thus o can be pruned for the candidate list.

Fig. 7. Refinement area defined by q_1, q_2 and o_1

Refinement. In the refinement step, each candidate c is verified by performing a dynamic skyline query using c as query point. The result should contain all $q_i \in Q$, otherwise c is dropped. The refinement step can be improved by the following observation (cf. Figure 7): for checking if a candidate o_1 has all $q_i \in Q$ in its dynamic skyline, it suffices to check whether there exists at least one other object $o_j \in \mathcal{D}$ which prevents one q_i from being part of the skyline. Such an object has to lie within the MBR defined by q_i and q_i' (which is obtained by reflecting q_i through o_1). If no point is within the $|Q|$ MBRs, then o_1 is reported as result.

6.2 Algorithm

The algorithm for $I\text{-}DSQ$, during the filter step, traverses the tree in a best first manner, where entries are accessed by their minimal distance (MinDist) to the farthest query object. For each entry e we check if e is completely contained in the union of pruning regions defined by all pairs of queries $(q_i, q_j) \in Q$; i.e., $\bigcup_{(q_i,q_j) \in Q} PR_{q_i}(q_j)$. In addition, for each accessed database object o_i and each query object q_j, the pruning region is extended by $PR_{q_j}(o_i)$. Analogously to the $Ik\text{-}NN$ case, lists for the candidates and pruned entries are maintained. Finally, the remaining candidates are refined using the refinement strategy described in Section 6.1.

7 Experiments

For each of the inverse query predicates discussed in the paper, we compare our proposed solution based on multi-query-filtering (MQF), with a naive approach (Naive) and another intuitive approach based on single-query-filtering (SQF). The naive algorithm (Naive) computes the corresponding reverse query for every $q \in Q$ and intersects their results iteratively. To be fair, we terminated Naive as soon as the intersection of results obtained so far is empty. SQF performs a Rk-NN (Rε-range / RDS) query using one randomly chosen query point as a filter step to obtain candidates. For each candidate an ε-range (k-NN / DS) query is issued and the candidate is confirmed if all query points are contained in the result of the query (refinement step). Since the pages accessed by the queries in the refinement step are often redundant, we use a buffer to further boost the performance of SQF. We employed R^*-trees ([1]) of pagesize 1Kb to index the data sets used in the experiments. For each method, we present the number of page accesses and runtime. To give insights into the impact of the different parameters on the cardinality of the obtained results we also included this number to the charts. In all settings we performed 1000 queries and averaged the results. All methods were implemented in Java 1.6 and tests were run on a dual core (3.0 Ghz) workstation with 2 GB main memory having windows xp as OS. The performance evaluation settings are summarized below; the numbers in **bold** correspond to the default settings:

parameter	values
db size	100000 (synthetic), 175812 (real)
dimensionality	2, **3**, 4, 5
ε	0.04, 0.05, **0.06**, 0.07, 0.08, 0.09, 0.1
k	50, **100**, 150, 200, 250
# inverse queries	1, 3, 5, **10**, 15, 20, 25, 30, 35
query extent	0.0001, 0.0002, 0.0003, **0.0004**, 0.0005, 0.0006

The experiments were performed using several data sets:

- Synthetic data sets: Clustered and uniformly distributed objects in d-dimensional space.
- Real Data set: Vertices in the Road Network of North America[2]. Contains 175,812 two-dimensional points.

The data sets were normalized, such that their minimum bounding box is $[0, 1]^d$. For each experiment, the query objects Q for the inverse query were chosen randomly from the database. Since the number of results highly depends on the distance between inverse query points (in particular for the $I\varepsilon$-RQ and Ik-NNQ) we introduced an additional parameter called *extent* to control the maximal distance between the query objects. The value of *extent* corresponds to the volume (fraction of data space) of a cube that minimally bounds all queries. For example in the 3D space the default cube would have a side length of 0.073. A small *extent* assures that the queries are placed close to each other generally resulting in more results. In this section, we show the behavior of all three algorithms on the uniform data sets only. Experiments on the other data sets can be found in the full version of the paper [2].

7.1 Inverse ε-Range Queries

We first compared the algorithms on inverse ε range queries. Figure 8(a) shows that the relative speed of our approach (MQF) compared to Naive grows significantly with increasing ε; for Naive, the cardinality of the result set returned by each query depends on the space covered by the hypersphere which is in $O(\varepsilon^d)$. In contrast, our strategy applies spatial pruning early, leading to a low number of page accesses. SQF is faster than Naive, but still needs around twice as much page accesses as MQF. MQF performs even better with an increasing number of query points in Q (as depicted in Figure 8(b)), as in this case the intersection of the ranges becomes smaller. The I/O-cost of SQF in this case remains almost constant which is mainly due to the use of the buffer which lowers the page accesses in the refinement step. Similar results can be observed when varying the database size (Figure 8(e)) and query extent (Figure 8(d)). For the data dimensionality experiment (Figure 8(c)) we set epsilon such that the sphere defined by ε covers always the same percentage of the dataspace, to make sure that we still obtain results when increasing the dimensionality (note, however, that the number of results is still unsteady). Increasing dimensionality has a negative effect on performance. However

---

[2] Obtained and modified from *http://www.cs.fsu.edu/~lifeifei/SpatialDataset.htm*. The original source is the *Digital Chart of the World Server (http://www.maproom.psu.edu/dcw/)*.

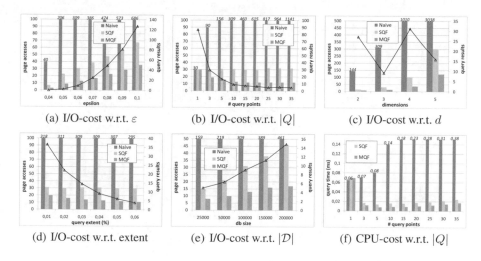

(a) I/O-cost w.r.t. ε (b) I/O-cost w.r.t. $|Q|$ (c) I/O-cost w.r.t. d

(d) I/O-cost w.r.t. extent   (e) I/O-cost w.r.t. $|\mathcal{D}|$ (f) CPU-cost w.r.t. $|Q|$

Fig. 8. $I\varepsilon$-Q algorithms on uniform data set

MQF copes better with data dimensionality than the other approaches. Finally, Figure 8(f) compares the computational costs of the algorithms. Even though Inverse Queries are I/O bound, MQF is still preferable for main memory problems.

7.2 Inverse k-NN Queries

The three approaches for inverse k-NN search show a similar behavior as those for the Iε-RQ. Specifically the behavior for varying k (Figure 9(a)) is comparable to varying ε and increasing the query number (Figure 9(b)) and the query extent (Figure 9(d)) yields the expected results. When testing on data sets with different dimensionality, the advantage of MQF becomes even more significant when d increases (cf. Figure 9(c)). In contrast to the Iε-RQ results for Ik-NN queries the page accesses of MQF decrease (see Figure 9(e)) when the database size increases (while the performance of SQF still degrades). This can be explained by the fact, that the number of pages accessed is strongly correlated with the number of obtained results. Since for the Iε-RQ the parameter ε remained constant, the number of results increased with a larger database. For Ik-NN the number of results in contrast decreases and so does the number of accessed pages by MQF. As in the previous set of experiments MQF has also the lowest runtime (Figure 9(f)).

7.3 Inverse Dynamic Skyline Queries

Similar results as for the Ik-NNQ algorithm are obtained for the inverse dynamic skyline queries (I-DSQ). Increasing the number of queries in Q reduces the cost of the MQF approach, while the costs of the competitors increase. Since the average number of results approaches 0 faster than for the other two types of inverse queries we choose 4 as the default size of the query set. Note that the number of results for I-DSQ intuitively increases exponentially with the dimensionality of the data set (cf. Figure 10(b)),

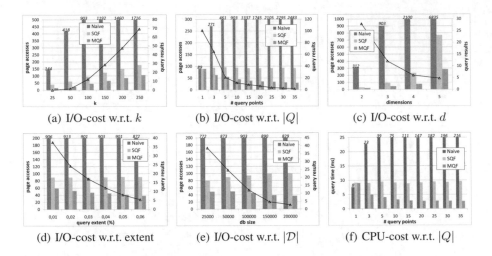

(a) I/O-cost w.r.t. k (b) I/O-cost w.r.t. $|Q|$ (c) I/O-cost w.r.t. d

(d) I/O-cost w.r.t. extent   (e) I/O-cost w.r.t. $|\mathcal{D}|$ (f) CPU-cost w.r.t. $|Q|$

Fig. 9. $Ik\text{-}NNQ$ algorithms on uniform data set

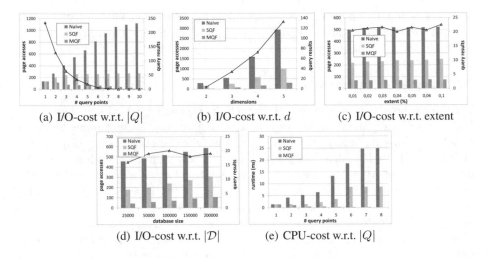

(a) I/O-cost w.r.t. $|Q|$ (b) I/O-cost w.r.t. d (c) I/O-cost w.r.t. extent

(d) I/O-cost w.r.t. $|\mathcal{D}|$ (e) CPU-cost w.r.t. $|Q|$

Fig. 10. $I\text{-}DSQ$ algorithms on uniform data set

thus this value can be much larger for higher dimensional data sets. Increasing the distance among queries does not affect the performance as seen in Figure 10(c); regarding the number of results in contrast to inverse range- and k-NN queries, inverse dynamic skyline queries are almost insensitive to the distance among the query points. The rationale is that dynamic skyline queries can have results which are arbitrary far from the query point, thus the same holds for the inverse case. The same effect can be seen for increasing database size (cf. Figure 10(d)). The advantage of MQF remains constant over the other two approaches. Like inverse range and k-NN queries, $I\text{-}DSQ$ are I/O bound (see Figure 10(e)), but MQF is still preferable for main-memory problems.

8 Conclusions

In this paper we introduced and formalized the problem for inverse query processing. We proposed a general framework to such queries using a filter-refinement strategy and applied this framework to the problem of answering inverse ε-range queries, inverse k-NN queries and inverse dynamic skyline queries. Our experiments show that our framework significantly reduces the cost of inverse queries compared to straightforward approaches. In the future, we plan to extend our framework for inverse queries with different query predicates, such as top-k queries. In addition, we will investigate inverse query processing in the bi-chromatic case, where queries and objects are taken from different data sets. Another interesting extension of inverse queries is to allow the user not only to specify objects that have to be in the result, but also objects that must not be in the result.

Acknowledgements. This work was supported by a grant from the Germany/Hong Kong Joint Research Scheme sponsored by the Research Grants Council of Hong Kong (Reference No. G_HK030/09) and the Germany Academic Exchange Service of Germany (Proj. ID 50149322).

References

1. Beckmann, N., Kriegel, H.-P., Schneider, R., Seeger, B.: The R*-Tree: An efficient and robust access method for points and rectangles. In: Proc. SIGMOD (1990)
2. Bernecker, T., Emrich, T., Kriegel, H.-P., Mamoulis, N., Zhang, S., Renz, M., Züfle, A.: Inverse queries for multidimensional spaces. In: The ACM Computing Research Repository, CoRR (2011), http://arxiv.org/abs/1103.0172
3. Dellis, E., Seeger, B.: Efficient computation of reverse skyline queries. In: VLDB, pp. 291–302 (2007)
4. Emrich, T., Kriegel, H.-P., Kröger, P., Renz, M., Züfle, A.: Boosting spatial pruning: On optimal pruning of mbrs. In: SIGMOD (June 6-11, 2010)
5. Hjaltason, G.R., Samet, H.: Ranking in spatial databases. In: Egenhofer, M.J., Herring, J.R. (eds.) SSD 1995. LNCS, vol. 951. Springer, Heidelberg (1995)
6. Korn, F., Muthukrishnan, S.: Influence sets based on reverse nearest neighbor queries. In: Proc. SIGMOD (2000)
7. Lian, X., Chen, L.: Monochromatic and bichromatic reverse skyline search over uncertain databases. In: SIGMOD Conference, pp. 213–226 (2008)
8. Papadias, D., Shen, Q., Tao, Y., Mouratidis, K.: Group nearest neighbor queries. In: Proceedings of the 20th International Conference on Data Engineering, ICDE 2004, Boston, MA, USA, March 30 -April 2, pp. 301–312 (2004)
9. Papadias, D., Tao, Y., Mouratidis, K., Hui, C.K.: Aggregate nearest neighbor queries in spatial databases. ACM Trans. Database Syst. 30(2), 529–576 (2005)
10. Singh, A., Ferhatosmanoglu, H., Tosun, A.S.: High dimensional reverse nearest neighbor queries. In: Proc. CIKM (2003)
11. Stanoi, I., Agrawal, D., Abbadi, A.E.: Reverse nearest neighbor queries for dynamic databases. In: Proc. DMKD (2000)
12. Tao, Y., Papadias, D., Lian, X.: Reverse kNN search in arbitrary dimensionality. In: Proc. VLDB (2004)
13. Vlachou, A., Doulkeridis, C., Kotidis, Y., Nørvåg, K.: Reverse top-k queries. In: ICDE, pp. 365–376 (2010)
14. Yang, C., Lin, K.-I.: An index structure for efficient reverse nearest neighbor queries. In: Proc. ICDE (2001)

Efficient Evaluation of k-NN Queries Using Spatial Mashups*

Detian Zhang[1,2,3], Chi-Yin Chow[2], Qing Li[2,3],
Xinming Zhang[1,3], and Yinlong Xu[1,3]

[1] Department of Computer Science and Technology,
University of Science and Technology of China, Hefei, China
[2] Department of Computer Science, City University of Hong Kong,
Hong Kong, China
[3] USTC-CityU Joint Advanced Research Center, Suzhou, China
tianzdt@mail.ustc.edu.cn, {chiychow,itqli}@cityu.edu.hk,
{xinming,ylxu}@ustc.edu.cn

Abstract. K-nearest-neighbor (k-NN) queries have been widely studied in time-independent and time-dependent spatial networks. In this paper, we focus on k-NN queries in time-dependent spatial networks where the driving time between two locations may vary significantly at different time of the day. In practice, it is costly for a database server to collect real-time traffic data from vehicles or roadside sensors to compute the best route from a user to an object of interest in terms of the driving time. Thus, we design a new spatial query processing paradigm that uses a spatial mashup to enable the database server to efficiently evaluate k-NN queries based on the route information accessed from an external Web mapping service, e.g., Google Maps, Yahoo! Maps and Microsoft Bing Maps. Due to the expensive cost and limitations of retrieving such external information, we propose a new spatial query processing algorithm that uses shared execution through grouping objects and users based on the road network topology and pruning techniques to reduce the number of external requests to the Web mapping service and provides highly accurate query answers. We implement our algorithm using Google Maps and compare it with the basic algorithm. The results show that our algorithm effectively reduces the number of external requests by 90% on average with high accuracy, i.e., the accuracy of estimated driving time and query answers is over 92% and 87%, respectively.

1 Introduction

With the ubiquity of wireless Internet access, GPS-enabled mobile devices and the advance in spatial database management systems, location-based services (LBS) have been realized to provide valuable information for their users based

---

* The work described in this paper was partially supported by a grant from City University of Hong Kong (Project No. 7200216) and the National Natural Science Foundation of China under Grant 61073185.

D. Pfoser et al. (Eds.): SSTD 2011, LNCS 6849, pp. 348–366, 2011.

on their locations [1, 2]. LBS are an abstraction of spatio-temporal queries. Typical examples of spatio-temporal queries include range queries (e.g., "*How many vehicles in a certain area*") [3, 4, 5] and k-nearest-neighbor (k-NN) queries (e.g., "*Find the k-nearest gas stations*") [4, 6, 7, 8].

The distance between two point locations in a road network is measured in terms of the network distance, instead of the Euclidean distance, to consider the physical movement constrains of the road network [5]. It is usually defined by the distance of their shortest path. However, this kind of distance measure would hide the fact that the user may take longer time to travel to his/her nearest object of interest (e.g., restaurant and hotel) than other ones due to many realistic factors, e.g., heterogeneous traffic conditions and traffic accidents. Driving time (or travel time) is in reality a more meaningful and reliable distance measure for LBS in road networks [9, 10, 11]. Figure 1 depicts an example where Alice wants to find the nearest clinic for emergency medical treatment. A traditional shortest-path based NN query algorithm returns Clinic X. However, a driving-time based NN query algorithm returns Clinic Y because Alice will spend less time to reach Y (3 mins.) than X (10 mins.).

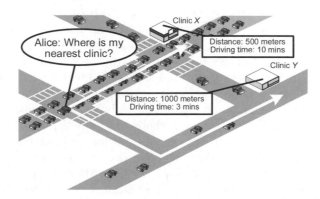

Fig. 1. A shortest-path based NN (X) versus a driving-time based NN (Y)

Since driving time is highly dynamic, e.g., the driving time on a segment of I-10 freeway in Los Angeles, USA between 8:30AM to 9:30AM changes from 30 minutes to 18 minutes, i.e., 40% decrease in driving time [9], it is almost impossible to accurately predict the driving time between two point locations in a road network based on their network distance. The best way to provide real-time driving time computation is to continuously monitor the traffic in road networks; however, it is difficult for every LBS provider to do so due to very expensive deployment cost and privacy issues.

A spatial mashup[1] (or GIS mashup), one of the key technologies in Web 2.0, provides a more cost-effective way to access route information in road networks from external Web mapping services, e.g., Google Maps, Yahoo! Maps, Microsoft

---

[1] A mashup is a web application that combines data, representation, and/or functionality from multiple web applications to create a new application [12].

Bing Maps and government agencies. However, existing spatial mashups suffer from the following limitations. (1) It is costly to access direction information from a Web mapping service, e.g., retrieving driving time from the Microsoft MapPoint web service to a database engine takes 502 ms while the time needed to read a cold and hot 8 KB buffer page from disk is 27 ms and 0.0047 ms, respectively [13]. (2) There is usually a limit on the number of requests to a Web mapping service, e.g., Google Maps allows only 2,500 requests per day for evaluation users and 100,000 requests per day for premier users [14]. (3) The use of retrieved route information is restricted, e.g., the route information must not be pre-fetched, cached, or stored, except only limited amount of content can be temporarily stored for the purpose of improving system performance [14]. (4) Existing Web mapping services only support primitive operations, e.g., the driving direction and time between two point locations. A database server has to issue a large number of external requests to collect small pieces of information from the supported simple operations to process relatively complex spatial queries, e.g., k-NN queries.

In this paper, we design an algorithm to processing k-NN queries using spatial mashups. Given a set of objects and a k-NN query with a user's location and a user specified maximum driving time t_{max} (e.g., "*Find the k-nearest restaurants that can be reached in 10 minutes by driving*"), our algorithm finds at most k objects with the shortest driving time and their driving time is no longer than t_{max}. The objectives of our algorithm are to reduce the number of external requests to a Web mapping service and provide query answers with high accuracy. To achieve our objectives, we use shared execution by grouping objects based on the road network topology and pruning techniques to reduce the number of external requests. We design two methods to group objects to adjust a performance trade-off between the number of external requests and the accuracy of query answers. We first present our algorithm in road networks with bidirectional road segments, and then adapt the algorithm to road networks with both one- and two-way road segments. In addition, we design another extension to further reduce the number of external requests by grouping users based on their movement direction and the road network topology for a system with high workloads or a large number of continuous k-NN queries.

To evaluate the performance of our algorithm, we build a simulator to compare it with a basic algorithm in a real road network. The results show that our algorithms outperform the basic algorithm in terms of the number of external requests and query response time. We also implement our algorithm using Google Maps [15]. The experimental results show that our algorithm provides highly accurate k-NN query answers.

The remainder of this paper is organized as follows. Section 2 describes the system model. Section 3 presents the basic algorithm and our algorithm. Section 4 gives two extensions to our algorithm. Simulation and experimental results are analyzed in Section 5. Section 6 highlights related work. Finally, Section 7 concludes this paper.

2   System Model

In this section, we describe our system architecture, road network model, and problem definition. Figure 2 depicts our system architecture that consists of three entities, users, a database server, and a Web mapping service provider. Users send k-NN queries to the database server at a LBS provider at anywhere and anytime. The database server processes queries based on local data (e.g., the location and basic information of restaurants) and external data (i.e., routes and driving time) accessed from the Web mapping service. In general, accessing external data is much more expensive than accessing internal data [13].

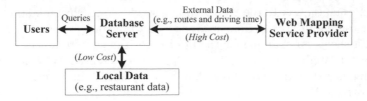

Fig. 2. System architecture

We use a graph $G = (V, E)$ to model a road network, where E and V are a set of road segments and intersections of road segments, respectively. For example, Figure 3a depicts a real road map that is modeled into an undirected graph (Figure 3b), where an edge represents a road segment (e.g., $I_1 I_2$ and $I_1 I_5$) and a square represents an intersection (e.g., I_1 and I_2). This road network model will be used in Section 3 because we assume that each road segment is bidirectional. In Section 4, we will use a directed graph to model a road network where each edge with an *arrow* or *double arrows* to indicate that the corresponding road segment is *one-way* or *two-way*, respectively (e.g., Figure 9).

Our problem is defined as follows. Given a set of objects O and a NN query $Q = (\lambda, k, t_{max})$ from a user U, where λ is U's location, k is U's specified maximum number of returned objects, and t_{max} is U's required maximum driving time from λ to a returned object, our system returns U at most k objects in O with the shortest driving time from λ and their driving time must be no longer than t_{max}, based on the routes and driving time accessed from a Web mapping service, e.g., Goolge Maps [15]. Since accessing the Web mapping service is expensive, our objectives are to reduce the number of external requests to the Web mapping service and provide highly accurate query answers.

3   Processing k-NN Queries Using Spatial Mashups

In this section, Section 3.1 describes a basic algorithm to process k-NN queries using spatial mashups. Then, Section 3.2 presents our efficient algorithm that aims to minimize the number of external requests to a Web mapping service by using shared execution and pruning techniques.

3.1 Basic Algorithm

Since there could be a very large number of objects in a spatial data set, it is extremely inefficient to issue an expensive external request to the Web mapping service to retrieve the route information and driving time from a user to each object in the data set. To reduce the number of external requests, the basic algorithm executes a range query in spatial networks [5] to prune the whole data set into a much smaller set of *candidate objects* that are within the maximum possible driving distance, $dist_{max}$, from the user. The most conservative way to compute $dist_{max}$ is to multiply the user-specified maximum driving time t_{max} by the maximum allowed driving speed of the road network. Since only the candidate objects can be reached by the user within the driving time of t_{max}, the database server only needs to issue one external request to the Web mapping service for each candidate object to access the route and driving time from the user to the object.

Figure 4 shows an example for the basic algorithm, where a user U is represented by a triangle, 10 objects R_1 to R_{10} are represented by circles, and the maximum allowed driving speed is 80 km/h. If U wants to find the nearest object within a driving time of five minutes (i.e., $k = 1$ and $t_{max} = 5$ mins.), $dist_{max}$ is 6.7 km. The database server executes a range query with a range distance of 6.7 km, where the road segments within the range distance are highlighted. Thus, the whole data set is pruned into a subset of seven candidate objects R_1, R_2, R_4, R_5, R_8, R_9 and R_{10}; the database server only needs to issue seven external requests rather than ten requests.

3.2 Our Efficient k-NN Query Processing Algorithm

Although the basic algorithm can reduce the whole data set into a much smaller subset, if the user-specified maximum driving time is very long or the object

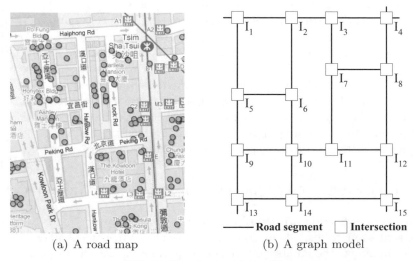

(a) A road map   (b) A graph model

Fig. 3. Road network model

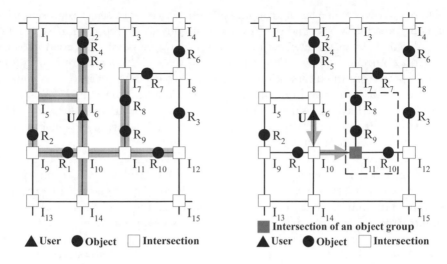

Fig. 4. Basic algorithm   **Fig. 5.** Object grouping

density in the vicinity of the user is very high, the database server may still need to issue a large number of external requests to the Web mapping service. To this end, we propose a new k-NN query processing algorithm that utilizes shared execution and pruning techniques to further reduce the number of external requests.

Overview. Our algorithm has four main steps. (1) Our algorithm takes a set of candidate objects computed by the basic algorithm as an input (Section 3.1). (2) It selects representative points in the road network and clusters objects to them to form groups (Section 3.2.1). (3) It issues an external request for each group to retrieve the route information from the user to the corresponding representative point and estimates the driving time from the user to each object in the group (Section 3.2.2). (4) The algorithm prunes candidate objects that cannot be part of a query answer (Section 3.2.3). The control flow of our algorithm is as follows. After it performs steps (1) and (2), it repeats steps (3) and (4) until all the candidate objects are processed or pruned.

3.2.1   Grouping Objects for Shared Execution

We observe that many spatial objects are generally located in clusters in real world. For example, many restaurants are located in a downtown area (Figure 3a). Thus, it makes sense to group nearby objects to a representative point. The database server issues only one external request to the Web mapping service for an object group. Then, it shares the retrieved route and driving time information from a user to the representative point among the objects in the group to estimate the driving time from the user to each object. This object grouping technique reduces the number of external requests from the number of candidate objects in the basic algorithm to the number of groups (at the worst case) for a k-NN query. Section 3.2.3 will describe another optimization to further reduce the number of external requests.

There are two challenges in grouping objects: (a) How to select representative points in a road network? and (b) How to group objects to representative points? Our solution is to select intersections as representative points in a road network and group objects to them. The reason is twofold: (1) Since there is only one possible path from the intersection of a group G to each object in G, it is easy to estimate the driving time from the intersection to each object in G. (2) A road segment should have the same conditions, e.g., speed limit and direction. The estimation of driving time would be more nature and accurate by considering a road segment as a basic unit [16].

Figure 5 gives an example for object grouping where objects R_8, R_9 and R_{10} are grouped together by intersection I_{11}. The database server only needs one external query from user U to I_{11} to the Web mapping service to retrieve the route information and driving time from U and I_{11}. Then, our algorithm estimates the driving time from I_{11} to each of its group members.

In the reminder of this section, we will present two methods for grouping objects. These methods have different performance trade-offs between the number of external requests and the accuracy of query answers. Section 3.2.2 (the third step) will discuss how to perform driving time estimation and Section 3.2.3 (the forth step) will discuss how to prune intersections safely to further reduce the number of external requests.

Method 1: Minimal intersection set (MinIn). To minimize the number of external requests to the Web mapping service, we should find a minimal set of intersections that cover all road segments having candidate objects, which are found by the basic algorithm. Given a k-NN query issued by a user U, we convert the graph G of the road network model into a subgraph $G_u = (V_u, E_u)$ such that E_u is an edge set where each edge contains at least one candidate object and V_u is a vertex set that consists of the vertices (intersections) of the edges in E_u. The result vertex set $V_u' \subseteq V_u$ is a minimal one covering all the edges in E_u. This problem is the vertex cover problem [17], which is NP-complete, so we use a greedy algorithm to find V_u'.

In the greedy algorithm, we calculate the number of edges connected to a vertex v as the degree of v. The algorithm selects the vertex from V_u with the largest degree to V_u'. In case of a tie, the vertex with the shortest distance to U is selected to V_u'. The selected vertex is removed from V_u and the edges of the selected vertex are removed from the E_u. After that, the degree of the vertices of the removed edges is updated accordingly.

Figure 6 gives an example for the MinIn method. After the basic algorithm finds a set of candidate objects (represented by black circles in Figure 7). The MinIn method first constructs a subgraph $G_u = (V_u, E_u)$ from G. Since edges I_2I_6, I_5I_9, I_9I_{10}, I_7I_{11} and $I_{11}I_{12}$ contain some candidate objects, these five edges constitute E_u and their vertices constitute V_u. Then, the MinIn method calculates the degree for each vertex in V_u. For example, the degree of I_{11} is two because two edges I_7I_{11} and $I_{11}I_{12}$ are connected to I_{11}. Figure 6a shows that I_{11} has the largest degree and is closer to U than I_9, I_{11} is selected to V_u' and the edges connected to I_{11} are removed from G_u. Figure 6b shows that the degree of the vertices I_7 and I_{12} of the two deleted edges is updated accordingly. Any vertex in G_u without any connected edge is removed from V_u immediately. Since I_9 has the largest degree, I_9 is selected to V_u' and the edges connected to I_9 are removed. Similarly, I_6 is selected to V_u' (Figure 6c). Since I_2 has no connected edge, I_2 is removed from V_u. After that, V_u becomes empty, so the MinIn method clusters the candidate objects into three groups (indicated by dotted rectangles), $I_6 = \{R_4, R_5\}$, $I_9 = \{R_1, R_2\}$, and $I_{11} = \{R_8, R_9, R_{10}\}$ (Figure 7).

Method 2: Nearest intersections (NearestIn). Although the MinIn method can minimize the number of external requests to the Web mapping service, it may lead to low accuracy in the estimated driving time for some objects. For example, consider object R_1 in Figure 7, since R_1 is grouped to intersection I_9 and I_{10} is not selected to V_u', the MinIn method will result in a path $P_1 : U \rightarrow I_6 \rightarrow I_5 \rightarrow I_9 \rightarrow R_1$. However, another path $P_2 : U \rightarrow I_{10} \rightarrow R_1$ may give shorter driving time than P_1. To this end, we design an alternative method, NearestIn, that groups objects to their nearest intersections. Figure 8 gives an example for NearestIn, where the candidate objects are represented by black circles. Since R_2 is closer to intersection I_9 than I_5, R_2 is grouped to I_9. The NearestIn method constructs six groups (indicated by dotted rectangles), $I_9 = \{R_2\}$, $I_{10} = \{R_1\}$, $I_2 = \{R_4, R_5\}$, $I_7 = \{R_8\}$, $I_{11} = \{R_9\}$, and $I_{12} = \{R_{10}\}$; thus, the database server needs to issue six external queries.

In general, NearestIn gives more accurate driving time than MinIn, but NearestIn needs more external requests than MinIn. As in our running examples, MinIn and NearestIn need to issue three and six external requests (one request per group), respectively. Thus, these methods provide different performance tradeoffs between the overhead of external requests and the accuracy of driving time estimation. We verify their performance tradeoffs in Section 5.

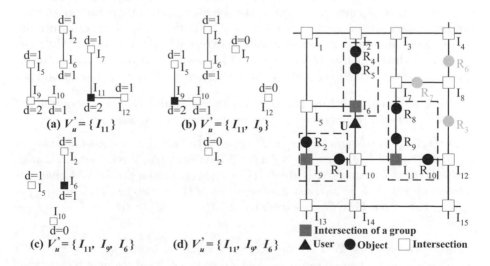

Fig. 6. Greedy algorithm for MinIn   **Fig. 7.** MinIn object grouping

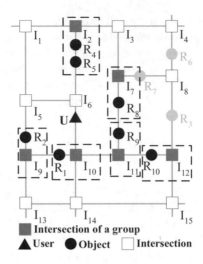

Fig. 8. NearestIn object grouping

3.2.2   Calculating Driving Time for Grouped Objects

Most Web mapping services, e.g., Goolge Maps, return turn-by-turn route information for a request. For example, if the database server sends a request to the Web mapping service to retrieve the route information from user U to intersection I_9 in Figure 7, the service returns the turn-by-turn route information, i.e., $Route(U \to I_9) = \{\langle d(U \to I_6), t(U \to I_6)\rangle, \langle d(I_6 \to I_5), t(I_6 \to I_5)\rangle, \langle d(I_5 \to I_9), t(I_5 \to I_9)\rangle\}$, where $d(A \to B)$ and $t(A \to B)$ are the Euclidean distance and driving time from location point A to location point B, respectively.

After grouping objects to intersections, each intersection in V'_u is processed based on its distance to the user in ascending order. The reason is that knowing the driving time of more objects closer to the user at an earlier stage gives more pruning power to the next step (Section 3.2.3). If a group contains only one object, the database server simply issues an external request to retrieve the route information from the user to the object. Otherwise, our algorithm calculates the driving time for each object and selects the best route (if appropriate) as follows.

Driving time calculation. Based on whether a candidate object is on the last road segment of a retrieved route, we can distinguish two cases.

Case 1: An object is on the last road segment. In this case, we assume that the speed of the last road segment is constant. For example, given the information of the route from U to I_9 (i.e., $Route(U \to I_9)$) retrieved from the Web mapping service, object R_2 is on the last road segment of the route, i.e., I_5I_9. Hence, the driving time from U to R_2 is calculated as: $t(U \to R_2) = t(U \to I_9) - t(I_5 \to I_9) \times \frac{d(R_2 \to I_9)}{d(I_5 \to I_9)}$.

Case 2: An object is NOT on the last road segment. In this case, a candidate object is not on a retrieved route, i.e., the object is on a road segment S connected to the last road segment S' of the route. We assume that the driving speed of

S is the same as that of S'. For example, given the retrieved information of the route from U to I_9 (i.e., $Route(U, I_9)$), object R_1 is not on the last road segment of the route, i.e., $I_5 I_9$. Hence, the driving time from U to R_1 is calculated as: $t(U \rightarrow R_1) = t(U \rightarrow I_9) + t(I_5 \rightarrow I_9) \times \frac{d(I_9 \rightarrow R_1)}{d(I_5 \rightarrow I_9)}$.

Route selection. The object grouping methods may select both the intersections of a road segment for a candidate object. In this case, the database server finds the route from the querying user to the object via each of the intersections. Thus, our algorithm selects the route with the shortest driving time. For example, since both the intersections of edge $I_9 I_{10}$, i.e., I_9 and I_{10}, are selected (Figure 8), the driving time from user U to R_1 is selected as $t(U \rightarrow R_1) = \min(d(U \rightarrow I_9 \rightarrow R_1), d(U \rightarrow I_{10} \rightarrow R_1))$.

3.2.3 Object Pruning

After the algorithm finds a current answer set A (i.e., the best answer so far) for a user U's k-NN query, this step keeps track of the longest driving time A_{max} from U to the objects in A, i.e., $A_{max} = \max\{t(U \rightarrow R_i) | R_i \in A\}$. It uses A_{max} to prune the candidate objects to further reduce the number of external requests to the Web mapping service. The basic idea is that a candidate object can be pruned safely if the smallest possible driving time from U to the object is not shorter than A_{max}, because the object cannot be part of a query answer. The smallest possible driving time of a candidate object is calculated by dividing the distance of the shortest path from the user to the object by the maximum allowed driving speed of the underlying road network. Whenever A_{max} is updated, this step checks all unprocessed candidate objects and prunes objects that cannot be part of the query answer. After removing an object from a group, if the group becomes empty, the corresponding intersection is also removed from the vertex set V'_u.

4 Extensions

In this section, we present two extensions to our advanced k-NN query processing algorithm. The first extension enables our algorithm to support one-way road segments (Section 4.1). Our algorithm with the second extension can group users for shared execution to further reduce the number of external requests (Section 4.2).

4.1 One-Way Road Segments

In real world, a street could be only one-way, so it is essential to extend our algorithm to support one-way road segments. The basic algorithm can be easily extended to support one-way streets by using a directed graph to model a road network. Figure 9 depicts an example where the objects on the highlighted road segments are probably reached from user U with the user-specified maximum driving time t_{max}. In this example, edges $I_{10} I_{14}$ and $I_{11} I_{12}$ are no longer considered, compared to the road network with only bi-directional road segments depicted in Figure 4.

To enable our algorithm to support one-way road segments, we only need to slightly modify the object grouping methods.

The MinIn method. After the basic algorithm finds a set of candidate objects, the MinIn method constructs a directed graph $G_u = (V_u, E_u)$ where E_u is a set of road segments containing some candidate objects and V_u is set of vertices of the edges in E_u. The degree of each vertex in V_u is calculated by the number of its outgoing edges in E_u. The basic idea of the greedy algorithm is that a vertex in V_u with the highest degree is selected to the result vertex set V'_u. Then, the edges of the selected vertex are removed from E_u and the degree of other adjacent vertices are updated accordingly. Any vertex with no edges is also removed from V_u.

Figure 12 depicts an example, where six candidate objects are found by the basic algorithm. Since there are four edges containing some candidate objects, we construct a directed graph G_u with $E_u = \{I_5 I_9, I_9 I_{10}, I_2 I_6, I_7 I_{11}\}$ and $V_u = \{I_2, I_5, I_6, I_7, I_9, I_{10}, I_{11}\}$. Since I_9 has two outgoing edges $I_9 I_5$ and $I_9 I_{10}$, the degree of I_9 is two (Figure 11a). Figure 11a shows that I_9 has the largest degree, I_9 is selected to V'_u and I_9 is removed from V_u. After the edges of I_9 are removed, I_5 and I_{10} have no more edges, so they both are deleted from V_u. Then, since I_6 is closer to U than I_{11}, I_6 is selected (Figure 11b). I_{11} is next to be selected to V'_u (Figure 11c). After deleting I_7, V_u becomes empty (Figure 11d), so the MinIn method is done. The candidate objects are grouped into three groups indicated by dotted rectangles, i.e., $I_9 = \{R_1, R_2\}$, $I_6 = \{R_4, R_5\}$, and $I_{11} = \{R_8, R_9\}$, as illustrated in Figure 12.

The NearestIn method. Since a candidate object on a one-way road segment can only be reached from a user through its starting intersection, the object is simply grouped to the starting intersection. For example, object R_1 is grouped

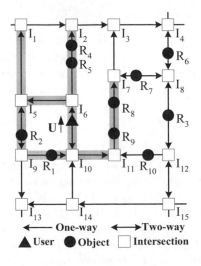

Fig. 9. Basic algorithm

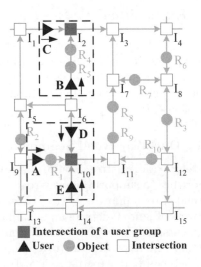

Fig. 10. User grouping

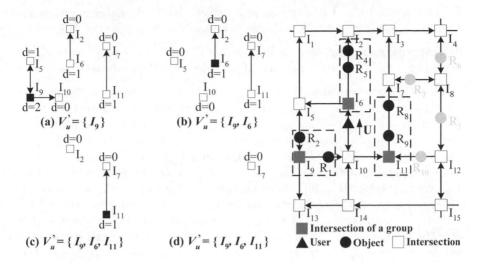

Fig. 11. Greedy algorithm for MinIn   **Fig. 12.** MinIn object grouping

to I_9, in Figure 12. However, a candidate object is grouped to the nearest intersection for a two-way road segment.

4.2 Grouping Users for Shared Execution

To further improve the system performance of a database server with a very high
workload, e.g., a large number of users or continuous queries, we design another
shared execution for grouping users. Similar to object grouping, we consider
intersections in the road network as representative points. The database server
only needs to issue one external request from the intersection of a user group $\mathcal{G}_u$
to the intersection of an object group $\mathcal{G}_o$ to estimate the driving time from each
user U_i in $\mathcal{G}_u$ to each object R_j in $\mathcal{G}_o$. In general, our algorithm with the user
grouping extension has five main steps.

User grouping. The key difference between user grouping and object grouping
is that users are moving. Grouping users has to consider their movement direction, so a user is grouped to the nearest intersection to which the user is moving.
Figure 10 depicts an example, where user A on edge I_9I_{10} is moving towards
I_{10}, so A is grouped to I_{10}. Similarly, users D and E are also grouped to I_{10}.
Users C and B on edges I_1I_2 and I_6I_2, respectively, are both moving to I_2, so
they are grouped to I_2.

Candidate objects. This step uses the basic algorithm (described in Sec-
tion 3.1) to find a set of candidate objects $\mathcal{R}$ for a user group $\mathcal{G}_u$. Since the
users in $\mathcal{G}_u$ may have different user-required maximum driving times, the basic algorithm finds the candidate objects that are within the largest required
maximum driving time of the intersection of $\mathcal{G}_u$. Consider the user group of intersection I_{10} in Figure 10, if the required maximum driving times of users A,

D, and E are 5, 10, and 20 minutes, respectively, the basic algorithm finds a set of candidate objects within 20 minutes driving time from I_{10}.

Object grouping. This step employs one of the object grouping methods presented in Section 3.2.1 to group candidate objects in $\mathcal{R}$ to intersections.

Driving time calculation. For each user group $\mathcal{G}_u$, our algorithm processes the object groups one by one based on the distance between their intersection and the intersection of $\mathcal{G}_u$ in ascending order. Such a processing order not only gives more pruning power to the object pruning step, but it also provides a fair response time for the users in $\mathcal{G}_u$. When our algorithm finds that a candidate object cannot be part of answers of any users in $\mathcal{G}_u$, the object is pruned. Similarly, when we guarantee that the remaining candidate objects in $\mathcal{R}$ cannot be part of a user's answer, the user's current answer is returned to the user without waiting the completion of processing all the queries issued by the users in $\mathcal{G}_u$. After the database server retrieves the route and driving time information from the intersection of $\mathcal{G}_u$ (I_u) to the intersection of an object group $\mathcal{G}_o$ (I_o), this step computes the driving time from I_u to each object R_j in $\mathcal{G}_o$, i.e., $t(I_u \rightarrow R_j)$, as presented in Section 3.2.2. Then, this step estimates the driving time from each user U_i in $\mathcal{G}_u$ to I_u, i.e., $t(U_i \rightarrow I_u)$. Since a user is grouped to an intersection to which the user is moving, the driving time from the user U_i to I_u has to be added to the driving time from I_u to each object R_j in $\mathcal{G}_o$. The user is either on the first road segment $I_u I_p$ of the route retrieved from the Web mapping service or on another road segment $I_u I_q$ connected to $I_u I_p$, so we use the driving speed of $I_u I_p$ to estimate the required diving time by $t(U_i \rightarrow I_u) = t(I_u \rightarrow I_p) \times \frac{d(U_i \rightarrow I_u)}{d(I_u \rightarrow I_p)}$; hence, the driving time from U_i to R_j is $t(U_i \rightarrow R_j) = t(U_i \rightarrow I_u) + t(I_u \rightarrow R_j)$.

Object pruning. After the algorithm finds a current answer set A_i (i.e., the best answer so far) for each user U_i in a user group $\mathcal{G}_u$, this step keeps track of the longest driving time A_{max_i} from U_i to the objects in A_i and the smallest possible driving time $A_{min_{\mathcal{G}}}(R_j)$ from any user U_i in $\mathcal{G}_u$ to an unprocessed candidate object $R_j \in \mathcal{R}$. $A_{min_{\mathcal{G}}}(R_j)$ is calculated by dividing the distance of the shortest path from U_i to R_j by the maximum allowed driving speed of the underlying road network. The step finds the largest value of A_{max_i} of $\mathcal{G}_u$, i.e., $A_{max_{\mathcal{G}}} = \max\{A_{max_i} | U_i \in \mathcal{G}_u\}$. For a candidate object R_j in $\mathcal{R}$, if $A_{min_{\mathcal{G}}}(R_j) \geq A_{max_{\mathcal{G}}}$, R_j is pruned from $\mathcal{G}_u$ because R_j cannot be part of any query answer. Whenever $A_{max_{\mathcal{G}}}$ is updated, this step checks the candidate objects in $\mathcal{R}$. If an object group becomes empty, the intersection of the object group is removed from V_u'. If $A_{max_i} < \min\{A_{min_{\mathcal{G}}}(R_j) | R_j \in \mathcal{R}\}$, any unprocessed candidate object cannot be part of U_i's query answer; thus, U_i's current answer is returned to U_i and U_i is removed from $\mathcal{G}_u$. The processing of $\mathcal{G}_u$ is done if all the intersections in V_u' have been processed/pruned or $\mathcal{G}_u$ becomes empty.

4.3   Performance Analysis

We now evaluate the performance of our algorithm. Let M and N be the number of users and the number of objects in the database server, respectively. Without any optimization, a naive algorithm issues N external requests for each user; hence, the total number of external requests is $Cost_N = N \times M$. By using the basic algorithm, the database server executes a range query to find a set of candidate objects for each user. Suppose the number of candidate objects for user U_i is α_i; the total number of external requests of M users is $Cost_B = \sum_{i=1}^{M} \alpha_i$. Since usually $\alpha_i \ll N$, $Cost_B \ll Cost_N$.

Our efficient algorithm further uses object- and user-grouping shared execution schemes to reduce the number of external requests. For each user group $\mathcal{G}_j$, our algorithm uses one of the object grouping methods to group its candidate objects. Suppose that the number of user groups is m and the number of object groups is β_j. The total number of required external requests is $Cost_E = \sum_{j=1}^{m} \beta_j$. In general, $m \ll M$ and $\beta_i \ll \alpha_i \ll N$, so $Cost_E \ll Cost_B \ll Cost_N$. We will confirm our performance analysis through the experiments in Section 5.

5   Performance Evaluation

In this section, we evaluate our efficient k-NN query processing algorithm using spatial mashups in a real road network of Hennepin County, MN, USA. We select a square area of 8×8 km^2 that contains 6,109 road segments and 3,593 intersections, and the latitude and longitude of its left-bottom and right-top corners are (44.898441, -93.302791) and (44.970094, -93.204015), respectively. The maximum allowed driving speed is 110 km per hour. In all the experiments, we compare our advanced algorithm with user grouping (UG) and object grouping, including MinIn and NearestIn, which are denoted as MI-UG and NI-UG, respectively, with the basic algorithm. We first evaluate the performance of our algorithm through a large-scale simulation (Section 5.1), and then evaluate its accuracy through an experiment using Google Maps [15] (Section 5.2).

5.1   Simulation Results

Unless mentioned otherwise, we generate 10,000 objects and 10,000 users that are uniformly distributed in the road network for all the simulation experiments. The default user required maximum driving time (t_{max}) is 120 seconds and the requested number of nearest objects (k) is 20. We measure the performance of our algorithm in terms of the average number of external requests per user to the Web mapping service and the average query response time per user.

Effect of the number of objects. Figure 13 depicts the performance of our algorithms with respect to increasing the number of objects from 4,000 to 20,000. Our algorithms, i.e., MI-UG and NI-UG, outperform the basic algorithm. The performance of our algorithms is only slightly affected by the increase of the number of objects (Figure 13a). The results confirm that our algorithms can scale up to a large number of objects. Figure 13b shows the average query response time of our algorithms. The average query response time is the sum of the average query processing time of the algorithm and the multiplication of the average

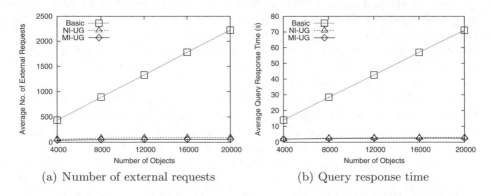

(a) Number of external requests   (b) Query response time

Fig. 13. Number of objects

number of external queries per user and the average response time per external request. The average response time per external request is 32 milliseconds that is derived from the experiments (Section 5.2).

Effect of the number of users. Figure 14 gives the performance of our algorithms with an increase of the number of users from 4,000 to 20,000. The results also show that our algorithms outperform the basic algorithm. It is expected that MI-UG needs smaller numbers of external requests than NI-UG. Since our algorithms effectively group users to intersections for shared execution, when there are more users, the number of external requests reduces.

Effect of the user-required maximum driving time. Figure 15 shows the performance of our algorithms with various user-required maximum driving times (t_{max}) that are increased from a range of [60, 120] to [60, 600] seconds. When t_{max} gets longer, more candidate objects can be reached by the user; thus, the number of external requests increases. The results also indicate that the increase rate of our algorithms is much smaller than that of the basic algorithm, so our algorithms significantly improve the system scalability.

In summary, all the simulation results consistently show that our algorithms outperform the basic algorithms, in terms of both the number of external requests to the Web mapping service and the query response time. The results also confirm that our algorithms effectively scale up to a large number of objects, a large number of users, and long user-required maximum driving times.

5.2   Experiment Results

To evaluate the accuracy of our query processing algorithm, we implement it with the proposed object and user grouping schemes using the Google Maps [15]. Because Google Maps allows only 2,500 requests per day for evaluation users, all the experiments in this section contain 100 users and 500 objects that are

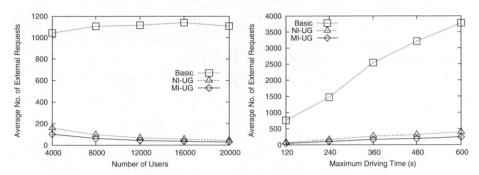

Fig. 14. Number of users   **Fig. 15.** Required maximum driving time

uniformly distributed in the underlying road network. We evaluate the accuracy of the driving time estimation and the accuracy of k-NN query answers.

Accuracy of the driving time estimation. The first experiment evaluates the accuracy of the driving time estimation used in object- and user-grouping methods with respect to varying the user required maximum driving time t_{max} from 120 to 600 seconds, as depicted in Figure 16. We compare the accuracy of the driving time estimation of our algorithms MI-UG and NI-UG with the basic algorithm which retrieves the driving time from each user to each object from the Google Maps directly and finds query answers. The accuracy of an estimated driving time is computed by:

$$Accuracy\ of\ estimated\ driving\ time\ (Acc_{time}) = 1 - \min\left(\frac{|\widehat{T} - T|}{T}, 1\right), \quad (1)$$

where T and $\widehat{T}$ are the actual driving time (retrieved by the basic algorithm) and the estimated one, respectively, and $0 \le Acc_{time} \le 1$. The results show that the accuracy of our algorithms is at least 0.92; our algorithms achieve highly accurate driving time estimation. Since MI-UG generates the smallest number of external requests, its accuracy is worse than NI-UG.

Accuracy of query answers. The second experiment evaluates the accuracy of k-NN query answers returned by our algorithms MI-UG and NI-UG with respect to increasing the required number of nearest objects (k) from 1 to 50. The accuracy of a query answer returned by our algorithms is calculated by:

$$Accuracy\ of\ a\ query\ answer\ (Acc_{ans}) = \frac{|\widehat{A} \cap A|}{|A|}, \quad (2)$$

where A is an exact query answer returned by the basic algorithm and $\widehat{A}$ is a query answer returned by our algorithms and $0 \le Acc_{ans} \le 1$. Figure 17 shows that our algorithms can provide highly accurate query answers. When $k = 1$, the accuracy of all our algorithms is over 87%. When k increases, the accuracy

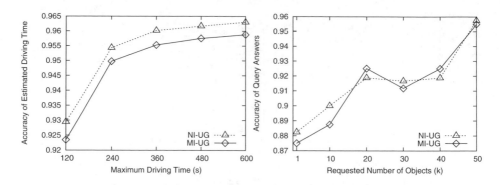

Fig. 16. Accuracy of driving time   **Fig. 17.** Accuracy of query answers

of our algorithms improves. Therefore, the results confirm that our algorithms not only effectively reduce the number of external requests to the Web mapping service, but they also provide highly accurate k-NN query answers.

6   Related Work

Existing location-based query processing algorithms can be categorized into two main classes: location-based queries in *time-independent spatial networks* and in *time-dependent spatial networks*. (1) *Time-independent spatial networks.* In this class, location-based query processing algorithms assume that the cost or weight of a road segment, which can be in terms of distance or travel time, is constant (e.g., [5, 18, 19]). These algorithms mainly rely on pre-computed distance or travel time information of road segments in road networks. However, the actual travel time of a road segment may vary significantly during different times of the day due to dynamic traffic on road segments [9, 10]. (2) *Time-dependent spatial networks.* The location-based query processing algorithms designed for time-dependent spatial networks have the ability to support dynamic weights of road segments and topology of a road network, which can change with time. Location-based queries in time-dependent spatial networks are more realistic but also more challenging. George et al. [11] proposed a time-aggregated graph, which uses time series to represent time-varying attributes. The time-aggregated graph can be used to compute the shortest path for a given start time or to find the best start time for a path that leads to the shortest travel time. In [9, 10], Demiryurek et al. proposed solutions for processing k-NN queries in time-dependent road networks where the weight of each road segment is a function of time.

In this paper, we focus on k-nearest-neighbor (NN) queries in time-dependent spatial networks. Our work distinguishes from previous work [9, 10, 11] is that our focus is not on modeling the underlying road network based on different criteria. Instead, we rely on third party Web mapping services, e.g., Google Maps, to compute the travel time of road networks and provide the direction and driving time information through spatial mashups. Since the use of external requests to access route information from a third party is more expensive than accessing local data [13], we propose a new k-NN query processing algorithm for spatial mashups that uses shared execution and pruning techniques to reduce the number of external requests and provide highly accurate query answers.

There are some query processing algorithms designed to deal with expensive attributes that are accessed from external Web services (e.g., [13, 20, 21]). To minimize the number of external requests, these algorithms mainly focused on using either some *cheap* attributes that can be retrieved from local data sources [13, 20] or sampling methods [21] to prune candidate objects, and they only issue absolutely necessary external requests. The closest work to this paper is [13], where Levandoski et al. developed a framework for processing spatial skyline queries by pruning candidate objects that are guaranteed not to be part of a query answer and only issuing an external request for each remaining candidate object to retrieve the driving time from the user to the object. However, all previous work did not study how to use shared execution techniques and the road network topology to reduce the number of external requests to external Web services that is exactly the problem that we solved in this paper.

7   Conclusion

In this paper, we proposed a new k-nearest-neighbor query processing algorithm for a database server using a spatial mashup to access driving time from a Web mapping service, e.g., Google Maps. We first designed two object grouping methods (i.e., MinIn and NearestIn) to group objects to intersections based on the road network topology to achieve shared execution and use a pruning technique to reduce the number of expensive external requests to the Web mapping service. We also extended our algorithm to support one-way road segments, and designed a user grouping method for shared execution to further reduce the number of external requests. We compare the performance of our algorithms with the basic algorithm through simulations and experiments. The results show that our algorithms significantly reduce the number of external requests and provide highly accurate of k-NN query answers.

References

[1] Jensen, C.S.: Database aspects of location-based services. In: Location-Based Services, pp. 115–148. Morgan Kaufmann, San Francisco (2004)

[2] Lee, D.L., Zhu, M., Hu, H.: When location-based services meet databases. Mobile Information Systems 1(2), 81–90 (2005)

[3] Gedik, B., Liu, L.: MobiEyes: A distributed location monitoring service using moving location queries. IEEE TMC 5(10), 1384–1402 (2006)

[4] Mokbel, M.F., Xiong, X., Aref, W.G.: SINA: Scalable incremental processing of continuous queries in spatio-temporal databases. In: ACM SIGMOD (2004)

[5] Papadias, D., Zhang, J., Mamoulis, N., Tao, Y.: Query processing in spatial network databases. In: VLDB (2003)

[6] Hu, H., Xu, J., Lee, D.L.: A generic framework for monitoring continuous spatial queries over moving objects. In: ACM SIGMOD (2005)

[7] Mouratidis, K., Papadias, D., Hadjieleftheriou, M.: Conceptual partitioning: An efficient method for continuous nearest neighbor monitoring. In: ACM SIGMOD (2005)

[8] Tao, Y., Papadias, D., Shen, Q.: Continuous nearest neighbor search. In: VLDB (2002)

[9] Demiryurek, U., Banaei-Kashani, F., Shahabi, C.: Efficient k-nearest neighbor search in time-dependent spatial networks. In: Bringas, P.G., Hameurlain, A., Quirchmayr, G. (eds.) DEXA 2010. LNCS, vol. 6261, pp. 432–449. Springer, Heidelberg (2010)

[10] Demiryurek, U., Banaei-Kashani, F., Shahabi, C.: Towards k-nearest neighbor search in time-dependent spatial network databases. In: International Workshop on Databases in Networked Systems (2010)

[11] George, B., Kim, S., Shekhar, S.: Spatio-temporal network databases and routing algorithms: A summary of results. In: Papadias, D., Zhang, D., Kollios, G. (eds.) SSTD 2007. LNCS, vol. 4605, pp. 460–477. Springer, Heidelberg (2007)

[12] Vancea, A., Grossniklaus, M., Norrie, M.C.: Database-driven web mashups. In: IEEE ICWE (2008)

[13] Levandoski, J.J., Mokbel, M.F., Khalefa, M.E.: Preference query evaluation over expensive attributes. In: Proc. of ACM CIKM (2010)

[14] Google Maps/Google Earth APIs Terms of Service, http://code.google.com/apis/maps/terms.html (last updated: May 27, 2009)

[15] Google Maps, http://maps.google.com

[16] Kanoulas, E., Du, Y., Xia, T., Zhang, D.: Finding fastest paths on a road network with speed patterns. In: IEEE ICDE (2006)

[17] Cormen, T.H., Leiserson, C.E., Rivest, R.L., Stein, C.: Introduction to Algorithms, 2nd edn. The MIT Press, Cambridge (2001)

[18] Samet, H., Sankaranarayanan, J., Alborzi, H.: Scalable network distance browsing in spatial databases. In: ACM SIGMOD (2008)

[19] Huang, X., Jensen, C.S., Šaltenis, S.: The islands approach to nearest neighbor querying in spatial networks. In: Anshelevich, E., Egenhofer, M.J., Hwang, J. (eds.) SSTD 2005. LNCS, vol. 3633, pp. 73–90. Springer, Heidelberg (2005)

[20] Bruno, N., Gravano, L., Marian, A.: Evaluating top-k queries over web-accessible databases. In: IEEE ICDE (2002)

[21] Chang, K.C.C., Hwang, S.W.: Minimal probing: Supporting expensive predicates for top-k queries. In: ACM SIGMOD (2002)

SeTraStream: Semantic-Aware Trajectory Construction over Streaming Movement Data

Zhixian Yan[1,*], Nikos Giatrakos[2,**], Vangelis Katsikaros[2],
Nikos Pelekis[2,**], and Yannis Theodoridis[2,**]

[1] EPFL, Switzerland
zhixian.yan@epfl.ch
[2] University of Piraeus, Greece
{ngiatrak,vkats,npelekis,ytheod}@unipi.gr

Abstract. Location data generated from GPS equipped moving objects are typically collected as streams of spatiotemporal $\langle x, y, t \rangle$ points that when put together form corresponding *trajectories*. Most existing studies focus on building ad-hoc querying, analysis, as well as data mining techniques on formed trajectories. As a prior step, trajectory construction is evidently necessary for mobility data processing and understanding, including tasks like trajectory data cleaning, compression, and segmentation so as to identify semantic trajectory episodes like stops (e.g. while sitting and standing) and moves (while jogging, walking, driving etc). However, semantic trajectory construction methods in the current literature are typically based on offline procedures, which is not sufficient for real life trajectory applications that rely on timely delivery of computed trajectories to serve real-time query answers. Filling this gap, our paper proposes a platform, namely SeTraStream, for online semantic trajectory construction. Our framework is capable of providing real-time trajectory data *cleaning, compression, segmentation* over streaming movement data.

1 Introduction

With the growth of location-based tracking technology like GPS, RFID and GSM networks, an enormous amount of trajectory data are generated from various real life applications, including traffic management, urban planning and geo-social networks. A lot of studies have already been established on trajectories, ranging from data management to data analysis. The focus of trajectory data management includes building data models, query languages and implementation aspects, such as efficient indexing, query processing, and optimization techniques [12][25]; whilst the analysis aims at trajectory data mining including issues like classification, clustering, outlier detection, as well as trajectory pattern discovery (e.g. sequential, periodic and convoy patterns) [9][13][20][21].

---

* This work was conducted during a Short Term Scientific Mission (STSM) of the author sponsored by the COST MOVE project.
** Nikos Giatrakos, Nikos Pelekis and Yannis Theodoridis were partially supported by the EU FP7/ICT/FET Project MODAP.

D. Pfoser et al. (Eds.): SSTD 2011, LNCS 6849, pp. 367–385, 2011.

Recently, semantic trajectory computation has attracted the research interest [1][3][29][30][31][32]. The focus of semantic trajectory construction is initially on the extraction of *meaningful* trajectories from the raw positioning data like GPS feeds. Moreover, sensory elements placed on vehicles can provide additional lower-scale information about their movement. Semantic trajectories manage to encompass both objects' spatiotemporal movement characteristics (at a certain level of abstraction) as well as useful information regarding objects' movement patterns (e.g dwelling, speeding, tailgating) and social activities (see Fig. 1) assigned to different time intervals throughout their lifespan. Current methods of such kind of trajectory construction are mainly offline [1][3][29][30][31][32], which is not enough for modern, real life applications, because positioning data of moving objects are continuously generated as streams and corresponding querying operations often demand result delivery in an online and continuous fashion.

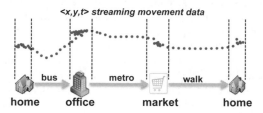

Fig. 1. From streaming movement data to semantic trajectory

Motivating Examples. Online semantic trajectory construction can be useful in many traffic monitoring scenarios where authorities are interested in identifying apart from recent (i.e., within a restricted time window) objects' trajectory representation, the behavior of the drivers by posing queries of the form: *"Report every τ secs the movement and driving behavior of the objects within area A during the last T minutes"*. In that, authorities are able to continuously diagnosing streets where the density of vehicles whose drivers tend to have aggressive (speeding, tailgating, driving at the edges of the lanes etc.) behavior has recently become high, thus enabling suitable placement and periodic rearrangement of traffic wardens and patrol cars. As another example, state-of-the-art navigation services (http://world.waze.com/) provide the potential for combining traditional routing functionality with social networking facilities. Online semantic trajectory construction allows users to acquire a compact picture of the movement and the social activities of interconnected friends around their moving area.

This paper proposes *SeTraStream*, a real-time platform that can progressively process raw mobility data arriving within a restricted time window and compute semantic-aware trajectories online. Before that, a number of data preparation steps need to be considered so as to render data easy to handle and ready to reveal profound movement patterns. The talk regards data cleaning and compression that precede the online segmentation and semantic trajectory computation procedures. Data cleaning is dealing with trajectory errors, including systematic errors (outlier removal) and random errors (smooth noise) [22][31]; compression considers data reduction because trajectory data grow rapidly and lack of compression sooner or later leads to exceeding system capacity [16][23]; segmentation is used for dividing trajectories into episodes where each episode is in some sense homogeneous (e.g. sharing similar velocity, direction etc.) [3] and thus expresses unchanged movement pattern; semantic computation can further extract high-level trajectory concepts like stops/moves [29], and even provide additional tagging support like the activity for stops (e.g. home, office, shopping) and the transportation mode (e.g. metro, bus, walking) for moves [1][30][31][33].

Challenges. It is non-trivial to establish a real-time semantic trajectory computation platform. There exist new technical challenges compared to the existing offline solutions: (1) *Efficient Computation:* Large amounts of movement data are generated continuously, therefore we need to come up with more efficient algorithms which can handle different levels of trajectories in an acceptable time – including all data processing aspects like data cleaning, compression, segmentation, and semantic tagging; (2) *Suitable Trajectory Segmentation Decision Making:* Algorithms in offline trajectory construction typically tune a lot of thresholds placed on movement features (like *acceleration, direction alteration, stop duration etc.*) to find their most suitable values, sometimes in a per object fashion. However, in the real-time context the movement attribute distribution may tremendously vary over time and continuous parameter tuning is prohibitive for real-time semantic trajectory construction. Thus, suitable techniques should not rely on many predefined thresholds on certain movement features but instead consider pattern alterations during the trajectory computation process. (3) *Semantic Trajectory Tagging:* After trajectory segmentation, the outcomes should provide the potentials for semantic tags to be explored, e.g. characterization of the activity (shopping, work) or means of movement that is taking place in episodes (e.g. car, metro, bus in Fig. 1).

Contributions. Towards the objective of real-time semantic trajectory construction, the core contributions of our paper are:

- *Online Trajectory Preprocessing.* As a prior step for constructing semantic trajectories, we significantly redesign trajectory data preprocessing in the real-time context, including *online cleaning* and *online compression*. Our cleaning includes an one-loop procedure for removing outliers and alleviating errors based on a *Kernel smoothing* method. SeTraStream's compression scheme uses a combination of the *Synchronized Euclidean Distance* (*sed*) and the novel definition of a *Synchronized Correlation Coefficient* (*scc*).
- *Online Trajectory Construction.* We design techniques for finding division points which infer trajectory episodes during online trajectory segmentation. SeTraStream's segmentation outcomes are later easy to handle and a *semantic tagging* classifier can then be applied for tag assignment on identified episodes, e.g. "driving", "jogging", "dwelling for shopping" etc.
- *Implementation Platform & Evaluation.* We implement SeTraStream's multilayer procedure for semantic trajectory construction and evaluate it, considering different real life trajectory datasets. The results demonstrate the

ability of SeTraStream to accurately provide computed semantic-aware trajectories in real-time, readily available for applications' querying purposes.

The rest of the paper proceeds as follows. In the upcoming section we discuss existing related works. Section 3 describes the preliminaries for semantic trajectory computation in SeTraStream, while in section 4 we present the data preparation procedures regarding incoming data cleaning and compression. In Section 5 we present SeTraStream's online segmentation algorithms and in Section 6 we experimentally evaluate our techniques. Eventually, section 7 includes concluding remarks and future work considerations.

2   Related Work

Trajectory construction is the procedure of reconstructing trajectories from the original sequence of spatiotemporal records of moving objects. Tasks involved in this procedure mainly include *data cleaning*, *data compression* and *data segmentation*. *Data cleaning* is dealing with trajectory errors which are quite common in GPS alike trajectory recordings. There are two types of errors: the outliers which are far away from the true values and need to be removed; the noisy data that should be corrected and smoothed. Several works [22][28][31] design specific filtering methods to remove outliers and smoothing methods to deal with small random errors. Regarding network-constrained moving objects, a number of map matching algorithms have been designed to refine the raw GPS records [2][16].

Trajectory data are generated continuously, in a high frequency and sooner or later grow beyond systems' computational and memory capacity. Therefore, *data compression* is a fundamental task for supporting scalable applications. The spatiotemporal compression methods for trajectory data can be classified into four types: i.e. *top-down*, *bottom-up*, *sliding window*, and *opening window*. The top-down algorithm recursively splits the trajectory sequence and selects the best position in each sub-sequence. A representative top-down method is the Douglas-Peucker (DP) algorithm [6], with many extended implementation techniques. The bottom-up algorithm starts from the finest possible representation, and merges the successive data points until some halting conditions are met. Sliding window methods compress data in a fixed window size; whilst open window methods use a dynamic and flexible window size for data segmentation. To name but a few methods: Meratnia et al. propose *Top-Down Time Ratio* (TD-TR) and *OPen Window Time Ratio* (OPW-TR) for the compression of spatiotemporal trajectories [23]. In addition, the work of [26] provides two sampling based compression methods: *threshold-guided sampling* and *STTrace* to deal with limited memory capacity.

Recently, semantic-based trajectory model construction has emerged as a hot topic for reconstructing trajectories, such as the stop-move concept in [29]. From a semantic point of view, a raw trajectory as a sequence of GPS points can be abstracted to a sequence of meaningful episodes (e.g. *begin, move, stop, end*). Yan et al. design a computing platform to progressively generate spatiosemantic trajectories from the raw GPS tracking feeds [31][32]. In that approach, different levels of trajectories are constructed, from *spatiotemporal trajectories, structured trajectories* to the final *semantic trajectories*, in four computational layers, i.e. *data preprocessing, trajectory identification, trajectory structure* and *semantic enrichment*.

Trajectory episodes like stops and moves can be computed with given geographic artifacts [1] or only depend on spatiotemporal criteria like density, velocity, direction etc. [24][27][31]. Alvares et al. develop a mechanism for the automatic extraction of stops that is based on the intersection of trajectories and geometries of geographical features considered relevant to the application [1]. In this approach the semantic information is limited to geographic data that intersect the trajectories for a certain time interval. This approach is restricted to applications in which geographic information can help to identify places visited by the moving object which play the essential role.

Recently, more advanced methods use spatiotemporal criteria to perform trajectory segmentation and identify episodes like stops/moves: Yan et al design a velocity-based method providing a dynamic velocity threshold on stop computation, where the minimal stop duration is used to avoid false positives (e.g. congestions) [31]; several clustering-based stop identification methods have been developed, e.g. using the velocity [24] and direction features [27] of movement. Finally, Buchin et al. provide a theoretical trajectory segmentation framework and claim that the segmentation problem can be solved in $O(nlogn)$ time [3].

Online segmentation concepts can be traced back to the time series and signal processing fields [17], but not initially for trajectories. Although, some of the above works are capable of adapting to an online context [2][16], none of them focuses on revealing the profound semantics present in the computed trajectories in real-time. To the best of our knowledge, online algorithms for semantic trajectory construction are significantly missing. Our objective is to design such online computation methods for real-time semantic trajectory construction.

3 Preliminaries

3.1 Data and Semantic Trajectory Models

In our setting, a central server continuously collects the status updates of moving objects that move inside an area of interest – monitoring area of moving objects. First, such updates involving an object O_i contain spatiotemporal $\langle x, y, t \rangle$ points forming its *"Raw Location Stream"*.

Definition 1 (Raw Location Stream). *The continuous recording of spatiotemporal points that update the status of a moving object O_i, i.e. $\langle Q_1^{\ell s}, Q_2^{\ell s}, \ldots, Q_n^{\ell s} \rangle$, where $Q_i^{\ell s} = \langle x, y, t \rangle$ is a tuple including moving object's O_i, position $\langle x, y \rangle$ and timestamp t.*

By means of the raw location stream, we can derive information of movement features such as acceleration, speed, direction etc., which make up a *"Location Stream Feature Vector"* ($Q^{\ell f}$). Moreover, depending on the application, updates include additional attributes such as heading, steering wheel activity, lane position, distance to headaway vehicle (e.g to assess tailgating), displacement and so on. These features formulate a *"Complementary Feature Vector"* (Q^{cf}). Consequently, the two types of feature vectors combined together are forming the *"Movement Feature Vector"* ($Q = \langle Q^{\ell f}, Q^{cf} \rangle$) of d dimension describing d attributes of O_i movement at a specific timestamp.

Definition 2 (Movement Feature Vector). *The movement attributes of object O_i at timestamp t can be described by a d-dimensional vector that is the concatenation of the location stream feature vector and the complementary feature vector $Q = \langle Q^{\ell f}, Q^{cf} \rangle$.*
– Location Stream Feature Vector ($Q^{\ell f}$): The movement features of object O_i that can be derived from the raw location stream tuple $Q^{\ell s}$.
– Complementary Feature Vector (Q^{cf}): The movement features that cannot be derived from the location stream but are explicitly included in O_i's status updates.

To provide better understanding and mobility data abstraction, in [29][31] the concept of *semantic trajectories* is introduced, where the trajectory is thought of as a sequence of meaningful episodes (e.g. stop, move, and other self-contained and self-correlated trajectory portions).

Definition 3 (Semantic Movement). *A semantic movement or trajectory consists of a sequence of meaningful trajectory units, called "episodes", i.e. $\mathcal{T}_{sem} = \{e_{first}, \ldots, e_{last}\}$.*
– An episode (e) groups a subsequence of the location stream (a number of consecutive $\langle x, y, t \rangle$ points) having similar movement features.
– From a semantic data compression point of view, an episode stores the subsequence's temporal duration as well as its spatial extent $e_i = (time_{from}, time_{to}, geometry_{bound}, tag)$.

The $geometry_{bound}$ is the geometric abstraction of the episode, e.g. the bounding box of a stop area or the shape trace of roads that the moving object has followed. The term *tag* in the last part of the previous definition refers to the semantics of the episode, i.e. characterization of the activity or means of movement that is taking place in an episode (see Fig. 1).

3.2   Window Specifications

The window specification is a fundamental concept in streaming data processing [8]. In our context, the time window size T expresses the most recent portion of semantic trajectories the server needs to be informed about. An additional parameter τ specifies a time interval in which client side devices, installed on moving objects, are required to collect and report batches of their time ordered status updates [8]. Thus, $\frac{T}{\tau}$ batches are included in the window. Obviously, posed prerequisites are: 1) $\tau \ll T$ and 2) $T \mod \tau = 0$. As the window slides, for each monitored object O_i, the most aged batch expires and a newly received one is appended to it. The size of τ may vary from a few seconds to minutes depending on the application's sampling frequency. Small τ values enable fine-tuned episode extend determination with the make-weight of increased processing costs, while larger τ values reduce the processing load by increasing the granules that are assigned to episodes.

3.3 SeTraStream Overview

Having presented the primitive concepts utilized by our framework, in this subsection we outline SeTraStream's general function. Details will be provided in the upcoming sections. The whole process is depicted in Fig. 2. Upon the receipt of a batch containing the status updates including $Q^{\ell s}, Q^{cf}$ vectors at different timestamps in τ, a cleaning and smoothing technique is applied on it (Step 1 on the right part of the figure). Consequently, a novel compression method (Step 2) is applied on the batch considering both $Q^{\ell s}, Q^{cf}$ characteristics while performing the load shedding. Finally, at a third step $Q^{\ell f}, Q^{cf}$ feature vectors are extracted, a corresponding matrix is formed and the batch is buffered until it is processed at the SeTraStream's segmentation stage. During the segmentation stage (left part of Fig. 2), a previously buffered batch is dequeued and compared with other batches' feature matrices in O_i's window. SeTraStream seeks both for short and long term changes in O_i's movement pattern, and identifies an episode whenever feature matrices are found to be dissimilar based on the RV-Coefficient (to be defined later) and a specified division threshold σ.

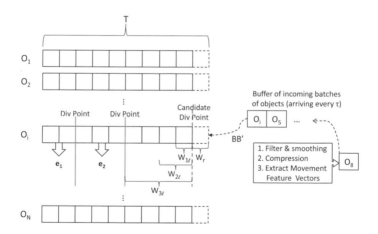

Fig. 2. The SeTraStream Framework

4 Online Data Preparation

As already described, arriving batches involving monitored objects contain their raw location stream, as well as complementary feature vectors. In this section, we discuss the initial steps of data preparation before proceeding to episode determination (i.e. trajectory segmentation). The talk regards three steps depicted in the right part of Fig. 2: (1) an *online cleaning* step that deals with noisy tuples, (2) an *online compression* stage that manages to reduce both the available memory usage and the processing cost in computing trajectories, and (3) extracting movement feature vectors, including both the location stream features and complementary features. Table 1 summarizes the symbology utilized in the current and the upcoming sections as well.

Table 1. Notations of symbols

Symbol	Description
N	Number of monitored objects
T, τ	Window size and batch interval
d	Number of movement features
O_i	The i-th monitored object id
B_i	The i-th batch from a candidate div. point
$Q^{\ell s}$	Tuple including $\langle x, y, t \rangle$ triplet of an objects' raw location stream
$Q^{\ell f}$	Feature vector derived from the raw location stream at t
Q^{cf}	Complementary feature vector at timestamp t
$\delta_{outlier}, \delta_{smooth}, \sigma$	Filtering, smoothing and segmentation thresholds respectively
res	The residual between the smoothed and the true value
sed, scc	Synchronous Euclidean Distance and Correlation Coefficient
W_ℓ, W_r	A left and right workpiece respectively
e_i	The i-th episode in an object's window

4.1   Online Cleaning

The main focus of trajectory data cleaning is to remove GPS errors. Jun et al. [14] summarize two types of GPS errors: *systematic errors* (i.e. the totally different GPS positioning from the actual location which is caused by low number of satellites in view, Horizontal Dilution Of Position HDOP etc.) and *random errors* (i.e. the small errors up to ± 15 meters which can be caused by the satellite orbit, clock or receiver issues). These systematic errors are also named "outliers", where researchers usually design *filtering* methods to remove them; whilst random errors are small distortions from the true values and their influences can be decreased by *smoothing* methods. Many offline GPS data cleaning works can be found such as [14][28][31].

In the context of streaming data, *online filtering & smoothing* of streaming tuples has become a hot topic [5][10][11][15][19]. Different from the focus of prior works on data accuracy and distribution estimation, our primary concern of cleaning streaming movement data is refining the data points that have substantial distortion of movement features for computing semantic trajectories[1].

For efficient data cleaning, we need to combine *online filtering* and *online smoothing* in a single loop. When a new batch B regarding object O_i arrives (right part of Fig.2), we do the following cleaning steps:

---

[1] Q^{cf} values are not examined as the micro-sensory devices of vehicles usually possess self-calibrating capabilities.

1. Build a kernel based smoothing model: $(\widehat{x}, \widehat{y}) = \frac{\sum_i k(t_i)(x_{t_i}, y_{t_i})}{\sum_i k(t_i)}$ where $k(t)$ is a function with the property $\int_0^{|B|} k(t)dt = 1$. The kernel function describes the weight distribution, with most of the weight in the area near the point. In our experiments, as in [28], we apply the Gaussian kernel $k(t_i) = e^{-\frac{(t_i-t)^2}{2\beta^2}}$, where ß refers to the bandwidth of the kernel.

2. Calculate the residual between the model prediction and the true value $\langle x, y \rangle$ of the examined point $Q_p^{\ell s}$, i.e. $res = \sqrt{(\widehat{x} - x)^2 + (\widehat{y} - y)^2}$.

3. By using a speed limit v_{limit} and the speed $v_{Q_{p-1}^{\ell s}}$ at the previous point $Q_{p-1}^{\ell s}$, respectively compute the outlier bound ($\delta_{outlier} = v_{limit} \times (t_{Q_p^{\ell s}} - t_{Q_{p-1}^{\ell s}})$) and the smooth bound ($\delta_{smooth} = v_{Q_{p-1}^{\ell s}} \times (t_{Q_p^{\ell s}} - t_{Q_{p-1}^{\ell s}}) \times 120\%^2$).

4. Filter out the point if the residual is more than the outlier bound, i.e. $res > \delta_{outlier}$, or replace the location of the point $\langle x, y \rangle$ with the smoothed value $\langle \widehat{x}, \widehat{y} \rangle$ if the residual is between the outlier bound and the smooth bound, i.e. $\delta_{smooth} < res < \delta_{outlier}$. Otherwise, we keep the original $\langle x, y \rangle$ of the point.

This cleaning method has taken both advantages of the distance based outlier removal and the local-weighted kernel smoothing method with linear memory requirements of $O(|B|)$, where $|B|$ is the size of a batch.

4.2   Online Compression

A primary concern when operating in a streaming setting regards the load shedding with respect to incoming tuples. In the context of semantic trajectory computation, this happens both for limiting the available buffer usage as well as to reduce the processing cost [4][16][23][26]. In our approach, as both Definitions 2, 3 imply, the approximation quality of the mere spatiotemporal trajectories is not our only concern. Semantic trajectories will be extracted based on additional features other than those derived from spatiotemporal $\langle x, y, t \rangle$ points. On the other hand, if we overlook the spatiotemporal trajectory approximation quality, the portion of the movement features that rely on the pure location stream will later be uncontrollably distorted. To cope with the previous requirements, we propose a method and define a *significance score* suitable to serve our purposes.

Assume that a batch regarding object O_i is processed (step. 2 at right part of Fig.2) and $(Q_{p-1}^{\ell s}, Q_p^{\ell s})$ is the last examined pair of points in it. When a new point $Q_{p+1}^{\ell s}$ is inspected, we first obtain the significance of $Q_p^{\ell s}$ from a spatiotemporal viewpoint by fostering the *Synchronous Euclidean Distance*, defined as [23][26]:

$$sed(Q_p^{\ell s}, Q_{p-1}^{\ell s}, Q_{p+1}^{\ell s}) = \sqrt{(x_{Q_p'^{\ell s}} - x_{Q_p^{\ell s}})^2 + (y_{Q_p'^{\ell s}} - y_{Q_p^{\ell s}})^2}, \text{ with } x_{Q_p'^{\ell s}} = x_{Q_{p-1}^{\ell s}} +$$

$v^x_{Q_{p-1}^{\ell s} Q_{p+1}^{\ell s}} \cdot (t_{Q_p^{\ell s}} - t_{Q_{p-1}^{\ell s}})$ and $y_{Q_p'^{\ell s}} = y_{Q_{p-1}^{\ell s}} + v^y_{Q_{p-1}^{\ell s} Q_{p+1}^{\ell s}} \cdot (t_{Q_p^{\ell s}} - t_{Q_{p-1}^{\ell s}})$ while v^x, v^y refer to the velocity vector (please refer to [26] for further details).

Nevertheless, *sed* constitutes an absolute number that lacks the ability to quantify the particular significance of a point with respect to other spatiotemporal

---

[2] Here, we increase the smooth bound by 20% of the location prediction provided by the speed of the previous point.

points within the current batch. In order to appropriately derive the aforementioned significance quantification, in SeTraStream's compression scheme we normalize sed and define the relative spatiotemporal significance Sig^{SP}:

$$Sig^{SP}(Q_p^{\ell s}) = \frac{sed(Q_p^{\ell s}, Q_{p-1}^{\ell s}, Q_{p+1}^{\ell s})}{max_{sed}} \tag{1}$$

with $0 \leq Sig^{SP}(Q_p^{\ell s}) \leq 1$. The denominator max_{sed} denotes the current maximum sed of points in the batch. Obviously, increased $Sig^{SP}(Q_p^{\ell s})$ estimations represent points of higher spatiotemporal significance.

Carefully inspecting sed's formula, we can conceive that the intuition behind its definition is to measure the amount of distortion that can be caused by pruning the spatiotemporal point $Q_p^{\ell s}$. That is, having omitted $Q_p^{\ell s}$ we could virtually infer the respective data point at timepoint $t_{Q_p^{\ell s}}$ using the preceding and succeeding ones ($Q_{p-1}^{\ell s}, Q_{p+1}^{\ell s}$). And calculating $Q_p^{\prime \ell s}$, $sed(Q_p^{\ell s}, Q_{p-1}^{\ell s}, Q_{p+1}^{\ell s})$ measures the incorporated distortion.

Thus, as regards the complementary feature vectors of O_i we choose to base the measure of their significance on the *Correlation Coefficient (corr)* metric. First, fostering an attitude similar to that in sed's calculation as explained in the previous paragraph, we estimate the value at the i-th position of vector $Q_p^{\prime cf}$ as: $[Q_p^{\prime cf}]_i = [Q_{p-1}^{cf}]_i + \frac{[Q_{p+1}^{cf}]_i - [Q_{p-1}^{cf}]_i}{t_{Q_{p+1}^{cf}} - t_{Q_{p-1}^{cf}}}(t_{Q_p^{cf}} - t_{Q_{p-1}^{cf}})$. Then, based on *corr* we define the *Synchronized Correlation Coefficient (scc)* between $(Q_p^{\prime cf}, Q_p^{cf})$ of complementary feature vectors:

$$scc(Q_p^{\prime cf}, Q_p^{cf}) = \frac{E(Q_p^{\prime cf} Q_p^{cf}) - E(Q_p^{\prime cf})E(Q_p^{cf})}{\sqrt{(E((Q_p^{\prime cf})^2) - E^2(Q_p^{\prime cf}))(E((Q_p^{cf})^2) - E^2(Q_p^{cf}))}} \tag{2}$$

where $E()$ refers to the mean and $-1 \leq scc(Q_p^{\prime cf}, Q_p^{cf}) \leq 1$.

The choice of scc is motivated by the fact that its stem, *corr*, possesses the ability to indicate the similarity of the trends that are profound in the examined vectors rather than relying on their absolute values [5][10][11][19]. Hence, it provides an appropriate way to identify (dis)similar patterns in the complementary vectors and can be generalized in order to detect similar patterns between movement feature vectors in their entirety. Values of scc that are close to -1 exhibit high dissimilarity between $(Q_p^{\prime cf}, Q_p^{cf})$, indicating that omitting Q_p^{cf} results in higher pattern distortion. Calculating $1 - scc$ enables higher measurements to account for more dissimilar patterns and taking one step further, min-max normalization on $1 - scc$ allows (dis)similarity values lie within $[0, 1]$. Thus, we eventually compute the relative significance of the complementary feature vector:

$$Sig^C(Q_p^{cf}) = \frac{1 - scc(Q_p^{\prime cf}, Q_p^{cf})}{2max\{(1 - scc)\}} \tag{3}$$

In the context of our compression scheme, the more dissimilar $(Q_p^{\prime cf}, Q_p^{cf})$ are, the higher the probability to be included in the window should be. As a result, the overall significance $Sig(Q_p)$ of Q_p can be estimated by the combination of both the location stream feature $Sig^{SP}(Q_p^{\ell s})$ and the complementary feature $Sig^C(Q_p^{cf})$. The weight balance between them is application dependent, though we choose to treat them equally important [20]:

$$Sig(Q_p) = \frac{1}{2}(Sig^{SP}(Q_p^{\ell s}) + Sig^C(Q_p^{cf})) \tag{4}$$

Eventually, for a threshold $0 \leq Sig_{thres} \leq 1$, Q_p remains in the batch when $Sig(Q_p) \geq Sig_{thres}$, or it is removed for compression purposes otherwise.

5   Semantic Trajectory Construction

We now describe the core of SeTraStream, the online trajectory segmentation stage. This stage comes after data cleaning and compression utilizing the *extracted feature vectors* of a batch (step. 3 at right part of Fig.2).

5.1   Online Episode Determination – Trajectory Segmentation

Upon deciding the data points of a batch that are to be included in the window as devised in the previous subsection, SeTraStream proceeds by examining episode existence in T. To start with, we assume the simple case of the current window consisting of a couple of τ-sized batches (i.e. $T = 2\tau$). We will henceforth refer to each part of the window composed of a number of compressed batches as *workpiece*. Intuitively, distinguishing episodes is equivalent to finding a *division point*, where the movement feature vectors on its left and right sides are uncorrelated and thus correspond to different movement patterns. In our simple scenario, a *candidate division point* is placed in the middle of the available workpieces.

Hence, we subsequently need to dictate a suitable measure in order to determine movement pattern change existence. We already noted the particular utility of the correlation coefficient on the discovery of trends [5][10][11][19], and thus (in our context) patterns in the movement data. In this processing phase movement feature vectors composing each workpiece essentially form a pair of matrices for which correlation computation needs to be conducted. As a result, we will reside to the *RV-coefficient* which constitutes a generalization of the correlation coefficient for matrix data. We organize W_ℓ into a $d \times m$ matrix, where d is the number of movement features and m represents a number of vectors (at different timestamps) that are the columns of the matrix. Similarly, W_r is organized in a $d \times n$ matrix i.e. n columns exist. The *RV-Coefficient* between $\langle W_\ell, W_r \rangle$ is defined as:

$$RV(W_\ell, W_r) = \frac{Tr(W_\ell W_\ell' W_r W_r')}{\sqrt{Tr([W_\ell W_\ell']^2)Tr([W_r W_r']^2)}} \tag{5}$$

where W_ℓ', W_r' refer to the transpose matrices, $Tr()$ denotes the trace of a matrix and $0 \leq RV \leq 1$. RV values closer to zero are indicative of uncorrelated movement patterns. Based on a division point threshold σ workpieces W_ℓ, W_r can be assigned to a pair of different episodes $e_\ell = (0, T - \tau, geometry_{bound})$, $e_r = (T - \tau + 1, T, geometry_{bound})$ when:

$$RV(W_\ell, W_r) \leq \sigma \qquad (6)$$

or to a single episode $e = (0, T, geometry_{bound})$ otherwise.

Now, consider the general case of T covering an arbitrary number of batches. It can easily be conceived that in a larger time window an alteration in the movement pattern may happen: (a) instantly as a sharp change, or (b) in a more smooth manner as time passes. As a result, upon the arrival of a new workpiece W_r, we initially check for short-term changes in the patterns of movement. We thus place a candidate division point between the newly received workpiece and the last of the existing ones. Then the correlation between the movement feature vectors present in $\langle W_{1\ell}, W_r \rangle$ is computed. Notice that $W_{1\ell}$ this time possesses an additional subscript which denotes the step of the procedure, as will be shortly explained. Similarly to our discussion in the previous paragraphs, when $RV_1(W_{1\ell}, W_r)$ is lower than the specified division threshold, a division point exists and signals the end of the previous episode e_ℓ and starts a new one e_r.

No short-term change existence triggers our algorithm to proceed by seeking long-term dis-correlations. For this purpose, we first examine $RV_2(W_{2\ell}, W_r)$ doubling the time scale of the left workpiece by going 2τ units back in the window from the candidate division point. In case RV_2 does not satisfy Inequality 6, this procedure continuous by *exponentially expanding* the time scale of the left workpiece in a way such that at the i-th step of the algorithm the size of $W_{i\ell}$ is $2^{(i-1)}\tau$ units and $RV_i(W_{i\ell}, W_r)$ is calculated. When Inequality 6 is satisfied the candidate division point is a true division point which bounds the previous episode $e_i = (time_{from}, time_{to}, geometry_{bound})$ and constitutes the onset of a new. Otherwise, W_r is rendered the current bound of the last episode by being appended to it. If no long-term change is detected, the aforementioned expansion ceases when either the beginning of the last episode or the start of T (in case all previous batches have been attributed to the same episode) is reached, i.e. no data points of the penultimate episode are considered since its extend has already been determined.

The exponential workpiece expansion fostered here is inspired by the *tilted time window* definition [8] as a general and rational way to seek movement pattern changes in different time granularities. Other expansion choices can also be applied. All of these options are orthogonal to our approaches and do not affect the generic function of SeTraStream. Our approach manages to effectively handle sliding windows as a slide of τ time units results in: (1) the expiration of the initial batch of the first episode e_{first} of O_i which affects its $(time_{from}, geometry_{bound})$ attributes and (2) the appendage of a newly received batch that either extends the last episode e_{last} (when no division point is detected) or starts a new episode. The outcome of the online segmentation consists of tuples $\mathcal{T}_{O_i} = \{e_{first}, \ldots, e_{last}\}$ representing objects' semantic trajectories.

5.2   Time and Space Complexity

The introduced trajectory segmentation procedure, premises that a newly appended batch will be compared with left workpieces that may be (depending on whether a division point is detected) exponentially expanded until either the previous episode end or the start of the window is reached. Based on this observation, the lemma below elaborates on the complexity of the checks required during candidate division point examination.

Lemma 1. *The time complexity of SeTraStream's online segmentation procedure, for N monitored objects, under exponential $W_{i\ell}$ expansion is $O(Nlog_2(\frac{T}{\tau}))$ per candidate division point.*

Proof. For a single monitored object, the current window is composed of $\frac{T}{\tau} - 1$ batches (excluding the one belonging to W_r). The worst case scenario appears when no previous episode exists in the window and the candidate division point is not proven to be an actual division point. By considering the exponential workpiece expansion, comparisons (i.e., σ checks) may reach a number of $k = min\{i \in \mathbb{N}^* : \frac{\frac{T}{\tau}-1}{2^{(i-1)}} \geq 1\}$ at most. Adopting logarithms on the previous expression and summing for N objects completes the proof.   □

Now, recalling the definition of the *RV-Coefficient* measure, it can easily be observed that its computation relies on the multiplication of the bipartite matrices with their transpose. Assume that the number of d-dimensional movement feature vectors in a cleaned and compressed batch are n. Based on the above observation we can see that instead of maintaining the original form of the vectors which requires $O(d \cdot n)$ memory space, we can reduce the space requirements during episode determination by computing the product of the $d \times n$ matrix of the batch with its transpose. This reduces the space requirements to $O(d^2)$ per batch since in practice $d \ll n$. So, to check a short-term change in the movement patterns we do not need to store the full matrices of $W_{1\ell}, W_r$ which in this case are composed of one batch each, but only the matrix products as described above.

However, this point may not be of particular utility since left workpieces are expanded during the long-term pattern alteration checks. A natural question that arises regards whether or not the product $W_{i\ell}W'_{i\ell}$ can be expressed by means of the multiplication of single batch matrices, with their transposes.

Lemma 2. *$W_{i\ell}W'_{i\ell}$ is the sum of batch matrix products with their transposes: $W_{i\ell}W'_{i\ell} = \sum_{j=1}^{2^{(i-1)}} B_j B'_j$, where B_j is used to notate the matrix formed by the vectors in the j-th batch (from a candidate division point to the end of $W_{i\ell}$).*

Proof. Let $W_{i\ell} = [B_1|B_2|\cdots|B_{2^{(i-1)}}]$ the matrix of the (i-th) left workpiece during the current division point check. B_js are used to denote sub-matrices belonging to individual batches that were appended to the workpiece. It is easy to see that the transpose matrix can be produced by transposing these submatrices: $W'_{i\ell} = [B'_1|B'_2|\cdots|B'_{2^{(i-1)}}]$. And then $W_{i\ell}W'_{i\ell}$ can be decomposed into $B_j B'_j$ products: $W_{i\ell}W'_{i\ell} = B_1 B'_1 + B_2 B'_2 + \cdots + B_{2^{(i-1)}} B'_{2^{(i-1)}} = \sum_{j=1}^{2^{(i-1)}} B_j B'_j$ □

Thus, for each batch we only need to store a square $d \times d$ matrix[3], which determines the space complexity of online segmentation leading to Lemma 3.

Lemma 3. *During the online episode determination stage of SeTraStream, the memory requirements per object O_i are $O(d^2 \frac{T}{\tau})$ and assuming N objects are being monitored the total space utilization is $O(d^2 N \frac{T}{\tau})$.*

5.3   Episode Tagging

Having detected an episode e_i, SeTraStream manages to specify in an online fashion the triplet $(time_{from}, time_{to}, geometry_{bound})$ describing its spatio-temporal extend. The final piece of information associated with an episode regards its *tag* as it was described in Section 3.1. Given application's context, possible *tag* instances form a set of movement pattern classes and notice that the instances of the classes are predetermined for the applications we consider (Section 1). Hence, the problem of episode tag assignment can be smelted to a trivial classification task, where the classifier can be trained in advance based on the collected episodes (with features like segment *distance, duration, density, avg. speed, avg. acceleration, avg. heading* etc.) and the detected episode e_i can be timely classified based on the trained model and the episode features. Suitable techniques include decision trees, boosting, SVM, neural or Bayesian networks [7]. Additional Hidden Markov Model based trajectory annotation can be referred to [30].

6   Experiments

In this section, we present our experimental results in real-time construction of semantic trajectories from streaming movement data.

Experimental Setup. We utilize two different datasets: *Taxi Data* - this dataset includes taxi trajectory data for 5 months with more than 3M GPS records, which do not have any complementary features. We mainly use taxi data to validate compression. It is non-trivial to get real life on-hand dataset with both complementary features and the underlying segment ground-truth tags. Therefore, we collect our own trajectory data by developing Python S60 scripts deployed in a Nokia N95 smartphone, which can generate both GPS data and accelerometer data from the embedded sensors. We calculate GPS features (e.g. *transformed longitude, latitude, speed, direction*) as the location stream vectors $(Q^{\ell f})$ and accelerometer features (e.g. *mean, variance, magnitude, covariance of the 3 accelerometer axis*) as the complementary feature vectors (Q^{cf}). We term the latter dataset as *Phone Data* within which, we also provide our own real segment tags (e.g. *standing, jogging, walking*) to validate the online segmentation accuracy. For *Phone Data*, we also work on the GPS data from the data campaign organized by Nokia Research Center - Lausanne, which has collected 185 users' phone data with about 7M records in total [18][30].

---

[3] We also keep the geometry bound of the batch that is utilized in the final episode geometry bound determination as well as some additional aggregate statistics, of minor storage cost, for classification and tag assignment in the next step.

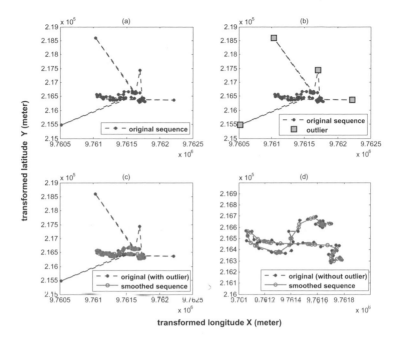

Fig. 3. Data cleaning (outlier removal and smoothing)

Data Cleaning. As described previously, our online data cleaning needs to consider two types of GPS data errors, i.e. filtering outliers as systematic errors and smoothing the random errors. The experimental cleaning results are shown in Fig. 3: (a) sketches the original trajectory data; (b) identifies the outliers during the online cleaning process; (c) and (d) present the original movement sequences together with the final smoothed trajectories, where (c) includes the outliers in the original sequences, whilst (d) removes them for better visualization.

Online Compression. Technically, compression makes sense when dealing with large data sets, however both *Taxi Data* and the big part of *Phone Data* have no complementary features (Q_p^{cf}) available but only the GPS features ($Q_p^{\ell s}$). Thus, our current experiment validates the sensitivity of data compression rate with respect to the spatiotemporal significance $Sig^{SP}(Q_p^{\ell s})$ on location streams, without considering the significance of the complementary features $Sig^C(Q_p^{cf})$. As shown in Fig. 4, we plot the compression rate sensitivity when applying different thresholds on $Sig^{SP}(Q_p^{\ell s})$. The results are proportional when using the *Phone Data* with respective $Sig(Q_p)$ thresholds.

Online Segmentation. SeTraStream's procedure in online trajectory segmentation relates to (1) initially computing the RV-coefficient between two workpieces $RV(W_\ell, W_r)$ and (2) expanding W_ℓ if $RV(W_\ell, W_r)$ is bigger than the given threshold σ (otherwise, we identify a division point between two episodes). Results are shown in Fig. 5, where for $T = 60s$ we can discover two main division

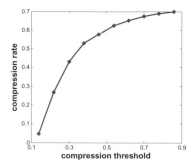

Fig. 4. Data compression rate w.r.t. different thresholds $Sig^{SP}(Q_p^{\ell s})$

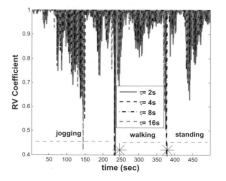

Fig. 5. Episode identification varying batch size, for $\sigma = 0.6$

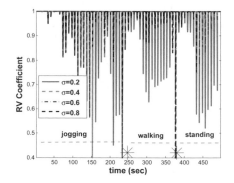

Fig. 6. Sensitivity of RV w.r.t. different σ at $\tau = 8s$

points (with RV-coefficient< 0.6 and batch size $\tau < 16$), which is consistent with the underlying ground-truth tags. The stars in the figures are the real division points in the streaming data, which indicate when user changes their movement behaviors e.g. from jogging to walking and finally to standing.

Fig. 5 analyzes the sensitivity of using different batch sizes, where the best outcome (i.e accurate episode extend determination) is $\tau = 8s$; when $\tau = 16s$, we actually identify three division points, which is partially correct, since as we can see there are only two real division points in the stream. Similarly, we also investigate the segmentation sensitivity regarding different division thresholds σ in Fig. 6. The best segmentation result is achieved when $\sigma = 0.6$.

Finally, we evaluate the time performance of SeTraStream's trajectory segmentation module. We measure the segmentation latency with 25 users in the *Phone Data*. In the experiments, we used a laptop with 2.2 Ghz CPU and 4 Gb of memory. From Fig.7 and Fig. 8, we can see the segmentation time is almost linear, in both situations with different batch sizes (τ) and different division thresholds (σ), which is quite consistent with Lemma 1.

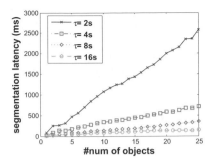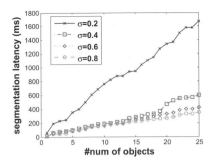

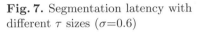

Fig. 7. Segmentation latency with different τ sizes (σ=0.6)

Fig. 8. Segmentation latency with different σ thresholds ($\tau = 8s$)

7 Conclusions and Future Work

In this paper, we proposed a novel and complete online framework, namely *SeTraStream* that enables semantic trajectory construction over streaming movement data. As far as we know, this is the first method proposed in the literature tackling with this problem in real-time streaming environments. Moreover, we considered challenges occurring in real world applications including data cleaning and load shedding procedures before accurately identifying trajectory episodes in objects' streaming movement data.

Our future work is to further evaluate this method with larger datasets including more complementary features and ground-truth tags. In addition, we are planning to extend SeTraStream to (1) handle multiple window types for online trajectory segmentation, and (2) perform real-time trajectory construction in distributed settings often encountered in large scale application scenarios.

References

1. Alvares, L.O., Bogorny, V., Kuijpers, B., Macedo, J., Moelans, B., Vaisman, A.: A Model for Enriching Trajectories with Semantic Geographical Information. In: GIS (2007)
2. Brakatsoulas, S., Pfoser, D., Salas, R., Wenk, C.: On map-matching vehicle tracking data. In: VLDB (2005)
3. Buchin, M., Driemel, A., Kreveld, M.V., Sacristan, V.: An Algorithmic Framework for Segmenting Trajectories based on Spatio-Temporal Criteria. In: GIS (2010)
4. Cao, H., Wolfson, O., Trajcevski, G.: Spatio-Temporal Data Reduction With Deterministic Error Bounds. The VLDB Journal 15(3) (2006)
5. Deligiannakis, A., Kotidis, Y., Vassalos, V., Stoumpos, V., Delis, A.: Another Outlier Bites the Dust: Computing Meaningful Aggregates in Sensor Networks. In: ICDE (2009)
6. Douglas, D., Peucker, T.: Algorithms for the reduction of the number of points required to represent a digitized line or its caricature. The Canadian Cartographer 10(2) (1973)

7. Fayyad, U.M., Piatetsky-Shapiro, G., Smyth, P., Uthurusamy, R. (eds.): Advances in Knowledge Discovery and Data Mining. AAAI/MIT Press (1996)
8. Giannella, C., Han, J., Pei, J., Yan, X., Yu, P.S.: Mining Frequent Patterns in Data Streams at Multiple Time Granularities. MIT Press, Cambridge (2002)
9. Giannotti, F., Nanni, M., Pinelli, F., Pedreschi, D.: Trajectory Pattern Mining. In: KDD (2007)
10. Giatrakos, N., Kotidis, Y., Deligiannakis, A.: PAO: Power-efficient Attribution of Outliers in Wireless Sensor Networks. In: DMSN (2010)
11. Giatrakos, N., Kotidis, Y., Deligiannakis, A., Vassalos, V., Theodoridis, Y.: TACO: Tunable Approximate Computation of Outliers in Wireless Sensor Networks. In: SIGMOD (2010)
12. Güting, R., Schneider, M.: Moving Objects Databases. Morgan Kaufmann, San Francisco (2005)
13. Jeung, H., Yiu, M.L., Zhou, X., Jensen, C.S., Shen, H.T.: Discovery of Convoys in Trajectory Databases. In: VLDB (2008)
14. Jun, J., Guensler, R., Ogle, J.: Smoothing Methods to Minimize Impact of Global Positioning System Random Error on Travel Distance, Speed, and Acceleration Profile Estimates. Transportation Research Record: Journal of the Transportation Research Board 1972(1) (January 2006)
15. Kanagal, B., Deshpande, A.: Online Filtering, Smoothing and Probabilistic Modeling of Streaming data. In: ICDE (2008)
16. Kellaris, G., Pelekis, N., Theodoridis, Y.: Trajectory Compression under Network Constraints. In: Mamoulis, N., Seidl, T., Pedersen, T.B., Torp, K., Assent, I. (eds.) SSTD 2009. LNCS, vol. 5644, pp. 392–398. Springer, Heidelberg (2009)
17. Keogh, E., Chu, S., Hart, D., Pazzani, M.: An Online Algorithm for Segmenting Time Series. In: ICDM (2001)
18. Kiukkoneny, N., Blom, J., Dousse, O., Gatica-Perez, D., Laurila, J.: Towards Rich Mobile Phone Datasets: Lausanne Data Collection Campaign. In: ICPS (2010)
19. Kotidis, Y., Vassalos, V., Deligiannakis, A., Stoumpos, V., Delis, A.: Robust management of outliers in sensor network aggregate queries. In: MobiDE (2007)
20. Lee, J.-G., Han, J., Whang, K.-Y.: Trajectory Clustering: a Partition-and-Group Framework. In: SIGMOD (2007)
21. Li, Z., Ding, B., Han, J., Kays, R., Nye, P.: Mining Periodic Behaviors for Moving Objects. In: KDD (2010)
22. Marketos, G., Frentzos, E., Ntoutsi, I., Pelekis, N., Raffaetà, A., Theodoridis, Y.: Building real-world trajectory warehouses. In: MobiDE (2008)
23. Meratnia, N., de By, R.A.: Spatiotemporal Compression Techniques for Moving Point Objects. In: Hwang, J., Christodoulakis, S., Plexousakis, D., Christophides, V., Koubarakis, M., Böhm, K. (eds.) EDBT 2004. LNCS, vol. 2992, pp. 765–782. Springer, Heidelberg (2004)
24. Palma, A.T., Bogorny, V., Kuijpers, B., Alvares, L.O.: A Clustering-based Approach for Discovering Interesting Places in Trajectories. In: SAC (2008)
25. Pelekis, N., Frentzos, E., Giatrakos, N., Theodoridis, Y.: HERMES: Aggregative LBS via a Trajectory DB Engine. In: SIGMOD (2008)
26. Potamias, M., Patroumpas, K., Sellis, T.: Sampling Trajectory Streams with Spatiotemporal Criteria. In: SSDBM (2006)
27. Rocha, J.A.M.R., Times, V.C., Oliveira, G., Alvares, L.O., Bogorny, V.: Db-Smot: a Direction-Based Spatio-Temporal Clustering Method. In: Intelligent Systems (2010)

28. Schüssler, N., Axhausen, K.W.: Processing GPS Raw Data Without Additional Information. Transportation Research Record: Journal of the Transportation Research Board 8 (2009)
29. Spaccapietra, S., Parent, C., Damiani, M.L., de Macedo, J.A., Porto, F., Vangenot, C.: A Conceptual View on Trajectories. Data and Knowledge Engineering 65(1) (2008)
30. Yan, Z., Chakraborty, D., Parent, C., Spaccapietra, S., Karl, A.: SeMiTri: A Framework for Semantic Annotation of Heterogeneous Trajectories. In: EDBT (2011)
31. Yan, Z., Parent, C., Spaccapietra, S., Chakraborty, D.: A Hybrid Model and Computing Platform for Spatio-Semantic Trajectories. In: Aroyo, L., Antoniou, G., Hyvönen, E., ten Teije, A., Stuckenschmidt, H., Cabral, L., Tudorache, T. (eds.) ESWC 2010. LNCS, vol. 6088, pp. 60–75. Springer, Heidelberg (2010)
32. Yan, Z., Spremic, L., Chakraborty, D., Parent, C., Spaccapietra, S., Karl, A.: Automatic Construction and Multi-level Visualization of Semantic Trajectories. In: GIS (2010)
33. Zheng, Y., Chen, Y., Li, Q., Xie, X., Ma, W.-Y.: Understanding transportation modes based on GPS data for web applications. Transactions on the Web (TWEB) 4(1) (2010)

Mining Significant Time Intervals
for Relationship Detection

Zhenhui Li, Cindy Xide Lin, Bolin Ding, and Jiawei Han

University of Illinois at Urbana-Champaign, Illinois, US

Abstract. Spatio-temporal data collected from GPS have become an important resource to study the relationships of moving objects. While previous studies focus on mining objects being together for a long time, discovering real-world relationships, such as friends or colleagues in human trajectory data, is a fundamentally different challenge. For example, it is possible that two individuals are friends but do not spend a lot of time being together every day. However, spending just one or two hours together at a location away from work on a Saturday night could be a strong indicator of friend relationship.

Based on the above observations, in this paper we aim to analyze and detect semantically meaningful relationships in a supervised way. That is, with an interested relationship in mind, a user can label some object pairs with and without such relationship. From labeled pairs, we will learn what time intervals are the most important ones in order to characterize this relationship. These significant time intervals, namely T-Motifs, are then used to discover relationships hidden in the unlabeled moving object pairs. While the search for T-Motifs could be time-consuming, we design two speed-up strategies to efficiently extract T-Motifs. We use both real and synthetic datasets to demonstrate the effectiveness and efficiency of our method.

1   Introduction

With the increasing popularity of GPS devices, the tracking of moving objects in general has become a reality. As a result, a vast amount of trajectory data is being collected and analyzed. Based on the temporal meeting pattern of objects, one of the most important and interesting problems in trajectory data analysis is *relationship detection.*

Previous studies of moving object relationships have been constrained to detecting moving object clusters. Studies such as flock [13], moving cluster [11], convoy [10], and swarm [16] focus on the discovery of a group of objects that move together. All these studies take the entire trajectory as a sequence of time points, and treat every time point or time interval *equally.* Therefore, the longer two moving objects are together, the better they are in terms of forming a cluster. However, all of these studies suffer from several drawbacks. On one hand, clusters discovered in this way usually do not carry any semantical meaning, such as friends, colleagues and families which naturally exist in human trajectory data. On the other hand, object pairs with certain relationship may not

D. Pfoser et al. (Eds.): SSTD 2011, LNCS 6849, pp. 386–403, 2011.

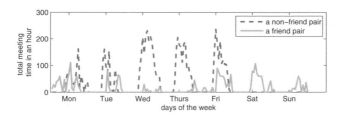

Fig. 1. Meeting frequency for a friend and a non-friend pair

necessarily meet more often than the other pairs, hence it leads to the failure of aforementioned methods in detecting such relationship. Considering the following example.

Example 1. Reality Mining project[1] collected 94 human trajectories of the 2004-2005 academic year and conducted a survey about their friendship to each other. In Figure 1, we plot the meeting frequencies for one friend pair and one non-friend pair. Comparing two frequency curves, the one has overall higher meeting frequency is the non-friend pair. Thus, longer overall meeting time does not necessarily indicate friend relationship. In addition, we observe that the friend pair shows significantly higher meeting frequency on weekends, which indicates that, in the friend relationship case, not every time point has equal importance. In other words, since two people who meet more frequently on weekends are more likely to be friends, the weekend time interval are considered more discriminative and should play a more important role in the friend relationship detection task.

The above example reveals an important problem when analyzing relationship for moving objects: some time intervals are more discriminative than the others for a particular relationship. In fact, besides friend relationship, many relationships have their unique temporal patterns. For example, if we want to examine whether or not two people are colleagues, daytime on weekdays becomes the discriminative time intervals. If two people are family members, they often gather on holidays. Therefore, to detect semantically meaningful relationships in moving objects, we cannot treat all time intervals equally, instead we need to learn from the data what time intervals are the most important ones to characterize a relationship.

Consequently, in this paper we aim to detect relationship for moving object in a *supervised* way. That is, given a set of labeled data consisting of positive pairs having such relationship and negative pairs not having such relationship, our job is to first find those discriminative time intervals, namely *T-Motifs*. Then, these T-Motifs are used as features to detect this relationship in the remaining unlabeled pairs. Consider the following example.

Example 2. In Reality Mining dataset, [21:56 Wed., 23:08 Wed.] is a T-Motif for friend relationship, because 37.3% friend pairs have meeting frequency more

---
[1] http://reality.media.mit.edu/

than 12 minutes in this time interval whereas only 3.17% non-friend pairs have meeting frequency that could reach 12 minutes.

According to the above example, we can interpret T-Motif as a time interval for which the meeting frequency between positive and negative pairs can be well split by a frequency value. Hence we propose to use *information gain* to measure the significance of a time interval. We need to calculate the significance score for any time interval and pick those intervals with high scores as T-Motifs. So the main technical challenge remains in the computation of significance score for huge number of T-Motif candidates.

To efficiently handle the large number of T-Motifs candidates, we design two efficient speed-up techniques for our algorithm. The first speed-up strategy is based on the observation that two similar time intervals, such as $[S, T]$ and $[S, T + 1]$, should have similar significance scores. Therefore, we propose to use time-indexed meeting pairs, so that when shifting the ending time from T to $T + 1$, we only need to update pairs who meet at time $T + 1$ and at the same time maintain the sorted list for all the pairs. The second speed-up technique takes advantage of skewed data. That is, positive pairs are only a small portion of all pairs. Based on a property of information gain, we could reduce the time to find the best split point from $O(|D|)$ to $O(|D^+|)$, where $|D^+|$ is the number of positive pairs. This further speeds up the computation when the positive pairs are only a small portion of all pairs, which is indeed the case for our problems.

In summary, the contributions of our work are as follows. (1) Our work is the first to detect semantically meaningful relationships in moving objects in a supervised way. This is done by introducing the concept of T-Motifs to properly represent the temporal characteristics for a relationship. (2) Two speed-up techniques are proposed to efficiently discover the T-Motifs. (3) The effectiveness and efficiency of our methods are demonstrated on both real and synthetic datasets.

The rest of this paper is organized as follows. Section 2 depicts the general framework. Section 3 describes the basic algorithm to mine T-Motifs. In Section 4, we introduce the speed-up techniques. Section 5 shows experimental results with respect to effectiveness and efficiency. Related work is discussed in Section 6, and the paper concludes in Section 7.

2   Problem Analysis

In this paper, the time intervals are defined in a *relative* time frame instead of an absolute one. This is because the movements of objects such as human usually have strong spatio-temporal regularities [8][6][19]. Therefore for human movements, for instance, it is more informative use "week" as the relative time frame. By default, we take minute as the basic time unit and consider any time point in a weekly time window (see Figure 1). Hence the total number of time points is $P = 7$ (days) $\times$ 24 (hours) $\times$ 60 (minutes) = 10080. Any minute in the original absolute frame can be mapped to an integer from 1 to P. Similarly, a time interval $[S, T]$ is also defined in the relative time frame and should be

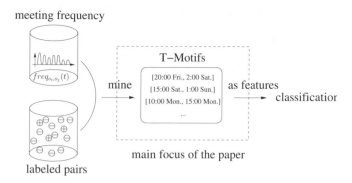

Fig. 2. Framework Overview

understood in a cyclic order when $S > T$. The maximum length of a time interval is P by definition.

Let $O_{DB} = \{o_1, o_2, \ldots, o_n\}$ be the set of all moving objects. The meeting frequency between any two objects could be inferred from their movements. For any object pair (o_i, o_j) and time $t, 1 \leq t \leq P$, the meeting frequency $freq_{o_i, o_j}(t)$ is defined as the total number of times they meet at t in the relative time frame. There are various ways to determine whether two objects *meet* at one time. The most common way is to see whether the distance of their spatial locations are within certain distance threshold. Such distance threshold is based on the property of moving objects and the specific application. Another way could be using bluetooth to detect nearby objects, such as the dataset used in our experiment.

With a particular relationship in mind, a user can label some pairs of objects having or not having such relationship. We use D^+ and D^- to denote the set of positive and negative pairs, respectively. For example, we write $(o_i, o_j) \in D^+$ if objects o_i and o_j are labeled by the user to have such a relationship. Further, we use $D = D^+ \bigcup D^-$ to denote the set of all labeled pairs.

Figure 2 shows an overview of our framework. Given the meeting frequencies of the labeled pairs, a set of significant time intervals, namely T-Motifs, are extracted to capture the temporal characteristics of the relationship. Then, a classification model is built using T-Motif features of training data. The remaining unlabeled pairs can be classified using the learned classification model. In this framework, the most challenging part is how to find T-Motifs. In the following sections, we will present the definition of T-Motifs and an efficient method to extract them.

3   Finding T-Motifs

A T-Motif is a significant time interval to characterize the relationship. In this section, we will first describe how to calculate the significance score for a time interval and then analyze the basic algorithm to extract top-k T-Motifs.

3.1   Significance of Time Intervals

To examine the significance of a time interval $[S,T]$, we need to first calculate the meeting frequency for every pair in this time interval. Meeting frequency $freq_{o_i,o_j}(S,T)$ is the amount of time that o_i and o_j meet within the time interval $[S,T]$:

$$freq_{o_i,o_j}(S,T) = \sum_{t\in[S,T]} freq_{o_i,o_j}(t).$$

In addition, for any set of pairs A consisting of positive pairs A^+ and A^-, its *entropy* is

$$H(A) = -\alpha \log \alpha - (1-\alpha)\log(1-\alpha),$$

where $\alpha = \frac{|A^+|}{|A|}$ is the fraction of positive numbers.

Intuitively, a time interval $[S,T]$ is significant if a large portion of positive (or negative) pairs have higher meeting frequencies than most of the negative (or positive) pairs. Once we select a meeting frequency value v as the *split point*, the pairs in D having meeting frequency in $[S,T]$ no less than split point form the set $D_{\geq}^{S,T}(v)$ and the rest form $D_{<}^{S,T}(v)$:

$$D_{\geq}^{S,T}(v) = \{(o_i,o_j)|freq_{o_i,o_j}(S,T) \geq v\}, D_{<}^{S,T}(v) = \{(o_i,o_j)|freq_{o_i,o_j}(S,T) < v\}.$$

The *information gain* of $[S,T]$ at split point v is:

$$IG^{S,T}(v) = H(D) - \frac{|D_{\geq}^{S,T}(v)|}{|D|}H(D_{\geq}^{S,T}(v)) - \frac{|D_{<}^{S,T}(v)|}{|D|}H(D_{<}^{S,T}(v)).$$

The *significance score* of $[S,T]$ is the highest information gain that any split point can achieve:

$$G(S,T) = \max_v IG^{S,T}(v).$$

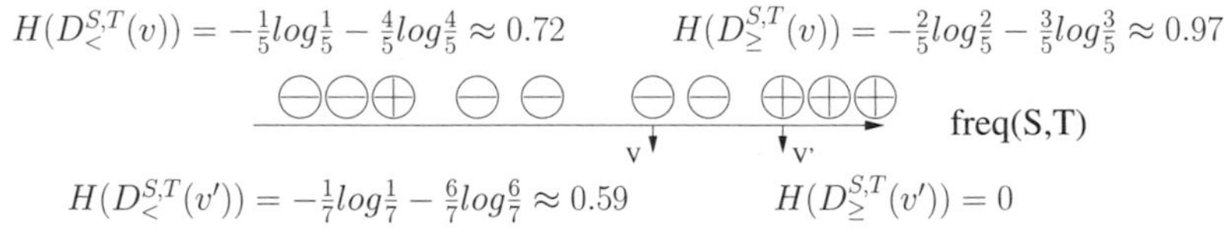

$$H(D_{<}^{S,T}(v)) = -\tfrac{1}{5}log\tfrac{1}{5} - \tfrac{4}{5}log\tfrac{4}{5} \approx 0.72 \qquad H(D_{\geq}^{S,T}(v)) = -\tfrac{2}{5}log\tfrac{2}{5} - \tfrac{3}{5}log\tfrac{3}{5} \approx 0.97$$

$$H(D_{<}^{S,T}(v')) = -\tfrac{1}{7}log\tfrac{1}{7} - \tfrac{6}{7}log\tfrac{6}{7} \approx 0.59 \qquad H(D_{\geq}^{S,T}(v')) = 0$$

Fig. 3. An example for calculation of $G(S,T)$

This concept is illustrated in Figure 3.

Example 3. Suppose the labeled set D contains 4 positive pairs and 6 negative pairs. Figure 3 shows the meeting frequency of each pair in a time interval $[S,T]$. We compute $H(D) = -\frac{4}{10}log\frac{4}{10} - \frac{6}{10}log\frac{6}{10} \approx 0.97$. At split point v, the information gain of $[S,T]$ is $IG^{S,T}(v) = 0.97 - \frac{5}{10} \times 0.72 - \frac{5}{10} \times 0.97 \approx 0.125$. The highest information gain is achieved at split point v': $G(S,T) = IG^{S,T}(v') = 0.97 - \frac{7}{10} \times 0.59 - \frac{3}{10} \times 0 \approx 0.557$.

3.2   Overview of Basic Algorithm

The basic algorithm is summarized in Algorithm 1. To find the T-Motifs, we first need to compute the significance score (i.e., information gain) for every time interval $[S, T]$. But sometimes it is unnecessary to consider time intervals which are too short or too long, such as one-minute time interval or the intervals with maximum length such as $[1, P]$. So the algorithm has an option to limit the length of time interval to $[\delta_{min}, \delta_{max}]$, where δ_{min} and δ_{max} are specified by the user. Now, for a time interval $[S, T]$, to get the meeting frequency of each pair takes $O(|D|)$ time (Line 5 in Algorithm 1). In order to calculate the significance score, the pairs will be sorted first (Line 6 in Algorithm 1). The sorting takes $O(|D| \log |D|)$ time. Taking each meeting frequency as split point v, the information gain $IG^{S,T}(v)$ can be calculated (Line 7-10 in Algorithm 1). The time complexity for this step is $O(|D|)$. And finally the maximal information gain value is set as the significance score for interval $[S, T]$. With the significance scores for all time intervals, we pick the top-k non-overlapped time intervals as T-Motifs. This procedure is similar to the selection of discriminative patterns in [3][17].

From Algorithm 1, we can see that the number of all time intervals is $O(P^2)$ in the worst case. And for each time interval $[S, T]$, it takes $O(|D| \log |D|)$ to compute the significance score. So the overall time complexity is $O(P^2|D| \log |D|)$.

Algorithm 1. Find T-Motifs

Input:

$freq$: meeting frequency for each pair;

D^+: positive pairs;

D^-: negative pairs.

Output: T-Motifs.

Algorithm:

1: $D \leftarrow D^+ \bigcup D^-$

2: **for** $S \leftarrow 1$ to P **do**

3:      **for** $len \leftarrow \delta_{min}$ to δ_{max} **do**

4:          $T \leftarrow S + len - 1$

5:          $freq\_arr \leftarrow \{freq_{o_i, o_j}(S, T), \forall (o_i, o_j) \in D\}$

6:          Sort $freq\_arr$

7:          **for** $i \leftarrow 1$ to $|D|$ **do**

8:              $v \leftarrow freq\_arr(i)$

9:              **if** $IG(D, v) > best\_IG$ **then**

10:                  $best\_IG = IG(D, v)$

11:          $G(S, T) = best\_IG$

12: Return top-k non-overlapped time intervals

---

4   Speed Up the Searh for T-Motifs

In this section, we propose two accelerating techniques to our basic algorithm. The first one is to build a time-indexed data structure, which allows us to quickly retrieve the pairs that meet at a certain time point T, and locally adjust the order of all pairs based on the changes in meeting frequencies. The second speed-up technique is based on an important property of information gain, which greatly reduces the number of split points one needs to examine for each time interval in order to compute its significance score.

4.1   Time-Indexed Meeting Pairs

In Algorithm 1, for a time interval $[S, T]$, we need to compute the meeting frequency for every pair (Line 5), which takes $O(|D|)$ time. However, it may be unnecessary to update *every* pair when expanding time interval from $[S, T]$ to $[S, T+1]$, since in the real data only a limited number of pairs meet at time $T+1$. For any time point t, to retrieve the pairs that meet at t, we use a time-based list $T\_list(t)$ to record those pairs,

$$T\_list(t) = \{(o_i, o_j) | freq_{o_i, o_j}(t) \neq 0\}.$$

With this data structure, Line 5 in Algorithm 1 can be replaced by retrieving every pair stored in $T\_list(t)$ and just update frequencies for those pairs. Even though T_list takes $\Omega(P \cdot d)$ additional memory, where d is the average number of meeting pairs per time point, it helps the updating step reduce its time complexity from $O(|D|)$ to $O(d)$. In real scenarios, as shown in our experiments (Section 5.3), d is usually much smaller than $|D|$

After updating frequencies, all the pairs need to be sorted according to their frequencies (Line 6 in Algorithm 1), which takes $O(|D| \log |D|)$ time. But when expanding T to $T + 1$, only a few pairs update their frequencies. Therefore, instead of doing sort all over again, we can update the sorted list with a small number of adjustments.

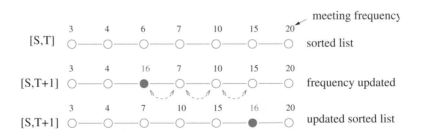

Fig. 4. Updating sorted list

Take Figure 4 for example. All the pairs are sorted in ascending order when ending time is T. When a pair increases its meeting frequency from 6 to 16 for time interval $[S, T + 1]$, it switches its position with the one on the right repeatedly, until it reaches the right-most position or the value on the right is larger than 16.

To update the position for one pair, it takes $O(|D|)$ in the worst case. In total, it takes $O(d|D|)$ to adjust the positions in the sorted list for all the updated pairs in T_list. Theoretically, this sorting strategy is no better than fast sort $(O(|D|\log|D|))$ because d may not be smaller than $\log|D|$. However, it takes much less time in practice since one pair will not increase its meeting frequency drastically by expanding ending time from T to $T+1$. We will verify this in experiment.

4.2   Finding the Best Split Point

Given the order of all pairs, it takes $O(|D|)$ time to consider each pair as a split point and calculate corresponding significance score. We next prove that it suffices to enumerate the values of all positive pairs, which takes $O(|D^+|)$ time. We observe that, in most real scenarios, the data have an important property: the number of pairs having the labeled relationship only takes a small portion of all the pairs (i.e., $|D^+| \ll |D|$).

For a split point v in time interval $[S, T]$, let $p(v)$ and $q(v)$ be the fractions of positive and negative pairs whose meeting frequencies are no less than v, respectively.

$$p(v) = \frac{|D^+ \cap D_{\geq}^{S,T}(v)|}{|D^+|}, \ q(v) = \frac{|D^- \cap D_{\geq}^{S,T}(v)|}{|D^-|}. \tag{1}$$

Given a pair $(p(v), q(v))$, we can write the information gain $IG(v)$ as a function of p and q:

$$IG(v) = IG(p(v), q(v)).$$

To introduce our second speed-up technique, we will find the following general property of information gain useful:

Lemma 1. *Given a pair of probabilities (p, q) as defined in (1), we have the following two properties of information gain:*

1. if $p > q$, then $\frac{\partial IG}{\partial p} > 0$, $\frac{\partial IG}{\partial q} < 0$,
2. if $p < q$, then $\frac{\partial IG}{\partial p} < 0$, $\frac{\partial IG}{\partial q} > 0$.

Basically, the above lemma states that if the frequency difference of positive pairs and negative pairs increases, then the split point v becomes more significant. In fact, in addition to information gain, it can be proven that many other popular statistical measures, such as the G-test score, also satisfy this good property. Interested readers are referred to [23] and the reference therein for the proof and more discussion of the lemma.

In the context of our work, Lemma 1 could be interpreted as follow.

Corollary 1. *For any two split points v_1 and v_2 and a time interval $[S, T]$,*

1. if $p(v_1) > q(v_1)$, $p(v_2) \geq p(v_1)$ and $q(v_2) \leq q(v_1)$ where the two equalities do not hold simultaneously, then $IG(v_2) > IG(v_1)$,

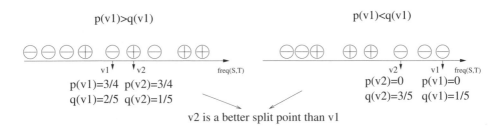

Fig. 5. Illustration for Corollary 1

2. *if $p(v_1) < q(v_1)$, $p(v_2) \leq p(v_1)$ and $q(v_2) \geq q(v_1)$ where the two equalities do not hold simultaneously, then $IG(v_2) > IG(v_1)$.*

The left subfigure of Figure 5 illustrates the first case in Corollary 1, i.e., $p(v_1) > q(v_1)$. As we can see, v_1 takes a negative pair as a split point. If we select v_2 as the split point, there is one less negative pair on the right side (i.e., $q(v_2) < q(v_1)$) but the number of positive pairs on the right side remains the same (i.e., $p(v_2) = p(v_1)$). Since the difference between positive pairs and negative pairs on the right side increases, $IG(v_2) > IG(v_1)$. In practice, we can observe the following two facts in the real data: (1) $|D^-| \gg |D^+|$ and (2) a considerable portion of negative pairs have very low or zero meeting frequency for any given time interval $[S, T]$. Therefore, for any time interval $[S, T]$, we can always assume $p(v) > q(v)$, where $0 < v \leq \max_{(o_i, o_j) \in D^+} \{freq_{o_i, o_j}(S, T)\}$. With this assumption, we can skip any negative pair whose meeting frequency is no larger than the maximum meeting frequency among all positive pairs.

Therefore, in addition to the positive pairs, we only need to examine those negative pairs which are on the right of the rightmost positive pair, as shown in the right subfigure of Figure 5. In fact, among all of these negative pairs, examining the leftmost one suffices. This is because in this case we have $p(v_1) = 0 < q(v_1)$. Therefore, by further shifting the split point to the left from v_1 to v_2, we always have $p(v_2) = p(v_1) = 0$ and $q(v_2) > q(v_1)$, thus $IG(v_2) > IG(v_1)$ by Corollary 1.

We summarize our second speed-up technique as the following theorem.

Theorem 1. *For a time interval $[S, T]$, assume the maximum meeting frequency of all positive pairs is v^*, then the best split point v must take one of following values:*

1. *the value of one positive pair: $v \in \{freq_{o_i, o_j}(S, T) : (o_i, o_j) \in D^+\}$,*
2. *the smallest value for the negative pairs that is larger than the maximum meeting frequency of all positive pairs:*

$$v = \min_{(o_i, o_j) \in D^-} \{freq_{o_i, o_j}(S, T) : freq_{o_i, o_j}(S, T) > v^*\}.$$

Therefore, for a time interval $[S, T]$, it takes $O(|D^+|)$ to compute significance score $G(S, T)$. Since $|D^+| \ll |D|$ in our problem, this saves a lot of time comparing with original time complexity $O(|D|)$.

5 Experiment

Our experiments are carried on both real and synthetic datasets. All the algorithms are implemented in C++, and all the experiments are carried out on a 2.13 GHz Intel Core machine with 4GB memory, running Linux with version 2.6.18 and gcc 4.1.2.

5.1 Dataset Description

To evaluate the effectiveness of our method, we use the Reality Mining dataset[2]. The Reality Mining project was conducted from 2004-2005 at MIT Media Laboratory. The study followed 94 subjects using mobile phones. We filtered the subjects with missing information and 86 subjects are left. The proximity between subjects can be inferred from repeated Bluetooth scans. When a Bluetooth device conducts a discovery scan, other Bluetooth devices within a range of 5-10 meters respond with their user-defined names. Therefore, instead of using the trajectory data, we take the Bluetooth data directly to generate the meeting frequencies for any two subjects. Although 86 subjects should form 3655 pairs, there are 1856 pairs which have zero meeting frequency. After filtering those pairs, our dataset has 1799 pairs in total.

We study two relationships on this dataset.

- **Friend relationship.** Subjects were asked about their friend relationship to the other individuals in the study. The survey question was "Is this person a part of your close circle of friends?" According to the survey, there are 22 reciprocal friend pairs, 39 non-reciprocal friend pairs and 1718 reciprocal non-friend pairs. In the survey, subjects were also asked about their physical proximity with the other individuals. In this analysis, we only use the pairs who have mutually reported some proximity. By doing so, we filter out those pairs with low meeting frequencies. The reason of doing this is because it is more trivial to achieve high accuracy if we include those pairs with few interactions. Similar pre-processing step was conducted in work [7] to study friend relationship. In the remaining pairs, we take reciprocal and non-reciprocal friend pairs as positive pairs (i.e., $|D^+| = 59$) and the non-friend pairs as negative ones (i.e., $|D^-| = 441$).
- **Colleague relationship.** One subject could belong to one of the affiliations such as media lab graduate student, media lab staff, professor, Sloan business school, etc.. To study colleague relationship, we take all the pairs belong to Sloan business school as the positive pairs and the remaining ones as the negative pairs. There are 218 positive pairs (i.e., $|D^+| = 218$) and 1561 (i.e., $|D^-| = 1561$) negative pairs in this case.

---

[2] http://reality.media.mit.edu/

5.2   Discovery of T-Motifs

In this section, we will show the T-Motifs for friend relationship and colleague relationship separately.

Table 1. Top-10 T-Motifs for friend relationship

T-Motif $[S, T]$	best split point v	$\dfrac{\lvert D_{\geq}^{S,T}(v) \cap D^+ \rvert}{\lvert D^+ \rvert}$	$\dfrac{\lvert D_{\geq}^{S,T}(v) \cap D^- \rvert}{\lvert D^- \rvert}$
[21:56 Wed., 23:08 Wed.]	12	0.372881	0.031746
[22:45 Tue., 23:39 Tue.]	55	0.305085	0.0181406
[19:07 Sat., 7:07 Sun.]	249	0.220339	0.00453515
[20:56 Tue., 22:44 Tue.]	1	0.508475	0.113379
[23:55 Tue., 1:42 Wed.]	10	0.355932	0.0453515
[23:22 Wed., 3:43 Thurs.]	53	0.220339	0.00680272
[7:08 Sun., 16:49 Sun.]	53	0.40678	0.0770975
[1:20 Fri., 5:12 Fri.]	12	0.20339	0.00680272
[21:52 Mon., 9:00 Tue.]	11	0.644068	0.240363
[18:12 Sun., 20:01 Sun.]	3	0.389831	0.0793651

Table 1 shows the top-10 T-Motifs mined for friend relationship. Among all time intervals, [21:56 Wed., 23:08 Wed.] plays the most important role, as 37.3% friends have meeting frequency more than 12 minutes whereas only 3.17% non-friends can exceed 12-minutes meeting frequency. As one can see, the interactions at night are more discriminative for friend relationship in general, with exceptions during the daytime on weekends, such as [7:08 Sun., 16:49 Sun.].

Table 2. Top-10 T-Motifs for colleague relationship

T-Motif $[S, T]$	best split point v	$\dfrac{\lvert D_{\geq}^{S,T}(v) \cap D^+ \rvert}{\lvert D^+ \rvert}$	$\dfrac{\lvert D_{\geq}^{S,T}(v) \cap D^- \rvert}{\lvert D^- \rvert}$
[9:20 Thurs., 10:30 Thurs.]	7	0.7201	0.0557
[9:42 Tue., 10:35 Tue.]	3	0.7431	0.0749
[10:36 Tue., 11:34 Tue.]	56	0.6376	0.0384
[10:34 Thurs., 11:04 Thurs.]	31	0.6055	0.0358
[11:05 Thurs., 11:40 Thurs.]	31	0.6146	0.0589
[7:31 Tue., 8:44 Tue.]	1	0.4449	0.0109
[21:16 Thurs., 9:10 Fri.]	7	0.6376	0.0723
[8:02 Thurs., 8:49 Thurs.]	2	0.4220	0.0128
[8:45 Tue., 9:19 Tue.]	22	0.3853	0.0070
[5:32 Wed., 10:28 Wed.]	2	0.5917	0.0749

Table 2 shows the top-10 T-Motifs for the colleague relationship. Interestingly, the colleague pairs (students from Sloan business school) usually have high meeting frequencies during the morning, especially on Tuesdays and Thursdays. It may suggest that these students have classes on Tuesday and Thursday mornings. Comparing to the friend relationship, it is obvious that colleagues have quite different temporal interaction patterns. T-Motifs provie us an insight in the uniqueness of each relationship.

5.3   Efficiency Study

In this section, we analyze the scalability issue w.r.t. different data sizes and parameter settings. We compare the T-Motif mining baseline method as shown in Algorithm 1 (denoted as baseline) with the one with speed-up techniques (denoted as speedup). We first present the comparison results on the friend relationship data. By default, we will use all the pairs in D to find T-Motifs and set $\delta_{min} = 1$ and $\delta_{max} = 6 \times 60$.

Figure 6 shows the time spent on each step when computing the significance scores for time intervals. In baseline method, to compute $G(S, T)$, there are three steps: updating meeting frequency with the time complexity $O(|D|)$, sorting all pairs with the time complexity $O(|D|log|D|)$ and finding the best split point with the time complexity $O(|D|)$. As shown in Figure 6, all the steps take roughly the same time. Compared to baseline, speedup compresses updating and sorting of the meeting frequencies into one step. As we mentioned in Section 4.1, even though this step takes $O(d|D|)$ time in theory, it is actually much faster in practice. In particular, updating and sorting by speedup together take 10.15 seconds whereas they take 175 seconds for baseline. The reason is illustrated in Figure 7. For many time

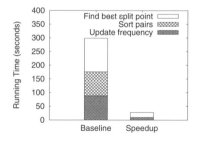

Fig. 6. Time spent on each step (friend relationship)

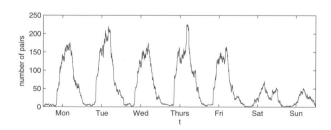

Fig. 7. Number of pairs meeting at each time point (friend relationship)

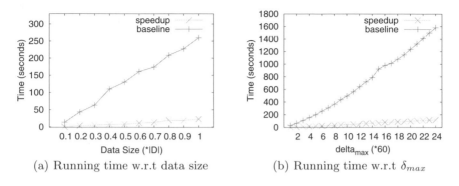

(a) Running time w.r.t data size   (b) Running time w.r.t δ_{max}

Fig. 8. Running time on friend relationship

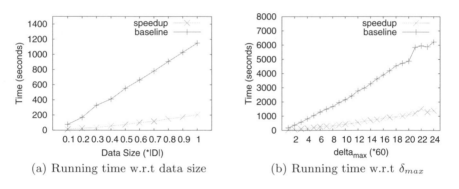

(a) Running time w.r.t data size   (b) Running time w.r.t δ_{max}

Fig. 9. Running time on colleague relationship

points, especially the ones at mid-night, very few pairs meet. On average, there are only 56.86 pairs meeting at each time point. Therefore, it is unnecessary to update the frequency for every pair and sort the frequencies all over again. Besides, the second speed-up technique reduces the best split point step from 299 seconds to 28 seconds since we only need to enumerate positive pairs as the split points.

Now we randomly select $p\%$ of the pairs from the entire dataset as the training samples and apply Algorithm 1. $p\%$ is enumerated from 10% to 100% with an increment of 10% in each trial. Figure 8(a) shows the running time w.r.t different data sizes. It is obvious that speedup techniques make the T-Motifs mining process much faster. The difference between speedup and baseline becomes bigger as the data size increases. When applying to the whole dataset, baseline takes 260 seconds whereas speedup only takes 23 seconds.

In Algorithm 1, we use δ_{min} and δ_{max} to limit the length of time intervals. When $\delta_{max} - \delta_{min}$ increases, the search space for T-Motifs also increases. By setting $\delta_{min} = 1$, we increase δ_{max} by hour, until it reaches 24 hours. The running times for baseline and speedup are plotted in Figure 8(b). Again, speedup is significantly faster than baseline, especially when δ_{max} is large.

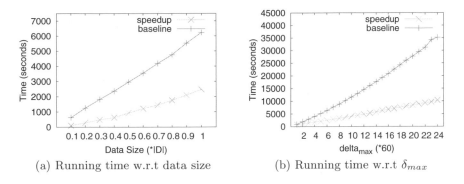

(a) Running time w.r.t data size   (b) Running time w.r.t δ_{max}

Fig. 10. Running time on synthetic dataset

We also conduct the same experiments on the dataset of colleague relationship. Figure 9 shows the running times w.r.t. different parameters, from which we can see that speedup is much faster than baseline.

Finally, we synthesize an even larger dataset based on the Reality Mining dataset. In our synthetic dataset, there are $1,000$ positive pairs and $10,000$ negative pairs. To generate a new positive pair, we randomly select one positive pair from the Reality Mining dataset and perturb its meeting frequency for each time point within 10% of the original value. We repeat this process to generate new negative pairs. We report the efficiency w.r.t. data size and δ_{max} in Figure 10. Compared to Figure 8, the difference in running time between the two methods is getting bigger in larger dataset.

5.4 Relationship Detection Using T-Motifs

Next we use the extracted T-Motifs to find the interested relationship in unlabeled data. In this experiment, we set $\delta_{min} = 1$ and $\delta_{max} = 12 * 60$ to mine T-Motifs and use the top-20 T-Motifs. A classification model is built using labeled pairs as training data and the meeting frequency within each T-Motif as the feature. We report the classification results using Support Vector Machine (SVM) and Gradient Boosting Machine (GBM) as learning methods. For comparison, we set the baseline as directly counting the meeting frequency over the entire time frame, which is equal to $freq_{o_i,o_j}(1, P)$ for a pair (o_i, o_j).

Using the classification model, we will get a score for each test sample indicating the probability to be a positive pair. In the top-k ranked test samples S_k, we use *precision* to examine the ratio of true positives, and *recall* to measure how many pairs having such relationship are retrieved. Precision and recall are defined as:

$$Prec@k = \frac{|D^+_{test} \cap S_k|}{k}, Rec@k = \frac{|D^+_{test} \cap S_k|}{|D^+_{test}|},$$

where D^+_{test} is the set of positive pairs in the test set.

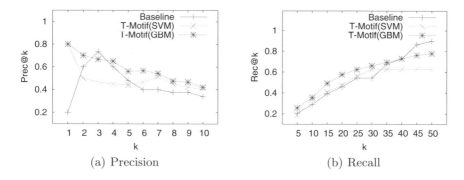

(a) Precision   (b) Recall

Fig. 11. Effectiveness comparison on friend relationship

We use 5-fold cross validation on the friend relationship data. Figure 11 shows the precision and recall of T-Motif based methods comparing with that of baseline. We can see that Prec@1=0.8 for both T-Motif(SVM) and T-Motif(GBM). It means that, when using T-Motifs, 80% top-1 pairs are real friends. In contrast, Prec@1=0.2 for baseline method, which indicates that the pair that has the highest meeting frequency does not necessarily have the friend relationship. In terms of recall measure in Figure 11(b), the methods based on T-Motif have higher recall value when k is smaller than 30. It means that T-Motifs can promote friend pairs to higher ranks. But it is worth noting that the baseline can retrieve 89% friend pairs at top-50 ranked pairs. This suggests that friends generally meet more frequently.

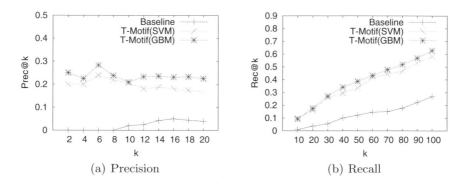

(a) Precision   (b) Recall

Fig. 12. Effectiveness comparison on colleague relationship

For colleague relationship, we use 10-fold cross validation. As one can see in Figure 12, baseline method performs poorly on this dataset, as it barely retrieves any colleague pair in the top-20 pairs. Even in the top-100 pairs, as shown in Figure 12(b), baseline can only retrieve less than 30% colleague pairs.

This indicates that the pairs meeting very often are not necessarily colleagues. As we have seen in Table 2, colleagues meet only at particular times. These patterns are well captured by T-Motifs, and we can retrieve 60% colleague pairs from top-100 pairs using T-Motifs.

6   Related Work

Previous studies mainly focus on the discovery of one specific relationship - moving object clusters, such as flock [13], moving cluster [11], convoy [10], and swarm [16]. They try to find a group of objects that move together for k consecutive or non-consecutive times. All these work simply count the timestamps that objects are being together. Laube et al. [14][9] define several spatio-temporal patterns, including flock, leadership, convergence, and encounter. However, each pattern is defined and studied individually. These patterns cannot be generalized to detect any real-world relationship for moving objects.

Many methods have been proposed to measure the similarity between two trajectories, such as Dynamic Time Warping (DTW) [25], Longest Common Subsequences (LCSS) [20], Edit Distance on Real Sequence (EDR) [2], and Edit distance with Real Penalty (ERP) [1]. The geometric trajectory distance plays an important role in determining the similarity between two objects. However, two objects having some relationship may not necessarily have similar trajectories. There are also studies to measure the similarities between two human trajectories [15][22]. But the similarity is measured by the travel sequence, such as shopping → movie → restaurant. Such similarity does not consider the times that two objects are being close.

There are several interesting studies showing the potential of using mobile or positioning technologies to study the human social behavior, such as mobility regularity [8][6][19] and interactions [18][7][4]. Miklas et al. [18] and Eagle et al. [7] focus on the analysis of the relationships between physical network and social network. [18] finds that "friends meet more regularly and for longer duration whereas the strangers meet sporadically" and [7] shows that "friend dyads demonstrate distinctive temporal and spatial patterns in their physical proximity and calling patterns". A more recent work by Cranshaw et al. [4] develops a technique to infer friendship in a supervised way, which is the most related work to ours. They design a set of spatial-temporal features and build a classification model for friend relationship. Their temporal features, such as the number of co-locations in evening/weekends, are heuristically designed. Our work is a much more general approach to detect any relationship with any temporal patterns.

Our idea of making use of T-Motifs is also motivated by a recent work [24] on time series classification problem. Different from other time series classification methods [12][21][5], Ye et al. [24] use shapelets, which are time series subsequences that can maximally represent a class. The pruning rules developed in [24] try to avoid the expensive time cost to compute the distance between a shapelet and a time series. Our problem is different from typical time series classification because our time dimension is fixed to a relative time frame, such as a week, and it only takes $O(1)$ to calculate the meeting frequency for each object pair. Our speed-up techniques aim to save time for computing significance scores.

7 Conclusion

In this paper, we introduce a supervised framework to detect relationships for moving objects from their meeting patterns. In this framework, the concept of T-Motifs is proposed to capture the temporal characteristics for relationships. A T-Motif is a time interval which has high information gain with respect to meeting frequencies for labeled pairs. We develop two speed-up techniques to enumerate T-Motif candidates and calculate their significance scores. In the experiments with real-world datasets, the proposed method is both efficient and effective in discovering the relationship for moving objects. Extensions to make use of spatial features to better detect the relationships could be interesting themes for future research.

Acknowledgment. The work was supported in part by the NSF IIS-1017362, NSF CNS-0931975, U.S. Air Force Office of Scientific Research MURI award FA9550-08-1-0265, Boeing company, and by the U.S. Army Research Laboratory under Cooperative Agreement Number W911NF-09-2-0053 (NS-CTA). The views and conclusions contained in this document are those of the authors and should not be interpreted as representing the official policies, either expressed or implied, of the Army Research Laboratory or the U.S. Government. The U.S. Government is authorized to reproduce and distribute reprints for Government purposes notwithstanding any copyright notation here on.

References

1. Chen, L., Ng, R.T.: On the marriage of lp-norms and edit distance. In: VLDB, pp. 792–803 (2004)
2. Chen, L., Özsu, M.T., Oria, V.: Robust and fast similarity search for moving object trajectories. In: SIGMOD Conference, pp. 491–502 (2005)
3. Cheng, H., Yan, X., Han, J., Hsu, C.-W.: Discriminative frequent pattern analysis for effective classification. In: ICDE, pp. 716–725 (2007)
4. Cranshaw, J., Toch, E., Hong, J.I., Kittur, A., Sadeh, N.: Bridging the gap between physical location and online social networks. In: UbiComp, pp. 119–128 (2010)
5. Ding, H., Trajcevski, G., Scheuermann, P., Wang, X., Keogh, E.J.: Querying and mining of time series data: experimental comparison of representations and distance measures. PVLDB 1(2), 1542–1552 (2008)
6. Eagle, N., Pentland, A.: Eigenbehaviors: identifying structure in routine. Behavioral Ecology and Sociobiology, 1057–1066 (2009)
7. Eagle, N., Pentland, A., Lazer, D.: Inferring friendship network structure by using mobile phone data. Proceedings of the National Academy of Sciences, 15274–15278 (2009)
8. González, M.C., Cesar, A., Hidalgo, R., Barabási, A.-L.: Understanding individual human mobility patterns. Nature 453, 779–782 (2008)

9. Gudmundsson, J., Laube, P., Wolle, T.: Movement patterns in spatio-temporal data. Encyclopedia of GIS, 726–732 (2008)
10. Jeung, H., Yiu, M.L., Zhou, X., Jensen, C.S., Shen, H.T.: Discovery of convoys in trajectory databases. PVLDB 1(1), 1068–1080 (2008)
11. Kalnis, P., Mamoulis, N., Bakiras, S.: On discovering moving clusters in spatio-temporal data. In: Anshelevich, E., Egenhofer, M.J., Hwang, J. (eds.) SSTD 2005. LNCS, vol. 3633, pp. 364–381. Springer, Heidelberg (2005)
12. Keogh, E.J., Kasetty, S.: On the need for time series data mining benchmarks: A survey and empirical demonstration. Data Min. Knowl. Discov. 7(4), 349–371 (2003)
13. Laube, P., Imfeld, S.: Analyzing relative motion within groups of trackable moving point objects. GIScience, 132–144 (2002)
14. Laube, P., van Kreveld, M.J., Imfeld, S.: Finding remo - detecting relative motion patterns in geospatial lifelines. In: Int. Symp. on Spatial Data Handling (2004)
15. Li, Q., Zheng, Y., Xie, X., Chen, Y., Liu, W., Ma, W.-Y.: Mining user similarity based on location history. In: GIS, p. 34 (2008)
16. Li, Z., Ding, B., Han, J., Kays, R.: Swarm: Mining relaxed temporal moving object clusters. PVLDB 3(1), 723–734 (2010)
17. Lo, D., Cheng, H., Han, J., Khoo, S.-C., Sun, C.: Classification of software behaviors for failure detection: a discriminative pattern mining approach. In: KDD, pp. 557–566 (2009)
18. Miklas, A.G., Gollu, K.K., Chan, K.K.W., Saroiu, S., Gummadi, P.K., de Lara, E.: Exploiting social interactions in mobile systems. In: Krumm, J., Abowd, G.D., Seneviratne, A., Strang, T. (eds.) UbiComp 2007. LNCS, vol. 4717, pp. 409–428. Springer, Heidelberg (2007)
19. Song, C., Qu, Z., Blumm, N., Barabasi, A.L.: Limits of predictability in human mobility. Science, 1018–1021 (2010)
20. Vlachos, M., Gunopulos, D., Kollios, G.: Discovering similar multidimensional trajectories. In: ICDE, pp. 673–684 (2002)
21. Xi, X., Keogh, E.J., Shelton, C.R., Wei, L., Ratanamahatana, C.A.: Fast time series classification using numerosity reduction. In: ICML, pp. 1033–1040 (2006)
22. Xiao, X., Zheng, Y., Luo, Q., Xie, X.: Finding similar users using category-based location history. In: GIS, pp. 442–445 (2010)
23. Yan, X., Cheng, H., Han, J., Yu, P.S.: Mining significant graph patterns by leap search. In: SIGMOD Conference, pp. 433–444 (2008)
24. Ye, L., Keogh, E.J.: Time series shapelets: a new primitive for data mining. In: KDD, pp. 947–956 (2009)
25. Yi, B.-K., Jagadish, H.V., Faloutsos, C.: Efficient retrieval of similar time sequences under time warping. In: ICDE, pp. 201–208 (1998)

A Uniform Framework for Temporal Functional Dependencies with Multiple Granularities

Carlo Combi[1], Angelo Montanari[2], and Pietro Sala[1]

[1] Dipartimento di Informatica, Università degli Studi di Verona
strada le Grazie 15, 37134 Verona Italy
{carlo.combi,pietro.sala}@univr.it
[2] Dipartimento di Matematica e Informatica, Università degli Studi di Udine
via delle Scienze 206, 33100 udine Italy
angelo.montanari@uniud.it

Abstract. Temporal functional dependencies (TFDs) add a temporal component to classical functional dependencies to deal with temporal data. As an example, while functional dependencies model constraints like *"employees with the same role get the same salary"*, TFDs can represent constraints like *"for any given month, employees with the same role get the same salary (but their salary may change from one month to the next one)"* or *"current salaries of employees uniquely depend on their current and previous roles"*. In this paper, we propose a general framework for specifying TFDs, possibly involving different time granularities, and for checking whether or not a given database instance satisfies them. The proposed framework subsumes existing formalisms for TFDs and it allows one to encode TFDs which are not captured by them.

1 Introduction

Temporal functional dependencies (TFDs) add a temporal dimension to classical functional dependencies (FDs) to deal with temporal data [2,11,12,13,14] (as a matter of fact, two temporal dimensions have been considered only in [6], where standard FDs are evaluated at every database snapshot). As an example, while FDs model constraints like *"employees with the same role get the same salary"*, TFDs can represent constraints like *"for any given month, employees with the same role have the same salary, but their salary may change from one month to the next one"* [2,13] or *"current salaries of employees uniquely depend on their current and previous roles"* [11]. Since temporal constraints may refer to different time units, e.g., university courses are organized on semesters, the scheduling of business activities usually refers to business months or weeks, follow-up visits are usually planned on working days, TFDs must allow one to express temporal constraints at different time granularities.

In this paper, we propose a general framework that makes it possible to formally specify TFDs, possibly involving multiple time granularities, and to check whether or not a given database instance satisfies them. We will prove that the proposed framework subsumes all existing formalisms for TFDs, and it allows

D. Pfoser et al. (Eds.): SSTD 2011, LNCS 6849, pp. 404–421, 2011.

one to express TFDs which are not captured by them. As an example, assuming months as the basic time unit, we can encode constraints such as *"employees with the same role, who will not change it from the current month to the next one, will get the same (unchanged) salary"*. Moreover, we will show the effectiveness of the approach by applying it to a real-world medical domain, related to the administration of chemotherapies.

The paper is organized as follows. In Section 2, we provide a motivating scenario. In Section 3, we describe the temporal data model the proposed framework relies on. In Section 4, we introduce a new formalism for the representation of TFDs, and we show that it allows one to capture a large variety of TFDs. Then, in Section 5, we explain how to reduce the problem of checking whether a temporal relation satisfies a given TFD to the problem of checking whether the evaluation of a suitable algebraic expression returns the empty relation. We conclude the section with a short discussion of computational aspects. In Section 6, we briefly survey existing systems for TFDs and we compare them with the framework we propose. Section 7 provides an assessment of the work and it outlines future research directions.

2   A Motivating Scenario from Clinical Medicine

Most health care institutions collect a large quantity of clinical information about patients and physicians' actions, such as therapies and surgeries, and health care processes, such as admissions, discharges, and exam requests. All these pieces of information are temporal in nature and thus the associated temporal dimension must be carefully modeled to make it possible to properly represent clinical data and to reason on them.

To illustrate the relevance of properly expressing and checking temporal constraints on data, we consider a real-world example taken from the domain of chemotherapies for oncology patients. Oncology patients undergo several chemotherapy cycles. Each one can be repeated several times, and it typically includes the administration of several drugs according to a predefined temporal pattern. The problem of managing chemotherapy plans has been extensively studied by the clinical research. Several clinical practice guidelines describe and recommend them in detail.

We consider the following chemotherapy recommendations related to FAC and CEF regimens [1,8] for the treatment of breast cancer.

Example 1. Recommended FAC and CEF regimens.

FAC regimen: *"The recommended FAC regimen consists of 5-fluorouracil on days 1 and 8, and doxorubicin and cyclophosphamide on day 1. This is repeated every 21 days for 6 cycles (that is, 6 cycles of 21 days each)"*.

CEF regimen: *"The recommended CEF regimen consists of 14 days of oral cyclophosphamide, and intravenous injection of epirubicin and 5-fluorouracil on days 1 and 8. This is repeated every 28 days for 6 cycles."*

A relation schema *Patient* to be used for storing information about patients who underwent chemotherapies can be structured as follows. For each patient, we store the therapy, the patient's identifier, the blood group, the physician who prescribed the therapy, the taken drug with its quantity, and the specific drug-taking time by means of attributes *Chemo*, *PatId*, *BG*, *Phys*, *Drug*, *Qty*, and *VT*, respectively. We assume that attribute *VT* specifies the valid time of a tuple in term of days (starting from an implicit day taken as the origin of the time domain).

Table 1 shows a possible instance *r* of *Patient* describing chemotherapy treatments for patients with *PatId* p1, p2, and p3. As an example, according to the prescription of the FAC treatment reported in Example 1, the first day of the cycle the patient has to take drugs *Flu*, *Dox*, and *Cyc*, that is, drugs containing 5-fluorouracil, doxorubicin, and cyclophosphamide, respectively.

Table 1. An instance *r* of *Patient* storing data about chemotherapy treatments

TUPLE#	*Chemo*	*PatId*	*BG*	*Phys*	*Drug*	*Qty*	*VT*
1	FAC	p1	0+	Smith	Flu	500	1
2	FAC	p1	0+	Hubbard	Dox	50	1
3	FAC	p1	0+	Verdi	Cyc	500	1
4	FAC	p1	0+	Smith	Flu	500	8
5	CEF	p2	AB-	Verdi	Cyc	600	1
6	CEF	p2	AB-	Hubbard	Flu	600	1
7	CEF	p2	AB-	Hubbard	Epi	60	1
8	CEF	p2	AB-	Verdi	Cyc	600	2
9	CEF	p2	AB-	Smith	Cyc	500	3
..	..	..	..	..	..	..	..
20	CEF	p2	AB-	Verdi	Cyc	600	8
21	CEF	p2	AB-	Hubbard	Flu	600	8
22	CEF	p2	AB-	Hubbard	Epi	60	8
..	..	..	..	..	..	..	..
33	CEF	p3	AB-	Verdi	Cyc	550	1

According to the clinical meaning of data, several requirements can be imposed on this relation schema. They stem from both clinical/medical reasons and organizational rules of the medical unit managing the chemotherapy administration. In the following, we list some typical requirements to be represented and managed by the database systems.

(1) *A patient may take any given drug at most one time per day*: such a requirement prevents any patient from taking two or more times the same drug during the same day. As the administration of a drug within a chemotherapy has relevant side effects on the patient status, it is a quite obvious requirement. Nevertheless, checking it prevents possible data insertion errors.

(2) *For any chemotherapy, the quantities of a given drug prescribed to a patient on two days which are at most 14 days far away cannot be different*: for any

patient, such a requirement forces drug quantities of chemotherapies to remain unchanged whenever the drug is taken several times within 14 days (notice that the relation depicted in Table 1 violates it). According to the considered chemotherapies, this amounts to impose that drug quantities cannot change during a chemotherapy cycle (obviously, they can change if the chemotherapy changes).

(3) *For any chemotherapy, the quantities of a given drug administered during the same month cannot be different*: this requirement states that, regardless of the patient, for any chemotherapy and any month, the administered quantity of a drug is always the same. A possible explanation of such a requirement is that there exists some form of synchronization among different administrations of the same chemotherapy, which forces changes in drug quantities to be done only when changes in month occur (for instance, to take into account different seasonal conditions).

(4) *For patients undergoing a specific chemotherapy regimen, with the same drug being prescribed on two consecutive days, the quantity of the drug administered on the latter day depends solely on the drug quantity administered on the former*: this requirement constrains different patients not to take different quantities of the drug on the next day, if they take the same quantity of it on the current day. A possible explanation of such a requirement is that, regardless of the patient and the chemotherapy, the physician must follow a predefined therapy plan, suggested by the clinical practice, for what concerns the quantities of drugs administrated on consecutive days.

(5) *For any pair of administrations of the same drug prescribed by the same physician to the same patient on two consecutive days, the quantity of the second administration uniquely depends on (the drug and) the quantity of the first administration*: this requirement slightly differs from the previous one as it further constrains the physician to be the same.

(6) *For any pair of successive administrations of the same drug to the same patient within the same chemotherapy, the quantity of the second administration uniquely depends on (the drug and) the quantity of the first administration*: such a requirement constrains successive administrations of a drug to a patient within an assigned chemotherapy. As a general rule, time delays between successive administrations may differ from one drug to another. As an example, oral cyclophosphamide in the CEF regimen is taken daily for 14 days, while there is an interval of 7 days between two successive intravenous injections of epirubicin.

(7) *For any pair of successive administrations of the same drug to the same patient within the same month, the quantity of the second administration uniquely depends on (the drug and) the quantity of the first administration*: such a requirement imposes suitable constraints on successive drug administrations only when they occur during the same month. We may assume the rationale of this requirement to be the same as that of Requirement (4).

(8) *For any given chemotherapy, the quantities of drugs taken by patients with successive administrations that take place one 7 days after the other cannot change*: for any chemotherapy, such a requirement basically imposes that (only) the quantities of drugs taken by patients every 7 days cannot change.

3   Temporal Data Model and Temporal Views

In the next two sections, we will outline a general framework for the specification and verification of different kinds of TFD. As a preliminary step, in this section, we describe the temporal data model we rely on.

To represent TFDs, we will make use of a simple data model based on the notion of *temporal relation*. We assume the time domain T to be isomorphic to the set of natural numbers $\mathbb{N}$ with the usual ordering $\leq$. Let U be a set of atemporal attributes and VT be a temporal attribute, called valid time attribute. A temporal relation r is a relation on a temporal relation schema R with attributes $U \cup \{VT\}$ ($r \in R$ for short). Given a tuple $t \in r$ and an attribute $A \in U \cup \{VT\}$, we denote by $t[A]$ the value that t assumes on A. The temporal attribute VT specifies the valid time of a tuple, and it takes its value over the time domain T, that is, $t[VT] \in T$.

In the following, we will exploit the *tuple relational calculus* [10] to define suitable temporal views on data, that will help us in specifying and analyzing TFDs without harming the simplicity of the data model. In particular, these views will allow us to easily "move through time" in order to establish a connection between corresponding tuples valid at different time points.

A special role will be played by the following two temporal views, respectively called *next* and *nexttuple*, that allow one to link tuples that satisfy a specific temporal relation in order to represent relevant cases of (temporal) evolution.

Given a temporal distance k, with $k \geq 1$, the temporal view *next* allows one to join pairs of corresponding tuples at distance k (for $k = 1$, it joins pairs of consecutive corresponding tuples). More precisely, given a temporal relation schema R, with attributes $U \cup \{VT\}$, $r \in R$, and $t, t' \in r$, the application of *next* to r, denoted $\chi_Z^k(r)$, with $Z \subseteq U$ and $k \geq 1$, joins t, t' if (and only if) $t[Z] = t'[Z]$ and $t'[VT] = t[VT] + k$ as formally stated by the following definition.

Definition 1. *Let R be a temporal relation schema with attributes $U \cup \{VT\}$, $Z \subseteq U$, $W = U - Z$, $\overline{W}$ be obtained from W by renaming each attribute $A \in W$ as $\overline{A}$, $r \in R$, and $t, t' \in r$. The relation $\chi_Z^k(r)$, with schema $ZW\overline{W} \cup \{VT, \overline{VT}\}$, is defined as follows:*

$$\chi_Z^k(r) \stackrel{def}{=} \{s \mid \exists t, t'(t \in r \land t' \in r \land t[Z] = t'[Z] \land t'[VT] = t[VT] + k \land$$
$$s[U] = t[U] \land s[VT] = t[VT] \land s[\overline{W}] = t'[W] \land s[\overline{VT}] = t'[VT])\}$$

Hereinafter, when no confusion may arise, we will write $\chi_Z(r)$ for $\chi_Z^1(r)$. It is worth pointing out that, in the definition of $\chi_Z^k(r)$, we make use of the non-standard (arithmetic) selection condition $t'[VT] = t[VT] + k$. However, the expression for $\chi_Z^k(r)$ can be turned into a standard relational calculus expression, as shown in detail in [4].

The temporal view *nexttuple* allows one to join pairs of successive (with respect to the values they take on attribute VT) tuples. More precisely, given a temporal relation schema R, with attributes $U \cup \{VT\}$, $r \in R$, and $t, t' \in r$, the application of *nexttuple* to r, denoted $\tau_Z(r)$, with $Z \subseteq U$, joins t, t' if (and only if) $t[Z] = t'[Z]$ and t' is the tuple immediately following t with respect to VT. The view *nexttuple* is formally defined as follows.

Definition 2. *Let R be a temporal relation schema with attributes $U \cup \{VT\}$, $Z \subseteq U$, $W = U - Z$, $\overline{W}$ be obtained from W by replacing each attribute $A \in W$ by $\overline{A}$, $r \in R$, and $t, t' \in r$. The relation $\tau_Z(r)$, with schema $ZW\overline{W} \cup \{VT, \overline{VT}\}$, is defined as follows:*

$$\tau_Z(r) \stackrel{def}{=} \{s \mid \exists t, t'(t \in r \land t' \in r \land t[Z] = t'[Z] \land t[VT] < t'[VT] \land$$
$$s[U] = t[U] \land s[VT] = t[VT] \land s[\overline{W}] = t'[W] \land s[\overline{VT}] = t'[VT] \land$$
$$\neg \exists t''(r(t'') \land t[Z] = t''[Z] \land t[VT] < t''[VT] \land t''[VT] < t'[VT]))\}$$

Table 2. The view $\chi_{PatId,Phys}(r)$ for the instance r of *Patient* given in Table 1

TUPLES#	Chemo	PatId	BG	Phys	Drug	Qty	VT	Chemo	BG	Drug	Qty	VT
5-8	CEF	p2	AB-	Verdi	Cyc	600	1	CEF	AB-	Flu	500	2

The temporal views *next* and *nexttuple* make it possible to represent non-trivial aspects of the temporal evolution of data. More precisely, *next* joins pairs of corresponding tuples, that is, tuples with matching values for the specified attributes, which are a given temporal distance apart; *nexttuple* joins pairs of corresponding tuples with one tuple being the next occurrence of the other (with no constraints on their temporal distance) in a data-dependent manner. As an example, let us consider the instance r of the relation schema *Patient* given in Table 1 (we restrict our attention to the tuples explicitly reported in Table 1). The view $\chi_{PatId,Phys}(r)$, depicted in Table 2, joins pairs of tuples that take the same values on $PatId$ and $Phys$ and are valid at two consecutive points t and $t+1$. The view $\tau_{PatId,Phys}(r)$, depicted in Table 3, joins pairs of tuples that take the same values on $PatId$ and $Phys$ and such that there are no tuples temporally located in between of them with the same values for $PatId$ and $Phys$.

Table 3. The view $\tau_{PatId,Phys}(r)$ for the instance r of *Patient* given in Table 1

TUPLES#	Chemo	PatId	BG	Phys	Drug	Qty	VT	Chemo	BG	Drug	Qty	VT
1-4	FAC	p1	0+	Smith	Flu	500	1	FAC	0+	Flu	500	8
5-8	CEF	p2	AB-	Verdi	Cyc	600	1	CEF	AB-	Flu	500	2
6-21	CEF	p2	AB-	Hubbard	Flu	600	1	CEF	AB-	Flu	600	8
8-20	CEF	p2	AB-	Verdi	Cyc	600	2	CEF	AB-	Cyc	600	8
7-21	CEF	p2	AB-	Hubbard	Epi	60	1	CEF	AB-	Flu	600	8
6-22	CEF	p2	AB-	Hubbard	Flu	600	1	CEF	AB-	Epi	60	8
7-22	CEF	p2	AB-	Hubbard	Epi	60	1	CEF	AB-	Epi	60	8

4   A Uniform Framework for TFDs

In this section, we propose a new formalism for the specification of TFDs. We first introduce its syntax and semantics; then, we show that it allows one to express all TFDs dealt with by existing formalisms as well as to cope with new classes of TFDs.

Definition 3. *Let R be a temporal relation schema with attributes $U \cup \{VT\}$. A TFD is an expression of the following form:*

$$[E\text{-}Exp(R), t\text{-}Group(i)]X \to Y,$$

where $E\text{-}Exp(R)$ is a relational expression on R, called evolution expression, *$t\text{-}Group$ is a mapping $\mathbb{N} \to 2^{\mathbb{N}}$, called* temporal grouping *($t\text{-}Group(i)$ denotes the image of a generic element $i \in \mathbb{N}$), and $X \to Y$ is a functional dependency. We distinguish two types of TFD:*

1. *if the schema of the expression $E\text{-}Exp(R)$ is $U \cup \{VT\}$, then, $X, Y \subseteq U$;*
2. *if the schema of the expression $E\text{-}Exp(R) \cap (\overline{U} \cup \{\overline{VT}\}) \neq \emptyset$, then $X, Y \subseteq U\overline{U}$ and, for each $A \in Y$, both $XA \cap U \neq \emptyset$ and $XA \cap \overline{U} \neq \emptyset$.*

Temporal grouping specifies how to group tuples on the basis of the values they take on the temporal attribute VT (and on the temporal attribute $\overline{VT}$, if present), when $X \to Y$ is evaluated.

For the sake of simplicity, we confined ourselves to TFDs involving at most two database states. However, Definition 3 can be easily generalized to the case of tuple evolutions involving n database states.

Evolution expressions $E\text{-}Exp(R)$ make use of temporal views to select those tuples, valid at different time points, that must be merged in order to track the evolution of domain objects over time.

In principle, there are no restrictions on the form that $t\text{-}Group$ may assume. However, such a generality is not necessary from the point of view of applications. In the following, we will restrict ourselves to specific mappings, as those captured by Bettini et al.'s granularities [2] and Wjisen's time relations [13]. A time granularity G is a partition of the time domain T in groups of indivisible, disjoint units, called *granules*. Formally, a time granularity G is defined as a mapping from an index set $\mathcal{I} \subseteq \mathbb{N}$ to subsets of T. Classical examples of granularity are Day, $\mathsf{WorkingDay}$, Week, and Month. Various formalisms for representing and reasoning about time granularities have been proposed in the literature, including Granular Calendar Algebra by Ning et al. [9] and Ultimately Periodic Automata, by Bresolin et al. [3]. Wijsen's time relations are subsets of the set $\{(i,j) \mid i \in \mathbb{N}, \; j \in \mathbb{N}, \; i \leq j\}$. As an example, we can define a time relation that, for any $i \in \mathbb{N}$, collects all and only the pairs of time points (i,j) such that $j - i \leq k$, for some fixed k. Granularities can be recovered as special cases of time relations, called chronologies (it is not difficult to see that there exist quite natural time relations that cannot be expressed by means of granularities).

We constrain $t\text{-}Group(i)$ to take one of the following two forms:

$$t\text{-}Group(i) \quad \overset{\text{def}}{=} \quad \mathsf{G}(i),$$

where $\mathsf{G}(i)$ is the i-th granule of granularity G;

$$t\text{-}Group(i) \quad \overset{\text{def}}{=} \quad \bigcup_{j=1}^{n} \{(i + \alpha_j)\} \text{ for some } n \geq 1,$$

where $\alpha_1 = 0, \forall j \in [1, n] \ (\alpha_j \in \mathbb{N})$, and $\forall k \in [1, n-1] \ (\alpha_k < \alpha_{k+1})$.

In the first case, given a granularity G, $t\text{-}Group(i)$ groups time points according to the granule $\mathsf{G}(i)$ (for some $i \in \mathcal{I}$) they belong to. In the second case, $t\text{-}Group$ groups time points into an infinite number of intersecting finite sets. For every $i \in \mathcal{I}$, the i-th set $t\text{-}Group(i)$ consists of the i-th time point plus other $n-1$ time points identified by their offset with respect to such a point (n is fixed and it does not depend on i).

As in the case of standard FDs, a TFD is a statement about admissible temporal relations on a temporal relation schema R. We say that $r \in R$ satisfies a TFD $[E\text{-}Exp(R), t\text{-}Group(i)]X \rightarrow Y$ if it is not possible that the relation obtained from r by applying the expression $E\text{-}Exp(R)$ (hereinafter, the evolution relation) features two tuples t, t' such that (i) $t[X] = t'[X]$, (ii) $t[VT]$ and $t'[VT]$ (the same for $t[\overline{VT}]$ and $t'[\overline{VT}]$, if present) belong to the same temporal group $t\text{-}Group(i)$, and (iii) $t[Y] \neq t'[Y]$. This amounts to saying that the FD $X \rightarrow Y$ must be satisfied by each relation obtained from the evolution relation by selecting those tuples whose valid times belong to the same temporal group $t\text{-}Group(i)$. We partition the set of relevant TFDs into four classes.

- *Pure temporally grouping TFDs.* $E\text{-}Exp(R)$ returns the given temporal relation r. Tuples are grouped on the basis of $t\text{-}Group$.
- *Pure temporally evolving TFDs.* $E\text{-}Exp(R)$ merges tuples modeling the evolution of a real-world object. There is no temporal grouping, that is, there is only one group collecting all tuples of the computed relation.
- *Temporally mixed TFDs.* First, $E\text{-}Exp(R)$ merges tuples modeling the evolution of a real-world object; then, temporal grouping is applied to the resulting set of tuples.
- *Temporally hybrid TFDs.* First, $E\text{-}Exp(R)$ selects those tuples of the given temporal relation that contribute to the modeling of the evolution of a real-world object, that is, it removes isolated tuples; then, temporal grouping is applied to the resulting set of tuples.

In the following, we will describe in some detail the above classes of TFDs and we will give some meaningful examples of TFDs belonging to them.

Pure temporally grouping TFDs. In these TFDs, $E\text{-}Exp(R)$ returns the (original) temporal relation r. This forces FD $X \rightarrow Y$, with $X, Y \subseteq U$, to be checked on every (maximal) subset of r consisting of tuples whose VT values belong to the same temporal group $t\text{-}Group(i)$.

Example 2. Let us consider the first three requirements of relation *Patient* reported in Section 2:

1. *a patient may take any given drug at most one time per day;*
2. *for any chemotherapy, the quantities of a given drug prescribed to a patient on two days which are at most 14 days far away cannot be different;*
3. *for any chemotherapy, the quantities of a given drug administered during the same month cannot be different.*

The first requirement is captured by the following TFD:

$$[Patient, \{i\}] PatId, Drug \rightarrow Chemo, BG, Phys, Qty$$

For each time point i, the FD $PatId, Drug \rightarrow Chemo, BG, Phys, Qty$ must be satisfied by the tuples of (the instance of the relation schema) *Patient* valid at time i. As a matter of fact, this forces attributes $PatId, Drug$ to be a *snapshot key* for relation schema *Patient*: the set of tuples of *Patient* valid at a given time point (snapshot) must have $PatId, Drug$ as a (standard) key [6].

The second requirement can be encoded as follows:

$$[Patient, \{i, i+1, \ldots, i+13\}] PatId, Chemo, Drug \rightarrow Qty$$

In such a way, we force the FD $PatId, Chemo, Drug \rightarrow Qty$ to be checked at each time point i on the tuples of (the instance of the relation schema) *Patient* valid at time points $i, i+1, i+2, \ldots,$ or $i+13$, that is, for each i, $Chemo, PatId, Drug \rightarrow Qty$ must be satisfied by the tuples belonging to the union of snapshots of the temporal relation at time instants $i, \ldots, i+13$.

The third requirement can be expressed by means of the following TFD:

$$[Patient, \mathsf{Month}(\mathsf{i})] Chemo, Drug \rightarrow Qty$$

Pure temporally evolving TFDs. In these TFDs, $E\text{-}Exp(R)$ returns a relation over (a subset of) attributes $U\overline{U} \cup \{VT, \overline{VT}\}$, which is computed by means of join operations on some subset of U. Temporal grouping considers all tuples together.

Example 3. Let us consider the following three requirements of relation *Patient* reported in Section 2:

4. *for patients undergoing a specific chemotherapy regimen, with the same drug being prescribed on two consecutive days, the quantity of the drug administered on the latter day depends solely on the drug quantity administered on the former;*
5. *for any pair of administrations of the same drug prescribed by the same physician to the same patient on two consecutive days, the quantity of the second administration uniquely depends on (the drug and) the quantity of the first administration;*
6. *for any pair of successive administrations of the same drug to the same patient within the same chemotherapy, the quantity of the second administration uniquely depends on (the drug and) the quantity of the first administration.*

Let Top(i) be the top granularity collecting all time points in a single nonempty granule [2]. Requirement (4) can be formalized as follows:

$$[\chi_{PatId,Chemo,Drug}(Patient), \mathsf{Top(i)}]Drug, Qty \rightarrow \overline{Qty}$$

while Requirement (5) is captured by the following TFD:

$$[\chi_{PatId,Chemo,Drug,Phys}(Patient), \mathsf{Top(i)}]Drug, Qty \rightarrow \overline{Qty}$$

TFDs in this class can be viewed as a generalization of Vianu's dynamic dependencies [11]: $E\text{-}Exp(R)$ defines an evolution mapping (update mapping, according to Vianu's terminology) that associates each tuple valid at a time i with its corresponding tuple (if any) valid at time $i + 1$ making use of the values of specific relation attributes. In the relational framework we propose, evolution mappings are thus expressed by means of suitable joins on a subset of U. Moreover, TFDs for Requirements (4) and (5) show the possibility of defining dynamic dependencies according to a number of evolution mappings.

Let us consider now Requirement (6). To cope with it, we need the ability to join tuples valid at non-consecutive time points. Successive administrations of the same drug to the same patient may indeed occur at consecutive time points (tuples #8 and #9 in Table 1), but they may also take place at time points which are far away from one another (tuples #1 and #4 in Table 1), and thus we must be able to deal with a kind of "asynchronous" temporal evolution. Requirement (6) is encoded by the following TFD:

$$[\tau_{PatId,Chemo,Drug}(Patient), \mathsf{Top(i)}]Drug, Qty \rightarrow \overline{Qty}$$

Temporally mixed TFDs. In these TFDs, $E\text{-}Exp(R)$ returns a relation over (a subset of) attributes $U\overline{U} \cup \{VT, \overline{VT}\}$, which is computed by means of join operations on some subset of U. Temporal grouping groups together tuples according to their VT values. In such a way, one can define evolving (dynamic) dependencies that must hold at all (and only) time points *belonging to the same temporal group*.

Example 4. Let us consider now the seventh requirement of relation *Patient* reported in Section 2:

7. *For any pair of successive administrations of the same drug to the same patient within the same month, the quantity of the second administration uniquely depends on (the drug and) the quantity of the first administration.*

Such a requirement is encoded by the following TFD:

$$[\tau_{PatId,Drug}(Patient), \mathsf{Month(i)}]Drug, Qty \rightarrow \overline{Qty}$$

Temporally hybrid TFDs. In these TFDs, $E\text{-}Exp(R)$ returns a relation over the schema $U \cup \{VT\}$, which is computed by means of join operations on some subset of U and further renaming, project, and union operations. Temporal grouping groups together tuples according to their VT values. These TFDs allow one to express requirements on evolving values that must be preserved by all evolutions. As we will show in Section 6, such a class of requirements is captured neither by Vianu's dynamic dependencies nor by the other "grouping" TFDs proposed in the literature.

Example 5. Let us consider the last requirement of the relation *Patient* reported in Section 2:

8. *For any given chemotherapy, the quantities of drugs taken by patients with successive administrations that take place one 7 days after the other cannot change.*

In this case, grouping tuples which "happen" every seven days and then checking dependency $Chemo, Drug \rightarrow Qty$ against them is not enough, and thus TFD $[r, \{i, i+7\}] Chemo, Drug \rightarrow Qty$ does not help us in representing this last requirement. A TFD that takes into consideration the evolution of tuples in an appropriate way is the following one:

$$[He^{Patient}, \mathsf{Top}(\mathrm{i})] Chemo, Drug \rightarrow Qty,$$

where

$$
\begin{aligned}
He^{Patient} \stackrel{def}{=} \{t| \; \exists t' (t' \in \tau_{PatId,Chemo,Drug}(Patient) \wedge t'[\overline{VT}] = t'[VT] + 7 \wedge \\
t[PatId, Chemo, Drug] = t'[PatId, Chemo, Drug] \wedge \\
((t[BG, Phys, Qty] = t'[BG, Phys, Qty] \wedge t[VT] = t'[VT]) \vee \\
(t[BG, Phys, Qty] = t'[\overline{BG}, \overline{Phys}, \overline{Qty}] \wedge t[VT] = t'[\overline{VT}]))) \}.
\end{aligned}
$$

The evolution expression $He^{Patient}$ first joins tuples representing successive administrations (of a given drug to a given patient), that is, pairs of administrations with no administrations in between. Then, two sets of tuples, respectively featuring the attribute values of the first and of the second drug administration, are computed by two existential subqueries. Finally, the two sets are merged (logical disjunction) to make it possible to specify (and check) the functional dependency. In such a way, the functional dependency is only checked on tuples referring to successive administrations of a given drug to a given patient that take place on time points (days) which are 7 unit (days) from one another. As all tuples have to be considered together, temporal grouping is $\mathsf{Top}(\mathrm{i})$.

We conclude the section by giving another example of this new kind of TFD. Let us consider the constraint "*employees with the same role, who will not change it from the current month to the next one, will get the same (unchanged) salary*" that we already mentioned in the introduction. We need both to join consecutive tuples modeling old and new values for a given employee (this can be done with Vianu's DFDs as well) and to compare these old and new values (no one of existing formalisms for TFDs support these comparisons). Given a relation *Employee* over the schema $U = \{empId, salary, role\}$ and $r \in Employee$, the above constraint can be encoded by means of the following temporally hybrid TFD:

$$[ev^{Employee}, \{i, i+1\}]role \rightarrow salary,$$

where

$$ev^{Employee} \stackrel{\text{def}}{=} \{t | \exists t'(t' \in \chi_{empId,role}(Employee) \wedge$$
$$t[empId, role] = t'[empId, role] \wedge$$
$$((t[salary] = t'[salary] \wedge t[VT] = t'[VT]) \vee$$
$$(t[salary] = t'[\overline{salary}] \wedge t[VT] = t'[\overline{VT}])))\}.$$

The evolution expression $ev^{Employee}$ joins tuples related to the same employee with the same role holding at two consecutive time points, taking months as the basic time unit. Then, two sets of tuples, respectively featuring the attribute values holding at the first time point and at the second one, are computed by two existential subqueries. Finally, the two sets are merged (logical disjunction) to make it possible to specify (and check) the functional dependency $role \rightarrow salary$. In such a way, tuples describing employees that change their role and, in the new role, get a salary different from that of the other employees already in this role are allowed, as they do not appear in $ev^{Employee}$.

5   TFD Checking

In this section, we show that the problem of checking whether a temporal relation $r \in R$ satisfies a given (set of) TFD(s) can be reduced to the problem of establishing whether the evaluation of a suitable relational query on r returns the empty relation [7]. To make the computational steps of the checking procedure explicit, we adopt the *named relational algebra* featuring selection, projection, natural join, set difference, set union, and renaming as its basic operations [7].

We can verify whether a given relation $r \in R$ satisfies the TFD $[E\text{-}Exp(R), t\text{-}Group(i)]X \rightarrow Y$ by checking the emptiness of the result of the following query:

$$\sigma_{Cnd}(E\text{-}Exp(R) \bowtie_{X=\widehat{X}} \rho_{W \rightarrow \widehat{W}} E\text{-}Exp(R))$$

where W is the set of attributes of $E\text{-}Exp(R)$ and Cnd stands for $\bigvee_{A \in Y}(A \neq \widehat{A}) \wedge SameTGroup$.

The predicate $SameTGroup$ verifies that all the given valid times belong to the same group, according to the expression $t\text{-}Group$. Thus, if $E\text{-}Exp(R)$ is defined over a schema $\subseteq U\overline{U} \cup \{VT, \overline{VT}\}$, then $SameTGroup \equiv VT \in t\text{-}Group(i) \wedge \overline{VT} \in t\text{-}Group(i) \wedge \widehat{VT} \in t\text{-}Group(i) \wedge \widehat{\overline{VT}} \in t\text{-}Group(i)$; if $E\text{-}Exp(R)$ is defined over the schema $U \cup \{VT\}$, then $SameTGroup \equiv VT \in t\text{-}Group(i) \wedge \widehat{VT} \in t\text{-}Group(i)$.

The predicate $SameTGroup$ can take two different forms depending on the way in which temporal grouping is defined: a granularity or a set of intersecting finite sets. For the sake of simplicity, we restrict our attention to the case of an evolution relation over the schema $U \cup \{VT\}$. The proposed solution can be easily adapted to the other cases.

As a preliminary remark, we observe that we are interested in checking TFDs on finite database instances, and thus temporal grouping returns a finite set of granules of a given granularity, each one consisting of a finite set of time points (resp., a finite set of intersecting finite sets of time points).

Let us consider first the case of granularities. Granularities are modeled by means of a relation $Gran$ over the attributes (G_Id, I, G_s, G_e), where G_Id is the granularity identifier, e.g., Month, I is the granule identifier (granule index), G_s and G_e are respectively the starting point and the ending point of the granule (in fact, the attribute I can be omitted). For the sake of simplicity, we restrict ourselves to granularities with no internal gaps. Let $t\text{-}Group(i)$ denote the i-th granule of granularity G ($G(i)$ for short). It can be easily shown that the problem of checking whether or not a temporal relation $r \in R$ satisfies a TFD $[E\text{-}Exp(R), t\text{-}Group(i)]X \to Y$ is equivalent to the problem of checking whether or not the following relational algebra query returns the empty set:

$$\sigma_{Cnd}(E\text{-}Exp(R) \bowtie_{X=\widehat{X}} \rho_{W \to \widehat{W}} E\text{-}Exp(R) \bowtie \sigma_{G\_Id=``G"} Gran),$$

where Cnd stands for $\vee_{A \in Y}(A \neq \widehat{A}) \wedge G_s \leq VT \wedge VT \leq G_e \wedge G_s \leq \widehat{VT} \wedge \widehat{VT} \leq G_e$.

Let us consider now the case of intersecting finite sets. Without loss of generality, we may assume (the finite set of) intersecting finite sets of time points to be represented by a relation $tGroups$ over the attributes (I, T_1, T_2), where I is the group index and (T_1, T_2) is an ordered pair of time points belonging to the same temporal group (identified by the value of the group index). It can be easily shown that the problem of checking whether or not a temporal relation $r \in R$ satisfies a TFD $[E\text{-}Exp(R), t\text{-}Group(i)]X \to Y$ is equivalent to the problem of checking whether or not the following query returns the empty set:

$$\sigma_{Cnd}(E\text{-}Exp(R) \bowtie_{X=\widehat{X}} \rho_{W \to \widehat{W}} E\text{-}Exp(R) \bowtie tGroups)$$

where Cnd stands for $\vee_{A \in Y}(A \neq \widehat{A}) \wedge VT = T_1 \wedge \widehat{VT} = T_2$.

As a matter of fact, in various practical cases the join with the relation $tGroups$ can be avoided by including temporal grouping into Cnd. Concrete examples are given below.

We now exemplify the checking-for-emptiness approach to TFD verification by providing a query for each class of TFDs (we refer to the TFDs encoding the requirements of chemotherapy recommendations given in the previous section).

Pure temporally grouping TFDs. The TFD $[Patient, \{i, \ldots, i+13\}] Chemo,$ $PatId, Drug \to Qty$ encodes requirement (2). We can establish whether or not $Patient$ satisfies it by checking for emptiness the following query:

$$\sigma_{Qty \neq \widehat{Qty} \wedge (\widehat{VT} - VT) \leq 13}(Patient$$
$$\bowtie_{Chemo = \widehat{Chemo} \wedge PatId = \widehat{PatId} \wedge Drug = \widehat{Drug}} \rho_{W \to \widehat{W}} Patient)$$

Pure temporally evolving TFDs. The TFD $[\chi_{PatId, Chemo, Drug}(Patient),$ $\text{Top}(i)] Drug, Qty \to \overline{Qty}$, encoding requirement (4), can be verified by checking for emptiness the following query:

$$\sigma_{\overline{Qty}\neq\widehat{\overline{Qty}}}(\chi_{PatId,Chemo}(Patient)$$
$$\bowtie_{Drug=\overline{Drug}\wedge Qty=\widehat{Qty}}\rho_{W\to\widehat{W}}\chi_{PatId,Chemo,Drug}(Patient))$$

Temporally mixed TFDs. To verify whether or not *Patient* satisfies the TFD $[\tau_{PatId,Drug}(Patient), \mathsf{Month}(i)]Drug, Qty \to \overline{Qty}$, encoding requirement (7), we can check for emptiness the following query:

$$\sigma_{\overline{Qty}\neq\widehat{\overline{Qty}}\wedge G_s\leq VT\wedge VT\leq G_e\wedge G_s\leq\widehat{VT}\wedge\widehat{VT}\leq G_e}(\tau_{PatId,Drug}(Patient)$$
$$\bowtie_{Drug=\overline{Drug}\wedge Qty=\widehat{Qty}}\rho_{W\to\widehat{W}}\tau_{PatId,Drug}(Patient)\bowtie\sigma_{G\_Id=\text{``}Month\text{''}}Gran)$$

Temporally hybrid TFDs. Finally, the TFD $[He^{Patient}, \mathsf{Top}(i)]Chemo, Drug \to Qty$, encoding requirement (8), can be verified by checking for emptiness the following query:

$$\sigma_{Qty\neq\widehat{Qty}}(He^{Patient}\bowtie_{Drug=\overline{Drug}\wedge Chemo=\widehat{Chemo}}\rho_{W\to\widehat{W}}He^{Patient})),$$

which can be expanded as follows:

$$\sigma_{Qty\neq\widehat{Qty}}((\pi_{U\cup\{VT\}}(\sigma_{\overline{VT}=VT+7}(\tau_{PatId,Chemo,Drug}(Patient)))\bigcup$$
$$\rho_{\overline{U},\overline{VT}\to U,VT}\pi_{\overline{U}\cup\{\overline{VT}\}}(\sigma_{\overline{VT}=VT+7}(\tau_{PatId,Chemo,Drug}(Patient))))$$
$$\bowtie_{Drug=\overline{Drug}\wedge Chemo=\widehat{Chemo}}$$
$$\rho_{W\to\widehat{W}}(\pi_{U\cup\{VT\}}(\sigma_{\overline{VT}=VT+7}(\tau_{PatId,Chemo,Drug}(Patient)))\bigcup$$
$$\rho_{\overline{U},\overline{VT}\to U,VT}\pi_{\overline{U}\cup\{\overline{VT}\}}(\sigma_{\overline{VT}=VT+7}(\tau_{PatId,Chemo,Drug}(Patient))))))$$

Let us briefly analyze the computational complexity of TFD verification based on the proposed checking-for-emptiness method. As TFDs are represented by expressions of the form $[E\text{-}Exp(R), t\text{-}Group(i)]X \to Y$, the cost of checking the functional dependency $X \to Y$ depends on the structure of the evolution expression $E\text{-}Exp(R)$ and on the nature of the temporal grouping condition $t\text{-}Group(i)$.

Let n, n_E, and n_G be the cardinalities of the temporal relation $r \in R$, of the corresponding instance of the evolution relation, and of the relation representing the temporal grouping, respectively. The emptiness check consists of the execution of the join:

$$E\text{-}Exp(R) \bowtie_{X=\hat{X}} \rho_{W\to\widehat{W}}E\text{-}Exp(R),$$

followed by the execution of the join of the resulting relation with the relation representing the temporal grouping, and the execution of an operation of selection on the resulting relation to verify the condition on the consequent(s) of the functional dependency.

The cost of computing the instance of the evolution relation is in $O(n^3)$ (the most complex case being that of the temporal view *nexttuple*). An upper bound to the cost of executing the two joins is given by the cost of executing two cartesian products, which is in $O(n_E^2 \cdot n_G)$ (we assume that no special indexing data structures are available). The cost of the subsequent selection operation is in $O(n_E^2 \cdot n_G)$ as well (it must be executed as many times as the tuples of the resulting relation are). Hence, the overall cost of the emptiness check is in

$O(n^3 + n_E^2 \cdot n_G)$. As n_E ranges from n (for pure temporally grouping TFDs) to $O(n^2)$ (for pure temporally evolving, mixed, and hybrid TFDs), the complexity of the emptiness check ranges from $O(n^3 + n^2 \cdot n_G)$, for pure temporally grouping TFDs, to $O(n^4 \cdot n_G)$, for pure temporally evolving, mixed, and hybrid TFDs.

Similarly to the atemporal case [5], if it is known that relation r is a legal state and t is a tuple to be inserted, then checking whether or not $r \cup \{t\}$ satisfies a TFD can be performed in time $O(n^3 + n_E \cdot n_G)$ (in $O(n_E \cdot n_G)$ if we assume the current value of the evolution expression to be cached).

6   Related Work

Various representation formalisms for TFDs have been developed in the literature [2,6,11,12,13], which differ a lot in their structure as well as in the underlying data model. All of them basically propose alternative extensions to the relational model, often introducing non-relational features (this is the case with Wijsen's objects [13] and Vianu's update mappings [11]), making it difficult to identify their distinctive features and to systematically compare them in order to precisely evaluate their relative strength and their limitations. In the following, we take the set of requirements given in Section 2 as a sort of benchmark for their evaluation. A systematic analysis is provided in [4], where we first describe the most significant TFD formalisms proposed in the literature, following as much as possible the original formulation given by the authors, and, then, we formally prove that our proposal actually subsumes all of them. In the following, we provide a short account of such an analysis. In particular, we show that existing formalisms significantly differ in the requirements they are able to express and that there exist meaningful requirements they are not able to cope with. As an example, existing TFD systems are not able to express Requirements (5-7), which (from the point of view of the conditions they impose) look like minor variations of Requirement (4).

Let us assume to have a representation of the patient database example in (the data model underlying) all the TFD systems we are going to analyze. We first show how to represent Requirements (1-4). As a matter of fact, not all these requirements can be encoded in all TFD systems.

In [6], Jensen et al. propose a bitemporal data model that allows one to associate both valid and transaction times with data. Jensen et al.'s TFDs make it possible to express conditions that must be satisfied at any (valid) time point taken in isolation. Requirement (1), which prevents any patient from having two or more administrations of the same drug during the same day, can be modeled by Jensen et al. 's TFDs as follows:

$PatId, Drug \rightarrow^T Chemo, BG, Phys, Qty$

A general formalism for TFDs on complex (temporal) objects has been proposed by Wijsen in [13]. It is based on a data model that extends the relational model with the notion of object identity, which is preserved through updates, and with the ability of dealing with complex objects, that is, objects that may have other objects as components. Wijsen's TFDs have the form $c : X \rightarrow_\alpha Y$. Their meaning can be intuitively explained as follows. Let t_1 and t_2 be two objects of class c valid at time points i and j, respectively, where (i, j) belongs to the time relation α. If t_1 and t_2 agree on X, then they must agree on Y as well.

For any patient, Requirement (2) forces drug quantities of chemotherapies to remain unchanged whenever administrations take place within 14 days. Such a requirement constrains the duration of the time span between two administrations and it can be modeled using Wijsen's TFDs as follows:

$Patient : PatId, Chemo, Drug \rightarrow_{14Days} Qty,$

where *14Days* is a time relation grouping 14 consecutive days.

Bettini, Jajodia, and Wang's notion of TFD takes advantage of time granularity [2]. Their TFDs allow one to specify conditions on tuples associated with granules of a given granularity and grouped according to a coarser granularity. It is not difficult to show that Wijsen's TFDs actually subsume Bettini et al.'s TFDs. More precisely, Bettini et al.'s TFDs are exactly all and only Wijsen's TFDs on chronologies (the class TFD-C in Wijsen's terminology).

Requirement (3) essentially states that, regardless of the patient, for any chemotherapy and any month, the administered quantity of a drug is always the same. This requirement can be expressed in Bettini et al.'s formalism by the following TFD on the temporal module schema (*Patient*, Day):

$Chemo, Drug \rightarrow_{\mathsf{Month}} Qty,$

where Month is the granularity grouping days of the same month.

Its representation in Wijsen's TFD formalism is as follows:

$Patient : Chemo, Drug \rightarrow_{Month} Qty,$

where *Month* is a time relation grouping days of the same month.

It is not difficult to show that pure temporally grouping TFDs are very close to TFDs proposed by Jensen et al., Wijsen, and Bettini et al., as witnessed by the above three temporal functional dependencies.

In [11], Vianu proposes a simple extension to the relational model in order to describe the evolution of a database over time. He defines a *database sequence* as a sequence of consecutive instances of the database, plus *"update mappings"* from one instance (the "old" instance) to the next one (the "new" instance). Constraints on the evolution of attribute values of tuples (objects) over time are expressed by means of dynamic functional dependencies (DFDs), that make it possible to define dependencies between old and new values of attributes on updates. For example, Requirement (4) constrains the administrations of a drug on two consecutive days by imposing different patients not to take different quantities of the drug on the next day, if they take the same quantity of it on the current day. Assuming that the update mapping represents the evolution of a tuple for a given patient and a given drug and chemotherapy, Requirement (4) can be expressed by the following DFD, where for each attribute A, $\overset{\vee}{A}$ represents its old value and $\hat{A}$ its new value:

$\overset{\vee}{Drug}, \overset{\vee}{Qty} \rightarrow \overset{\wedge}{Qty}$

We conclude the section by an intuitive account of the fact that the last four requirements cannot be dealt with by existing TFD systems. This is true, in particular, for Requirements (5-7) that present some similarities with Requirement (4), which can be easily encoded using Vianu's DFDs.

Consider, for instance, Requirement (5). Such a requirement is based on an update mapping which differs from the one of Requirement (4) in an essential way. Such a mapping must indeed associate any tuple involving a patient, a drug, a chemotherapy, and a physician, valid at a given day, with a tuple involving the same patient, drug, chemotherapy, and physician, valid at the next day (if any). Unfortunately, Vianu's DFDs cannot help, as they are based on a fixed update mapping (that does not allow one to constrain the physician to be the same). Moreover, Vianu's update mappings are partial one-to-one mappings, and thus they cannot be exploited to deal with the case where a tuple, valid at a given day, must be associated with more than one tuple, valid at the next day. Requirement (6) constrains successive administrations of a drug possibly involving different delays: Vianu's update mappings cannot cope with "asynchronous" updates of different tuples. Requirement (7) cannot be fulfilled by existing TFDs as well. It indeed requires a sort of combination of Vianu's DFDs and tuple temporal grouping supported by Wijsen's TFDs (and by Bettini et al.'s TFDs).

Requirement (8) deserves a deeper analysis. Basically, it imposes that, for any chemotherapy, (only) the quantities of drugs taken by patients every 7 days cannot change. On the one hand, this constraint cannot be expressed by Vianu's DFDs, as they do not allow one to formulate a condition of the form: "something cannot change in the evolution of the database". On the other hand, the other TFD systems have no the capability of "mapping tuples' evolution" (the evolution of a database can only be modeled through the union of consecutive states). As an example, Wijsen's TFD $Patient : Chemo, Drug \rightarrow_{7Days} Qty$, where the time relation $7Days$ groups days which are exactly 7 days far from each other, does not capture the intended meaning. Consider, for instance, the tuples belonging to the relation depicted in Table 1. The relation violates such a TFD as the tuple for patient 3 at time 1 violates it with respect to drug Cyc and chemotherapy CEF, with respect to time points 1 and 8. On the contrary, according to its intended meaning, requirement (8) is actually fulfilled by the relation depicted in Table 1, as the drugs that are taken by patients every 7 days (Flu by patient p1, and Flu and Epi by patient p2) do not change their quantities from one administration to the successive one (7 days later).

7   Conclusions

In this paper, we focused our attention on the specification and checking of temporal functional dependencies (TFDs), possibly involving multiple time granularities. To overcome the limitations of existing TFDs, we have proposed a new *general notion* of TFD, that subsumes all of them and allows one to cope with temporal requirements they cannot deal with. The simplest TFDs are directly brought back to atemporal FDs; to manage the most complex ones some additional machinery is needed. As for the problem of checking whether a temporal relation satisfies a given set of TFDs, we have shown how to uniformly reduce it to the problem of checking for emptiness a suitable relational algebra expression over the considered temporal relation.

References

1. Assikis, V., Buzdar, A., Yang, Y., et al.: A phase iii trial of sequential adjuvant chemotherapy for operable breast carcinoma: final analysis with 10-year follow-up. Cancer 97, 2716–2723 (2003)
2. Bettini, C., Jajodia, S., Wang, X.: Time granularities in Databases, Data Mining, and Temporal Reasoning. Springer, Heidelberg (2000)
3. Bresolin, D., Montanari, A., Puppis, G.: A theory of ultimately periodic languages and automata with an application to time granularity. Acta Informatica 46(5), 331–360 (2009)
4. Combi, C., Montanari, A., Sala, P.: A uniform framework for temporal functional dependencies with multiple granularities. Technical Report RR 81/2011, Department of Computer Science, University of Verona, Verona, Italy (2011)
5. Hegner, S.J.: The relative complexity of updates for a class of database views. In: Seipel, D., Torres, J.M.T. (eds.) FoIKS 2004. LNCS, vol. 2942, pp. 155–175. Springer, Heidelberg (2004)
6. Jensen, C., Snodgrass, R., Soo, M.: Extending existing dependency theory to temporal databases. IEEE Transactions on Knowledge and Data Engineering 8(4), 563–581 (1996)
7. Kanellakis, P.C.: Elements of relational database theory. In: van Leeuwen, J. (ed.) Handbook of Theoretical Computer Science, Volume B: Formal Models and Semantics (B), pp. 1073–1156. Elsevier and MIT Press (1990)
8. Levine, M., Sawka, C., Bowman, D.: Clinical practice guidelines for the care and treatment of breast cancer: 8. Adjuvant systemic therapy for women with node-positive breast cancer (2001 update). Canadian Medical Association Journal, 164 (2001)
9. Ning, P., Jajodia, S., Wang, X.S.: An algebraic representation of calendars. Annals of Mathematics and Artificial Intelligence 36, 5–38 (2002)
10. Ullman, J.D.: Principles of Database and Knowledge-Base Systems. I. Computer Science Press, Rockville (1988)
11. Vianu, V.: Dynamic functional dependency and database aging. Journal of the ACM 34(1), 28–59 (1987)
12. Wijsen, J.: Design of temporal relational databases based on dynamic and temporal functional dependencies. In: Clifford, J., Tuzhilin, A. (eds.) International Workshop on Temporal Databases. Recent Advances in Temporal Databases, pp. 61–76. Springer, Heidelberg (1995)
13. Wijsen, J.: Temporal FDs on complex objects. ACM Transactions on Database Systems 24(1), 127–176 (1999)
14. Wijsen, J.: Temporal dependencies. In: Liu, L., Özsu, M.T. (eds.) Encyclopedia of Database Systems, pp. 2960–2966. Springer US, Heidelberg (2009)

Quality of Similarity Rankings in Time Series

Thomas Bernecker[1], Michael E. Houle[2], Hans-Peter Kriegel[1], Peer Kröger[1],
Matthias Renz[1], Erich Schubert[1], and Arthur Zimek[1]

[1] Ludwig-Maximilians-Universität München
Oettingenstr. 67, 80538 München, Germany
{bernecker,kriegel,kroegerp,renz,schube,zimek}@dbs.ifi.lmu.de
http://www.dbs.ifi.lmu.de
[2] National Institute of Informatics
2-1-2 Hitotsubashi, Chiyoda-ku, Tokyo 101-8430, Japan
meh@nii.ac.jp
http://www.nii.ac.jp/en/

Abstract. Time series data objects can be interpreted as high-dimensional vectors, which allows the application of many traditional distance measures as well as more specialized measures. However, many distance functions are known to suffer from poor contrast in high-dimensional settings, putting their usefulness as similarity measures into question. On the other hand, shared-nearest-neighbor distances based on the ranking of data objects induced by some primary distance measure have been known to lead to improved performance in high-dimensional settings. In this paper, we study the performance of shared-neighbor similarity measures in the context of similarity search for time series data objects. Our findings are that the use of shared-neighbor similarity measures generally results in more stable performances than that of their associated primary distance measures.

1 Introduction

One of the most fundamental operations in data mining applications is that of similarity search. The retrieval of similar objects during a given data mining task may be facilitated using a 'k-nearest-neighbor' (k-NN) search with an appropriate distance or similarity measure. For data representable as real-valued feature vectors, many similarity measures are in common usage, such as the cosine distance measure and L_p norms — which include the Euclidean distance ($p = 2$) and Manhattan distance ($p = 1$).

The effectiveness of similarity measures in data mining applications depends on their ability to discriminate among the various groupings of the data that arise from different generation mechanisms or statistical processes. These groupings or subsets, although usually unknown in practical settings, may be regarded as 'classes' in the context of classification, as 'clusters' in the context of clustering, as 'usual' vs. 'conspicuous' data objects in outlier detection, or simply as 'similar' vs. 'dissimilar' or 'relevant' vs. 'irrelevant' objects in the context of similarity search and information retrieval applications. Generally speaking, k-NN queries

D. Pfoser et al. (Eds.): SSTD 2011, LNCS 6849, pp. 422–440, 2011.

should return for a query object generated by a given mechanism other objects generated by the same mechanism.

For data mining applications involving time-series data, the presence of such spatio-temporal phenomena generally prevents the use of simple distance measures (such as cosine similarity and L_p norms) for meaningful similarity measurement. To cope with spatio-temporal data, several specialized distance measures have been developed that attempt to determine the best matching of events along the time axis. One of the most prominent of these is the *Dynamic Time Warping (DTW)* distance [1], used extensively in speech recognition. *DTW* supports asynchronous matching — matches with shifts along the time dimension — by extending each sequence with repeated elements, and applying L_p norms to the extended time series. The advantages of *DTW* are invariance to (local) phase-delays, the acceleration or deceleration of signals along the time dimension, and the ability to support matches between series of differing lengths. Like the L_p norms, *DTW* requires a complete matching of both time series, in that each value from one time series must be matched with at least one value from of the other time series. For this reason, *DTW* is sensitive to noise and outliers.

In general, distance measures that are robust to extremely noisy data typically violate the triangle inequality [2], and thus are inapplicable for most indexing methods. Well-known distance measures for sequence data that fall into this category are the *Longest Common Subsequence (LCSS)* distance [2], the *Edit Distance on Real sequence (EDR)* [3] and the *Edit distance with Real Penalty (ERP)* [4]. In contrast with *DTW*, *LCSS* and *EDR*, the measure *ERP* has the advantage of satisfying the triangle inequality and is therefore a distance metric. The aforementioned measures are adaptations of the edit distance, a commonly-used distance measure for matching strings that can accommodate gaps in the matching. Unlike the L_p norms and *DTW*, these measures are able to ignore noise and outliers. As such, edit distance variants are better at coping with different sampling rates, different time rates, and different series lengths; they can also be computed more efficiently.

The high computational cost associated with distance measures for time-series data has led to the development of many methods for dimensionality reduction, in which distance measures are applied to subsets of features extracted from objects in the series [5,6,7,8,9]. Distance measures based on dimensionality reduction can be regarded as specialized similarity measures, in that they process the full set of spatio-temporal features to ultimately produce similarity values for the original objects.

A rather different approach to the design of similarity measurement is that of 'shared nearest neighbor' (*SNN*) dissimilarity, in which 'secondary' similarity measures are derived from any 'primary' similarity measure supplied for the objects. Given two objects for which the similarity is to be computed, secondary similarity measures consider the extent to which the neighbor sets of these objects resemble each other, where the neighbors are determined according to the primary similarity measure. In principle, any primary similarity measure can be used — including L_p norms and any specialized time series distance measures —

although the effectiveness of the secondary measure does depend on that of the primary measure. *SNN* measures have found many applications for computing *k*-NN sets in high-dimensional data, and have been reported to be less prone to the 'curse of dimensionality' than conventional distance measures. The 'curse of dimensionality' is a general phenomenon that drastically limits the performance of search, clustering, classification, and other fundamental operations in applications of data mining for high-dimensional data (see for example [10]).

The earliest known use of *SNN* dissimilarity was as the merge criteria for agglomerative clustering algorithms [11]. *SNN* was also used for clustering high-dimensional data sets [12,13], and for finding outliers in subspaces of high-dimensional data [14]. In these applications, the preference for *SNN* dissimilarity was due to the perception of it being more stable and robust than conventional distance measures for high-dimensional data; however, in all of these early studies, any such claims were intuitive and unsubstantiated (when articulated at all). The first detailed study of the merits of *SNN* dissimilarity for high-dimensional real-valued feature vector data appeared in [15], which assessed the effects of the curse of dimensionality on L_p distances, the cosine distance, and *SNN* measures based upon these primary distance measures. For this context, the study came to the following general conclusions:

1. The quality of a ranking, and thus the separability of different groupings (such as classes or clusters), depends less on the data dimensionality, and more on the proportion of data elements that are relevant to the grouping.
2. The use of *SNN* dissimilarity with L_p distances or the cosine distance significantly boosts the quality of neighbor ranking, compared to the use of the corresponding primary distances.
3. The performance of similarity search and related problems in data mining becomes more accurate and more reliable (less variable) when using *SNN* distance measures in place of their associated primary distance measures.

As time-series data are representable as high-dimensional feature vectors, they too are susceptible to the curse of dimensionality. Although the study did not directly consider the effect of *SNN* dissimilarity for time-series data, the improvements observed for high-dimensional vector data for L_p norms and the cosine distance do suggest the possibility of improvements for time series. On the other hand, specialized time series distance measures perform rather differently from L_p norms or the cosine distance: assumptions that hold for Euclidean vector spaces do not necessarily apply to the spaces within which time-series data reside. *SNN* dissimilarity measures work best when the groupings (classes or clusters) form compact, 'spherical' structures. Classes and clusters of time-series data objects are generally not compact in terms of Euclidean distance. It is possible, however, that these groupings may be made more compact by means of a transformation to some suitable space or geometry.

The potential impact of transformation on the performance of search serves as an interesting motivation for the investigation of *SNN* measures for time-series data. In this paper, we extend the study of [15] to time series of varying lengths (dimensionalities), by comparing the accuracy of common primary distance measures with that of their associated secondary *SNN* dissimilarity measures. Previously, no comprehensive study has been made of the effects of shared-neighbor distance measures on the alleviation of the curse of dimensionality for the specialized distance measures used for typical time-series data sets.

Distance measures based on dimensional reduction techniques will not be considered in our study. The dimensional reduction effectively transforms the original time-series data domain into a second time-series data domain, one whose nature is harder to characterize than the original set. It is important that the nature of the data sets for the experimental study be transparently understood, so that the effects of the secondary similarity methods can be properly assessed. For this reason, we consider the effects of *SNN* dissimilarity only on simple benchmark data sets, and higher-dimensional extensions of these data sets, where the characteristics of the data can be clearly understood. The interaction of *SNN* dissimilarity and specific dimensional-reduction techniques is beyond the scope of this paper.

The structure of the remainder of this paper is as follows. We will first introduce the concept of *SNN* in more detail in Section 2. Section 3 presents the experimental setup for this study. We compare the performance of 'secondary' *SNN* measures with their respective 'primary' measures in Section 4. Finally, we summarize and discuss our findings in Section 5.

2   Shared Nearest Neighbor Similarity

The most basic form of the shared-nearest-neighbor (*SNN*) similarity measure is that of the 'overlap'. Given a data set S consisting of $n = |S|$ objects and $s \in \mathbb{N}^+$, let $NN_s(x) \subseteq S$ be the set of s nearest neighbors of $x \in S$ as determined using some specified primary similarity measure. The overlap between objects x and y is then defined to be the intersection size

$$SNN_s(x, y) = |NN_s(x) \cap NN_s(y)|. \tag{1}$$

In addition to the overlap, a number of similarity measures have been proposed in the research literature, including:

– The 'cosine measure', given as:

$$simcos_s(x, y) = \frac{SNN_s(x, y)}{s}, \tag{2}$$

so called as it is the cosine of the angle between the characteristic vectors for $NN_s(x)$ and $NN_s(y)$. This normalization of the overlap was used in [12,16] as a local density measure for clustering.

– The 'set correlation', given as:

$$simcorr_s(x, y) = \frac{n}{n - s} \left(\frac{SNN_s(x, y)}{s} - \frac{s}{n} \right), \tag{3}$$

which results when the standard Pearson correlation formula

$$r = \frac{\sum_{i=1}^{n} x_i y_i - n\bar{x}\bar{y}}{\sqrt{(\sum_{i=1}^{n} x_i^2 - n\bar{x}^2)(\sum_{i=1}^{n} y_i^2 - n\bar{y}^2)}}$$

is applied using the coordinates of the characteristic vectors of $NN_s(x)$ and $NN_s(y)$ as variable pairs. Objects of S that appear in both $NN_s(x)$ and $NN_s(y)$, or neither of $NN_s(x)$ and $NN_s(y)$, support the correlation of the two neighborhoods (and by extension the similarity of x and y); those objects that appear in one neighborhood but not the other detract from the correlation. Note that the set correlation value tends to the cosine measure as $\frac{s}{n}$ tends to zero. Set correlation was introduced in [13] for the purpose of assessing the quality of cluster candidates, as well as ranking the cluster objects according to their relevance (or centrality) to the cluster.

It should be noted that when s and n are fixed (as they typically are in practice), the rankings determined by each of these similarity measures are identical. The functions only differ in the contrast of distance values achieved over different neighborhood subranges.

Dissimilarity measures can generally be derived from similarity measures in straightforward fashion. For the SNN similarity $simcos$ (Equation 2) with a given choice of s, the following distance measures have been proposed [15]:

$$dinv_s(x,y) = 1 - simcos_s(x,y) \tag{4}$$

$$dacos_s(x,y) = \arccos(simcos_s(x,y)) \tag{5}$$

$$dln_s(x,y) = -\ln simcos_s(x,y) \tag{6}$$

3   Experimental Setup

3.1   Distance Measures

We study the behavior of several representative distance measures for time series, including L_1 (Manhattan distance) and L_2 (Euclidean distance) as baseline measures, and DTW, LCSS, EDR, and ERP as distance measures specialized for time series data (see Section 1). Note that we will not compare and discuss the accuracies of different primary distance functions on time-series data, since the 'best' primary distance function in any case depends heavily on the characteristics of the data set. Rather, we will limit our investigations to the effects of the use of the secondary (SNN) distance measure $dinv_s$ (as defined in Equation 4) relative to their corresponding primary distance measures.

All primary and secondary distance measures were implemented in the ELKI framework [17].

3.2   Evaluation Criteria

The purpose of a distance function or similarity measure is to discriminate between relevant and irrelevant data. Relevant objects should be closer or more similar to the reference object than irrelevant objects. In time-series data sets whose objects are of multiple classes, 'relevant' objects are those objects belonging to the same class as the query object, whereas 'irrelevant' objects belong to a different class. The ability of a distance function to separate out relevant from irrelevant data can be best evaluated by computing a nearest neighbor ranking of all data objects with respect to a given query object. Ideally, we would find all objects of the same class as the query object at the highest ranks, followed by the objects belonging to the remaining classes.

As a visual presentation of the effectiveness of dissimilarity measures in discriminating between different classes within a data set, we make use of grey-scale similarity matrices, where black values indicate high similarity and white values indicate low similarity. Along each dimension, the items are sorted according to their class labels. Since all items are self-similar, the diagonal entries are always black.

The quality of a particular distance function is also presented numerically in the form of a pair of histogram plots, where one plot displays distances between members sharing the same class (intra-class distances), and the other displays distances between members of differing classes (inter-class distances). If the two histograms do not overlap, a single distance threshold would be sufficient to discriminate between the cluster and all others. However, in most real applications these histograms would interpenetrate substantially. The visualization gives a good impression of the over-all separability of classes over different ranges of distance values.

The discriminative ability of a dissimilarity measure can be rated over the full range of distance values by means of 'received operator characteristic' (ROC) curves, which plot the ratio between the true positive rate and the false positive rate against neighborhood size s. For each query, the objects are ranked according to their similarity to the query object, and objects are said to be true positives when they belong to the same class as the query object. For each ranked list of results, we compute an ROC curve and the corresponding area under the curve (AUC). We shall denote the AUC values by $AUC(o, D, d)$, where o is the query object, D the database and d the distance function used. An AUC of 1.0 indicates a perfect separation using the distance function, whereby all relevant objects are ranked ahead of all irrelevant objects. The expected AUC value for a random ordering of objects is 0.5.

The ROC curve and its AUC value summarize the quality of a query result ranking for a single reference object. By performing a query based at each object of the database, we can generate an ROC curve and an AUC value for each object in the data set and aggregate these either using histograms, or more compactly by computing the mean AUC value and variance. The mean AUC value can be used to rate the quality of a particular distance measure for the entire data set:

$$\text{MAUC}(D, d) := \frac{1}{|D|} \sum_{o \in D} \text{AUC}(o, D, d))) \tag{7}$$

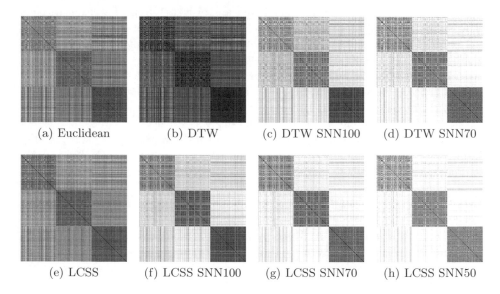

(a) Euclidean   (b) DTW   (c) DTW SNN100   (d) DTW SNN70

(e) LCSS   (f) LCSS SNN100   (g) LCSS SNN70   (h) LCSS SNN50

Fig. 1. Similarity matrices for *CBF*: All items are sorted according to their class label and pairwise distances are plotted. Similarity values are shown in grey scale, where black indicates high similarity and white indicates low similarity.

Note again that primary distance measures will be evaluated only against their corresponding secondary distance measures.

The difference $1 - \mathrm{MAUC}$ can be regarded as the 'mean ranking error'. By comparing the mean error of a primary distance (the reference distance, referred to here as d_{ref}) and a secondary distance (the distance to be evaluated, referred to here as d_{eval}) we can evaluate the gain in ranking quality achieved using the *SNN* distance:

$$\mathrm{Gain}(D, d_{ref}, d_{eval}) := 1 - \frac{1 - \mathrm{MAUC}(D, d_{eval})}{1 - \mathrm{MAUC}(D, d_{ref})} \qquad (8)$$

3.3 Data Sets

For our experiments we used two synthetic data sets derived from the data used in [18], as well as two real-world data sets. The first and most basic synthetic data set, the *Cylinder-Bell-Funnel* (*CBF*) set, describes an artificial problem first proposed by [19] where the task is to classify an instance of a data stream into one of three pattern classes *Cylinder*, *Bell* or *Funnel*, where each instance is bracketed by a non-class-specific base signal. The original data set consists of 930 objects per class, with each object represented as a 128-dimensional vector. In our experiments, the evaluation was performed over the first 100 objects of each class.

The *Synthetic Control* data set from [20] features synthetically generated control charts. This data set consists of 600 time series of length 60. There are six classes with different shape characteristics: Normal, Cyclic, Increasing Trend, Decreasing Trend, Upward Shift and Downward Shift. Each class contains exactly 100 objects.

The *Leaf* data set from [21] is in fact not a time series, but rather a spatial series taken from the outlines of leaves from 6 different types of tree (such as maple). It has been shown that time warping distances work well for such data, in that they are capable of adapting to variations in natural leaf growth. This set has 442 instances with 150 dimensions.

The real-world *Lightning-7* data set is the largest (7-class) variant of the FORTE ('Fast On-orbit Rapid Recording of Transient Events') time series data used in [22]. The objects represent information on lightning strokes as captured by a low-orbit satellite. While this data set has only 143 objects, the series are rather long, with 3181 dimensions. An additional challenge to classification arises from one class containing 'Off-Record' objects that could potentially belong to the other classes.

3.4   Dimensional Scaling

Some of the experiments presented here are specifically designed to assess the impact of dimensionality on the quality of similarity rankings for time series data, using different primary distance measures and the corresponding secondary distance measures. The evaluation framework is similar to that of [15], here using sets that exhibit the typical characteristics of time series data, together with specialized distance measures for time series.

In order to assess the performance of the similarity measures over a wide range of dimensions, we extend the original time series data in several different ways. Ideally, any extension of time series should be done in a way that is semantically meaningful. It should also be noted that extension has the potential to improve or weaken the discriminative ability of similarity measures, depending on whether it reinforces the relevant data content, or acts as noise. To extend the dimensionality of the time series for our experimentation, we modify the data set in one of two ways (always keeping the original data as a subseries unless explicitly stated otherwise):

Relevant attributes. To extend a time series so as to increase the amount of information that relates to its class, we extend the series by appending other time series data objects randomly selected from the same class.

Irrelevant attributes. To extend a time series so as to decrease the amount of information that relates to its class, we append noise values to the time series. Although there are many conceivable ways of introducing noise into a time series, we adopted different methods of noise generation for each data set, each method specifically tailored to the data domain. For the data sets *CBF* and *Leaf*, we generate noise values as follows. Let the average value amplitude of the data set be denoted by *mean*, let *ext* be the maximum possible deviation from the mean, and let $0 < p \leq 1$ be a reduction factor. The noise points are uniformly distributed around *mean* in a reduced range based on p and *ext*. The creation of a noise point N can be formalized as:

$$N = mean \pm Uniform[-1, 1] \cdot p \cdot ext$$

In our experimentation, p is set to 0.1 so as to emulate the noise observed in the original data sets. Increasing this type of noise leads to distortion of the original class information, which in turn makes it more difficult to discriminate between classes. In the *CBF* data set, this type of noise resembles the 'base signal' — that is, the values that bracket the cylinder, bell and funnel time-series patterns associated with each data object. This is true since these bracketing values predominate over all time series of the data set, and thus contribute highly to the mean value at which the generation of noise is based. As a result, increasing the amount of noise reduces the proportion of information associated with the three class-distinctive pattern types. In the *Leaf* data set, the noise simply results in a much lower amplitude compared to the original data, since all amplitudes of the data points contribute to the mean value with a similar impact. The noise values for the *Synthetic Control* data set were generated in a similar manner. However, the center of the generation interval was set at a baseline value far lower than the data values of patterns from any of the six classes, so as to be fully distinct from any of these.

For the second real-world data set, *Lightning-7*, noise elements are introduced in an entirely different way, since the original data clearly follows a more complex distribution: all series begin with an initial peak, followed by an interval of relative low values ('silence'), and then by a region with a higher baseline containing spikes that coincide with the main part of the lightning stroke. Here, we chose to use sampling to extend the series: assuming that we will be appending a noise series of length n, the value of the i-th entry of the noise series is decided by selecting the i-th entry from a randomly-chosen time-series data object, where $i = 1 \ldots n$. As a consequence of appending noise series generated in this way, the distinctions among the classes are blurred, and discrimination between them becomes more difficult.

Dimensional Scaling. Dimensional scaling is effectuated by appending to the original time series one or more additional series of the same length (either class-relevant or noise, as described above). Each of these (the original and all appended series) can be regarded as a 'block' within the expanded time series. The total length of the expansion is then an integer multiple of the original time series length (the original dimensionality, denoted by d). For our experimentation, the first block contains the original data, and subsequent blocks consist of either relevant information or noise. The total time-series lengths considered are of the form $m = 2^n \times d$ with $n \in \{1, 2, 3, 4\}$. Since none of the data sets under consideration, and since none of our algorithms take advantage of *periodic* information in the time series, we opt for a straightforward design in which the first

k blocks of the series consists of class-relevant data, and the remaining $m - k$ blocks consist of noise. We vary the choice of parameter k so as to produce time series with various ratios between 'signal' (the class-relevant information at the head of the series) and 'noise'; these ratios include $1/2^l$ for $l = 0 \ldots 4$ as well as $3/4$.

4   Performance of Secondary Measures on Time Series

4.1   Discriminability of Classes

The similarity matrices plotted in Figure 1 give an indication of the compactness of the different classes in *CBF* for a given distance measure. The classes are compact for the measure if there are distinct dark squares along the diagonal with a side-length roughly corresponding to the number of class members.

Perhaps surprisingly, under DTW the classes in this data set can be seen to be less compact than with the Euclidean distance. Nevertheless, SNN based on DTW (first row), as well as based on LCSS (second row), considerably increases the compactness and separability of the different classes. The choice of parameter s — the number of neighbors considered for the computation of SNN — does have an impact on the degree of compactness and separability (as will be seen in the next subsection).

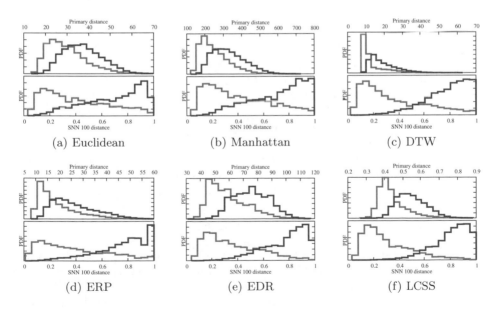

Fig. 2. Distance histograms for *CBF*: distribution of intra-class distances (red) vs. inter-class distances (blue). The y-axis has been normalized to facilitate the comparison. Shared-nearest-neighbor distances (lower plots) allow for much better discrimination of classes than their corresponding primary distances (upper plots).

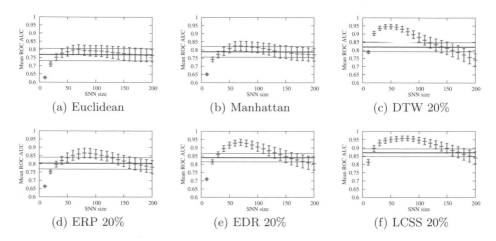

Fig. 3. Performance on the *CBF* data set for varying choices of the SNN neighborhood size s. Horizontal lines indicate the corresponding mean primary distance plus/minus one standard deviation.

The compactness of individual classes and separability of pairs of classes can be assessed by means of a visualization of the (normalized) distributions of intra-class distances and of inter-class distances. If both distributions overlap strongly, the discrimination power of the corresponding distance measure is generally poor. We see in Figure 2 the results on the *CBF* data set. The upper plots indicate the primary distances, and the lower plots indicate the derived SNN distances. Together, they show that the numerical contrast improves significantly when the SNN measures are adopted, as the pairs of distributions are much more widely separated in all of the lower plots. Consider the plots that would result after vertical scaling by the number of intra-class and inter-class distances, respectively. When the discrimination boundary is placed at the intersection of these scaled plots, the classification error would be proportional to the intersection of the areas under the scaled curves. The strong increase in the inter-class distance means for the SNN measures indicates that these areas would be substantially smaller than those arising from their corresponding primary measures.

4.2   Impact of the Choice of s on SNN Distances

In the case of traditional spatial vector data representations, it has been already observed that SNN performs well when s is chosen to be of the same approximate size as the natural cluster to which the query object belongs (or larger) [15]. For the data sets used in our experimentation, as with most realistic data sets, the classes are fragmented into several smaller natural clusters, which explains why performance is best when s is chosen to be slightly less than the expected class size. However, as a general rule we chose s to be equal to the class size in the experiments described below.

Table 1. Mean ROC AUC scores for unmodified data sets

CBF						leaf shapes					
Distance	Primary	SNN100	Gain	SNN70	Gain	Distance	Primary	SNN80	Gain	SNN60	Gain
Euclid.	0.769	0.792	9.9%	0.800	13.5%	Euclid.	0.550	0.556	1.3%	0.570	4.3%
Manh.	0.789	0.815	12.5%	0.825	17.3%	Manh.	0.578	0.594	3.7%	0.613	8.2%
DTW 20%	0.820	0.879	33.1%	0.930	61.1%	DTW 20%	0.713	0.777	22.3%	0.810	33.6%
ERP 20%	0.804	0.860	28.4%	0.861	28.7%	ERP 20%	0.751	0.823	28.7%	0.826	30.2%
EDR 20%	0.840	0.905	40.7%	0.932	57.8%	EDR 20%	0.716	0.761	15.7%	0.778	21.9%
LCSS 20%	0.871	0.947	59.3%	0.958	67.5%	LCSS 20%	0.766	0.823	24.4%	0.838	30.8%

synthetic control						FORTE lightning 7-class					
Distance	Primary	SNN100	Gain	SNN80	Gain	Distance	Primary	SNN30	Gain	SNN20	Gain
Euclid.	0.898	0.929	29.7%	0.933	34.2%	Euclid.	0.656	0.688	9.3%	0.684	8.1%
Manh.	0.902	0.932	30.2%	0.937	35.4%	Manh.	0.673	0.697	7.3%	0.692	5.7%
DTW 20%	0.961	0.972	27.1%	0.979	46.2%	DTW 10%	0.661	0.677	4.9%	0.665	1.1%
ERP 20%	0.934	0.941	10.8%	0.945	16.7%	ERP 10%	0.668	0.677	2.7%	0.670	0.7%
EDR 20%	0.930	0.967	53.1%	0.970	57.6%	EDR 10%	0.624	0.652	7.4%	0.641	4.5%
LCSS 20%	0.948	0.979	59.2%	0.980	62.4%	LCSS 10%	0.687	0.771	27.0%	0.775	28.2%

First, we evaluate the impact of the choice of s over broad range of values. Figure 3 shows the results obtained for the *CBF* data set when s is varied, with vertical error bars indicating the extent of one standard deviation from the mean. The plots indicate that SNN improves over the corresponding primary distance measure (the straight horizontal lines) over a fairly wide range of s. Only very small or very large values of s lead to poor performance. Similar results were obtained for the other data sets as well. For each data set, Table 1 displays the gain of SNN over the primary measure for various choices of s. From this table, it can also be observed that across the various choices of data sets and primary similarity measures, SNN measures can be expected to perform relatively well whenever the associated primary measures perform relatively well. Combinations where the primary distance performed relatively poorly (such as the L_p norms on the *Leaf* data set) see the smallest improvement when using SNN. For most of our experiments, the relative improvement of SNN over the primary measure (as expressed by the gain) is significant.

4.3   Results on Dimensionally Scaled Data Sets

In Figure 4, we show the normalized intra-class distances and inter-class distances (in the same format as in Figure 2) for the *CBF* data set, after 3-fold extension with noise blocks. Compared to the original data set, the overlap in the primary distances is larger, diminishing the separability of the classes. The SNN distances remain largely unaffected.

We next examine the effect of noise objects on performance. Figure 5 shows the results obtained when the data set is augmented with objects comprising a single noise block each. Although one would perhaps expect the performance of the Euclidean distance to drop rapidly as the data set size increases, it in fact remained relatively stable, with only a slight quality drop observed. Despite having performed well before the introduction of noise data, the distance functions LCSS and EDR showed the biggest sensitivity to the added noise, while the other distance measures appear to be largely unaffected. The SNN-based

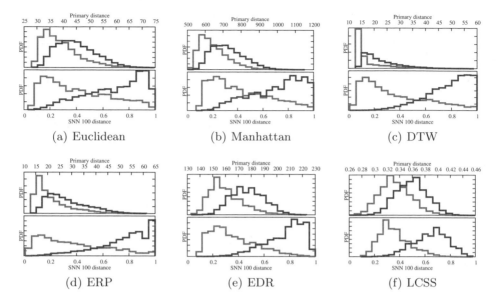

(a) Euclidean   (b) Manhattan   (c) DTW

(d) ERP   (e) EDR   (f) LCSS

Fig. 4. Distance histograms for *CBF* after extension with 3-fold noise: the relative numerical contrast for the primary distance functions decreases, but the classes remain well-differentiated under SNN. Compare with Figure 2.

distances remained stable, handling more than twice the amount of noise as their corresponding primary measures, before eventually deteriorating as well.

The reason why most of the distance functions considered here are largely unaffected by noise can be attributed to the characteristics of the added noise. When adding hundreds of independently and identically distributed attributes to a data set, their total effect accumulates to an almost constant value with a low variance. As long as this variance is lower than the variance within the relevant attributes, the added noise does not strongly affect the resulting ranking.

Note that the EDR distance on the *CBF* data set shows an anomaly, since the results improve even for the primary distance function after the addition of the first block of noise. This however is an artifact of the way this synthetic data set was created. Essentially, the noise appended has values similar to the baseline mean value of the data set, and as such the extended records can indeed be more similar to each other for those situations in which a series has only few bracketing values before or after the main pattern (cylinder, bell, or funnel). In realistic data sets, such effects are not likely to occur. They can be regarded as record-boundary effects wherein one series has only a limited range remaining for matching. Such effects have previously been noted in the research literature, for example in [2]. Since they affect the comparative performance of primary distance measures, they fall outside the scope of this study. With the synthetic control set, the effect is exactly the opposite: despite its low variance, the added noise is sufficiently different from the original data as to negatively affect the performance of EDR.

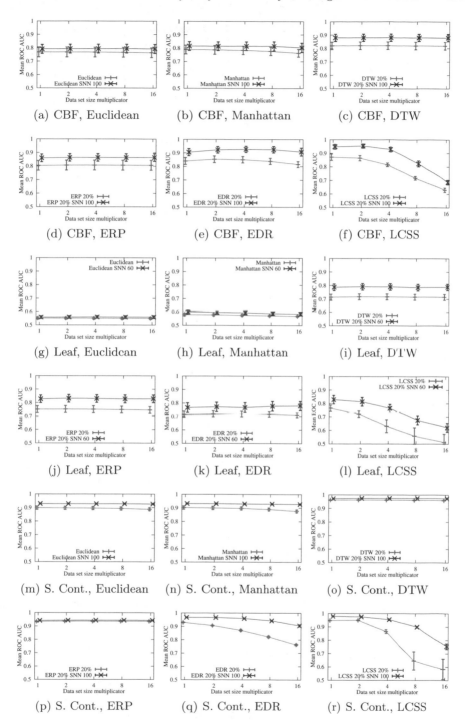

Fig. 5. Result quality after extension with noise blocks, shown with the x-axis in log scale

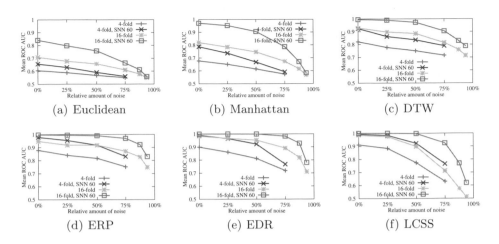

Fig. 6. Result quality in relation to noise ratio for the Leaf data set expanded 4-fold and 16-fold with noise blocks. As the number of relevant dimensions increases, a higher ratio of noise is tolerated. With a ratio of 1.0, the data would consist only of noise, and all ROC curves would be expected to have an AUC of 0.5.

For the next experiment, we fix the total number of dimensions and vary the proportion of noise added. Figure 6 shows the performance of distance measures

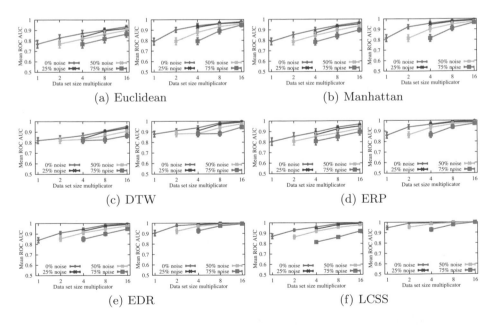

Fig. 7. Result quality with increasing dimensionality and fixed noise ratio for the *CBF* data set. The left-hand plots show the performances of the primary distance, the right-hand plots show the performances of the corresponding SNN $s = 100$ distances.

for the Leaf data set, when expanded 4-fold and 16-fold with class-relevant blocks and noise blocks. The performance remains relatively stable when the amount of noise is moderate. With the 16-fold expanded data set, the drop in quality occurs later, indicating that longer time series may be more robust to noise than smaller ones. Apart from improving the general result quality, SNN also is apparently able to handle a larger share of noise in the data. The other data sets show essentially the same behavior.

To assess the relative performance of primary and secondary similarity with respect to the absolute dimensionality, we maintain fixed proportions of noise and vary only the total dimensionality. The results for the *CBF* data set can be seen in Figure 7. The trend in these results holds throughout all settings: increasing the data dimensionality for a fixed amount of noise improves the performance, with higher dimensionality leading to good performance for higher proportions of noise. This confirms the findings of [15] also for time series data and time series distance measures. The performance of SNN measures again essentially depends only on the performance of the primary measure, not on the actual characteristics of the data set.

5   Discussion and Conclusions

Studies of the infamous curse of dimensionality (such as [23,24,25]) have long been used as a rationale for practitioners and researchers to justify their avoidance of large data sets, or for algorithms failing to find meaningful results in such data, or for the motivation of new heuristics (see [15] for a discussion of examples). The same has also been alleged for time series or sequence data [5,26,27,28,29]. Other researchers (such as [30,23], and more recently [15]) have argued that the findings of such studies should be interpreted with a degree of caution.

The essential point made in [15] is that the effects of the curse of dimensionality largely depend on the presence or absence of clusters in a dataset, and similarity measures by which they may be distinguished. In the presence of data clusters (or localized distributions of any kind), the curse applies to the discrimination between pairs of neighbors within clusters (in that the nearest and farthest neighbors within a cluster may become indistinguishable with increasing dimensionality), but not between pairs of neighbors from different clusters. As long as the proportion of class-relevant information remains sufficiently high, the contrast between different clusters may even improve with increases in dimensionality. The experimental study of this paper confirms the findings of [15], and shows that they also apply to distance measures specific to time series data.

Aside from the (expected) confirmation of earlier findings, we demonstrated the capability of shared-neighbor-based secondary distance measures to improve on the quality of similarity rankings provided by primary measures, in the context of time series data and time series distance measures. Furthermore, we showed that SNN similarity measures are capable of withstanding larger proportions of noise than their associated primary distances. Even in cases where the contrast between clusters has deteriorated due to noise, provided that the primary distance measure still produces meaningful ranking information, secondary SNN measures often remain capable of discrimination between the clusters.

Although extensive within the scope outlined early on, this study could not account for all situations that may be of interest to practitioners or researchers. For example, we did not evaluate the use of approximations such as wavelets and Fourier transformation for dimensional reduction. We also did not work with periodic signals that could be better analyzed in a frequency domain. Only a few common distance functions could be considered, and the patterns of dimensional expansion and noise generation were also of necessity very limited — and doubtlessly there are instances for which even the best distance functions (both primary and secondary) must fail. Any of these topics could conceivably be greatly expanded into a large study in its own right.

However, the goal of this study was not to debate the issue as to which primary distance measure is generally better than another, but instead to support three key conclusions:

1. High dimensionality of a data set does not in itself lead to the effect known as the curse of dimensionality. The proportion of data attributes that is relevant to the query has far more influence on quality and performance.
2. The relative ranking of data objects according to a similarity measure can remain meaningful even when the absolute distance values become difficult to distinguish. Poor distance contrast can be boosted by using an approriate secondary distance measure as SNN.
3. Rank-based similarity measures such as SNN can be more discriminative than their associated primary distance measures.

These points relate to conclusions 1-3 in [15], and establish these observations also in the context of time series data and time series distance measures.

References

1. Berndt, D., Clifford, J.: Using dynamic time warping to find patterns in time series. In: KDD Workshop (1994)
2. Vlachos, M., Kollios, G., Gunopulos, D.: Discovering similar multidimensional trajectories. In: Proc. ICDE (2002)
3. Chen, L., Özsu, M.T., Oria, V.: Robust and fast similarity search for moving object trajectories. In: Proc. SIGMOD (2005)
4. Chen, L., Ng, R.: On the marriage of Lp-norms and edit distance. In: Proc. VLDB (2004)
5. Agrawal, R., Faloutsos, C., Swami, A.: Efficient similarity search in sequence databases. In: Lomet, D.B. (ed.) FODO 1993. LNCS, vol. 730. Springer, Heidelberg (1993)
6. Yi, B.K., Jagadish, H., Faloutsos, C.: Fast time sequence indexing for arbitrary Lp norms. In: Proc. VLDB (2000)
7. Ahmed, N., Natarajan, T., Rao, K.R.: Discrete cosine transform. IEEE TC 23(1), 90–93 (1974)

8. Chan, K.P., Fu, A.W.C.: Efficient time series matching by wavelets. In: Proc. ICDE, pp. 126–133 (1999)
9. Cai, Y., Ng, R.: Index spatio-temporal trajectories with Chebyshev polynomials. In: Proc. SIGMOD (2004)
10. Kriegel, H.P., Kröger, P., Zimek, A.: Clustering high dimensional data: A survey on subspace clustering, pattern-based clustering, and correlation clustering. IEEE TKDD 3(1), 1–58 (2009)
11. Jarvis, R.A., Patrick, E.A.: Clustering using a similarity measure based on shared near neighbors. IEEE TC C-22(11), 1025–1034 (1973)
12. Ertöz, L., Steinbach, M., Kumar, V.: Finding clusters of different sizes, shapes, and densities in noisy, high dimensional data. In: Proc. SDM (2003)
13. Houle, M.E.: The relevant-set correlation model for data clustering. Stat. Anal. Data Min. 1(3), 157–176 (2008)
14. Kriegel, H.P., Kröger, P., Schubert, E., Zimek, A.: Outlier detection in axis-parallel subspaces of high dimensional data. In: Theeramunkong, T., Kijsirikul, B., Cercone, N., Ho, T.-B. (eds.) PAKDD 2009. LNCS, vol. 5476, pp. 831–838. Springer, Heidelberg (2009)
15. Houle, M.E., Kriegel, H.P., Kröger, P., Schubert, E., Zimek, A.: Can shared-neighbor distances defeat the curse of dimensionality? In: Gertz, M., Ludäscher, B. (eds.) SSDBM 2010. LNCS, vol. 6187, pp. 482–500. Springer, Heidelberg (2010)
16. Houle, M.E.: Navigating massive data sets via local clustering. In: Proc. KDD (2003)
17. Achtert, E., Bernecker, T., Kriegel, H.P., Schubert, E., Zimek, A.: ELKI in time: ELKI 0.2 for the performance evaluation of distance measures for time series. In: Mamoulis, N., Seidl, T., Pedersen, T.B., Torp, K., Assent, I. (eds.) SSTD 2009. LNCS, vol. 5644, pp. 436–440. Springer, Heidelberg (2009)
18. Keogh, E., Xi, X., Wei, L., Ratanamahatana, C.A.: The UCR time series classification/clustering homepage (2006), http://www.cs.ucr.edu/~eamonn/time_series_data/
19. Saito, N.: Local feature extraction and its application using a library of bases. PhD thesis, Yale University (1994)
20. Pham, D.T., Chan, A.B.: Control chart pattern recognition using a new type of self-organizing neural network. Proceedings of the Institution of Mechanical Engineers, Part I: Journal of Systems and Control Engineering 212(2), 115–127 (1998)
21. Gandhi, A.: Content-based image retrieval: Plant species identification. Master's thesis, Oregon State University (2002)
22. Davis, S., Jacobson, A., Suszcynsky, D., Cai, M., Eads, D.: FORTE forte time series dataset (2005), http://nis-www.lanl.gov/~eads/datasets/forte
23. Beyer, K., Goldstein, J., Ramakrishnan, R., Shaft, U.: When is "nearest neighbor" meaningful? In: Beeri, C., Bruneman, P. (eds.) ICDT 1999. LNCS, vol. 1540, pp. 217–235. Springer, Heidelberg (1998)
24. Hinneburg, A., Aggarwal, C.C., Keim, D.A.: What is the nearest neighbor in high dimensional spaces? In: Proc. VLDB (2000)
25. Aggarwal, C.C., Hinneburg, A., Keim, D.: On the surprising behavior of distance metrics in high dimensional space. In: Van den Bussche, J., Vianu, V. (eds.) ICDT 2001. LNCS, vol. 1973, p. 420. Springer, Heidelberg (2000)
26. Chakrabarti, K., Keogh, E., Mehrotra, S., Pazzani, M.: Locally adaptive dimensionality reduction for indexing large time series databases. ACM TODS 27(2), 188–228 (2002)

27. Verleysen, M., François, D.: The curse of dimensionality in data mining and time series prediction. In: Cabestany, J., Prieto, A.G., Sandoval, F. (eds.) IWANN 2005. LNCS, vol. 3512, pp. 758–770. Springer, Heidelberg (2005)
28. Radovanovic, M., Nanopoulos, A., Ivanovic, M.: Time-series classification in many intrinsic dimensions. In: Proc. SDM (2010)
29. Radovanovic, M., Nanopoulos, A., Ivanovic, M.: Hubs in space: Popular nearest neighbors in high-dimensional data. J. Mach. Learn. Res. 11, 2487–2531 (2010)
30. Bennett, K.P., Fayyad, U., Geiger, D.: Density-based indexing for approximate nearest-neighbor queries. In: Proc. KDD (1999)

Managing and Mining Multiplayer Online Games

Hans-Peter Kriegel, Matthias Schubert, and Andreas Züfle

Institute for Informatics, Ludwig-Maximilians-Universität München, Oettingenstr. 67, D-80538 Munich, Germany
{kriegel,schubert,zuefle}@dbs.ifi.lmu.de

Abstract. Modern massive-multiplayer-online games (MMOs) allow thousands of users to simultaneously interact within a virtual spatial environment. Developing an MMO servers yields a lot of new challenges in managing and mining spatial temporal data. In this paper, we describe problems of handling spatial data in an MMO and outline why analyzing spatial patterns is an important task for keeping an MMO successful over a long period of time. Though many of these problems share a close connection to well-known tasks in temporal and spatial databases, there are many significant differences as well. Thus, we will describe the similarities to established problems and outline new challenges for managing and mining spatial temporal data in MMOs.

1 Introduction

In recent years, computer games have become a driving factor in the entertainment industry having a business volume exceeding even the movie business and the music business. An important factor in modern computer games is the possibility to link the gaming experience of thousands of players via the internet. Massive-Multiplayer-Online games (MMOs) allow several thousand players to simultaneously participate within the same virtual spatial environment. Since each game entity in the game, e.g. a player character, has a spatial position on a virtual landscape, MMO servers often have to provide similar functionalities as spatio-temporal databases. For example, a player needs information about all other game entities being located in his current area of sight which can be mapped to an ϵ-range query. Thus, computing the result of spatial queries is an important functionality within processing the game state in an MMO server. However, the requirements for game server w.r.t. query processing, data management or persistency are often quite different from those in a database system yielding new challenges to data structures and algorithms. Beyond efficient processing of the game state there is another interesting relation between data mining in temporal spatial databases and monitoring player behavior which can be modeled by trajectory data. Monitoring the player behavior is an important tool in maintaining the interest of players over a long period of time. Tracking and analyzing the player behavior has two main purposes. The first purpose is to detect cheating: A cheating player may either use unallowed tools like bot-programs, directly modify the gaming client or employ other unallowed actions violating the rules of the game. In most cases, the goal of cheating is to gain an unfair advantage. If cheating becomes a common practice, large groups of fair playing customers might lose their interest in the game which causes a large economic damage.

D. Pfoser et al. (Eds.): SSTD 2011, LNCS 6849, pp. 441–444, 2011.

Furthermore, many modern MMOs rely on the concept of *Micro Transactions* which allows players to gain game advantages by directly paying the game operator. Thus, cheating players may gain these advantages for free for instead of paying for them. Due to these reasons, game companies treat the problem of cheating very seriously.

The second purpose of analyzing the player behavior is to adjust the game design to be interesting for large number of players over a long period of time. A game should be challenging, but it should allow gaming progress for players having a wide variety of playing skill. Furthermore, their should be a balance between the powers of the classes, units or fractions a player can select. When designing game content, the developers imagine a certain portfolio on strategies a player or a team might pursue to solve a particular situation. Since the challenge of this situation is adjusted around these strategies, a game provider should automatically detect unexpected new strategies which do not fit to the intended difficulty level. Since playing an identical game for up to several years eventually gets boring, the game providers change rules and add new content from time to time. However, changing the rules in a running game where instantly thousands of players a confronted with new situations often leads to unexpected results. Again monitoring player actions is essential to detect unexpected effects and quickly react accordingly.

The rest of the paper is structured as follows. Section 2 discusses tasks related to managing spatial temporal data in a game. Afterwards section 3 discusses the challenges in mining player behavior for cheat detection and general player behavior. The paper concludes with a brief summary in section 4.

2   Spatial Temporal Data Management

As mentioned above, many MMOs have to offer similar functionalities as temporal spatial databases managing the movement of real persons w.r.t. GPS coordinates. Thus, we might encounter similar types of spatial queries, e.g. ϵ-range queries. An MMO server employs spatial queries for several purposes. For example, if a player drops a water bomb into a crowd the server needs to find out who is getting wet. A further example are monsters which may always attack the closest player, requiring a bi-chromatic nearest neighbor join.

The most important requirement of a game server is to consistently achieve very low response times. An ordinary real time game is processed in so called ticks. In each tick the game state is processed one step further. To run a game fluidly the server must process the complete tick in at most 100 ms. If the server starts to exceed this time limit too often the game begins to stutter noticeable and cannot by played fluidly anymore. For the data structure managing the spatial locations this yields the challenge of processing potentially thousands of updates within the fraction of the tick being reserved for spatial data management. Furthermore, we need to consider the worst case insertion time to make sure that the update does not cause regular lags. Let us note that there are first solutions for achieving persistency [1] on game servers considering this effect. However, these methods do not especially cover the spatial requirements but with generally saving a game state without generating lags. To further emphasize the importance of low response times, we want to point out that the problem can be expected to increase because game companies keep pushing the maximum amount of players being simultaneously online on the same server.

Another important requirement is spatio-temporal accuracy. While systems tracking real world objects via GPS often have to allow positioning errors and unsynchronized position updates, spatial management in a gaming server should be accurate and consistent in time. In other words, the server must synchronize to the time on a client to prevent confusing the temporal order of events. However, the network delay (lag) often represents a source of uncertainty which must be dealt with. If a game server sends positioning information belonging to time t_k to a player who submitted his last update at time $t_0 << t_k$, the server has to make sure that an update having a time stamp $t_l < t_k$ is processed in a way not compromising the causality of the gaming world.

A further challenge of managing spatial information in a game server is to optimize the amount of queries the server has to process. Though it is necessary to determine a list of all other entities being within interaction range, it might not be required to completely process this list at each point of time. Further directions for reducing processing costs are the use of join processing and distributing the required queries as uniform as possible over multiple ticks.

A final aspect that should be mentioned being a further source of challenging new problems is that game servers are more and more implemented within a cloud environment. Thus, there is a need for large distributed algorithms maintaining the high requirement for temporal synchronization. First solution can be found in [2].

3   Monitoring Player Behavior

Monitoring player behavior is an important aspect when trying to maintain a successful MMO over a long period of time. The behavior of a player can be modeled by two components: A spatio-temporal trajectory, describing the position of a player at each point of time; and the set of actions performed by the player each associated with its respective time-stamp. In the following, we will survey tasks that arise in the field of MMOs and relate them to the field of Spatio-Temporal data management and data mining.

A first very important purpose for monitoring player behavior is detecting cheating players. There are currently two common ways of cheating: hacks and bots. A hack comprises every action that circumvents the rules of the games. Hacks often aim at causing spatial and temporal effects. Common examples are speed hacks (a characters moves faster than allowed), teleport hacks (a character instantly changes its location to a far off place) and map/wall hacks(the player accesses location information which should not be available to him). The other well established form of cheating is using bots. A bot is a computer program controlling a player character. Using a bot is useful if game progress can be achieved by spending time in performing primitive actions or being online at a regular basis. Since most bots usually act according to static scripts determining their actions, they often generate specific temporal patterns which are distinguishable from human behavior. A further purpose of monitoring players is analyzing their success and their strategy. This information can be used to adjust the difficulty of a game depending to the skill level of a player. Furthermore, it is necessary to check whether the game is played as intended by the developers. In many games, proper movement is an essential part of the game-play. In many situations, it is required to move closer or farther away from your team mates to be successful. Furthermore, a skilled player simultaneously moves and acts differently than a beginner. Thus, the movement patterns represent important information about players and strategies.

To monitor players, the necessary data has to be collected and stored within server logs. As mentioned in the previous section, processing time on the game server is a limiting factor. Thus, the first problem is to write server logs while straining the server resources as little as possible. A further problem is the constantly growing amount of logged data which is caused by the high frequency of ticks and the large amount of game entities. Thus, running data analysis on complete server log may not be a viable option. Instead, it is necessary to select a fraction of the log containing the behavior of a subset of players in certain time periods. However, identifying interesting players and time intervals for a particular purpose is not a trivial task. For example, how can a game provider detect as many cheaters as possible while processing as little logged data as possible?

Analyzing gaming data requires multiple data mining tasks. Checking logged info for well-known cheats or bots is a classification task, while analyzing team strategies requires clustering techniques. For many tasks, outlier detection might reveal important information. Since a game provider does not exactly know which types of cheats are currently employed, finding unusual behavior is essential. Furthermore, outliers often reveal unintended gaming strategies circumventing the intended design.

A major challenge in data mining in game logs is their highly dynamic nature. First of all, having several thousand players yields a large potential for unexpected behavior. Thus, new playing strategies will emerge frequently. Furthermore, cheats are often developed by highly skilled programmers and thus, hacks and bots are often updated to counter possible detection methods. Finally, player communities are usually highly connected. Thus, yesterdays outliers often become todays clusters.

4   Conclusion

In this paper, we briefly pointed out the usefulness of methods from the area of managing and mining temporal spatial data to massive multiplayer online (MMO) games. While many tasks in MMOs are quite similar to current research problems, there still exists multiple new challenges in this interesting application area offering a rewarding direction for future work. Let us note that we do not consider the list challenges as complete. Especially, there is a wide variety of further problems in designing the artificial intelligence of computer opponents.

References

1. Salles, M.V., Cao, T., Sowell, B., Demers, A., Gehrke, J., Koch, C., White, W.: An Evaluation of Checkpoint Recovery for Massively Multiplayer Online Games. In: 35th international conference on very large databases (VLDB 2009), Lyon, France (2009)
2. Wang, G., Salles, M.A.V., Sowell, B., Wang, X., Cao, T., Demers, A.J., Gehrke, J., White, W.M.: Behavioral simulations in mapreduce. PVLDB 3(1), 952–963 (2010)

Citizens as Database: Conscious Ubiquity in Data Collection

Kai-Florian Richter and Stephan Winter

Department of Infrastructure Engineering
The University of Melbourne
{krichter,winter}@unimelb.edu.au

1 Users as Data Providers

Crowd sourcing [1], *citzens as sensors* [2], *user-generated content* [3,4], or *volunteered geographic information* [5] describe a relatively recent phenomenon that points to dramatic changes in our information economy. Users of a system, who often are not trained in the matter at hand, contribute data that they collected without a central authority managing or supervising the data collection process. The individual approaches vary and cover a spectrum from conscious user actions ('volunteered') to passive modes ('citizens as sensors'). *Volunteered* user-generated content is often used to replace existing commercial or authoritative datasets, for example, Wikipedia as an open encyclopaedia, or OpenStreetMap as an open topographic dataset of the world. Other volunteered content exploits the rapid update cycles of such mechanisms to provide improved services. For example, fixmystreet.com reports damages related to streets; Google, TomTom and other dataset providers encourage their users to report updates of their spatial data. In some cases, the database itself is the service; for example, Flickr allows users to upload and share photos. At the *passive* end of the spectrum, data mining methods can be used to further elicit hidden information out of the data. Researchers identified, for example, landmarks defining a town from Flickr photo collections [6], and commercial services track anonymized mobile phone locations to estimate traffic flow and enable real-time route planning.

In short, user-generated content drastically reduces the costs of data collection and time to next update. Users expect free access: the data is user-generated and shared, reproduction and distribution costs are close to zero [7]. Further, user-generated data has proven to match the quality of authoritative datasets [8]. In studying the motivation of people to contribute, Benkler demonstrates "that the diverse and complex patterns of behavior observed [...] are perfectly consistent with much of our contemporary understanding of human economic behavior" [9, p. 91]. The phenomenon of user-generated content is not going away soon. The traditional economy of spatial information is fundamentally challenged.

Thinking this emergent field further into the future, some limitations become immediately obvious. User-generated spatial data consists of a combination of sensor data (e.g., coordinates) and user-added semantics (e.g., place descriptions). While databases are good at collecting and interpreting sensor data, the management and use of the user-added semantics is still in its infancy.

D. Pfoser et al. (Eds.): SSTD 2011, LNCS 6849, pp. 445–448, 2011.

Traditionally, semantics is inferred by data mining methods on sensor data, and the same has been done for user-added semantics. We believe that this traditional approach will fall short, since, fundamentally, adding the kind of semantics discussed in this paper is a human intelligence task. Human Intelligence Tasks (HIT) are simple, often menial tasks (e.g., identifying objects on a photo), which appear to be ideally suited to be solved automatically, but are in fact easier and cheaper to be solved through crowd-sourcing. The term was introduced by Amazon's Mechanical Turk (https://www.mturk.com).

The kind of semantic information envisioned here goes well beyond the simple tasks of Mechanical Turk, so it is bound to be contributed by humans. Accordingly, the true challenge is collecting highly redundant user-added semantics and developing novel information inference processes for semantically rich data.

2   The Volunteered Image of the City

Crowd sourcing in its various forms is ideally suited to collect *semantic information* rather than just data. In the spatial domain, locals provide not only their tracks or locations, but also their local expertise for free. For example, users map their neighbourhoods in OpenStreetMap, or annotate places in CityFlocks [10], which can then be accessed by others for local decision making. Since contributors are in the environment when they trace geographic features' geometries and describe their semantics, they are ideal candidates to provide their insights and experience for others to use: physical presence increases their credibility when reporting on the experience of spatial and social structures of environments.

People's local experience is captured in the idea of the *image of the city* [11]. This 'image' is a metaphor for people's mental conceptualisation of the environment they are living in—its spatial layout in a narrow sense and, more broadly, how they conceive the environment's social structure. It evolves over time through interaction with the environment. It is this image that lets us find our way around, know which places to avoid at night, recommend places to buy specialist items to others, and figure out that our favourite restaurant and favourite coffee place are really only two blocks away from each other.

We all carry this 'image of the city' with us all the time—in our heads: the *citizen as a database.* Imagine it could be externalised! Suddenly, everybody else could benefit from our experiences. Partially, this information is already available today of course. But it is fragmented, distributed, and often only implicitly inferable. It is scattered on various web sources, such as restaurant reviews, Flickr image collections, local news, or social networks. If we could store each individual's 'image of their city' in electronic form in a holistic, comprehensive way, i.e., represent it in a common data structure, this would open up opportunities for location based services and man-machine communication that are unheard of today. Such future services would replace asking the 'local expert.' In fact they would access the collective wisdom of multiple local experts at any location, which would filter our individual eccentricities. These externalised, holistic images would deliver topographic data of the neighbourhood and place annotations as in CityFlocks [10], and provide access to local and global landmarks for navigation and gazetteers of official and vernacular place names, to name just a few examples. More generally, these services could communicate like humans [12], which would be vastly beneficial since humans are not necessarily experts when it comes to communicating with machines.

This vision of volunteering the image of the city has implications for data collection (how to capture the images), for data maintenance and integration (how to represent and map their semantics), for querying and analysis (how to exploit their semantics), and for communication (how to express semantics). Further, capture and communication require careful design of the human-computer interaction. For reasons of space in this paper, we will discuss only capturing of such images of the city in more detail. Already this limited discussion will illustrate some of the long-term research questions that come with this vision.

The problem can be phrased as: How to externalise the experientiable and experienced image of the city? And how then to put it into a (crowd-sourced) content platform for sharing with others? In contrast to general user-generated content, for spatial content users have to be or been in-situ to document their experiences; they report in real-time or from memory. Both ways have their advantages. In-situ collection may be more reliable and less distorted, but collection from memory had already a filter on selection and relevance. This interplay of capturing of both sensor and semantic information of corporeal experiences through crowd-sourcing methods is summarised as *conscious ubiquity*.

3 Conscious Ubiquity in Data Collection

Great advances in sensor technology and mobile computing allow for running powerful software while on the move. The ubiquity of mobile Internet provides means for data communication. The technologies of Web 2.0 enable users to contribute content using sophisticated (web-based) interfaces. These are the foundations for the vision of smart mobile devices capturing the image of the city.

Collecting data for this image requires a balance between ubiquity of service and dedicated human-computer interaction. People will contribute such content in large numbers and over longer periods of time only if this collection is facilitated unobtrusively, casually, or, as Weiser put it, calmly [13]. Data collection needs to be supported by automated capturing processes in order for users to accept this service as they cannot be constantly involved. Then again, externalising the image of the city requires significant human interaction; it is a human intelligence task. Adding semantic information will always be an active process, at least partly. Thus, communication between users and devices must become as natural and unobtrusive as possible. Interfaces need to disappear; contributing semantic information must become a negligible task, supported by intelligent, sensor-rich devices, such that people stay "tuned into what is happening around" [13].

At the same time, the collected data is highly sensitive, many privacy issues are involved. It is crucial that users are informed about which data gets collected and how it is distributed. They need to be in absolute control of their data. Keeping this control may require breaks in the seamless collection process outlined above, in order to make users aware of potential breaches of privacy.

Just as the image of the city develops over time in our heads, it will develop and change over time in external representations. New semantic information must be integrated in these representations in a consistent and coherent manner. Often, this may make existing information obsolete. But as this information is semantic and highly individual, it may not be obvious which information to replace. Issues of semantic similarity and automatic conflict resolution need to be solved, even more so if images of multiple users are to be integrated.

In summary, technologies for volunteering the image of the city need to be smart enough to collect sensor observations, provide disappearing interfaces for collection of semantic information, report to the user on request and in critical situations, contribute the collected observations to a content platform, and smoothly integrate these observations into the platform's databases. This interplay between human intelligence, calm technology, and autonomy leads to *conscious ubiquity* in data collection, and, ultimately, to the *citizen as a database*.

References

1. Surowiecki, J.: The Wisdom of Crowds. Doubleday, New York (2004)
2. Goodchild, M.: Citizens as sensors: the world of volunteered geography. Geo. Journal 69(4), 211–221 (2007)
3. Krumm, J., Davies, N., Narayanaswami, C.: User-generated content. Pervasive Computing 7(4), 10–11 (2008)
4. Haklay, M., Weber, P.: OpenStreetMap: User-generated street maps. Pervasive Computing 7(4), 12–18 (2008)
5. Elwood, S.: Volunteered geographic information: Key questions, concepts and methods to guide emerging research and practice. GeoJournal 72(3-4), 133–135 (2008)
6. Crandall, D.J., Backstrom, L., Huttenlocher, D., Kleinberg, J.: Mapping the world's photos. In: Proceedings of the 18th International Conference on World Wide Web, pp. 761–770. ACM, New York (2009)
7. Shapiro, C., Varian, H.R.: Information Rules: A Strategic Guide to the Network Economy. Harvard Business Press, Boston (1998)
8. Haklay, M.: How good is volunteered geographical information? A comparative study of OpenStreetMap and Ordnance Survey datasets. Environment and Planning B: Planning and Design 37(4), 682–703 (2010)
9. Benkler, Y.: The Wealth of Networks. Yale University Press, New Haven (2006)
10. Bilandzic, M., Foth, M., De Luca, A.: Cityflocks: designing social navigation for urban mobile information systems. In: DIS 2008: Proceedings of the 7th ACM Conference on Designing Interactive Systems, pp. 174–183. ACM, New York (2008)
11. Lynch, K.: The Image of the City. The MIT Press, Cambridge (1960)
12. Winter, S., Wu, Y.: Intelligent spatial communication. In: Navratil, G. (ed.) Research Trends in Geographic Information Science, pp. 235–250. Springer, Berlin (2009)
13. Weiser, M., Brown, J.S.: The coming age of calm technology. In: Denning, P.J., Metcalfe, R.M. (eds.) Beyond Calculation: The Next Fifty Years of Computing. Springer, Berlin (1997)

Spatial Data Management over Flash Memory

Ioannis Koltsidas[1] and Stratis D. Viglas[2]

[1] IBM Research, Zurich, Switzerland
iko@zurich.ibm.com
[2] School of Informatics, University of Edinburgh, UK
sviglas@inf.ed.ac.uk

Abstract. We present desiderata for improved I/O performance of spatial data structures over flash memory and hybrid flash-magnetic storage configurations. We target the organization of the data structures and the management of the in-memory buffer pool when dealing with spatial data. Our proposals are based on the fundamentals of flash memory, thereby increasing the likelihood of being immune to future trends.

1 Introduction

Flash memory has moved from being used for short-term, low-volume storage and data transfer, to becoming the primary alternative to magnetic disks for long-term, high-volume and persistent storage. Appearing in systems as disparate as PDAs, handheld tablets, laptops and desktops, enterprise servers and clusters, it is one of the most ubiquitous storage media on the market. At the same time, the proliferation of location-based services means that spatial data management techniques acquire increasing traction. As a result, access to and processing of spatial data moves from highly customized application scenarios to commodity ones. The combination of these two trends necessitates revisiting spatial data management for flash memory both at the server and at the client levels. In this paper we will present ideas that we posit are imperatives when it comes to high-performing techniques for flash-resident spatial data.

Solid state drives, or SSDs, are arrays of flash memory chips packaged in a single enclosure with a controller. An SSD is presented to the operating system as a single storage device using the same interface as traditional HDDs (*e.g.*, SATA). The similarities between the two types of storage medium, however, stop here. The key characteristics of SSDs are (*a*) the lack of mechanical moving parts; (*b*) the asymmetry between their read and write latencies; and (*c*) their *erase-before-write* limitation. The first characteristic means that there is no difference between sequential and random access latencies. Performance is not penalized by being dependent on the seek time or the rotational delay, as is the case for HDDs. The second characteristic has to do with the physical characteristics of flash memory, which make reading the value of a flash memory cell faster than changing it; therefore, reads are in general faster than writes. The discrepancy between read and write latencies is further influenced by the underlying technology of flash memory and, in particular, by how many bits are stored in each

D. Pfoser et al. (Eds.): SSTD 2011, LNCS 6849, pp. 449–453, 2011.
© Springer-Verlag Berlin Heidelberg 2011

cell: a *single-level cell* (SLC) device can store a single bit, while a *multi-level cell* (MLC) can store two bits. SLC devices inherently have higher performance but lower density; on the other hand, MLC devices have higher density but lower performance. Finally, the third characteristic stems for the inherent properties of SSDs: to update an already written flash page, the controller must first erase it and then overwrite it. Erasures are performed at an *erase unit* granularity, where each erase unit is a flash block, *i.e.*, a number of contiguous flash pages. A garbage collection mechanism is required to reclaim flash blocks as pages become invalid due to user overwrites. This is especially detrimental to the performance of random write workloads: random writes result in continuously erasing blocks and moving data at the flash level. It is not uncommon for the random write throughput to be five times less than the random read throughput for SLC devices; or even up to two orders of magnitude less for MLC devices. Though approaches for efficient spatial data structures have been proposed [1,5], clearly, data management over flash memory requires rethinking our priorities and not improving performance in specific cases. To complicate the situation, existing HDD-based hardware is not going to disappear any time soon; the trend is towards augmenting HDD storage with flash memory. It is therefore imperative to design towards hybrid storage configurations employing both SSDs and HDDs.

We propose ways to tackle spatial data management in flash and hybrid setups for the applications of the foreseeable future. We cannot claim that these ideas will be future-proof; however, we solidly ground them on the inherent characteristics of flash memory, thereby increasing their potential for applicability. We do not require any flash features be exposed apart from a standard I/O interface. Our proposals revolve around two axes: the organization of spatial data structures in SSD-only as well as in hybrid configurations, and the management of pages belonging to spatial data structures once the pages have been brought in main memory. In what follows we will present each axis in turn.

2   Data Structure Organization

One of the key goals of secondary storage data structures is performance guarantees. For instance, the best performing spatial data structures are balanced trees. Their guarantee is that all paths from root to leaves are of equal length. This is achieved through bottom-up management algorithms: insertions and deletions cause splits and merges at the leaves; these splits and merges are recursively applied in the bottom-up traversal and propagate higher up in the tree; resulting in the tree having its height increased or decreased. Such algorithms make perfect sense for structures over HDDs where the read and write costs are uniform.

Consider, however, the read/write discrepancy of SSDs and an insertion like the one of Fig. 1. Typically, as shown in the left part, an insertion to a full leaf L will cause a split of the leaf into itself and a sibling S; and an update at the original leaf's parent P with the new bounding box for L; and the insertion of the bounding box for S. To balance the tree we must perform three disk writes. An alternative, instead of splitting L into L and S, is to allow L to overflow into

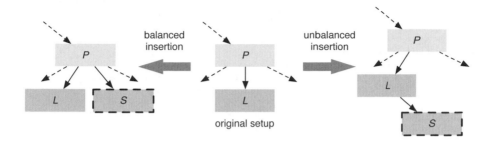

Fig. 1. Balanced *vs.* unbalanced insertions

S and we not update P, since L's bounding box has not changed (right part of Fig. 1). Thus, instead of performing three writes, we perform one: only for S.

Assume now that the ratio between write and read is x, *i.e.*, reading is x times faster than writing. Depending on the application, the cost associated with each type of operation may be any combination of the latency and throughput of the SSD for that type of operation. Saving two writes during insertion and causing the imbalance means that we have saved $2x$ time units: any operation that will read L would have read it anyway; any operation having to access S too, would have accessed it anyway, but going through a direct child pointer from P. Thus, the structure overall has x time units to spare with respect to its optimal form (*i.e.*, the fully balanced one). Then, we allow this imbalance and maintain a counter c at L that measures how many times the overflow pointer to S has been traversed. As long as $c < x$, the data structure still is more efficient; once $c = x$ we can rebalance it by propagating the bounding box from S to P, thereby expensing the write we saved in the first place. This idea can work across all levels of the tree, though one must be careful to avoid long overflow chains. In such cases rebalancing can be prioritized using global rather than local metrics.

In a typical scenario, the savings will be substantial. As the tree structure grows, the likelihood of a single leaf being "bombarded" with read or write requests decreases for all typical scenarios. Thus, the likelihood of the tree's imbalance paying off increases. Though generally applicable to all tree structures, catering for imbalance is more likely to pay off for spatial data and location-based applications: the update rate of the spatial data is low, while the point queries themselves are highly volatile as users move about the area.

Observations like the one that lead to unbalanced structures can be generalized for hybrid configurations. In more detail, and for a tree structure, the likelihood of having to write data increases as we descend the tree. One might then consider placing the different parts of the tree structure on different media. The top levels of the tree have a read-intensive workload, used to direct searches. Whereas the bottom levels of the tree might have an update-intensive workload; moreover, the likelihood of an update being propagated upstream decreases with the height of the tree. It might then be conducive to place the write-intensive leaves of the tree on the HDD and the read-intensive index nodes on the SSD [3].

Coupled with the previous idea of keeping the flash-resident portion of the tree unbalanced, this will result in further performance boosts.

3   Buffer Management

The next issue to focus on is buffering spatial data from the SSD into main memory. As is usually the case, we assume *demand paging*: data pages can only be processed in main memory and they are brought from the SSD into main memory only on reference. Main memory is of limited capacity and smaller than the capacity of the SSD; thus, only a subset of the flash-resident data pages can be kept in main memory at any point in time. Once the memory is full, data pages will be evicted to make room for new ones.

There has been a substantial body of research targeting general buffer management for SSDs [2,4]. Tailoring the approaches to spatial data, in conjunction with the I/O asymmetry of SSDs, offers further room for improvement. Consider a location-based application which continuously accesses a spatial data index for information (*e.g.*, points of interest close to the user's location). Such an application will generate multiple hot paths in the index, tracking the user's motion. As the user moves about, the paths followed from the root of the tree to the leaves will be constantly changing. Therefore, using simple reference counts as the eviction decision metric, though correct, can be further improved. By predicting the motion we can make more informed decisions regarding which pages to keep in the buffer pool and which have a lower probability of being referenced in the future. By selectively caching the parts of the indexed space that are more likely to receive updates, a lot of flash writes, and thereby, erasures can be avoided. Coupled with cost-based replacement policies [3], such an approach can improve I/O performance. One might even envision horizontal or vertical tree partitioning and placement based on access frequency and storage capabilities.

4   Conclusions and Outlook

While SSDs are attracting more traction from the research community and industry alike, surprisingly enough, one of the areas that is under-represented is spatial data management over flash memory. We have presented some initial ideas on improving the performance of flash-resident spatial data structures.

One of the aspects we did not address is endurance and wear-leveling. More precisely, each flash cell can only be written to a fixed number of times. Thus, the controller spreads writes evenly throughout the flash chips and maintains a mapping between logical block identifiers and physical ones, through a software layer called the *flash translation layer*, or FTL. While FTL algorithms are typically inaccessible to the user, some SSD manufacturers are now pushing more of that functionality to the OS driver level; in addition, SSD-specific commands have lately been introduced, giving direct control to the filesystem for some flash operations (*e.g.*, the TRIM command for explicitly erasing blocks). In light of that, the ideas presented in the previous sections can be more efficiently coupled with the SSD and result in further I/O savings.

References

1. Emrich, T., et al.: On the impact of flash SSDs on spatial indexing. In: DAMON 2010 (2010)
2. Kim, H., Ahn, S.: BPLRU: a buffer management scheme for improving random writes in flash storage. In: FAST (2008)
3. Koltsidas, I., Viglas, S.D.: Flashing up the storage layer. Proc. VLDB Endow. 1(1), 514–525 (2008)
4. Park, S.-Y., et al.: CFLRU: a replacement algorithm for flash memory. In: CASES (2006)
5. Wu, C.-H., et al.: An efficient R-tree implementation over flash-memory storage systems. In: GIS (2003)

Tipping Points, Butterflies, and Black Swans:
A Vision for Spatio-temporal Data Mining Analysis

James M. Kang and Daniel L. Edwards

National Geospatial-Intelligence Agency
7500 GEOINT Dr., Springfield, VA 22150
{James.M.Kang,Daniel.L.Edwards}@nga.mil

Abstract. Tipping points represent significant shifts that change the general understanding or belief of a given study area. The recent late winter 2011 events in the Mid-East and climate-level changes raise issues of whether such events are the result of random factors, tipping points, chaos theory or completely unpredicted black swans. Our vision is to understand how spatio-temporal data mining analysis can discover key variables and relationships involved in spatial temporal events and better detect when mining may give completely spurious results. One of the main challenges in discovering tipping point-like events is that the general assumptions inherent in any technique may become violated after an event occurs. In this paper, we explore our vision and relevant challenges to discover tipping point-like events in spatio-temporal environments.

Keywords: Data Mining, Spatial and Temporal, Tipping Points, Black Swans, Chaos Theory.

1 Vision

Though much research remains to be done, the Federal Government looks to the tools of data mining and spatio-temporal data mining to assist in identifying and understanding key patterns proactively, rather than reactively. Diverse events such as the climate changes occurring over differing parts of the world (e.g., El Niño) [8], or the spatially uneven economic recovery within the U.S. since 2008, or the recent late 2011 winter events in the Mid-East all represent key problems that need to be better understood and quantified so that strategies for managing can be formulated. Data mining of spatial, temporal and non-spatial data can help reveal patterns from assembled data that either reinforce or refute hypotheses, or quantify novel relationships. The trouble is that the three examples above represent case studies where additional forces may have played a role to further complicate the discovery and understanding of key underlying patterns.

A number of authors raise important questions about our human tendency to "rationalize by hindsight" [10] and have described examples in which accepted patterns of normality fall apart and may challenge accepted practices within the data mining community. One of these is the tipping point, which Gladwell defines as "the

D. Pfoser et al. (Eds.): SSTD 2011, LNCS 6849, pp. 454–457, 2011.

moment of critical mass, the threshold, the boiling point" [4]. For example, in climate-change, tipping-points are seen as significant shifts from one state to another, such as a fertile landscape shifting to a barren desert due to a drought in the area [12]. In a differing explanation, Taleb characterizes unpredicted surprise events having a major impact such as "Black Swans" which involve unpredictable patterns that do not appear to be Gaussian with an exponential diminishing tail, but a flatter curve with tails that are fatter with elevated probabilities [10]. Finally, a third possibility from Chaos theory seeks to explain the behavior of dynamical systems that are highly sensitive to initial conditions and/or their topological mixing; popularly known as the "butterfly effect" [5].

This paper envisions overcoming these challenges in spatio-temporal data mining so that more powerful tools are able to surface key patterns and identify possible tipping point-like events. Equally important is to better quantify during mining when patterns are either overly sensitive to their inputs (e.g., chaotic systems) or result in such spurious answers that attempting to understand tipping points is futile.

2   Challenges

Discovering "tipping points" (TP) in any analysis of a spatio-temporal dataset brings in three significant conceptual and computational challenges. First, assumptions on a dataset or phenomena may differ before and after a TP occurs. Second, characterizing a "tipping point" may be a challenge in itself based on how they are observed. Finally, an outside force that may not be known in advance nor represented in the observable data, e.g., natural disaster, may cause TPs.

One of the first challenges in discovering TPs is the significant shift of assumptions when identifying spatio-temporal patterns. For example, assumptions over a spatial region may differ such as Tobler's first law of geography, which states that "Everything is related to everything else, but near things are more related than distant things" [11] and the property of teleconnections where there may be a strong interaction between paired events that are spatially distant from each other [6]. For example, climate maps often illustrate contiguous surface areas where colder regions may appear in the northern hemisphere and warmer areas may appear in the south. In contrast, weather patterns due to El Niño exhibit a different relationship between a pair of linked events involving the eastern pacific region and unusual weather patterns occurring throughout the world [8]. The main challenge is to account for multiple or contrasting assumptions on a dataset to identify any indicators or warning signs that a significant shift or tipping point is imminent.

The second main challenge in discovering TPs may be characterizing the event itself. A TP may be seen as a single transient event such as a major traffic accident causing the normal patterns of rush hour in a city to change. Or a TP may be seen as a gradual persistent event such as the transition to a recession economy causing spending habits to change all over the world. For example, in this late winter of 2011, numerous significant events are ongoing in the Mid-East. Numerous reporting of the events try to connect underlying causes across differing countries such as the Washington Post Feb 25th "Turmoil in the Middle East" article [9] which had a graphic entitled "Behind the uprising: Oil wealth, widespread poverty." While Taleb's "rationalize by hindsight" will certainly apply to ultimately understanding these events in the Mid-East, from a data and spatial data mining perspective, having tools that mine for patterns to pull out obvious and non-obvious factors or indicators of change would be incredibly useful. In addition, a quick compilation of statistics from countries experiencing current protests from CIA's "the-world-factbook" [2] reveals a greater diversity across key variables than one would expect. For example, estimates of percent of population below the poverty line range from 4% in Tunisia to 45% in Yemen. Similarly, the literacy ranges from 83% in Libya to 50% in Yemen. Unemployment in Egypt is 10%; in Yemen it's 35%. Percentage of population owning cells phones ranges from 33% in Yemen to 176% in Saudi Arabia. The question related to tipping points is whether you would need a lot of historical trend datasets by country to detect and characterize tipping points or whether mining of data of all 267 countries in the world would suddenly make these variables appear less diverse creating patterns that exhibit tipping points.

In considering the final challenge of the need for tools to understand novel spatial and temporal events and trends, a greater question emerges. Are mining techniques robust enough to reject mined patterns that are overly sensitive to their inputs (chaotic systems)? Can mining someday detect when the input data is too insufficient to characterize a significant event as in the case of Black Swans? Clearly, the rule concerning garbage in-garbage out applies. In reality, we are often inept at differentiating good data from bad data or which data are key and which are irrelevant or understanding when we have sufficient data bounding a problem and when we do not. A prime example of this quandary is the internet-enabled plethora of data quantifying all aspects of the stock market over the last two decades and the realization of how little difference this has made in the performance of both professional and private stock portfolio performance during the same period. One of the points that Taleb [10] makes about Black Swans that after its first recording, the event is rationalized by hindsight, as if it could have been expected. In other words, some events may be so complex that they cannot possibly be detected beforehand by humans or powerful tools such as data mining and spatial data mining. For example, in the early days of March 2011, as the Mid-East events continue, some experts are categorizing these as black swans [7].   The reality is that we really do not and cannot know at this time. It will only be in hindsight, that we have the wisdom to know what data to assemble for mining to quantify the patterns and tipping points so we may properly write history and, possibly, in some way be better prepared for a future black swan.

3 Summary

In this paper, we first introduced our vision for the need to identify significant changes in the patterns discovered that may lead to the quantification of detected tipping points to further understand our environment. We illustrated this vision with several real-world examples of possible tipping-point-like events occurring today. A set of hard challenges was posed to discover these tipping-points in the dataset. These challenges showed how current assumptions on spatial datasets may be violated when observing the change or shift by a tipping-point. We also raise challenges that as mining gets more sophisticated and attempts to detect and quantify tipping points that we might also detect instability in variables (maybe better than sensitivity analysis) and/or detect when statistically we have too few variables relative to the size of the problem. We implore the data mining community to explore our vision and challenges of discovering these complex, yet interesting events.

References

1. Beaumont, P., Can social networking overthrow a government?, WAtoday (February 25, 2011),
   http://www.watoday.com.au/technology/technology-news/
   can-social-networking-overthrow-a-government-20110225-
   1b7u6.html?from=watoday_sb
2. CIA, The World Factbook,
   https://www.cia.gov/library/publications/the-world-factbook/
3. Elliott, M.: Learn to Love The Revolution. TIME Magazine, 30–35 (March 7, 2011)
4. Gladwell, M.: The Tipping Point: How Little Things Can Make a Big Difference. Back Bay Books (2002)
5. Gleick, J.: Chaos: Making a New Science. Penguin Books (2008)
6. Kang, J., Shekhar, S., Henjum, M., Novak, P., Arnold, W.: Discovering Teleconnected Flow Anomalies: A Relationship Analysis of spatio-temporal Dynamic (RAD) neighborhoods. In: Mamoulis, N., Seidl, T., Pedersen, T.B., Torp, K., Assent, I. (eds.) SSTD 2009. LNCS, vol. 5644, pp. 44–61. Springer, Heidelberg (2009)
7. Keyes, D.: The Middle East's "Black Swan", Reuters (January 28, 2011),
   http://blogs.reuters.com/great-debate/2011/01/28/
   the-middle-east%E2%80%99s-black-swan/
8. Pastor, R.: El niño climate pattern forms in pacific ocean (2006),
   http://www.usatoday.com/weather/climate/
   2006-09-13-el-nino_x.htm
9. Sly, Schneider, Wilson, Faiola: Turmoil in the Middle East, Washington Post, pp A9 (Febuary 25 2011)
10. Taleb, N.: The Black Swan: The impact the Highly Improbable, Random House Trade Paperbacks, 2nd edn. (2010)
11. Tobler, W.: A Computer Movie Simulating Urban Growth in the Detroit Region. Economic Geography 46(2), 234–240 (1970)
12. Walsh, B.: Is There a Climate-Change Tipping Point?, TIME (September 4, 2009),
   http://www.time.com/time/health/article/
   0,8599,1920168,00.html

On User-Generated Geocontent

Dieter Pfoser

Institute for the Management of Information Systems
Research Center Athena
G. Mpakou 17, 11524 Athens, Greece
pfoser@imis.athena-innovation.gr

Abstract. Spatiotemporal reasoning is a basic form of human cognition for problem solving. To utilize this potential in the steadily increasing number of mobile and Web applications, significant amounts of spatiotemporal data need to be available. This paper advocates user-generated content and crowdsourcing techniques as a means to create rich, both, in terms of quantity and quality, spatiotemporal datasets.

1 Problem Description

Effective discovery, integration, management and interaction with spatiotemporal knowledge is a major challenge in the face of the somewhat recent discovery of the spatial "dimension" of the World Wide Web. In many contexts, space in connection with time is used as the primary means to structure and access information simply because *spatiotemporal* (ST) reasoning is essential to everyday problem solving. In combination with the already staggering number of spatially-aware mobile devices, we are faced with a tremendous growth of user-generated content and demand for spatiotemporal knowledge in connection with novel applications and challenges.

With the proliferation of the Internet as the primary medium for data publishing and information exchange, we have seen an explosion in the amount of online content available on the Web. Thus, in addition to professionally-produced material being offered free on the Internet, the public has also been allowed, indeed encouraged, to make its content available online to everyone. The volumes of such User-Generated Content (UGC) are already staggering and constantly growing. Our goal has to be to tame this data explosion, which applied to the spatial domain translates to massively collecting and sharing knowledge to ultimately digitize the world. Our ambition has to be to go beyond considering traditional ST data sources and to include any type of available content such as narratives in existing Web pages in a ST data collection effort. We view all available content that has a ST dimension as a potential data source that can be used for computation. When utilized, this vast amount of data will lead to a digitized ST world view beyond mere collections of co-ordinates and maps.

One could argue that as early maps were traces of people's movements, i.e., view representations of people's experiences, digitizing the world in the this context relates to collecting pieces of knowledge gained by a human individual

D. Pfoser et al. (Eds.): SSTD 2011, LNCS 6849, pp. 458–461, 2011.

tied not only to space and time, but also to her context, personal cognition, and experience.

To realize this vision, (i) a proper understanding of ST language expressions will allow us to utilize non-traditional sources, e.g., narratives containing spatial objects and their relationships, (ii) appropriate representations of the collected data and related data management techniques will enable us to relate and integrate data, (iii) data fusion techniques will combine individual observations into an integrated ST dataset and (iv) adequate algorithms and queries will take (the continuously changing) uncertainty of the data into account. In addition, ST data is typically delivered and queried through map-based interfaces. With a thorough ST language understanding, (v) novel user interfaces that abstract and represent ST information in qualitative human terms might become a reality and provide for effortless and naive natural language interaction.

Overall, these ideas promote the GeoWeb 2.0 vision and advance the state of the art in collecting, storing, analyzing, processing, reconciling, and making large amounts of semantically rich user-generated geospatial information available.

2   Research Directions

In order for the proposed vision to become a reality, several research challenges spanning multiple disciplines have to be addressed.

User experience related to ST information is currently directly linked to the representation of the data; geographic co-ordinates pinpoint locations on maps, routing algorithms determine the best route based on distance, etc. In contrast, human interaction with the world is based on experience, learning and reasoning upon loosely coupled, qualitative entities, e.g., spatial relationships such as "near/far". The challenge will be on devising means to better understand people's perception of space by *deciphering how people express* ST *concepts in natural language terms* by means of a Rosetta-stone-equivalent tool for deciphering the ST component of (a range of) natural languages and, if possible, define the underlying building blocks, i.e., cognitive concepts inherent in ST reasoning. To improve *natural language processing*, a key aspect will be to engage the user in providing her conceptions of space in natural language terms. For example, *games-with-a-purpose* (GWAP) [10] can be used as a vehicle to record and analyze natural language descriptions of known ST scenarios, i.e., to provide spatial descriptions of scenes by means of text or audio during the course of the game. GWAP will be the motivator for crowdsourcing of spatial scene descriptions for a multitude of languages and to finally produce a ST language corpus. This corpus can then be used to extract ST knowledge from non-traditional content sources such as narratives in travel blogs.

Data capture focusses then on amassing user-generated ST data from various sources. Existing *attentional information* can be exploited by data mining user-generated ST content, e.g., point cloud data such geocoded flickr images [2] and by extracting ST data from text/audio narratives using NLP techniques, e.g., translating the phrase "the hospital is next to the church" to two spatial objects and respective relationship. In addition, *specifically designed tools* can support the user in the creation of geospatial content (cf. "Geoblogging" [7]). A means for the collection of *un-attentional data* is *ubiquitous positioning*, i.e., using a large number of complementary positioning solutions (GPS, WiFi, RFIDs, indoor positioning) to relate content to absolute spatial (coordinates) and temporal (time stamps) values. Data capture will not only produce quantitative data, i.e., spatial objects and their locations, but also qualitative data in terms of spatial relationships.

Efficient *data management* techniques will be of outmost importance when tapping into large amounts of geospatial data streams. The focus in this research will be on distributed data management schemes such as cloud computing. Issues to be addressed are spatial indexing and query processing (cf. geospatial data management using Apache Cassandra [5]). In addition, one needs to investigate novel concepts such as dataspaces [4] and linked data [1] for the specific case of geospatial information. *Mobile devices* are increasingly used as Web infrastructure nodes and, hence, will play an important role not only in data collection but also in distributed geospatial data management.

Data Fusion presents us with the problem of processing diverse incoming information and to specify the *relationships and correspondences between data and/or metadata* of different sources so as to reconcile them. Such a framework for matching and mapping different data streams of user-generated content involves identifying related data and generating better mappings by developing specific tools that involve the user in the process. The final goal is then to reconcile and fuse user-generated data to arrive at a single *ST* dataset. *Spatial uncertainty* has been described as "the Achilles' Heel of GIS, the dark secret that once exposed will bring down the entire house of cards" [3]. Uncertainty will be essential to the process of correlating and fusing previously unrelated *ST* data sources. In part, data fusion can be seen similar to the problem of adjustment computation in surveying engineering, in which for a number of observations the best fitting (mathematical) model and its parameters are determined, i.e., to derive the true values based on observations. Research needs to focus on relating and mapping qualitative data (relative *ST* data as denoted by topological, metric and directional spatial relationships) to uncertainty. To achieve this task, techniques based on graph similarity, spatial reasoning [11] and Bayesian statistics [8] need to be investigated.

Working with user-generated spatiotemporal data sources has the drawback that there are no "final" datasets, i.e., all datasets are affected by a varying degree of uncertainty, which any kind of *computation* has to take into consideration. An additional aspect is the evolving nature of the data. Given that each observation increases the scope and/or reduces the uncertainty, algorithms have to accommodate this fact (cf. Canadian Traveller Problem [6]).

Current *spatial information visualization and interaction* is typically map-based. Even in novel geographic fusion services over the Web, e.g., Google Maps, the dependence on map-based interfaces for querying and delivering information is dominant and quite often lacking in expressiveness and usability. While recently systems based on augmented reality concepts have been developed, improved ST language understanding may lead to alternative text and audio-based interfaces to consume ST information (cf. [9]).

3   Summary

The presented vision targets user-generated geocontent and turning it into a viable data source that will complement ST knowledge created by expert users. New means to harness, aggregate, fuse and mine massive amounts of ST data are needed in order to achieve best possible coverage and extract hidden knowledge. The task at hand is nothing less than taming this semantically rich user-generated geodata tsunami and addressing the challenge of transforming the data into meaningful chunks of information obtained with simplicity and speed comparable to that of Web-based search.

Acknowledgements. The research leading to these results has received funding from the European Union Seventh Framework Programme - Marie Curie Actions, Initial Training Network GEOCROWD (http://www.geocrowd.eu) under grant agreement No. FP7-PEOPLE-2010-ITN-264994. The author also thanks Spiros Athanasiou for many productive discussions on the various subjects.

References

1. Bizer, C., Heath, T., Berners-Lee, T.: Linked data - the story so far. Int'l J. on Semantic Web and Information Systems 5(3), 1–22 (2009)
2. Cope, A.: The Shape of Alpha. Where 2.0 conf. presentation (2009), http://where2conf.com/where2009/public/schedule/detail/7212
3. Goodchild, M.: Uncertainty: the Achilles Heel of GIS? Geo Info Systems 8(11), 50–52 (1998)
4. Halevy, A., Franklin, M., Maier, D.: Principles of Dataspace Systems. In: Proc. 25th PODS Conf., pp. 1–9 (2006)
5. Malone, M.: Building a Scalable Geospatial Database on top of Apache Cassandra (2010), http://www.youtube.com/watch?v=7J61pPG9j90
6. Papadimitriou, C., Yannakakis, M.: Shortest paths without a map. Journal Theoretical Computer Science 84(1), 127–150 (1989)
7. Pfoser, D., Lontou, C., Drymonas, E., Georgiou, S.: Geoblogging: User-contributed geospatial data collection and fusion. In: Proc. 18th ACM GIS Conf., pp. 532–533 (2010)
8. Richardson, H., Stone, L.: Operations Analysis during the underwater search for Scorpion. Naval Research Logistics Quarterly 18(2), 141–157 (1971)
9. Strachan, S., Williamson, J., Murray-Smith, R.: Show me the way to Monte Carlo: density-based trajectory navigation. In: Proc. CHI 2007 Conf., pp. 1245–1248 (2007)
10. von Ahn, L.: Games with a purpose. IEEE Computer 39(6), 92–94 (2006)
11. Wallgruen, J., Wolter, D., Richter, K.-F.: Qualitative matching of spatial information. In: Proc. 18th ACM GIS Conf., pp. 300–309 (2010)

Localizing the Internet: Implications of and Challenges in Geo-locating Everything Digital

Michael R. Evans and Chintan Patel

University of Minnesota
Computer Science and Engineering
Minneapolis, MN, USA
{mevans,cpatel}@cs.umn.edu

Abstract. The Internet allows for near-total anonymity in online discussions and transactions; leading to challenges such as identity theft, spear phishing, cyber-bullying and lack of user-retailer trust for small online retailers. In an effort to restore prosperity, security and trust to our increasingly digitally interconnected society, it is important to tie some sense of the physical world in with the cyber world, e.g., creating a geo-coded Internet. However, geo-coding the vast amount of content currently on the web, along with providing reliable geo-location authentication services against the difficulties of spatial analysis and the inherent probabilistic estimate of current location-identification technology.

Keywords: GIS, Geo-coding, Moving Objects, Spatial Databases.

1 Vision

Location information for everything on the Internet has the potential to impact our prosperity, security and civility. Location is fast becoming an essential part of Internet services, with HTML 5 providing native support for locating browsers and GPS-enabled phones locating people on the move. The Internet currently provides anonymity to a large degree, which in turn inhibits trust and leads to rampant problems such as spam, phishing, identity theft, banking fraud, etc. In the physical world, we are surrounded by context clues for making decisions, especially when it comes to safety. Could such problems be reduced if we knew the physical location of every packet, document, computer, server and person on the Internet?

Geo-coding the Internet will facilitate prosperity. In December 2010, daily deal website Groupon turned down a $6 billion dollar buyout offer from Google. In Groupon's wake is a sea of copycat startups, all aiming to make money out-'localizing' Groupon's daily deals [1]. The suggested weakness of Groupon's model is in the broadness of its daily deals. Already in 500 cities, Groupon targets entire metro areas, some with populations in the millions. Copycat startups are aiming for hyperlocal advertising, targeting down to the street or even block level, ensuring that each person finds a deal close to where they live and work.

D. Pfoser et al. (Eds.): SSTD 2011, LNCS 6849, pp. 462–466, 2011.

Websites like Facebook and eBay utilize user profiles to establish trust between users [2]. eBay builds reputations via seller feedback, Facebook through friendship networks. When users build these reputations, trust levels increase as people become less random and anonymous. This leads to more direct sharing, collaboration and commerce. A geo-coded Internet would increase this trust by connecting an online persona to a physical real-world location.

Geo-coding the Internet will improve security. In the physical world, we are surrounded by context clues for making decisions, especially when it comes to safety. Shoppers walking into flea markets and department stores naturally have different expectations of service and trust. Would it be possible to transition some of these physical clues into the cyber world? When visiting an unfamiliar online retailer, a user has few clues to determine a site's credibility - for example, if the site looks 'cheap', as compared to polished, it might affect a user's decision to purchase from the site. Being able to tie an online retailer to a physical location allows for context clues and implicit laws protecting customers based on the location. The ability to geo-locate online retailers will help build trust and enable commerce.

Although the security of online banking has increased dramatically over the years, there is still much work to be done. Currently, most banks authenticate users only through something they know (e.g., passwords, secret questions), as hardware-based tokens remain unused by the general public. Phishing attacks by fraudulent websites posing as major banking websites are becoming ever more sophisticated, targeting specific sets of users with more detailed attacks, known as spear phishing. Adding geo-location information to both the client connecting to the online banking server, and the online banking service itself will reduce fraud on both sides of the equation.

In 2010, 15% of global Internet traffic was briefly routed through China, including some US Military traffic [3]. This raises questions regarding privacy and security, and indicates a need for location validation technology. While it would likely involve modifying internet protocols and perhaps new hardware, adding verified location information to network level components, such as routers and servers, could provide crucial security guarantees.

Geo-coding the Internet will build civility. Online anonymity has been show in the social sciences to increase aggression and uninhibited, sometimes dangerous, behavior [4]. The ability to authenticate spatio-temporal location of users, client-server devices, documents and message sources, etc, would provide a dramatically improved Internet experience. While civilized communication is sometimes difficult to achieve even in the physical world, it can be even more difficult in online situations when widespread anonymity results in a lack of accountability. Enforcing geo-coded Internet technology will help bridge the divide between the cyber and physical world, hopefully allowing for transfers of empathy and humanity to an otherwise hostile online environment.

2   Challenges

There are major technical and societal challenges to fully realizing a geo-coded Internet.

Geo-coding Internet content is challenging. Early work in automated geo-coding information dealt with algorithms for textual content, such as news stories, textual driving directions, pictures, etc. The next step, of high value to law enforcement and US Department of Defense, is inferring location directly from picture images or videos. Repeated cases of terrorists and kidnappers using video and pictures to make threats is motivating urgent work in this area. An initiative was started by Dr. Beth Driver at the National Geospatial-Intelligence Agency to fund and develop various video and image geo-coding techniques. Simple things such as background textures, noises (e.g., in *The Fugitive*, where a call location is placed due to a background train whistle), landscapes, etc, can help give some bounds to the media's origin location.

This sort of technology will require massive databases of geo-located images and sounds, along with fast and efficient ways to query for partial matches. In addition, spatial reasoning is required to correctly deduct possible locations via these clues. Lastly, denial and deception attempts by our foes will increase the difficulty in these sort of geo-coding problems.

Geo-locating content for security, or using location as a part of an authentication scheme, will increase security on the Internet. New hardware is needed to enable authenticated personal location reporting through location-aware hardware tokens, combining location-determination technology (e.g., GPS or cellphone tracking) with traditional hardware tokens (e.g., RSA SecurID [5]). This location-aware token could report to a server, sending spatio-temporal information along with entity identifier to ensure continued entity authorization.

Research needs include investigation into threat-relevant movements for development expiration policies (e.g., timeout, geo-fence violation) using movement parameters (e.g., trajectory, speed, direction, acceleration) and authorized geometry (e.g., building-size). Geo-coded Internet authentication will require challenge-response protocols; and could be based on local broadcast, i.e. location-based signals accessible only from designated places. For example, a fixed kiosk may display CAPTCHA-like [6] message to challenge users. Alternatively, Wi-Fi transmitter power levels or variants of cell towers may be varied to challenge devices similar to work mentioned in [7].

Current authentication techniques are boolean in that they provide yes or no authentication responses. However, spatio-temporal location sensors and service provide a probabilistic estimate of location. For example, global positioning system (GPS) provides a root mean square error with each location-estimate. This 'fuzziness' presents difficulties for current querying models, such as the OGIS SQL extensions [8].

Geo-locating advertising, known as hyperlocal advertising, is the next step in the popular wave of 'daily deal' websites like Groupon.com [1]. However, a fully realized geo-coded Internet presents scalability issues. Tracking and tailoring content to location-providing mobiles, for example, will require fast and efficient indexing and querying techniques in database systems [9]. Hundreds of millions of cell phones sending real-time spatial join queries will quickly overload existing technology. Web Content servers will need to quickly do spatial queries to localize information for a user [10].

Spatial data mining algorithms need to be developed to take advantage of the exponential rise of location-aware content. Spatio-temporal hotspots of like-minded users could be clustered for more specific location-based advertising (e.g., target only the hipsters in the southeast section of town about an upcoming bar special). New user- and location-specific associations can be mined out of detailed geo-located transaction information [11].

3   Conclusion

The incorporation of authenticated and continuous location information for internet entities such as users, documents and servers will allow a flourishing of services designed around enhanced security and trust. Technology that allowed for universal authentication and location-determination services for permitted parties would allow a person to restrict online banking access to their own homes, or a government entity to require that classified information be accessed within pre-determined spatial boundaries. New avenues of research would open in efficient challenge/response protocols to ensure validity of a reported location, digital rights management algorithms for important documents or services that can only be accessed from certain locations. We believe that the development of spatio-temporal location authentication services may reduce or even prevent online fraud and facilitate Internet growth.

References

1. Bruder, J.: In Groupons $6 Billion Wake, a Fleet of Start-Ups (2011), http://goo.gl/EsIhR
2. Guha, R., Kumar, R., Raghavan, P., Tomkins, A.: Propagation of trust and distrust. In: Proceedings of the 13th International Conference on World Wide Web, pp. 403–412. ACM, New York (2004)
3. The Washington Times, Internet traffic was routed via Chinese servers (2010), http://goo.gl/ZMais
4. Wikipedia, Social psychology (psychology) (2011), http://goo.gl/Aa46N
5. Wikipedia, SecurID (2011), http://goo.gl/aPAOY
6. Wikipedia, CAPTCHA (2011), http://goo.gl/zkY53
7. Ferreres, G., Alvarez, R., Garnacho, A.: Guaranteeing the authenticity of location information. IEEE Pervasive Computing 7(3), 72–80 (2008)

8. Shekhar, S., Chawla, S.: Spatial databases: a tour. Prentice Hall, Englewood Cliffs (2003)
9. Güting, R.H., Schneider, M.: Moving objects databases. Morgan Kaufmann Pub., San Francisco (2005)
10. Mokbel, M., Ghanem, T., Aref, W.: Spatio-temporal access methods. IEEE Data Engineering Bulletin 26(2), 40–49 (2003)
11. Bogorny, V., Shekhar, S.: Spatial and spatio-temporal data mining. In: ICDM, p. 1217 (2010)

From Geography to Medicine: Exploring Innerspace via Spatial and Temporal Databases*

Dev Oliver[1] and Daniel J. Steinberger[2]

[1] Department of Computer Science, University of Minnesota, USA
[2] Department of Radiology, Medical School, University of Minnesota, USA
oliver@cs.umn.edu, stein012@umn.edu

Abstract. Spatial and temporal (ST) databases have traditionally been used to manage geographic data. However, the human body is another important low dimensional physical space which is extensively measured, queried and analyzed in the field of medicine. Health care is fundamental to the lives of all people, young or old, rich or poor, healthy or ill. This is an opportune time for research in this area due to the international priority to improve human health and the recent passage of health care legislation to provide electronic medical records and databases in many countries such as the United States. ST datasets in medicine include 3-D images (e.g., CT and MRI), spatial networks in the body (e.g., circulatory system), recorded for each patient at various times (per visit or frequently during hospitalization). We envision a spatio-temporal framework for monitoring health status over the long term (via dental X-rays, mammograms, etc.) or predicting when an anomalous decay or growth will change in size. An ST framework may play an important role in improving health care quality by providing answers on the progression of disease and the treatment of many pathologies (e.g., cancer). However, realizing such a framework poses significant challenges for researchers, each of which is a non-trivial task that has not been addressed by previous work. Taking on these challenges, therefore, would mark the beginning of the next fantastic voyage [2] in spatial and temporal databases.

Keywords: Spatial and Temporal Databases, Medical Data, Radiology.

Vision. Spatial and temporal (ST) data in medicine is available in many forms such as spatial networks formed by bodily systems, 3-D medical images, etc. [9,3]. Spatial networks in the body include the circulatory system or blood vessels, the network of nerves, the network of bronchi and bronchioles in the lungs, and the skeletal system. An example source of a spatial network in this context is an angiogram showing blood vessels with blockage. 3-D images, on the other hand, include computer aided tomography (CT), ultrasound, etc.; they allow visualization of important structures in great detail and are therefore an important tool for the diagnosis and surgical treatment of many pathologies [4].

---

* Innerspace [1] is a Spielberg remake of Fantastic Voyage [2], a science fiction movie about a journey inside the human body in a miniaturized submarine exploring the brain, blood vessels, eyes, ears, etc.

D. Pfoser et al. (Eds.): SSTD 2011, LNCS 6849, pp. 467–470, 2011.

A patient's spatial network or 3-D image data taken over time can be used in new ways such as long term study in which crucial monitoring, predictive, and routing questions may be answered algorithmically. Examples of monitoring questions include discovering how an anomalous growth (e.g., cancer) is changing over time or detecting the narrowing of blood vessels. Predictive questions involve determining therapy effect on tumors across a population as a guide for future therapies. Routing questions are asked to find a route through the body for minimally invasive surgery to remove a tumor.

A spatio-temporal framework for answering long term questions plays a critical role in improving health care quality by providing answers regarding the progression of disease and the comparative effectiveness of interventions. Such a framework (in conjunction with other information technologies) may also assist individuals to stay healthier by helping those with chronic or acute conditions to manage their disease outside of acute care settings [12]. Providers are empowered with a means of simplifying the tracking of multi-focal disease based on 3D images or spatial networks taken over time and this might go a long way in reducing expenses such as the $2 trillion a year that the United States spends on health care [5].

Previous work such as PACS [6] focuses on image and graphics processing to produce accurate 4-D images or models (3-D + time) of clinical quality data [15, 13]. Current commercially available tools in advanced image processing include applications for finding and measuring lesions on previous studies and applications for identifying parts of the brain automatically for comparison with a database of normals to determine if there has been relative mass loss in certain areas [7]. However, these tools have limited exploratory, associative and predictive analysis capabilities. Exploratory analysis enables the user to construct completely new queries on the data set (beyond a fixed list of canned queries), predictive analysis uses a time series of snapshots to determine future behavior using data driven techniques and associative analysis correlates a patient's history and context (e.g., age, race, comorbidities). Additional limitations include the fact that they are largely based on the raster data model (e.g., pixel, voxel) and do not adequately support vector data models (e.g., points, line strings, polygons, networks) and queries (e.g., topological operations, shortest path, etc.). They also do not provide a general frame of reference similar to geographic-based latitude/longitude or postal addresses. Applying solutions from spatial and temporal databases and data mining may well help answer monitoring, predictive and routing questions in the human body. There are, however, several challenges that must be overcome in order to make this vision a reality.

Challenges. Answering long term questions based on ST medical data sets gathered over time raises five conceptual and computational challenges. First, a reference frame analogous to latitude/longitude must be developed for the human body. Second, location determination methods are needed to know where we are in the body. Third, routing in a continuous space where no roads are defined is required to reduce the invasiveness of certain procedures. Fourth, defining and capturing change across two images is crucial for understanding trends. Fifth, scalability to potential petabyte and exabyte-sized data sets is essential.

Developing a reference frame for the human body entails defining a coordinate system to facilitate looking across snapshots. Rigid structures in the body such as bone landmarks provide important clues as to the current spatial location in relation to soft tissues. This has been used in Stereotactic surgery to locate small targets in the body for some action such as ablation, biopsy or injection [10, 8]. Identifying nodes (e.g., the start and end of branches), edges (e.g., vessels linking nodes), and locations on branches (e.g., using distance from end-nodes) might be useful for pin-pointing vessel blockage and tracking changes over time. For spatial networks in the body, using identified nodes and edges is equivalent to the use of street addresses by the US Post Office. However, the resolution of this coordinate system is important in automatically aligning certain structures in the body across snapshots; it may be difficult to accomplish this if the coordinate system's resolution is too coarse.

A related challenge is location determination. Although the reference frame might be useful in defining a coordinate system, location determination is needed to pinpoint specific coordinates in the body. Analogies include using global positioning systems to determine one's location on the earth or taking snapshots of a street network so that traffic on a certain street can be monitored across different times. In medicine, the challenge lies in aligning each structure in the body across multiple 3-D images so that it can be guaranteed that the same structure is being observed.

If we know our location in the body, it becomes possible to answer routing questions. Routing based on the body's spatial network over time is a difficult task given that the space is continuous. Defining "roads" in the human body for the purposes of routing is an interesting challenge. Analogous structures to roads are blood vessels and anatomic divisions of organs. Using these structures to discretize the space might be a useful way to attack the routing issue. An example of this problem is to find the shortest path to a brain tumor that minimizes tissue damage. What is unclear are corresponding definitions of shortest path weight [11] and paths for routing in the human body.

Once a patient's 3-D image is taken several different times, it is necessary to define change across the snapshots. An example of this is change detection in tumors where images at different times need to be observed. This is challenging because the images may be taken with different instruments across several visits and so calibration becomes a problem. Changes in the body's chemistry, for example, fasting status pre-PET Scan which alters physiologic distribution of the radiotracer, can impact imaging and thus should be accounted for. Hence defining changes across snapshots given a coarse resolution might be problematic.

Finally, we need to be able to scale up to potentially petabyte and exabyte-sized data sets. Large amounts of data are produced from medical imaging techniques [4] and replicating this data across different snapshots makes long term analysis prohibitive. Compression techniques (lossy and lossless) have been used to enable fast retrieval of static 3-D data (i.e., a single snapshot) but they are not adequate for dynamic 3-D data with features like interactive zoom in and out across the time dimension. For example, each snapshot of a large image might be approximately 8 - 16 gigabytes [14]; when this is multiplied by number of visits, number of images/visit and number of patients, scale increases to exabytes.

Summary. In this paper, we presented our vision of a spatio-temporal framework capable of answering long term questions based on 3-D medical images and spatial networks in the human body taken across several snapshots. This framework could play an important role in improving health care quality by providing answers about the comparative effectiveness of interventions. It could also provide doctors with a means of more quickly diagnosing and characterizing disease progression. We articulated the challenges associated with realizing such a vision including defining a reference frame for the human body, location determination, routing in a continuous space, observing change across snapshots and scalability. We encourage the Data Mining community to explore the vision and challenges we have proposed and we welcome future collaboration.

Acknowledgment. This material is based upon work supported by the National Science Foundation under Grant No. 1029711, III-CXT IIS-0713214, IGERT DGE-0504195, CRI:IAD CNS-0708604, and USDOD under Grant No. HM1582-08-1-0017, HM1582-07-1-2035, and W9132V-09-C-0009.

References

1. Innerspace, Wikipedia, http://en.wikipedia.org/wiki/Innerspace
2. Fantastic Voyage Movie (1966), www.imdb.com/title/tt0060397
3. Google Body, http://bodybrowser.googlelabs.com
4. Medical Imaging, Wikipedia, http://en.wikipedia.org/wiki/Medical_imaging
5. President Obama's speech to American Medical Association, USA Today (June, 15, 2009), http://goo.gl/bXK8o
6. PACS, Wikipedia, http://goo.gl/Tk9Kl
7. Siemens, www.siemens.com
8. Stereotactic surgery, Wikipedia, http://goo.gl/dDa6C
9. Banvard, R.: The Visible Human Project® Image Data Set From Inception to Completion and Beyond. In: Proceedings CODATA (2002)
10. Chin, L., Regine, W.: Principles and practice of stereotactic radiosurgery. Springer, Heidelberg (2008)
11. Cormen, T.: Introduction to algorithms. The MIT press, Cambridge (2001)
12. Graham, S., et al.: Information technology research challenges for healthcare: From discovery to delivery. Computing Community Consortium (2010), http://goo.gl/LuKQH
13. Lekadir, K., et al.: An Inter-Landmark Approach to 4-D Shape Extraction and Interpretation: Application to Myocardial Motion Assessment in MRI. IEEE Transactions on Medical Imaging 30(1), 52–68 (2011)
14. Peng, H., et al.: V3D enables real-time 3D visualization and quantitative analysis of large-scale biological image data sets. Nature Biotechnology (2010)
15. Zhang, H., et al.: 4-D cardiac MR image analysis: Left and right ventricular morphology and function. IEEE Transactions on Medical Imaging 29(2), 350–364 (2010)

Smarter Water Management: A Challenge for Spatio-Temporal Network Databases

KwangSoo Yang[1], Shashi Shekhar[1],
Jing Dai[2], Sambit Sahu[2], and Milind Naphade[2]

[1] Department of Computer Science, University of Minnesota, MN 55455
{ksyang,shekhar}@cs.umn.edu
[2] IBM T.J. Watson Research Hawthorne, NY 10532
{jddai,sambits,naphade}@us.ibm.com

Abstract. Developing intelligent water resource management systems is necessary for a sustainable future. Many aspects of these systems are highly related to spatio-temporal (ST) databases, particularly spatio-temporal network databases (STNDB). However, this domain poses several challenges. In this paper we present our view of the important research issues, including the challenges of modeling of spatio-temporal networks (STN) and data access methods.

Keywords: Water Management System, Network Tomography, Spatio Temporal Network Database System, Lagrangian Reference Framework.

Vision. Water is one of our most important natural resources and water scarcity may be the most underestimated resource issue facing the world today. The United Nations Millennium project reported, "By 2025 about 3 billion people could face water scarcity due to the climate change, population growth, and increasing demand for water per capita" [1,12]. Water scarcity is not only an issue of enough water but also of access to safe water [9,13]. About 80 percent of diseases in the developing world are attributed to lack of access to safe drinking water and basic sanitation [1,12]. Developing intelligent water resource management systems is necessary to remedy the problem.

According to a recent report [14], developing infrastructure for water distribution is one of the main issues for a sustainable future because water is not evenly distributed in the world. One technological approach to remedy this problem is to implement fully integrated systems which allow monitoring, analyzing, controlling, and optimizing of all aspects of water flows.

As a result, IBM Smarter Planet emphasizes smarter water management for planning, developing, distributing, and managing optimal use of limited water resources under relevant policies and regulations [2,3]. IBM smarter water projects explore sensor networks and smart water meters, integrated with the water network, to gather and integrate spatio-temporal network (STN) datasets [4,5]. Spatio-temporal network databases (STNDB) will likely be a key component of smarter water management since effectiveness of decision depends on the quality of information (e.g., current, past, and future data about water availability,

D. Pfoser et al. (Eds.): SSTD 2011, LNCS 6849, pp. 471–474, 2011.
© Springer-Verlag Berlin Heidelberg 2011

quality, usage, distribution, and management). Such data are inherently spatial and temporal. For example, smart water meters have a geographic location (e.g., home address) and their readings are reported at a specific time instant [11,10]. In addition, water distribution networks may be modeled as spatial graphs with nodes (e.g., sinks and branching points) and edges (e.g., pipes and channels). Many common queries on such datasets are spatio-temporal as well. For example, data warehouse reading from a smart meter may be queried to identify hot moments (e.g. work-day morning) of water consumption and hot pipes (a.k.a. water mains) to identify pipes with highest consumer demand. In addition to ST pipe access analysis, planning and design of distribution networks needs to consider location, demand, leakage, pressure, pipe-size, pressure loss, fire fighting flow requirements, etc [6]. Particularly, location must include spatial factors, such as distance between source and faucet, elevation of water tower and home, depth of water pipe to prevent freeze damage, as well as topological factors, such as loop or branch network, and zone.

The key issue is the quality of the datasets used to fully understand, model, and predict water flows in the network. Access and analysis of STN datasets is one of most important parts of these systems. In this paper, we mainly focus on STNDB, which is applicable for a variety of water network applications. In the following section, we discuss research challenges for STN.

Challenges. Storing and accessing STN datasets poses a number of challenges. First, STN datasets are imperfect due to errors of measurement and losses of data. Inference approach, therefore, is used to extrapolate the missing data. Consider, for example, a nuclear power plant and its water cooling system. The water cooling system delivers fresh water to cool the reactor and stabilize it. To cope with a malfunction such as Japan's nuclear crisis, the system should be monitored, upgraded, and tested based on various scenarios. However, the internal structure and status of the system may not be directly observable due to radioactive emissions from the damaged nuclear reactors. Network tomography is one solution to understand the internal characteristic of the network using end-to-end measurements, without needing the cooperation of internal nodes [15]. This approach is relatively easy and scalable, but inherently suffers from accuracy problem. General network tomography assumes that all links are independent although some links are correlated. Furthermore, network tomography simplifies the problem to linear-programming or convex optimization. These limitations lead to the need for novel network tomography techniques to facilitate rich analysis and use of STN datasets to further understanding of STN phenomena.

The second challenge concerns the complexity of STN datasets that arises from spatial, temporal, and network connectivity properties. To keep track of changes through time series, temporal attributes are associated with both geometric and topological data. As the number of dimensions increases, the size of the datasets grows massively while the density of the datasets becomes sparse. This situation creates other challenges to collect and analyze the activities in the network space. Consider, for example, a leakage problem. In this case, a water supply network transmits water using pipelines and water pressurizing components. Breaks occur frequently in these pipelines, causing a large amount of water leakage, reduced pressure, and service disruption. These leaks accelerate the deterioration of cracked pipelines, resulting in corroding of neighboring pipes and cascading effect. The problem is the detection of flow anomalies to prevent the loss of leakage. A general approach for identifying the water leaks is to measure subsurface moisture conditions or to detect changes in soil temperature. A sonic leak-detection technique, which identifies the sound of water escaping a pipe, can be used in small leaks. However, these methods do not focus on the network distribution and indirectly estimate the source of leaks. By associating with sensors which measure pressures and flow rates inside pipelines along with time series, the detection of unusual network flows help to pinpoint exact location of these leakages. The challenge here is that the detection algorithm would be intractable due to the size of the STN datasets. Furthermore, the complexity of the STN datasets will be increased by additional network constraints, such as direction, capacity, pressure, and flow rate.

Third, access of STN datasets requires a Lagrangian frame of reference which coordinates STN datasets with STN connectivity [8]. For instance, when describing a moving fluid, the motion of a particle is represented by space and time. To retrieve the path of the movement, an access operation should trace and follow the particle along STN connectivity. Consider a water quality specialist monitoring flow anomalies. If he retrieves information at a stationary point, he may miss the change due to ignoring of spatial variance. Tracking the flow of water through Lagrangian paths could show the overall flow status. The key challenge is that the Lagrangian frame of reference requires new data types, storage models, and query operations to efficiently store and query STN datasets.

Finally, STNDB lack general frameworks to analyze STN datasets. A mathematical approach uses differential equations to represent network-based phenomena (e.g., fluid flow). However, in many applications the underlying network model is unknown or too complex to be mathematically described. Statistics approaches use a statistical hypothesis and spatial framework to find interesting patterns. The key concept of spatial statistical analysis is that spatial data are highly self correlated; Datasets are gathered and analyzed using the notion of spatial relationships and similarity measures. In a STN model, relationships are defined by connectivity, centrality, and shortest-path. The challenge of designing analytical frameworks for STN is that the complexity would be increased by spatial correlation and network connectivity as well as flow constraints. In addition, temporal attributes change these relationships over time, which leads to the need for even more complex frameworks. Consider, for example, Nokia water supply contamination; a cross-connection between clean and sewage pipeline caused a massive contamination of drinking water distribution network and an epidemic with thousands of cases of diarrhea and vomiting [7]. This contamination is fundamentally occurred by misunderstanding of flow direction and water pressure. The STN model will help to find the origin of outbreak and ban all use of water. Also, the pipelines will be cleaned according to the STN connectivity. Therefore network connectivity and time-varying properties should be necessarily coupled together and modeled as interdependent networks. As an extension of this issue, multi-modal networks lead to other challenges. Multimodal networks are constructed from multiple features or environments. In order to provide real world network models, the integration of heterogeneous networks is necessary (e.g., network of clean and sewage water, and water distribution network and power grid). However, current STN models are limited in their ability.

Conclusion. In this paper we focused on spatio-temporal networks (STN), which are applicable for water network applications. We discussed the research issues that need to be addressed in order to retrieve useful information from STN datasets. We believe that the need of STN storage and access methods have been widely acknowledged through various water network models. The demand for STN increasingly impacts the societal and environmental applications. Therefore, new paradigms will be needed to meet the challenges posed by research in these areas.

Acknowledgement. This work was supported by NSF grant (grant number NSF III-CXT IIS-0713214) and USDOD grant (grant number HM1582-08-1-0017 and HM1582-07-1-2035).

References

1. Water: How can everyone have sufficient clean water without conflict?(2010), United Nations, http://goo.gl/pdsr0 (retrieved March 17, 2011)
2. Smarter planet, Wikipedia, http://goo.gl/ay5W8 (retrieved March 17, 2011)
3. Water Management, Wikipedia, http://goo.gl/aNllg (retrieved March 17, 2011)
4. Let's Build a Smarter Planet: Smarter Water Management, Dr. Cameron Brooks, IBM (September 22, 2010), http://goo.gl/XvCIB
5. IBM Smarter Water Keynote, IBM (2010), http://goo.gl/4zRMw
6. Water supply network, Wikipedia, http://goo.gl/cF4IG (retrieved March 17, 2011)
7. Nokia water supply contamination, Wikipedia, http://goo.gl/fDfVx (retrieved May 30, 2011)
8. Batchelor, G.K.: An introduction to fluid dynamics. Cambridge Univ. Pr., Cambridge (2000)
9. Chartres, C., Varma, S.: Out of Water: From Abundance to Scarcity and How to Solve the World's Water Problems. Ft Pr (2010)
10. Chen, F., et al.: Activity analysis based on low sample rate smart meters. In: Proceedings of the 16th ACM SIGKDD International Conference on Knowledge Discovery and Data Mining. ACM, New York (to appear, 2011)
11. Dai, J., Chen, F., Sahu, S., Naphade, M.: Regional behavior change detection via local spatial scan. In: Proceedings of the 18th SIGSPATIAL International Conference on Advances in Geographic Information Systems, pp. 490–493. ACM, New York (2010)
12. Marien, M.: Jc glenn, tj gordon and e. florescu, 2010 state of the future, the millennium project, washington (2010), http://www.stateofthefuture.org, Futures
13. Molden, D.: Water for food, water for life: a comprehensive assessment of water management in agriculture. Earthscan/James & James (2007)
14. Perry, W.: Grand challenges for engineering. Engineering (2008)
15. Vardi, Y.: Network Tomography: Estimating Source-Destination Traffic Intensities from Link Data. Journal of the American Statistical Association 91(433) (1996)

FlexTrack: A System for Querying Flexible Patterns in Trajectory Databases

Marcos R. Vieira[1], Petko Bakalov[2], and Vassilis J. Tsotras[1]

[1] UC Riverside
[2] ESRI
{mvieira,tsotras}@cs.ucr.edu, pbakalov@esri.com

Abstract. We describe the *FlexTrack* system for querying trajectories using *flexible pattern queries*. Such queries are composed of a sequence of simple spatio-temporal predicates, e.g., range and nearest-neighbors, as well as complex motion pattern predicates, e.g., predicates that contain *variables* and constraints. Users can interactively select spatio-temporal predicates to construct such pattern queries using a hierarchy of regions that partition the spatial domain. Several different query processing algorithms are currently implemented and available in the *FlexTrack* system.

1 Introduction

In this paper we describe *FlexTrack*, a system that allows users to query, in a very intuitive way, trajectory databases using *flexible patterns* [1,2]. A flexible pattern query (or pattern query for short) is specified over a fixed set of areas that partition the spatial domain and is defined as a combination of predicates that allow the end user to focus on specific parts of the trajectories that are of interest. For example, the pattern query "Find all trajectories that first were in downtown LA, later passed by Santa Monica, and then were closest to LAX" provides a mixture of range and Nearest-Neighbor (NN) predicates that have to be satisfied in the specific order. Essentially, flexible patterns cover that part of the query spectrum between the single predicate spatio-temporal queries, such as the range predicate that covers certain time instances of the trajectory life (e.g. "Find all trajectories that passed by area A at 11pm"), and similarity/clustering based ones, such as extracting similar movement patterns and periodicities from a trajectory archive that cover the whole lifespan of the trajectory (e.g. "Find all trajectories that are similar to a given query trajectory according to some similarity measure").

In order to provide more expressive power, flexible pattern queries can also include *variables* as predicates. An example of a query with a variable is "Find all taxi cabs that visited the same city district twice in the last 1 hour". Here the area of interest is not known in advance but it is specified by its properties (visited twice in the last 1 hour). We term these variable-enabled pattern queries as "flexible" as they provide a powerful way to query trajectories. Both the fixed and variable spatial predicates can express explicit temporal constraints

D. Pfoser et al. (Eds.): SSTD 2011, LNCS 6849, pp. 475–480, 2011.
© Springer-Verlag Berlin Heidelberg 2011

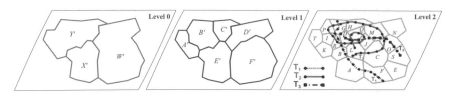

Fig. 1. Example of a set of regions defined using a hierarchy of 3 levels

(e.g., "between 10am and 11am") and/or implicit temporal ordering between them ("anytime later"). Flexible predicate queries can also include "numerical" constraints (NN and their variants) to provide "best fit" capabilities to the query language. Using this general and powerful querying framework, the user can "focus" the search only on the portions/events in a trajectory's lifetime that are of interest.

2   The Flexible Pattern Query Language

In this section we provide the definition of key elements in the *FlexTrack* system, as well as the description of the query language syntax.

A trajectory T_{id} is defined as a list of locations collected for a specific moving object over an ordered sequence of timestamps, and is stored as a sequence of w pairs $\{(ls_1, ts_1), \ldots (ls_w, ts_w)\}$, where $ls_i \in \mathbb{R}^d$ is the object location recorded at timestamp ts_i ($ts_{i-1} < ts_i$). In the *FlexTrack* system, the spatial domain is partitioned by a leveled hierarchy, where at each level l the spatial domain is divided by a fixed set Σ_l of non-overlapping regions, as shown in Figure 1. A region in level l is formed by the union of regions in the previous level $l - 1$. Regions correspond to areas of interest (e.g. *school districts*, *airports*) and form the alphabet $\Sigma = \bigcup_l \Sigma_l = \{A, B, C, ...\}$. Note the non-overlapping property between regions at a given level (e.g., W", X", Y" in level 0), while regions from different levels can overlap (e.g., regions W" in level 0 and F' in level 1).

In the *FlexTrack* query language, a spatio-temporal predicate $\mathcal{P}$ is defined by a triplet $\langle op, \mathcal{R}[, t] \rangle$, where $\mathcal{R}$ corresponds to a predefined spatial region in Σ or a *variable* in Γ ($\mathcal{R} \in \{\Sigma \cup \Gamma\}$), op describes the topological relationship (e.g. *meet*, *overlap*, *inside*) that the trajectory and the spatial region must satisfy over the (optional) time interval t ($t := (t_{from} : t_{to}) \mid t_s \mid t_r$). A predefined spatial region is explicitly specified by the user in the query predicate (e.g. "the convention center"). In contrast, a *variable*, e.g. "@x", denotes an arbitrary region using the symbols in $\Gamma = \{@a, @b, @c, ...\}$. Unless otherwise specified, a *variable* takes a single value (instance) from a given level Σ_l (e.g. @a=C), where the level l is specified in the query. Conceptually, *variables* work as placeholders for explicit spatial regions and can become instantiated (bound to a specific region) during the query evaluation.

Such spatio-temporal predicates $\mathcal{P}$ however cannot be used to specify distance based constraints (e.g., "best-fit" type of queries, like NN, that find trajectories which best match a specified pattern). This is because topological predicates involved are binary in nature and thus cannot capture distance based properties of the trajectories. To solve this problem we introduce the optional $\mathcal{D}$ part of a pattern query $\mathcal{Q}$ which allows us to describe distance-based or other constraints among the *variables* in $\mathcal{S}$ and the predefined regions (for more details, see [1]).

Having defined spatio-temporal predicates and the distance based constraints, we can now define a pattern query $\mathcal{Q} = (\mathcal{S} \, [\cup \, \mathcal{D}])$ as a combination of a sequential pattern $\mathcal{S}$ and (possibly) a set of constraints $\mathcal{D}$, where a trajectory matches $\mathcal{Q}$ if it satisfies both $\mathcal{S}$ and $\mathcal{D}$ parts. Here $\mathcal{S} := \mathcal{S}.\mathcal{S} \mid \mathcal{P} \mid !\mathcal{P} \mid \mathcal{P}^{\#} \mid ?^{+} \mid ?^{*}$ corresponds to a sequence of spatio-temporal predicates, while $\mathcal{D}$ represents a collection of distance functions (e.g. *NN*) and constraints (e.g. @x!=@y, @z={A,D,F}) that may contain regions defined in $\mathcal{S}$. The wild-card "?" is also considered a variable, however it refers to any region without occurring multiple times within a $\mathcal{S}$.

The use of the same set of *variables* in describing both the topological predicates and the numerical conditions provides a very powerful language to query trajectories. To describe a query in *FlexTrack*, the user can use fixed regions for the parts of the trajectory where the behavior should satisfy known (strict) requirements, and *variables* for those sections where the exact behavior is not known but can be described by *variables* and the constraints between them.

3   Pattern Query Evaluation

We continue with a description of the system architecture, its major components and evaluation algorithms.

In order to efficiently evaluate *flexible pattern queries*, the *FlexTrack* system employs two lightweight index structures in the form of ordered lists that are stored in addition to the raw trajectory data. There is one *region-list* (*R-list*) per region and one *trajectory-list* (*T-list*) per trajectory. The *R-list* $\mathcal{L}_{\mathcal{I}}$ of a given region $\mathcal{I} \in \Sigma$ acts as an inverted index that contains all trajectories that passed by region $\mathcal{I}$. Each entry in $\mathcal{L}_{\mathcal{I}}$ contains a trajectory identifier T_{id}, the time interval (*ts-entry*:*ts-exit*] during which the trajectory was inside $\mathcal{I}$, and a pointer to the *T-list* of T_{id}. Entries in a *R-list* are ordered first by T_{id} and then by *ts-entry*.

The only requirement for the region partitioning is that regions should be non-overlapping. In practice, there may be a difference between the regions presented to the end user as Σ and what is used internally for space partitioning. In the *FlexTrack* system we use a uniform grid to partition the space and we overestimate the regions in Σ by approximating each one of them with the smallest collection of grid cells that completely encloses the region. Because of the overestimation, false positives may be generated from regions that do not completely fit the set of covering grid cells. They, however, can be removed with a verification step using the original trajectory data.

In order to fast prune trajectories that do not satisfy $\mathcal{S}$, the *FlexTrack* system uses the *T-list*, where each trajectory is approximated by the sequence of regions it visited in each level of the partitioning space. A record in the *T-list* of T_{id} contains the region and the time interval (*ts-entry*:*ts-exit*] during which this region

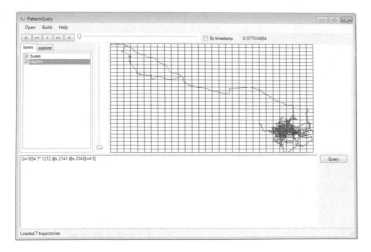

Fig. 2. The main interface of the *FlexTrack* System

was visited by T_{id}, ordered by *ts-entry*. In addition, entries in *T-list* maintain pointers to the *ts-entry* part in the original trajectory data. Given those index structures available, we propose four different strategies for evaluating flexible pattern queries (for the details on how $\mathcal{D}$ is evaluated, see [1]):

1. ***Index Join Pattern*** (*IJP*): this method is based on a merge join operation performed over the *R-lists* for every fixed predicate in $\mathcal{S}$. The *IJP* uses the *R-lists* for pruning and the *T-lists* for the *variable* binding;
2. ***Dynamic Programming Pattern*** (*DPP*): this method performs a subsequence matching between every predicate in $\mathcal{S}$ (including *variables*) and the trajectory approximations stored as the *T-lists*. The *DPP* uses mainly the *T-lists* for the subsequence matching and performs an intersection-based filtering with the *R-lists* to find candidate trajectories based on the fixed predicates in $\mathcal{S}$;
3. ***Extended-KMP*** (*E-KMP*): this method is similar to *DPP*, but uses the Knuth-Morris-Pratt algorithm [3] to find subsequence matches between the trajectory representations and the query pattern;
4. ***Extended-NFA*** (*E-NFA*): this is an NFA-based approach to deal with all predicates of our proposed language. This method also performs an intersection-based pruning on the *R-lists* to fast prune trajectories that do not satisfy the fixed spatial predicates in S.

4   Demonstration

For our demonstration we will use the *Trucks* and *Buses* datasets that contain moving object trajectories collected from the greater metropolitan area of Athens, Greece (www.rtreeportal.org). The *Trucks* dataset contains 112,203 locations generated from 276 moving objects. The *Buses* dataset has 66,096 locations from 145 moving objects. For the purposes of the demonstration we partition the spatial domain into regions using uniform grid with three levels. The granularity at levels 0, 1 and 2 is, respectively, 100×100, 50×50 and 25×25.

The first step in the query evaluation is to load the trajectory dataset from secondary storage. The next step is to create the index structures (*R-list* and *T-list*) used by our evaluation algorithms. During this process the users can tune several parameters (e.g. grid size, number of levels) for optimal performance. Using the system main interface, shown in Figure 2, users can visualize the trajectories in the spatial domain for a particular time interval. This property allows users to inspect, navigating in space and time, which regions have high concentration of trajectories. The system also has the property to "replay" the movement of the trajectories timestamp-by-timestamp.

After the data is loaded and the index structures are created, the user can create pattern queries using the Σ alphabet. The user can zoom in/out to select a lower/higher level of interest in the hierarchy. This allows the user to form a query with mixed size predicates where more detailed, lower level regions correspond to areas of particular interest, and less detailed, higher level regions are used otherwise. The user can also select variables or distance-based constraints at any level of the hierarchy. In addition to that, the user can create predicates that contain a set of regions or is defined by a maximum bounding rectangle (i.e. range predicate).

After the user's query Q is composed using the GUI it is then translated into the system's internal representation, as described in Section 2, and passed to the query engine. The pattern query is then evaluated using one of the four query evaluation algorithms available in the *FlexTrack* system (*IJP*, *DPP*, *E-KMP* or *E-NFA*). The trajectories in the result set are then plotted on the visualization canvas. Users can then zoom in/out and select parts of the trajectories by specifying the time interval of interest. The system also allows users to "replay" the movement of all the trajectories in the result set. Upon request, the system can provide textual description of trajectories using the regions in Σ.

5   Conclusion

This paper describes the *FlexTrack* system, which allows users to intuitively query trajectory databases by specifying complex motion pattern queries. Using the system GUI, users can easily construct those pattern queries that are further translated into a regular expression-like representation, which is then evaluated by the query evaluation module. Because of its expressive power, fast performance and intuitive user interface, the system can be of great help for users that work with large spatio-temporal archives.

Acknowledgements. This research was partially supported by NSF IIS grants 0705916, 0803410 and 0910859. Vieira's work was funded by a CAPES/Fulbright Ph.D fellowship.

References

1. Vieira, M.R., Bakalov, P., Tsotras, V.J.: Querying trajectories using flexible pat
   terns. In: EDBT, pp. 406–417 (2010)
2. Vieira, M.R., Martínez, E.F., Bakalov, P., Martínez, V.F., Tsotras, V.J.: Querying
   spatio-temporal patterns in mobile phone-call databases. In: MDM, pp. 239–248
   (2010)
3. Knuth, D., Morris, J., Pratt, V.: Fast pattern matching in strings. SIAM J. on
   Computing (1977)

A System for Discovering Regions of Interest from Trajectory Data

Muhammad Reaz Uddin, Chinya Ravishankar, and Vassilis J. Tsotras

University of California, Riverside, CA, USA
{uddinm,ravi,tsotras}@cs.ucr.edu

Abstract. We show how to find regions of interest (ROIs) in trajectory databases. ROIs are regions where a large number of moving objects remain for at least a given time interval. Our implementation allows a user to quickly identify ROIs under different parametric definitions without scanning the whole database. We generalize ROIs to be regions of arbitrary shape of some predefined density. We also demonstrate that our methods give meaningful output.

Keywords: Spatio-temporal database, Trajectory, Region of Interest.

1 Introduction

The widespread use of GPS-enabled devices has enabled many applications that generate and maintain data in the form of *trajectories*. Novel applications allow users to manage, store, and share trajectories in the form of GPS logs, and find travel routes, interesting places, or other people interested in similar activities.

This demonstration is based on the paper [1], where we give a novel and more intuitive definition of ROIs and propose a framework for identifying them. Recent works on discovering ROIs from trajectory data [2] define ROI as an (x, y) average of the points of a subtrajectory in which the object moves less than a prespecified distance threshold δ and takes longer than a prespecified time threshold τ. If either δ or τ changes, the entire trajectory database must be re-scanned. In contrast, our work removes this important limitation.

It is more intuitive to define ROIs in terms of speed. If an object takes at least time τ to travel at most distance δ, it maintains an average speed no more than $\frac{\delta}{\tau}$ for at least time τ. In our framework, we actually use a speed *range* to define ROIs, as this leads to a more generic definition. Further, we introduce the notion of *trajectory density* to define ROIs. In summary, our ROI definition uses (1) a range of speed that an object maintains while in an ROI (2) a minimum duration of staying in an ROI area and (3) the density of objects in that area.

We build an index on object speeds to avoid scanning the whole database. Given a range or a particular speed, we first retrieve trajectory segments with that speed using this index. We then verify the minimum stay duration condition. Objects that fulfill the speed and duration condition are *candidate objects*. Finally, we identify dense regions of candidate objects.

D. Pfoser et al. (Eds.): SSTD 2011, LNCS 6849, pp. 481–485, 2011.

2   Defining Regions of Interest

Conceptually, an ROI is intended to be a region where moving objects pause or wait or move slow in order to complete activities that are difficult or impossible to carry out while in fast motion. Examples of ROIs are restaurants, museums, parks, places of work, and so on. Generally, individual trajectories display idiosyncrasies, so ROIs are best defined in terms of collective behaviors of a collection of trajectories. That is, a collection of trajectories is needed to identify a location as an ROI.

The duration of an object's stay in a location is important in filtering out spurious ROIs, e.g. busy road intersections. So we will require a *minimum stay duration* for objects at ROIs. Nevertheless, if an object spends a long time in a large spatial region, a city, say, then that large region should not be considered as an ROI either. Hence, we must also consider the geographic extent of the object's movement, that is, the maximum area within which an object remains (or the maximum distance traveled by an object) during the minimum stay duration. Finally, to capture the collective behavior we consider the density of candidate objects in such a region. We identify dense regions adapting the *point-wise dense region* approach of [3].

Definition 1. *A region R is a **region of interest** if every point $p \in R$ has an l-square neighborhood containing segments from at least N distinct trajectories with object speeds in the range $[s_1, s_2]$, and where each such object remains in R for at least time τ before leaving R. The parameters l, N, τ, s_1, s_2 are user-defined.*

3   Indexing Trajectory Segments by Speed

Typically, objects in an ROI will maintain very low (or zero) speed. Hence, if we can quickly retrieve and analyze low speed trajectory segments, we can reduce query costs significantly.

Let s_{max} and s_{min} be the maximum and minimum speeds specifiable in an ROI query. We partition the speed values into *index ranges* $\mathcal{R} = [s_{min}, s_1), [s_1, s_2), \ldots, [s_{n-1}, s_{max})$. These ranges can be of arbitrary length. We maintain one bucket for each index range, with bucket B_i holding trajectory segments with speed range $[s_i, s_{i+1})$.

We consider the segments of a trajectory sequentially, and compute speeds assuming linear motion between two successive timestamps. If a series of consecutive segments fall within the same speed range, we combine them into one subtrajectory, and insert it into the index as one entry. Thus each entry in an index bucket points to a subtrajectory all of whose segments fall into within the speed range of the bucket.

We assume trajectories are sorted according to TID, so that subtrajectories in the buckets are also sorted according to TID. Having TID sorted entries in the buckets allows to perform a merge join to reconstruct trajectories from these buckets. When new trajectories are added to the database the index can easily be updated using the above algorithm.

4 Finding Regions of Interest

We find ROIs in three steps. First, we retrieve the appropriate buckets from the index. In the second step, we collect subtrajectories spanning multiple buckets by performing a merge-join, and check the stay durations. In the third step, we find regions with line segment density N/l^2, where each of N segments has to be from different trajectories.

It is straightforward to retrieve the segments falling into a given speed range $[s_1, s_2)$ using the speed index. No further discussion is needed.

4.1 Step 2: Verifying the Duration Condition

In this step, we consider only the buckets obtained from the previous step. To verify the duration condition for each trajectory we must join subtrajectories with same TID from different buckets. Let the query speed range include buckets B_i and B_j, and let $S_i \in B_i$ and $S_j \in B_j$ be subtrajectories. Let the start and end timestamps for S_i and S_j be $[t_{i1}, t_{i2}]$ and $[t_{j1}, t_{j2}]$ respectively. If S_i and S_j have the same TID and $t_{i2} = t_{j1}$ or $t_{i1} = t_{j2}$, then S_i and S_j should be merged into a single subtrajectory. The object's stay duration is the interval between the first and the last timestamps of the merged subtrajectory. We discard all subtrajectories with stay duration less than τ after merge, since they do not fulfil the stay duration condition.

In addition to minimum stay duration, our implementation also supports other temporal conditions, such as time intervals and weekdays/weekends. For example, ROIs during any weekday with $\tau = 15$ to 30 minutes, carry different semantics than those found in the afternoon or evening of any weekend, with a few hours of stay duration.

4.2 Step 3: Finding Dense Regions

This step involves finding points p whose l^2-neighborhood contains at least N distinct trajectories. For our purpose we use the Pointwise Dense Region (PDR) method [3] which was originally presented for point objects. The work in [3] describes two variations: (1) an exact, and (2) an approximate method. We use both of them. [3] uses Chebyshev polynomials to approximate the density of 2D points. We take the middle point p_m of each trajectory segment, and update the Chebyshev coefficient for the l-square neighborhood of p_m. A trajectory segment is a straight line between two points of a trajectory recorded at consecutive timestamps.

5 Demonstration

We develop a user interface where a user can specify values of query parameters e.g., speed, stay duration, other temporal conditions etc. Users can also select their data or use datasets on which we test our implementation. We show the

Table 1. Description of real data set

Description	Time of collection	♯trajectories
GeoLife Data: Beijing, China [4].	Apr 2007 to Aug 2009	165
TaxiCab Data: San Francisco, USA [5].	2008-05-17 to 2008-06-10	536

ROIs found by our methods using Google Maps API. The user can change any parameter value leaving others same and see the change. For example after identifying all ROIs for a region user can select only weekends or weekdays ROIs and see the difference. Figure 1 shows the user interface of our system.

Table 1 provides the description of the real datasets that we use to test our implementation. Using a short stay duration (15 to 30 min) for GeoLife data, we found bus stops, railway and subway stations, the Tsinghua University canteen, etc. We then considered weekends and a longer stay duration (1.5 to 4 hr). This resulted in ROIs in (1) the Sanlitun area which houses many malls, bars and is a very popular place, (2) the Wenhua square which contains churches, theaters, and other entertainment places, and (3) Zhongguancun, referred to as 'China's Silicon Valley', having a lot of IT and electronics markets. Figure

Figure 2(a) shows all the ROIs found using the TaxiCab dataset. We further zoomed in to ROIs and found (b) The San Francisco international airport, (c) a car rental, (d) the main downtown, union square, (e) San Francisco Caltrain station (f) the yellow cab access road. We also found hotels e.g. Star Wood, Westin, Mariott, Radisson, Ramada Plaza, Regency hotel, etc. These were found for short stay duration of 10 minutes. When the stay duration was increased to 12 hours we found only yellow cab access road, while for $2-3$ hours of stay duration we also found the airport.

Figure 2(g) and (h) shows Sanlitun and Zhongguancun area respectively in Beijing. When considering lunch and dinner time we found places that contain

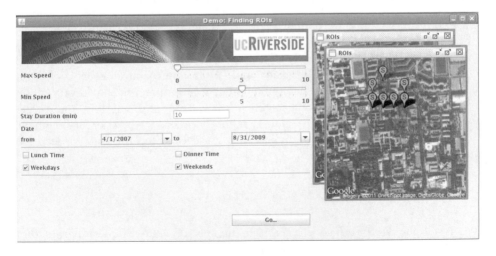

Fig. 1. The User Interface

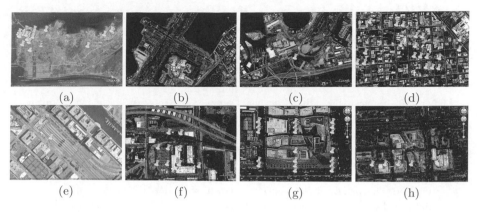

(a)   (b)   (c)   (d)

(e)   (f)   (g)   (h)

Fig. 2. ROIs identified for the TaxiCab and GeoLife data

many restaurants. Interestingly ROIs found at lunch time contain regions near the Microsoft China head quarters which are absent in dinner time ROIs. Finally, we identified ROIs on each individual day from April 2007 to August 2009. These resulted in (1) the Olympic media village, the Olympic sports center stadium during the Olympics 2008, (2) Peking University when the 'Regional Windows Core Workshop 2009 - Microsoft Research' was taking place in the PKU campus, (3) areas near the Great Wall in a weekend, (4) the Beijing botanical gardens, (5) the Celebrity International Grand Hotel, Beijing, etc.

References

1. Uddin, R., Ravishankar, C., Tsotras, V.J.: Finding regions of interest from trajectory data. In: MDM (to appear, 2011)
2. Cao, X., Cong, G., Jensen, C.S.: Mining significant semantic locations from gps trajectory. In: VLDB, pp. 1009–1020 (2010)
3. Ni, J., Ravishankar, C.V.: Pointwise-dense region queries in spatio-temporal databases. In: IEEE ICDE, pp. 1066–1075 (2007)
4. http://research.microsoft.com/en-us/projects/geolife/
5. http://crawdad.cs.dartmouth.edu

MARiO: Multi-Attribute Routing in Open Street Map

Franz Graf, Hans-Peter Kriegel, Matthias Renz, and Matthias Schubert

Institute for Informatics,
Ludwig-Maximilians-Universität München, Oettingenstr, 67, D-80538 Munich, Germany
{graf,kriegel,renz,schubert}@dbs.ifi.lmu.de

Abstract. In recent years, the Open Street Map (OSM) project collected a large repository of spatial network data containing a rich variety of information about traffic lights, road types, points of interest etc.. Formally, this network can be described as a multi-attribute graph, i.e. a graph considering multiple attributes when describing the traversal of an edge. In this demo, we present our framework for Multi-Attribute Routing in Open Street Map (MARiO). MARiO includes methods for preprocessing OSM data by deriving attribute information and integrating additional data from external sources. There are several routing algorithms already available and additional methods can be easily added by using a plugin mechanism. Since routing in a multi-attribute environment often results in large sets of potentially interesting routes, our graphical frontend allows various views to interactively explore query results.

1 Introduction

The Open Street Map (OSM)[1] project collects rich and up-to-date information about road networks and the landscape surrounding them. Combining this information with other publicly available information about the spatial landscape allows us to derive a large variety of information that previously has not been considered in routing systems. For example, a network might contain information about the distance, the speed limit, the altitude difference or the number of traffic lights for each road segment. Thus, a driver looking for the route which fits best to his personal preferences might want to consider various cost criteria at the same time. When employing ordinary shortest path routing, multiple attributes can be integrated by selecting a preference function combining cost criteria. For example, a user might enter that his major preference is driving the fastest path with a weight of 80%, but still wants to consider driving distance with a minor weight of 20%. By still considering travel distance with a minor weight, the selected route might be considerable shorter and only slightly slower than the fastest path. Thus, the gas consumption and the risk of getting into a congestion should be considerably smaller. However, finding an appropriate weighting is not intuitive and thus, better solutions should be found.

To conclude, considering multiple attributes has the potential to improve the usability of routing but raises a lot of further research questions requiring new problem specifications and solutions. First works in the area where proposed in [1] and [2]. While [1]

---

[1] http://www.openstreetmap.org

D. Pfoser et al. (Eds.): SSTD 2011, LNCS 6849, pp. 486–490, 2011.

ranks possible destinations w.r.t. to multiple cost attributes, [2] introduced route skyline queries. The result of a route skyline query consists of all routes connecting one starting point and one destination having an optimal cost value w.r.t. any linear combination of cost values.

In our demonstration, we want to present our framework for Multi-Attribute-Routing in OSM data (MARiO). MARiO is an open source project combining functionalities for data integration and preprocessing, implementation of new algorithms and performance evaluation. Our graphical frontend provides methods for posing queries and interactively exploring result routes. Since there is a number of queries computing multiple result routes, handling a result set of potentially hundreds of routes requires sophisticated tools. Thus, we integrated various interconnected views on the potentially multidimensional cost space and perform post processing in the form of clustering result routes.

The rest of this paper is organized as follows. In section 2, we provide an overview of the framework and its functionalities. Section 3 describes the already implemented algorithms. Afterwards, we sketch the content of the demonstration in section 4. Section 5 briefly summarizes the demonstrated system features.

2 System Overview and Functionalities

In this section, we want to give an overview of the functionalities of MARiO. We implemented our framework in Java 1.6 to be independent from a particular hardware platform.

A first functionality is importing map data from OSM. In order to apply multi-attribute routing, we cannot rely on the rich map representation provided by OSM. First of all, the OSM format contains a lot of unnecessary information for route computation. A second more important reason is that several of the employed optimization criteria are not directly maintained in the maps. For example, we have information about traffic lights and altitudes connected to the nodes which have to be reassigned and post processed into edge attributes of a multi-attribute graph. Furthermore, there is publicly available data from other sources than OSM that provide further useful information. Therefore, we allow to add topographic data from the SRTM[2] program. Another reason making preprocessing of the map information advisable is that available maps often contain a lot of nodes which are not required for routing purposes, e.g. nodes that are integrated to display turns in an edge. In order to allow efficient path computation, deleting these nodes and combining the neighboring edges can significantly reduce the number of considered routes.

After loading network data into an internal adjacency list representation, it is possible that additional preprocessing steps are required. An important functionality for many routing algorithms, e.g. A*-Search, is to compute an approximation for the minimal cost of a path between two nodes. A common approximation for the shortest path w.r.t. network distance is the Euclidian distance between the spatial coordinates of both nodes. However, the same idea is not applicable for general attributes. For example, the number of traffic lights on a route cannot be estimated based on distance. Therefore, we

---

[2] http://www2.jpl.nasa.gov/srtm/

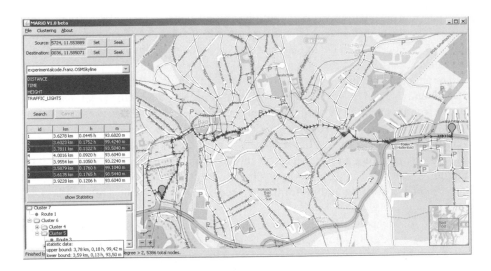

Fig. 1. Screen shot of the MARiO Frontend

implemented a reference node embedding storing at each node the distance to each of a well selected set of reference nodes. The advantage of this approach is that it is viable to arbitrary positive edge attributes. The drawback of the approach is the large memory consumption because it is necessary to store a distance value for each node, each reference node and each attribute type. To significantly lower the memory consumption of this method, we implemented a sparse variant of the embedding being proposed in [3].

To integrate various query types and compare algorithms solving the same problem, we designed our framework in a way allowing the fast and flexible integration of new algorithms. Therefore, new algorithms are integrated by employing a plugin mechanism. As a result, it is possible to add further query types or algorithms without altering the original code of the framework. After adding the algorithm the framework lists the algorithms in the frontend and automatically generates a dialog to select parameter values. The result is expected to be a list of result routes which can be displayed in the user interface. A further generalized feature of the framework is the possibility to analyze the performance of the algorithms. Therefore, is is possible to monitor and report general performance measures for route planning algorithms like query time, result size, the number of accessed network nodes or the number of extended routes.

A final component of the MARiO framework is its frontend which is displayed in figure 1. The frontend allows to display the OSM map data by using the map view component of SwingX-WS[3] which contains versatile viewing controls. Furthermore, the frontend allows the user to pose queries using various algorithms and provides multiple methods for displaying the result set. The first view on the result set consists of a grid control containing the cost w.r.t. each of the selected cost attributes. There exists further views visualizing the cost values for the case of two and three attributes. To handle the particularly large number of result routes that sometimes occur in

---

[3] http://swinglabs.org

multi-attribute routing, we can display the result in the form of a clustering tree. The clustering is derived by single link clustering which is based on a weighted variant of Hamming distance. Thus, the result is clustered w.r.t. the visited nodes instead of the cost attributes. The resulting clustering can be seen in the lower left corner of figure 1.

3   Implemented Algorithms

In the previous section, we described the general functionalities that can be used when implementing and testing a routing algorithm. In this section, we shortly review the already available algorithms. For basic shortest path computation based on a single cost attribute, the framework implements Dijkstra's algorithm and A*-search. The A*-search is based on the reference point embedding named above.

A second type of query being already implemented is a route skyline query. To calculate the route skyline for a given set of quality criteria, we employ the ARSC algorithm described in [2]. The basic idea of this algorithm is a best first traversal of the graph beginning with the starting position. During query procession the algorithm maintains two data structures. The first is a priority queue containing all nodes that still must be visited to find all skyline paths. The second structure consists of a table storing the already encountered pareto-optimal sub-routes for each visited node. Due to the monotonicity of local sub-routes, it can be shown that each sub route of a skyline route ending at the destination must be a skyline route between the starting location and its ending location. Thus, extending any path which is not part of the local skyline of its ending location cannot lead to a skyline route to the destination. To further speed up skyline computation, we additionally compare the lower bound approximation for any path to the current skyline of paths of the destination. If the lower bound approximation is already dominated by a member of the current skyline of the destination, the path can be pruned as well. The algorithms terminates when there is no path left that could be extended into a member of the route skyline to the destination node. For a more detailed description of the algorithm please refer to [2].

4   Demonstration

To demonstrate the functionalities of the MARiO framework, we will focus on query processing and result browsing in the frontend.

To pose a query, the user has to select an available query algorithm. Depending on this selection, the system can now generate a query dialog requesting the required input parameters from the users. For example, a route skyline query being processed by the ARSC algorithm requires a set of cost attributes, a starting point and a destination. The cost attributes are selected as a subset of the attributes being supported by the currently loaded graph. To select spatial locations the system allows to mark the coordinates directly on the map view. As an alternative, MARiO supports an address search to pinpoint locations. After parameter selection, the search is being started and the system collects the statistical information about query times, visited nodes and extended routes.

The result is a set of routes in the network which are characterized by a trajectory and a cost vector describing the cost of each of the selected attribute types. A basic view of this result set is a grid control containing a row for each result route and a column for each type of selected cost attribute. When clicking one or several routes in the control the corresponding route is marked in the map view. Furthermore, it is possible to sort the result set by any type of selected cost in the result set. A further view on the result data that is being made available for two attributes is a 2D vector view. For the route skyline query, this view always displays the well-known step function of a skyline. For 3D data, there exists a further view displaying the result set in a simplex control.

A final feature being extremely useful for rather large result sets is to view the result routes by browsing its cluster tree. The tree is displayed in a tree control and thus, a user can navigate deeper into the cluster by expanding the nodes. To get an impression of the contents of a cluster, it is possible to select a node in the tree and simultaneously display all contained routes in the map view. Furthermore, the tool tip of the node displays upper and lower bounds for each cost value of the clustered routes. For example, a cluster might be described by 4 routes having a travel time between 0.25 and 0.5 hours and a distance between 10 and 12 km. By clustering result routes w.r.t. the visited nodes in the graph, the routes within a cluster do not have to minimize the displayed intervals. However, the clusters display similar trajectories on the map view. Thus, top-level clusters distinguish rather general areas a trajectory is visiting while low-level clusters rather represent local variations. Thus, examining the top level can be employed to explore general directions and by traversing the tree the user can stepwise decide which route fits best to her particular preferences.

5   Conclusion

In this proposal, we introduced MARiO a framework for Multi-Attribute Routing in OSM data. Our framework, has three main functionalities. The first is data integration and preprocessing in order to construct multi-attribute graphs from OSM data. The second is the simple implementation and integration of new algorithm via a plugin mechanism. Finally, we provide a frontend for posing queries and exploring query results. Since the result set being generated by a multi-attribute routing algorithm can be rather large, there exists several interconnected views displaying result routes on the map, in the cost space or summarize the result with a clustering algorithms.

References

1. Mouratidis, K., Lin, Y., Yiu, M.: Preference queries in large multi-cost transportation networks. In: Proceedings of the 26th International Conference on Data Engineering (ICDE), Long Beach, CA, USA, pp. 533–544 (2010)
2. Kriegel, H.P., Schubert, M., Renz, M.: Route skyline queries: A multi-preference path planning approach. In: Proceedings of the 26th International Conference on Data Engineering (ICDE), Long Beach, CA, USA (2010)
3. Graf, F., Kriegel, H.P., Renz, M., Schubert, M.: Memory-efficient a*-search using sparse embeddings. In: Proc. ACM 17th International Workshop on Advances in Geographic Information Systems (ACM GIS), San Jose, CA, US (2010)

TiP: Analyzing Periodic Time Series Patterns

Thomas Bernecker, Hans-Peter Kriegel, Peer Kröger, and Matthias Renz

Institute for Informatics,
Ludwig-Maximilians-Universität München,
Oettingenstr, 67, 80538 München, Germany
{bernecker,kriegel,kroeger,renz}@dbs.ifi.lmu.de
http://www.dbs.ifi.lmu.de

Abstract. Time series of sensor databases and scientific time series often consist of periodic patterns. Examples can be found in environmental analysis, where repeated measurements of climatic attributes like temperature, humidity or barometric pressure are taken. Depending on season-specific meteorological influences, curves of consecutive days can be strongly related to each other, whereas days of different seasons show different characteristics. Analyzing such phenomena could be very valuable for many application domains. Convenient similarity models that support similarity queries and mining based on periodic patterns are realized in the framework *TiP*, which provides methods for the comparison of similarity query results based on different threshold-based feature extraction methods. In this demonstration, we present the visual and analytical methods of *TiP* of detecting and evaluating periodic patterns in time series using the example of environmental data.

1 Introduction

In a large range of application domains, e.g. environmental analysis, evolution of stock charts, research on medical behavior of organisms, or analysis and detection of motion activities, we are faced with time series data featuring activities which are composed of regularly repeating sequences of activity events. For that purpose, existing periodic patterns that repeatedly occur in specified periods over time have to be considered. Though consecutive motion patterns show similar characteristics, they are not equal. We can observe changes of significant importance in the shape of consecutive periodic patterns.

TiP utilizes the dual-domain representation of time series and the threshold-based approach [1] to extract periodic patterns from them. Beyond the interest of [2], the temporal location and the evolution of consecutive patterns are focused. For efficient similarity computation, relevant feature information is extracted so that data mining techniques can apply. By visualization, *TiP* provides first information about the existence and the location of periodic patterns in the time domains in the 3D space. Further knowledge about large datasets and the choice of adequate parameter settings for similarity queries and mining can be obtained by diverse analysis methods.

D. Pfoser et al. (Eds.): SSTD 2011, LNCS 6849, pp. 491–495, 2011.

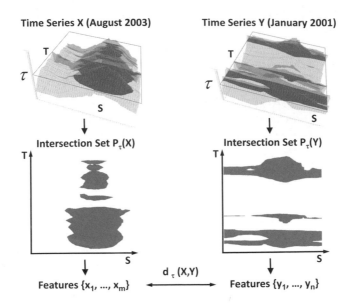

Fig. 1. Feature extraction and similarity computation on dual-domain time series

Overall, *TiP* serves as a framework to effectively and efficiently manage dual-domain time series. Furthermore, it provides several methods to process similarity queries on this type of time series based on user-defined features as well as an intuitive graphical user interface. Theoretical background of the applied techniques in *TiP* will be given in Section 2. Details of the system and the architecture will be explained in Section 3. Section 4 describes the planned demo tour in more detail.

2   Theoretical Background

Time Series Representation. A time series X can be split into a sequence of subsequences of fixed length by adding an additional time domain [1]. This yields a *dual-domain* representation having a 3D surface. A dual-domain time series represents the temporal behavior along two time axes. Consider an example of environmental research. The trend of temperature within one month of a year having one value for each hour of a day is then represented by the evolution of the temperature within one day (first time domain), and, additionally, over consecutive days of the entire time period (second time domain). Formally, a dual-domain time series is defined as

$$X_{dual} = \langle \langle x_{1,1}, \ldots, x_{1,N-1}, x_{1,N} \rangle, \ldots, \langle x_{M,1}, \ldots, x_{M,N-1}, x_{M,N} \rangle \rangle$$

where $x_{i,j}$ denotes the value of the time series at time slot i in the first (discrete) time domain $T = \{t_1, \ldots, t_N\}$ and at time slot j in the second (discrete) time domain $S = \{s_1, \ldots, s_M\}$.

Time Series Similarity. *TiP* compares dual-domain time series based on their periodic patterns that we call *intersection sets*. An intersection set $P_\tau(X)$ of a time series X is created by a set of polygons that are derived according to [2], where time series with one time domain are represented as a sequence of intervals w.r.t. to a threshold value τ. In this case, τ corresponds to a 2D plane which intersects the 3D time series X. The polygons of an intersection set represent the evolution of amplitude-level patterns as spatial objects, as they deliver all information about the periods of time during which the amplitudes of the time series exceed τ. In the temperature example, the polygons contain only values beyond a certain temperature level. Thus, the patterns emerge from temperature values that are greater than τ. However, patterns of summer and winter months are different in their occurrence and their dimensions, as depicted in Figure 1: the patterns of August 2003 are more concentrated within the days, but they are more constant over the whole month, whereas January 2001 contains patterns having a lower variance over consecutive days but that hardly change within the days. The degree of periodicity of a time series is reflected by the extent of the polygons, so even non-periodic patterns can be detected and addressed by their spatial location. The distance $d_\tau(X,Y)$ of two time series X and Y reflects the dissimilarity of the intersection sets $P_\tau(X)$ and $P_\tau(Y)$ for a certain τ. This step reduces the comparison of time series to comparing the sets of polygons such that spatial similarity computation methods can apply. To save computational cost while computing the distance as well as to allow for the usage of index structures like the R^*-*tree* [3], *TiP* derives local features (e.g. approximations or numerical values) from the polygons and global features (e.g. characteristics) from the intersection sets and computes the distance based on them (cf. Figure 1). The number of polygons and, thus, the number of local features varies for different intersection sets, so *TiP* employs the Sum of Minimal Distances (SMD) measure [4]. The distance based on global features is simply calculated as the L_p-distance between the associated features, as, for each intersection set, *TiP* derives the same amount of global features. For similarity calculation and further analysis, intersection sets for different threshold values can be pre-computed for a dataset and stored in an underlying database. Thus, relevant feature information of intersection sets can simply be reloaded for similarity queries. Furthermore, this threshold-based method is very efficient as, contrary to simple measures like the Euclidean distance, it does not need to consider the originally high-dimensional structure of time series.

3   Architecture

TiP has been implemented using Java and the Java3D API. The framework is able to load datasets of time series following the ARFF format[1] into the application. For each time series, the user can set several threshold planes to compute intersection sets w.r.t. these threshold values. The polygons can be approximated by simple conservative bounds like minimum bounding rectangles

---

[1] http://weka.wiki.sourceforge.net/ARFF

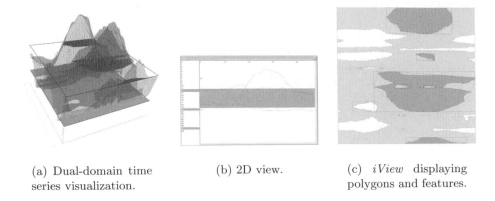

(a) Dual-domain time
series visualization.

(b) 2D view.

(c) *iView* displaying
polygons and features.

Fig. 2. Visual exploration of dual-domain time series with *TiP*

(MBRs) and thus be stored efficiently in the internal database by the support of
internal index structures. This also accelerates similarity queries, as multi-step
query processing can be applied. For a time series X, each intersection set $P_\tau(X)$
w.r.t. a specific value for the threshold τ does only have to be computed once.
Once stored in the internal database, the ID of $P_\tau(X)$ can simply be associated
with the ID of the time series X from which was derived when X is opened by *TiP*
for another time. The user is able to perform an extensive evaluation using several
techniques to mine databases of time series that are based on periodic patterns,
such as k-nearest-neighbor (kNN) queries with distance ranking, precision-recall
analysis and kNN classification.

TiP provides an intuitive graphical user interface. Exemplary elements are
depicted in Figure 2. The core elements include two Java3D applets displaying two time series simultaneously in their dual-domain representation (cf. Fig.
2(a)). As several user-defined threshold planes can be set for each time series,
TiP offers two additional applets called *iView* to depict the polygons and the
presentable features (i.e. approximations) of the corresponding intersection sets.
Additionally, the user is able to split a time series along each axis. Then each
subsequence of a dual-domain time series can be displayed in an external 2D
chart (cf. Fig. 2(b)).

4   Demo Tour

In this section, a short overview of the nature of the demonstration of *TiP* is
given, which will be performed applying the example of environmental research
where the time series are built of normalized temperature measurements. The
recorded and prepared data, labelled corresponding to the four seasons of a year,
was provided by the Environmental State Office of Bavaria, Germany[2].

---

[2] http://www.lfu.bayern.de/

In the demo, we first present how the dual-domain time series representation can discover a higher content of information compared to the original, sequential representation. For example, depending on the season, the temperature curves show different characteristics in their course, so representative patterns can already be derived visually from this 3D surface. When setting threshold planes arbitrarily, the user can materialize those patterns. By setting several threshold planes, the user can examine how to find suitable values for τ that should be used while performing similarity queries. In general, the higher a threshold is, the less is the number of amplitude values of a time series that exceed the threshold and, thus, contribute to a polygon of the intersection set. By defining a value or a range for τ, we show how to define the relevant time series values for the queries. A high τ value, e.g. $\tau = 80°$F, in the temperature example corresponds to the query: "Given the curve of a query month showing temperature values higher than 80°F on the first days at one certain time of day, please return all months that also show values higher than 80°F on their first days at this time of day.". If the query time series contains measurements of a summer month, it is likely to get summer months returned in the result set, as winter months show a different behavior, even if the amplitudes of the dataset are normalized. As an additional feature, the system supports the automatic detection of the value for τ that yields the best results without specifying any minimum query temperature level. For that purpose, a kNN classification is performed in order to find a τ value to obtain the best accuracy results. Exemplarily for the normalized temperature dataset, choosing $\tau = 0.6$ yields an accuracy value of 0.65, where by applying the Euclidean distance we get an accuracy of 0.56. To get an overview of the quality of the results that can be achieved with different values for τ, a parameter analysis shows the classification accuracies and average precision values for every possible τ within the amplitude range of the time series. Of course the result strongly depends on the selection of the features of the intersection sets. Therefore, we show that features can be combined to improve the expressiveness of the results.

References

1. Aßfalg, J., Bernecker, T., Kriegel, H.-P., Kröger, P., Renz, M.: Periodic pattern analysis in time series databases. In: Zhou, X., Yokota, H., Deng, K., Liu, Q. (eds.) DASFAA 2009. LNCS, vol. 5463, pp. 354–368. Springer, Heidelberg (2009)
2. Aßfalg, J., Kriegel, H.-P., Kröger, P., Kunath, P., Pryakhin, A., Renz, M.: Threshold similarity queries in large time series databases. In: Proceedings of the 22nd International Conference on Data Engineering, ICDE 2006, Atlanta, GA, USA, April 3-8, p. 149 (2006)
3. Beckmann, N., Kriegel, H.-P., Schneider, R., Seeger, B.: The R*-tree: An efficient and robust access method for points and rectangles. In: Proceedings of the 1990 ACM SIGMOD International Conference on Management of Data, Atlantic City, NJ, May 23-25, pp. 322–331 (1990)
4. Eiter, T., Mannila, H.: Distance measures for point sets and their computation. Acta Informatica 34(2), 109–133 (1997)

An Extensibility Approach for Spatio-temporal Stream Processing Using Microsoft StreamInsight

Jeremiah Miller[1], Miles Raymond[1], Josh Archer[1], Seid Adem[1], Leo Hansel[1],
Sushma Konda[1], Malik Luti[1], Yao Zhao[1], Ankur Teredesai[1], and Mohamed Ali[2]

[1] Institute Of Technology, University of Washington, Tacoma, WA, USA
`{jeremmi,reukiodo,archerjb,seidom,lhansel,skonda,`
`mluti,yaozerus,ankurt}@u.washington.edu`
[2] Microsoft StreamInsight, Microsoft Corporation, Redmond, WA, USA
`mali@microsoft.com`

Abstract. Integrating spatial operators in commercial data streaming engines has gained tremendous interest in recent years. Whether to support such operators natively or to enable the operator through an extensibility framework is a challenging and interesting debate. In this paper we leverage the Microsoft StreamInsight[TM] extensibility framework to support spatial operators enabling developers to integrate their domain expertise within the query execution pipeline.

We first justify our choice of adopting an extensibility approach over a native support approach. Then, we present an example set of spatiotemporal operations, e.g., KNN search, and range search; implemented as user defined operators using the extensibility framework within Microsoft StreamInsight. More interestingly, the demo showcases the how embedded devices and smartphones are shaping the future of streaming spatiotemporal applications. The demo scenario specifically features a smartphone based input adapter that provides a continuous stream of moving object locations as well a continuous stream of moving queries. To demonstrate the scalability of the implemented extensibility framework, the demo includes a simulator that generates a larger set of stationary/moving queries and streams of stationary/moving objects.

Keywords: Microsoft StreamInsight, extensibility, spatiotemporal data streaming, geostreaming.

1 Introduction

It has been a debate in the database community whether to support spatial operators natively or to provide an extensibility framework capable of integrating user defined operators across multiple domains within the query execution pipeline. This same debate has naturally evolved to include the data streams domain as more and more business scenarios demand spatiotemporal stream processing. In this demo, we explore the utility of the second approach (extensibility) to design a set of spatiotemporal streaming operators.

D. Pfoser et al. (Eds.): SSTD 2011, LNCS 6849, pp. 496–501, 2011.

1.1 Motivation

Spatial queries are becoming increasingly popular and are a necessary building block for a number of location-enabled applications (e.g. find coffee shops near me as I travel down the freeway, or alert the police cars nearest to a fleeing suspect.) Our approach utilizes a commercial data stream management system (DSMS), Microsoft StreamInsight, to handle streams of spatiotemporal location data. We leverage the extensibility features to update and query our spatial indexes. StreamInsight natively handles the details of the streaming data, so that we can focus our development efforts on the business logic implementation of real-time spatial queries.

1.2 The Case for Extensibility in a DSMS

Microsoft StreamInsight (StreamInsight, for brevity) is a commercial DSMS that adopts the semantics of the *Complex Event Detection and Response* (CEDR) project [1-2]. The underlying basis for processing long-running continuous queries in StreamInsight is a well-defined temporal stream model and operator algebra. It correlates stream data from multiple sources and executes standing queries on a low-latency query processor to extract meaningful patterns and trends [3-4].

Several business domains have explored the value that could be gained by using DSMSs to process real-time workloads. In previous work, StreamInsight has been used as a platform for web click analysis and online behavioral targeting [5], computational finance [6], and spatiotemporal query processing [7-9]. Because of the wide applicability of data streaming in many domains, and because of the domain expertise involved in each domain, the extensibility model provides a way to make DSMS capabilities more readily accessible to developers within their domain of expertise. For this demo, we leverage StreamInsight's extensibility model in the spatiotemporal domain. We discuss StreamInsight's extensibility model [10] in Section 2.3.

1.3 Applying StreamInsight to the Spatial Domain

Ali et al present two approaches (the extensibility approach and the native support approach) to enable spatiotemporal query processing in DSMSs and, more specifically, in StreamInsight [7]. Some have also investigated the extensibility approach [8-9] to extend StreamInsight with the capabilities of the SQL Server Spatial Library [11]. However, the SQL Server Spatial Library is tuned for non-streaming data and, hence, performance at real time remains an issue. Using the extensibility framework proposed by Ali et al [10], incremental streaming-oriented versions of spatial operators can be developed and integrated with the query execution pipeline. In particular, we utilize user defined operators (UDOs) in our demo.

This demo implements a set of K- nearest neighbor (KNN) search and range search operators. These operators receive a stream of location updates. Note that there are various flavors of the KNN search and range search problems. Typically, a fixed KNN search or range search can be posed against a stream of continuously moving objects (i.e. querying device is not moving, the query answers are derived from a stream of moving entities). Alternatively, the KNN search or the range search center can be moving while the objects are stationary – e.g. show me locations of coffee shops as I drive along on a freeway. More interestingly, both the objects and the search center can be moving; e.g. a user and the friends nearest to them at any given time in a public location such as a shopping mall. We support all these flavors in the demo.

2   StreamInsight Overview

This section summarizes the major features of StreamInsight and gives an overview of its developer's interface with an emphasis on the extensibility framework.

2.1   Capabilities of StreamInsight

Speculation and consistency levels: StreamInsight handles imperfections in data delivery and provides consistency guarantees on the resultant output. Such consistency guarantees place correctness measures on the output that has been generated so far, given that late and out-of-order stream events are still in transit.

Windowing Semantics: Windowing is achieved by dividing the time-axis into a set of possibly overlapping intervals, called windows. An event belongs to a window if and only if the event's lifetime overlaps with the window's interval (time span). The desired operation (e.g., sum or count) is applied over every window as time moves forward. The output of a window is the computation result of the operator over all events in that window, and has a lifetime that is equal to the window duration.

Scalability: Scalability is achieved by both stream partitioning and query partitioning. Stream partitioning clones a query into multiple queries (of the same operator tree) such that each query operates on a portion of the stream. Query partitioning divides a query into many sub queries, each deployed on an instance of StreamInsight.

Debugging: StreamInsight provides a graphical tool (the Event Flow Debugger) for the inspection of event flow in a query as a means of debugging and performing root cause analysis of problems. The Event Flow Debugger reports the per-query memory and CPU usage, latency, throughput, and other runtime statistics as well.

2.2   Developing a Streaming Solution with StreamInsight

To develop a streaming application using StreamInsight, a set of modules have to be written to interact with the system. These modules are classified as:

Input/output Adapters: Input data streams are fed to the streaming engine through the appropriate input adapters. Input adapters have the ability to interact with the stream source and to push the stream events to the streaming engine. The engine processes the queries issued by the user and streams the resultant output to the consumer through output adapters. For the demo, we wrote two input adapters – one that generates test data and one for real data streamed from smartphone apps. We also created an output adapter that provides the output data as a service consumed by our display applications on a PC or smartphone.

Declarative Queries in LINQ: Language Integrated Query (LINQ) [12] is the approach taken by StreamInsight to express continuous queries. The LINQ that invokes our spatial query, for example, might look like this:

```
var outputStream = inputStream.Scan(
  new RTreeUDO( queryType, xRange, yRange));
```

This scans the input stream with our R-tree UDO, and provides us with the output stream. The code inside our UDO will accept a set of input event to produce a set of output events. We define the payload of our input and output events to include geographic location, phone ID, and IP address.

User Defined Operators (UDOs): The extensibility framework enables domain experts to extend the system's functionality beyond relational algebra. Domain experts package their logic as libraries of UDOs that are invoked by the continuous query.

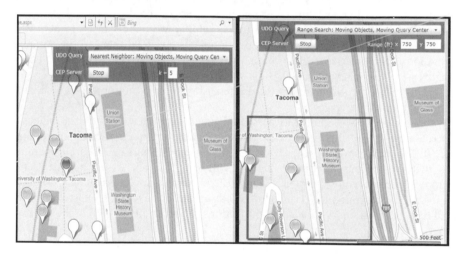

Fig. 1. Query output visualization of KNN search and range search query using the dashboard

StreamInsight's extensibility framework addresses two types of UDO developers. The first type is software developers who are not trained to think under the data streaming paradigm with its temporal attributes: for them there is the non-incremental model which provides a relational view of the world. The second type is developers of streaming applications where temporal attributes are first class citizens in their business logic. These developers seek maximum achievable performance through incremental query processing and may require full control over the temporal attributes of events as well. For them there is the incremental model, which provides the deltas or changes in the input to the UDO since the UDO's last invocation.

3 Demo Scenario

The demo scenario features a dashboard (Figure 1) that simultaneously visualizes a set of stationary objects (e.g., landmarks, coffee shops, shopping malls) and a set of moving objects (e.g., streamed from handheld devices and smartphones) using Bing Maps [13]. The dashboard is also used to compose and issue continuous queries against these objects. There are two versions of the dashboard: a PC-based dashboard and a smartphone-based dashboard. Note that a query that is issued using the smartphone based dashboard is assumed to have a continuously moving query center (as described in Section 1). The queries are processed using a data streaming engine and the query results are visualized using either version of the dashboard.

The demo utilizes Microsoft StreamInsight as the underlying data streaming engine. To interface with StreamInsight, a set of input/output adapters, a set of LINQ queries and a set of user defined operators (UDOs) are developed. We provide two types of input adapters. The first type of input adapters streams, at real time, GPS locations from handheld devices (e.g. smartphones) to the data streaming engine. The second type of input adapters simulates a larger set of moving objects/queries for the sake of demonstrating scalability. The simulator moves the device locations by random displacement vectors to simulate the variability we expect in real data. Moreover, it can replay historical logs of moving objects and feed them to the streaming engine to process queries over historical data. On the output side, the output adapter visualizes the query results using Bing Maps over the PC-based dashboard or sends the result back to be visualized at the dashboard of the moving object (say, a smartphone) that issued the query.

We implement R-tree and M-tree spatial index structures to track the objects as they roam the space. Our UDOs incrementally update and query these indexes at real time. In this demo, we present UDOs for KNN search and range search operations such that the various flavors of stationary/moving queries issued against stationary/moving objects are supported. Finally, LINQ queries are automatically composed and instantiated through the dashboard. These queries invoke the UDOs along with a set of relational operators (filters and projections) to declaratively define the requested output.

References

1. Barga, R.S., Goldstein, J., Ali, M., Hong, M.: Consistent Streaming Through Time: A Vision for Event Stream Processing. In: Proceedings of CIDR, pp. 412–422 (2007)
2. Goldstein, J., Hong, M., Ali, M., Barga, R.: Consistency Sensitive Streaming Operators in CEDR. Tech. Report, MSR-TR-2007-158, Microsoft Research (2007)
3. Chandramouli, B., Goldstein, J., Maier, D.: On-the-fly Progress Detection in Iterative Stream Queries. In: VLDB (2009)
4. Chandramouli, B., Goldstein, J., Maier, D.: High-Performance Dynamic Pattern Matching over Disordered Streams. In: VLDB (2010)
5. Ali, M., et al.: Microsoft CEP Server and Online Behavioral Targeting. In: VLDB (2009)
6. Chandramouli, B., Ali, M., Goldstein, J., Sezgin, B., Sethu, B.: Data Stream Management Systems for Computational Finance. IEEE Computer 43(12), 45–52 (2010)

7. Ali, M., Chandramouli, B., Raman, B.S., Katibah, E.: Spatio-Temporal Stream Processing in Microsoft StreamInsight. IEEE Data Eng. Bull. 33(2), 69–74 (2010)
8. Kazemitabar, J., Demiryurek, U., Ali, M., Akdogan, A., Shahabi, C.: Geospatial Stream Query Processing using Microsoft SQL Server StreamInsight. In: VLDB (2010)
9. Ali, M.H., Chandramouli, B., Raman, B.S., Katibah, E.: Real-time spatio-temporal analytics using Microsoft StreamInsight. In: ACM GIS (2010)
10. Ali, M., Chandramouli, B., Goldstein, J., Schindlauer, R.: The Extensibility Framework in Microsoft StreamInsight. In: ICDE (2011)
11. SQL Server Spatial Library, http://www.microsoft.com/sqlserver/2008/en/us/spatial-data.aspx (last accessed in March 2011)
12. Pialorsi, P., Russo, M.: Programming Microsoft LINQ. Microsoft Press, Redmond (May 2008)
13. Bing Maps, http://www.bing.com/maps

Efficient Spatio-temporal Sensor Data Loading
for a Sensor Web Browser

Chih-Yuan Huang, Rohana Rezel, and Steve Liang

Department of Geomatics Engineering, University of Calgary,
Calgary, Alberta, Canada
{huangcy,rdrezel,steve.liang}@ucalgary.ca

Abstract. We present a Sensor Web browser with an efficient spatio-temporal data loading mechanism as a client-side application in GeoCENS project. The same way the World Wide Web needs a web browser to load and display web pages, the World Wide Sensor Web needs a Sensor Web browser to access distributed and heterogeneous sensor networks. However, most existing Sensor Web browsers are just *mashups* of sensor locations and base maps that do not consider the scalability issues regarding transmitting large amount of sensor readings over the Internet. While caching is an effective solution to alleviate transmission latency and bandwidth problems, a method for efficiently loading spatio-temporal sensor data[1] from Sensor Web servers is currently missing. Therefore, we present LOST-Tree, a new spatio-temporal structure, intended to be the sensor data loading component on a Sensor Web browser. By applying LOST-Tree, redundant transmissions are avoided and consequently enables efficient loading with cached sensor data.

Keywords: sensor data management, spatio-temporal indexing, Sensor Web.

1 Introduction

The World-Wide Sensor Web [1] is increasingly attracting interests for a wide range of applications, including: large-scale monitoring of environment [2], roadways [3], etc. Thanks to the international sensor web standards (e.g., Open Geospatial Consortium (OGC)), it is possible to access heterogeneous sensor networks and their data with standard web service interfaces. The same way the World Wide Web needs a web browser to load and display web pages, the World-Wide Sensor Web needs a coherent frontend to access distributed and heterogeneous sensor networks. We call this kind of coherent frontend a *Sensor Web browser*. EarthScope[2], Sensorpedia[3] and SensorMap[4] are some example Sensor Web browsers.

---

[1] By sensor data, we mean the observations collected by sensors. They can be time series collected by stationary *in-situ* sensors, or a collection of single readings collected by transient sensors.
[2] http://www.earthscope.org/
[3] http://www.sensorpedia.com/
[4] http://atom.research.microsoft.com/sensewebv3/sensormap/

D. Pfoser et al. (Eds.): SSTD 2011, LNCS 6849, pp. 502–506, 2011.
© Springer-Verlag Berlin Heidelberg 2011

Nath et al. [4] discussed the challenges they encountered when building Sensor-Map. They highlighted transmitting large amount of sensor data efficiently over the network as one major challenge. They addressed this challenge by aggregating and sampling sensor data with COLR-Tree [5]. However, data uncertainty and quality degradation issues could occur with that approach. Therefore, we take a different approach in this research. One common solution for such challenge is to employ a caching mechanism that stores server responses on clients' local disks. The clients then only need to request for data in the case of cache miss. Take today's earth browser systems (e.g., Google Earth) as examples. They use a quadtree-based tiling scheme to store and manage the cached image tiles at different level of details. Before sending requests to servers, these systems check cache first. In the case of cache hit, no request needs to be sent. Otherwise (cache miss), the request will be sent, and the returned image tiles will then be inserted into the cache for future uses.

However, the same tiling and caching method cannot be directly applied to sensor data. There are two major reasons. First, sensor data is spatio-temporal in nature. Comparing to static map images, there is an additional temporal dimension to consider. Second, sensor data may be distributed sparsely in space and even more sparsely in time (e.g., transient sensors or sensors with very different sampling frequency). As a result, many spatio-temporal requests for sensor data of a particular phenomenon have server responses without any sensor data (empty hits). In order to prevent redundant empty hits, not only the responses need to be stored and managed in a cache, the requests also need to be stored and managed as a separate cache.

This work presents LOST-Tree, which stands for LOading Spatio-Temporal Tree. LOST-Tree uses predefined hierarchical spatial and temporal frameworks to manage requests. By applying LOST-Tree as a data loading management layer between a Sensor Web browser and servers, redundant requests can be prevented. Consequently, we can reduce server load, optimize data transmission, and save bandwidth.

2 LOST-Tree

The LOST-Tree manages a Sensor Web browser's requests. A typical request R from a Sensor Web browser to a Sensor Web server (e.g., an OGC Sensor Observation Service server [6]) can be defined by three parameters: R_{bbox}, R_{t_period}, and R_{obs}, where R_{bbox} is the minimum bounding box of the request's spatial extent, R_{t_period} is the request's temporal extent defined by a start time t_1 and an end time t_2, and finally R_{obs} is the observed phenomenon of interest (e.g., air temperature). In fact, we can further define the combination of R_{bbox} and R_{t_period} as a spatio-temporal cube R_{STCube}. Thus, a request R and its corresponding server response can be defined as follows: R (R_{STCube}, R_{obs}): $\{o_1, o_2, ..., o_i\}$, where o_i is an observation collected by a sensor that fulfill the request R.

LOST-Tree is a data loading management layer between a Sensor Web browser and servers. Its overall objective is to prevent sending unnecessary requests to Sensor Web servers. When our Sensor Web browser sends a request R through LOST-Tree, it consists four steps: (1) Decompose, (2) Filter, (3) Update, and (4) Aggregate. (Fig.1)

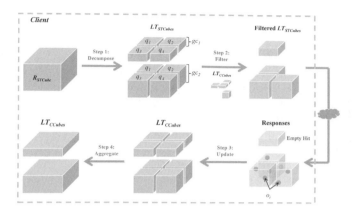

Fig. 1. Workflow of LOST-Tree

Decompose step: The purpose of the decompose step is to convert an *ad-hoc* R_{STCube} into one or many non-overlapping LOST-Tree-based requests: $LT_{STCubes}$. The decomposition is based on two predefined and hierarchical spatial and temporal frameworks. In this demo paper, we implement LOST-Tree with quadtree [7] as the spatial framework and with Gregorian calendar as the temporal framework. We use the LT_{STCube} *key*, that is the combination of a quadkey q and a calendar string gc (e.g., YYYY, YYYYMM, YYYYMMDD, YYYYMMDDHHMMSS) to represent a LT_{STCube}. A quadkey q represents a bounding box, and a calendar string gc represents a time period. One important characteristic in both quadkey q and calendar string gc is that they are hierarchical in nature. That means the length of q and gc represents its level-of-detail. That also means we can simply use a prefix matching method to identify whether $LT_{STCube\_A} \subseteq LT_{STCube\_B}$. Consequently we can easily and efficiently manage $LT_{STCubes}$ in a client's local cache by manipulating the LT_{STCube} *key*.

Filter step: The objective of the filter step is to filter out the requests that have been sent to servers previously. LT_{CCubes} are previously loaded $LT_{STCubes}$ found in the cache, and the filtering process is defined in Algorithm 1.

Update step: The aim of this step is to keep LT_{CCubes} up-to-date. Once the client receives responses from servers, the corresponding $LT_{STCubes}$ are inserted into LT_{CCubes}. What makes LOST-Tree algorithm new is that even if the server responses contain no sensor data (*i.e.*, empty hits), their corresponding $LT_{STCubes}$ will still be cached as LT_{CCubes}. In this way, LOST-Tree avoids sending repeated empty-hit requests.

Aggregate step: This step is intended to minimize the memory footprint of LT_{CCubes}. With the hierarchical characteristic of the spatial and temporal frameworks in LOST-Tree, when all *sub-LT_{CCubes}* (e.g., the eight small green cubes) of an LT_{CCube} (e.g., the two large green cubes) are loaded, we can replace all the *sub-LT_{CCubes}* with just one LT_{CCube}. Therefore, whenever the loaded $LT_{STCubes}$ are inserted in the update step, LOST-Tree first identifies aggregatable LT_{CCubes} and aggregates them into one LT_{CCube}. In this way, the smaller number of LT_{CCubes} allows better filtering performance in LOST-Tree. In addition, since LOST-Tree only maintains a quadkey and a calendar string for each LT_{CCube}, the tree size is small enough (<25Kbytes) to fit into memory for efficient filtering processing.

As can be seen from Fig. 2, which depicts the end-to-end latency with and without proposed scheme, LOST-Tree and local cache improves the system performance.

Algorithm 1. The filter function

Function	$Filter(LT_{STCubes}, LT_{CCubes})$: $FilteredLT_{STCubes}$

$FilteredLT_{STCubes} = \{\}$
1:    **FOREACH** $LT_{STCube} \in LT_{STCubes}$
2:        $previously\_loaded \leftarrow false$
3:        **FOREACH** $LT_{CCube} \in LT_{CCubes}$
4:            **IF** LT_{STCube} is contained by LT_{CCube} **THEN**
5:                $previously\_loaded \leftarrow true$
6:                **BREAK**
7:            **ELSE IF** LT_{STCube} contains LT_{CCube} **THEN**
8:                $LT_{STCube} \leftarrow LT_{STCube} - LT_{CCubes}$
9:            **END IF**
10:        **END FOREACH**
11:        **IF NOT** $previously\_loaded$ **THEN**
12:            $FilteredLT_{STCubes} \leftarrow FilteredLT_{STCubes} \cup$
13: LT_{STCube}
14:        **END IF**
15:    **END FOREACH**
16:    **RETURN** $FilteredLT_{STCubes}$

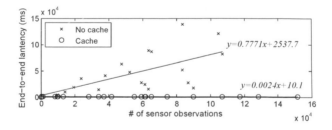

Fig. 2. End-to-end latency with and without proposed scheme

3 Demonstration

We will demonstrate the Sensor Web browser in the Geospatial Cyberinfrastructure for Environmental Sensing (GeoCENS) project (Fig. 3). GeoCENS Sensor Web browser is an OGC Sensor Web standard-based frontend, and users can browse sensor data from various sources within a coherent virtual globe geographical interface. We will first load sensor data through Internet then load the same data after restarting the Sensor Web browser. This will clearly show the contribution of this work. Then we will show other features that also benefit from the proposed scheme, such as point-to-surface interpolation.

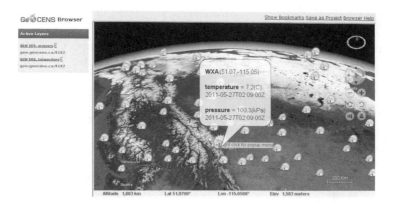

Fig. 3. The GeoCENS Sensor Web browser

References

[1] Liang, S.H.L., Croitoru, A., Tao, C.V.: A Distributed Geospatial Infrastructure for Sensor Web. Computers and Geosciences 31(2), 221–231 (2005)

[2] Hart, J.K., Martinez, K.: Environmental Sensor Networks: A Revolution in the Earth System Science? Earth Science Reviews 78, 177–191 (2006)

[3] Hsieh, T.T.: Using Sensor Networks for Highway and Traffic Applications. IEEE Potentials 23(2), 13–16 (2004)

[4] Nath, S., Liu, J., Zhao, F.: Challenges in Building a Portal for Sensors World-Wide. In First Workshop on World-Sensor-Web: Mobile Device Centric Sensory Networks and Applications, Boulder CO United States (2006)

[5] Ahmad, Y., Nath, S.: Colr-tree: Communication-efficient spatio-temporal indexing for a sensor data web portal. In: ICDE (2008)

[6] Open Geospatial Consortium: Sensor Observation Service (2007), http://www.opengeospatial.org/standards/sos (accessed March 18, 2011)

[7] Finkel, R., Bentley, J.L.: Quad Trees: A Data Structure for Retrieval on Composite Keys. ActaInformatica 4(1), 1–9 (1974)

A Visual Evaluation Framework for Spatial Pruning Methods

Tobias Emrich, Hans-Peter Kriegel, Peer Kröger,
Matthias Renz, Johannes Senner, and Andreas Züfle

Institute for Informatics, Ludwig-Maximilians-Universität München,
Oettingenstr, 67, D-80538 München, Germany
{emrich,kriegel,kroeger,renz,senner,zuefle}@dbs.ifi.lmu.de

Abstract. Over the past years, several pruning criteria for spatial objects have been proposed that are commonly used during the processing of similarity queries. Each of these criteria have different properties and pruning areas. This demo offers a visual interface for comparing existing pruning criteria under various settings and in different applications allowing an easy integration of new criteria. Thus, the proposed software helps to evaluate and understand the strengths and weaknesses of pruning criteria for arbitrary spatial similarity queries.

1 Introduction

During the processing of spatial similarity queries such as distance-range (a.k.a. ε-range) queries, k-nearest neighbor (NN) queries, reverse-kNN queries, etc., an important efficiency aspect is early pruning of objects which can not be part of the result set. Typically, in an early stage of the processing, the objects are approximated by suitable representations like minimum bounding rectangles (MBRs) or bounding spheres (BS). These representations are used in various applications in the spatial and/or temporal domain. In addition, using a feature-based similarity model, any type of objects (e.g. text documents) can be transferred into feature vectors (e.g. using term frequency) ending up in spatial objects.

Approximations like MBRs and BSs may be used in different contexts. For example, if the data objects have a complex spatial representation (e.g. polygons or probability density functions), MBRs serve as an approximation of the objects. During query execution costly distance functions can be avoided by using lower/upper bound distances on the approximations. Furthermore, spatial index structures use MBRs or BSs to approximate the area covered by the children of a node of the index structure.

Effective pruning criteria are among the most important components for efficiently answering spatial similarity queries. For decades, the minimum and maximum distance [1] between the approximations have been used to decide about pruning. Recently it has been shown that theses metrics can be improved under various settings (cf. [2,3,4]). In this work we describe and implement a system which visualizes the impact of several pruning criteria in different applications under various settings. The system should be helpful for giving insights on different pruning strategies under user specified settings.

D. Pfoser et al. (Eds.): SSTD 2011, LNCS 6849, pp. 507–511, 2011.

2   Domination Criteria

Depending on the underlying application and the type of query to be answered, pruning is performed differently. Thus, we will rely on the generalized concept of *spatial domination* introduced in [3] that can be used for any of the mentioned types of spatial similarity queries. Given the approximation A^*, B^*, R^* of three objects (A, B, R) then A^* dominates B^* w.r.t. R^*, written as $Dom(A^*, B^*, R^*)$, if

$$Dom(A^*, B^*, R^*) \Leftrightarrow \forall a \in A^*, b \in B^*, r \in R^* : dist(a, r) < dist(b, r),$$

where a, b and r are points within the corresponding approximations (and not necessarily within the objects) and $dist$ is a distance function defined on points (e.g. the L_p-norm).

For a kNN query with query object Q, we can prune an object O_i if we find k objects $O_j \neq O_i$ for which $Dom(O_j, O_i, Q)$ holds. For a RkNN query with query object Q, we can prune an object O_i if we find k objects $O_j \neq O_i$ for which $Dom(O_j, Q, O_i)$ holds.

In the following, we will review three pruning techniques and explain how to derive the domination concept using them. Figure 1 illustrates these techniques.

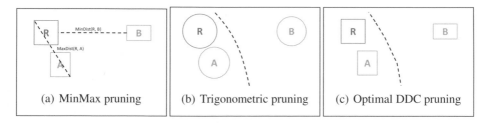

(a) MinMax pruning   (b) Trigonometric pruning   (c) Optimal DDC pruning

Fig. 1. Pruning Criteria for NN-queries

2.1   MinMax Domination

In [1] the minimum (MinDist) and the maximum (MaxDist) distance between a pair of objects is used for pruning in the context of kNN queries and MBR-based approximations (see Figure 1(a) for illustration). It can easily be used for other query types and spherical approximations (BSs), too. In relation to MBRs the minimum/maximum distance between two MBRs $X^\square$ and $Y^\square$ is the minimal/maximal distance that two points $x \in X^\square, y \in Y^\square$ can possibly have. Domination can be derived as follows:

$$Dom_{MinMax}(A^\square, B^\square, R^\square) \Leftarrow MaxDist(A^\square, R^\square) < MinDist(B^\square, R^\square)$$

Although the MinMax criterion is correct and efficiently computable it is not optimal (by means of pruning always the maximum amount of objects).

2.2   Trigonometric Pruning

Trigonometric pruning has been developed independently for all-nearest-neighbour queries [2] and for probabilistic reverse nearest neighbour queries [4] assuming spherical BS approximations. The hyperbola $H_{X,Y}$ between two approximating spheres $X°$ and $Y°$ which consists of all points e for which it holds that $MaxDist(X°, e) = MinDist(Y°, e)$ is used for pruning (see Figure 1(b) for illustration). Trigonometric pruning extends MinMax pruning since it is optimal in case of spherical approximations. However, it works for euclidean distance only and it cannot be applied to MBR-based approximations. Domination is defined as follows:

$$Dom_{Trig}(A°, B°, R°) \Leftarrow R° \text{ is completely on the same side as } A° \text{ of } H_{A,B}$$

The check involves complex trigonometric computations (see [4] for more details).

2.3   Optimal DDC

In [3] an optimal pruning criterion based on MBRs was developed. It also utilizes the geometric structure which consists of all points e for which it holds that $MaxDist(A, e) = MinDist(B, e)$ (cf. Figure 1(c)). Instead of materializing this complex structure (even in the two dimensional case), the criterion efficiently decides on which side of the structure a third object is.

$$Dom_{Opt}(A^{\square}, B^{\square}, R^{\square}) \Leftarrow \sum_{i=1}^{d} \max_{r_i \in \{R_i^{min}, R_i^{max}\}} (MaxDist(A_i, r_i)^2 - MinDist(B_i, r_i)^2) < 0$$

where X_i ($X \in \{A^{\square}, B^{\square}, R^{\square}\}$) denotes the projection interval of the rectangular region of X on the ith dimension, X_i^{min} (X_i^{max}) denotes the lower (upper) bound of the interval X_i, and $MaxDist(I, p)$ ($MinDist(I, p)$) denotes the maximal (minimal) distance between a one-dimensional interval I and a one-dimensional point p.

3   Visual Evaluation of Pruning Criteria

The choice of an adequate pruning criterion for a given application is not always obvious as the pruning power of a criterion depends on the character of the individual objects as well as the data distribution of the whole data set. For a deeper understanding which criterion to choose it is important to separately consider these two aspects. Thus, our visualisation tool offers two visualisation modes. In both modes, any L_p-norm can be applied as basic distance measure.

3.1   Individual Object View

In this mode only three objects (Q, A and B) are visualized in order to examine the concept of domination. For the Trigonometric and the Optimal domination decision criterion (DCC) pruning, the areas of domination are colored in dark gray. The three different pruning techniques can be evaluated against each other in this basic relation using a split screen. In order to evaluate variations and changes of the data objects on the domination relationship, objects can be re-sized as well as moved easily and the domination areas (if displayed) are adjusted accordingly. For each split screen, a signal light shows whether B is dominated (green light) or not dominated (red light) by A w.r.t. Q using the selected pruning technique.

Figure 2 illustrates a sample snapshot of this object view. It displays the MinMax pruning (left) and the Optimal DCC pruning (right) with query object Q and two data objects A, B. The red light in the screen showing the MinMax criterion indicates, that object B is not dominated by A w.r.t. Q when using the MinMax method (and cannot be pruned in a NN query setting with Q as the query). Contrary, the green light in the screen showing the Optimal DCC criterion indicates, that object B is dominated by object A w.r.t. Q and can be pruned when issuing a NN query around Q.

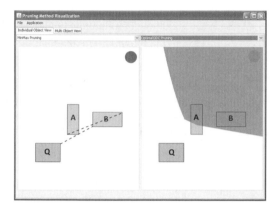

Fig. 2. Individual Object View

3.2 Data Set View

While the Object View is intended to illustrate the domination relationships among three objects, in the Data Set View, an entire data set can be visually analyzed and pruning can be visualized for different types of queries. Again, different pruning criteria can be compared using split screens. In particular, after loading a data set from a file, the user has the following options:

- The user can chose the pruning technique. Note that the Trigonometric pruning only works for BS approximations.
- The user can specify a query type, i.e., kNN or RkNN, and can set the parameter k.
- Finally, the user needs to pick a query object by clicking on an object in one of the split screens. The query object is highlighted in red color.

For each pruning criterion, the software highlights the true hits in green, the pruned true drops in blue. Objects that cannot be decided based on the approximations are colored in orange. In addition, a quantitative summary of the results is given for each criterion listing the number of true hits, true drops and candidate objects. Analogously to the Individual Object View, objects can be re-sized and moved to evaluate the impact of changes of the data objects on the pruning.

Figure 3 displays a sample snapshot showing true hits, true drops, and candidates computed by the Trigonometric pruning method for a R1NN query (query object in red) on a data set of 50 objects approximated with BS. In this setting, we can obtain one true hit (green), three undecidable objects (orange) and 46 pruned true drops (blue).

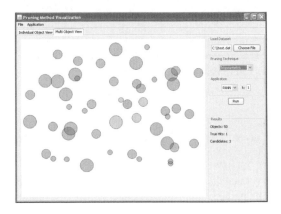

Fig. 3. Data Set View

4   Conclusions

In this demo, we enable visual comparison of different pruning techniques supporting rectangular and spherical object approximations that can be in spatial similarity query processing. The framework allows to use different L_p norms as basic distance measures and implements different types of query predicates. Finally, query parameters and individual data objects can be adjusted. Thus, this demo supports the evaluation of advantages and limitations of different pruning strategies that are an important ingredient in efficient processing of spatial similarity queries.

References

1. Roussopoulos, N., Kelley, S., Vincent, F.: Nearest neighbor queries. In: Proc. SIGMOD, pp. 71–79 (1995)
2. Emrich, T., Graf, F., Kriegel, H.-P., Schubert, M., Thoma, M.: Optimizing all-nearest-neighbor queries with trigonometric pruning. In: Gertz, M., Ludäscher, B. (eds.) SSDBM 2010. LNCS, vol. 6187, pp. 501–518. Springer, Heidelberg (2010)
3. Emrich, T., Kriegel, H.P., Kröger, P., Renz, M., Züfle, A.: Boosting spatial pruning: On optimal pruning of MBRs. In: Proc. SIGMOD (2010)
4. Lian, X., Chen, L.: Efficient processing of probabilistic reverse nearest neighbor queries over uncertain data. The VLDB Journal 18(3), 787–808 (2009)

Spatial Outlier Detection:
Data, Algorithms, Visualizations

Elke Achtert, Ahmed Hettab, Hans-Peter Kriegel,
Erich Schubert, and Arthur Zimek

Ludwig-Maximilians-Universität München,
Oettingenstr, 67, 80538 München, Germany
{achtert,hettab,kriegel,schube,zimek}@dbs.ifi.lmu.de
http://www.dbs.ifi.lmu.de

Abstract. Current Geographic/Geospatial Information Systems (GIS) and Data Mining Systems (DMS) so far are usually not designed to interoperate. GIS research has a strong emphasis on information management and retrieval, whereas DMS usually have too little geographic functionality to perform appropriate analysis. In this demonstration, we introduce an integrated GIS-DMS system for performing advanced data mining tasks such as outlier detection on geo-spatial data, but which also allows the interaction with existing GIS and this way allows a thorough evaluation of the results. The system enables convenient development of new algorithms as well as application of existing data mining algorithms to the spatial domain, bridging the gap between these two worlds.

1 Introduction

Geo Information Systems (GIS) are commonly associated with tasks such as the fast retrieval of cartography data at varying resolutions as seen in popular applications such as Google Maps and Google Earth. This poses a wide array of challenges, ranging from multi-resolution indexing, handling of polygon and 3D raster data types to fast route searching in dynamic traffic networks. However, many of these problems can now efficiently be handled by a wide array of software ranging from open source products such as PostGIS to commercial applications such as Oracle Spatial. The role of these systems closely resembles that of a traditional RDBMS system with strong query capabilities but few integrated advanced analysis methods.

From the initial challenges of efficiently managing these data, recent research often focuses on analyzing these data in detail, often in combination with non-spatial information. While the current systems are powerful in managing large route networks, they offer little assistance when analyzing geo-annotated information often found in health, government and marketing situations beyond the ability to select a subset of the data. Current systems are often specialized for specific tasks such as pareto-optimal route search (e.g. [1]).

In spatial data mining, instead of partitioning the data according to some more or less arbitrary geographical boundary – administrative zones and grid

D. Pfoser et al. (Eds.): SSTD 2011, LNCS 6849, pp. 512–516, 2011.

partitions are very often used – the spatial and non-spatial attributes are processed as two sides of the same coin. In the most basic view, objects that are spatially close are expected to have similar values in the non-spatial attributes as well. The spatial attributes are used to define a neighborhood an object can be compared to without having to partition the data and potentially introducing artefacts by this partitioning. Artefacts however are a major obstacle in many data mining applications, in particular when working with real data.

For this demonstration, we will focus on the data mining discipline known as outlier detection or anomaly detection which has seen an increasing interest over several years [2,3,4,5,6,7,8]. As opposed to traditional outlier detection (which can trivially be applied on the non-spatial attributes only of a data set), the objects of interest for spatial outlier detection are those who deviate from a "local trend" in the data, although they might be globally unremarkable. For example, a temperature low of $-10°C$ might globally be common, for an outdoor sensor near Los Angeles where temperatures barely drop to the freezing point it clearly is an outlier. Our demonstration uses traditional GIS features to compute the neighborhood sets for the database objects, then apply spatial outlier detection methods to identify trends and unusual behavior. However we do not just apply the algorithm in a standalone way, but we leverage a complete tool-chain starting from data import, running multiple algorithms, evaluating and comparing the results and inspecting the results in a traditional GIS front-end.

2   Workbench: System, Functionality, Work-Flow

The integrated GIS and Data-Mining system introduced here is written in Java employing a modular architecture, to allow the easy extension with additional modules. As it comes along with release 0.4 of ELKI[1] [9,10,11], modularization is not limited to algorithms: there are also modules for additional data types, distance functions, neighborhoods, input parsing, index structures, evaluation, visualization and output formats. A generic parametrization tool assists the user in choosing and configuring the modules as needed. Figure 1 gives a schematic overview of the information flow in the system.

2.1   Data – Typical Formats and Conversions

A classic but rather limited format for spatial data is the ESRI Shapefile format. While this is sufficient for representing simple shapes such as the outline of a lake, it does not store much more than 2D polygons. Another format in widespread use is the Keyhole Markup Language (KML) used by Google Earth. However, this format is designed with a strong focus on presentation. The most flexible format, published by the Open Geospatial Consortium, is the Geography Markup Language (GML). It still has a strong focus on "map features", but also can store observations and measurements. As of March 2011, this part of GML is still in flux with a 2.0 standard nearby. As such, it is not yet in widespread use.

---

[1] http://elki.dbs.ifi.lmu.de

Fig. 1. System overview for the integrated GIS Data Mining system

In order to obtain a database with both spatial and non-spatial attributes it will often be necessary to join multiple data sets. A very popular data set is published by the U.S. Census Bureau and contains various demographic values such as age, ethnicities, family and household sizes.[2] The whole data set consists of various Shapefiles that represent districts at varying resolution ranging from state to city district levels and a large set of tables that contain the summarized census results ("summary files"). These files can be joined using the FIPS codes which serve as geographic object ID. Given the amount of attributes and redundancy within attributes of the summary file, the analyst also needs to choose the interesting attributes and only load these parts of the data set.

2.2   Spatial Outlier Detection Algorithms

The workbench includes many specialized spatial outlier algorithms, for example SLOM [5,6], SOF [12], Trimmed-mean-approach [13], Random-Walk based Outlier Detection [7], and GLS-SOD along with its predecessors [2,3,4,8]. Additionally, the system allows the application of some non-specialized outlier detection methods such as LOF [14] (and many more) for comparison. This is particularly useful for evaluation of the effects of the actual spatial locality of the results, since global outliers will often also be local outliers. A fundamental motivation for developing a framework integrating re-implementations of the algorithms of different groups for a certain research topic (like spatial outlier detection) is to aim at consolidation of a research area that has shown some steps of innovation already but also is still ongoing. In an active research area, newly proposed algorithms are often evaluated in a sloppy way taking into account only one or two partners for comparison of efficiency and effectiveness, presumably because for most algorithms no implementation is at hand. And if an implementation is provided by the authors, a *fair* comparison is nonetheless all but impossible due to different performance properties of different programming languages,

---

[2] Available at the United States Census Bureau. http://www.census.gov/

 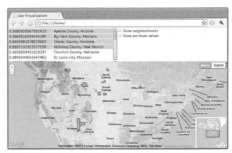

(a) Google Earth showing KML file   (b) Direct webbrowser interface

Background map data ©2011 Google, Europa Technologies, Geocentre Consulting, INEGI, Tele Atlas

Fig. 2. Visual analysis of detected outliers

frameworks, and, last but not least, implementation details. Eventually, an evaluation based on implementations of different authors is more likely to be a comparison of the efforts of different authors in efficient programming rather than truly an evaluation of algorithmic merits. But also w.r.t. effectiveness, using an integrated system allows the fair comparison of different algorithms within the same context to eliminate effects that result for example from data normalization, indexing and other implementation details.

2.3 Visualizations

Visualization components of the workbench include interfaces to KML (that can for example be read by Google Earth or NASA World Wind) and a webbrowser based application that uses map overlays and allows for greater interactivity than it is currently possible with KML files. Examples are depicted in Figure 2.

3 Conclusion

The software described in this paper comes along with release 0.4 of ELKI [9,10,11] as an application of the framework. Both ELKI and this application can be downloaded at

http://elki.dbs.ifi.lmu.de

The framework ELKI itself provides much more possibilities as are used by the application introduced here. The integrated nature of the ELKI workbench allows for rapid development of new methods, since existing components can be reused. The published source code of the methods ensures reproducible results. In addition, it allows for fair comparison of existing and new methods that eliminates bias introduced by implementation details.

References

1. Graf, F., Kriegel, H.P., Renz, M., Schubert, M.: PAROS: Pareto optimal route selection. In: Proceedings of the ACM International Conference on Management of Data (SIGMOD), Indianapolis, IN (2010)
2. Shekhar, S., Lu, C.T., Zhang, P.: A unified approach to detecting spatial outliers. GeoInformatica 7(2), 139–166 (2003)
3. Lu, C.T., Chen, D., Kou, Y.: Algorithms for spatial outlier detection. In: Proceedings of the 3rd IEEE International Conference on Data Mining (ICDM), Melbourne, FL (2003)
4. Kou, Y., Lu, C.T., Chen, D.: Spatial weighted outlier detection. In: Proceedings of the 6th SIAM International Conference on Data Mining (SDM), Bethesda, MD (2006)
5. Sun, P., Chawla, S.: On local spatial outliers. In: Proceedings of the 4th IEEE International Conference on Data Mining (ICDM), Brighton, UK (2004)
6. Chawla, S., Sun, P.: SLOM: A new measure for local spatial outliers. Knowledge and Information Systems (KAIS) 9(4), 412–429 (2006)
7. Liu, X., Lu, C.T., Chen, F.: Spatial outlier detection: Random walk based approaches. In: Proceedings of the 18th ACM SIGSPATIAL International Conference on Advances in Geographic Information Systems (ACM GIS), San Jose, CA (2010)
8. Chen, F., Lu, C.T., Boedihardjo, A.P.: GLS-SOD: A generalized local statistical approach for spatial outlier detection. In: Proceedings of the 16th ACM International Conference on Knowledge Discovery and Data Mining (SIGKDD), Washington, DC (2010)
9. Achtert, E., Kriegel, H.P., Zimek, A.: ELKI: a software system for evaluation of subspace clustering algorithms. In: Proceedings of the 20th International Conference on Scientific and Statistical Database Management (SSDBM), Hong Kong, China (2008)
10. Achtert, E., Bernecker, T., Kriegel, H.P., Schubert, E., Zimek, A.: ELKI in time: ELKI 0.2 for the performance evaluation of distance measures for time series. In: Mamoulis, N., Seidl, T., Pedersen, T.B., Torp, K., Assent, I. (eds.) SSTD 2009. LNCS, vol. 5644, pp. 436–440. Springer, Heidelberg (2009)
11. Achtert, E., Kriegel, H.P., Reichert, L., Schubert, E., Wojdanowski, R., Zimek, A.: Visual evaluation of outlier detection models. In: Kitagawa, H., Ishikawa, Y., Li, Q., Watanabe, C. (eds.) DASFAA 2010. LNCS, vol. 5982, pp. 396–399. Springer, Heidelberg (2010)
12. Huang, T., Qin, X.: Detecting outliers in spatial database. In: Proceedings of the 3rd International Conference on Image and Graphics, Hong Kong, China, pp. 556–559
13. Hu, T., Sung, S.: A trimmed mean approach to finding spatial outliers. Intelligent Data Analysis 8(1), 79–95 (2004)
14. Breunig, M.M., Kriegel, H.P., Ng, R., Sander, J.: LOF: Identifying density-based local outliers. In: Proceedings of the ACM International Conference on Management of Data (SIGMOD), Dallas, TX (2000)

Author Index